IET ENERGY ENGINEERING SERIES 131

Metaheuristic Optimization in Power Engineering

Other volumes in this series:

Volume 1	**Power Circuit Breaker Theory and Design** C.H. Flurscheim (Editor)
Volume 4	**Industrial Microwave Heating** A.C. Metaxas and R.J. Meredith
Volume 7	**Insulators for High Voltages** J.S.T. Looms
Volume 8	**Variable Frequency AC Motor Drive Systems** D. Finney
Volume 10	**SF_6 Switchgear** H.M. Ryan and G.R. Jones
Volume 11	**Conduction and Induction Heating** E.J. Davies
Volume 13	**Statistical Techniques for High Voltage Engineering** W. Hauschild and W. Mosch
Volume 14	**Uninterruptible Power Supplies** J. Platts and J.D. St Aubyn (Editors)
Volume 15	**Digital Protection for Power Systems** A.T. Johns and S.K. Salman
Volume 16	**Electricity Economics and Planning** T.W. Berrie
Volume 18	**Vacuum Switchgear** A. Greenwood
Volume 19	**Electrical Safety: A guide to causes and prevention of hazards** J. Maxwell Adams
Volume 21	**Electricity Distribution Network Design, 2nd Edition** E. Lakervi and E.J. Holmes
Volume 22	**Artificial Intelligence Techniques in Power Systems** K. Warwick, A.O. Ekwue and R. Aggarwal (Editors)
Volume 24	**Power System Commissioning and Maintenance Practice** K. Harker
Volume 25	**Engineers' Handbook of Industrial Microwave Heating** R.J. Meredith
Volume 26	**Small Electric Motors** H. Moczala *et al.*
Volume 27	**AC–DC Power System Analysis** J. Arrillaga and B.C. Smith
Volume 29	**High Voltage Direct Current Transmission, 2nd Edition** J. Arrillaga
Volume 30	**Flexible AC Transmission Systems (FACTS)** Y.-H. Song (Editor)
Volume 31	**Embedded generation** N. Jenkins *et al.*
Volume 32	**High Voltage Engineering and Testing, 2nd Edition** H.M. Ryan (Editor)
Volume 33	**Overvoltage Protection of Low-Voltage Systems, Revised Edition** P. Hasse
Volume 36	**Voltage Quality in Electrical Power Systems** J. Schlabbach *et al.*
Volume 37	**Electrical Steels for Rotating Machines** P. Beckley
Volume 38	**The Electric Car: Development and future of battery, hybrid and fuel-cell cars** M. Westbrook
Volume 39	**Power Systems Electromagnetic Transients Simulation** J. Arrillaga and N. Watson
Volume 40	**Advances in High Voltage Engineering** M. Haddad and D. Warne
Volume 41	**Electrical Operation of Electrostatic Precipitators** K. Parker
Volume 43	**Thermal Power Plant Simulation and Control** D. Flynn
Volume 44	**Economic Evaluation of Projects in the Electricity Supply Industry** H. Khatib
Volume 45	**Propulsion Systems for Hybrid Vehicles** J. Miller
Volume 46	**Distribution Switchgear** S. Stewart
Volume 47	**Protection of Electricity Distribution Networks, 2nd Edition** J. Gers and E. Holmes
Volume 48	**Wood Pole Overhead Lines** B. Wareing
Volume 49	**Electric Fuses, 3rd Edition** A. Wright and G. Newbery
Volume 50	**Wind Power Integration: Connection and system operational aspects** B. Fox *et al.*
Volume 51	**Short Circuit Currents** J. Schlabbach
Volume 52	**Nuclear Power** J. Wood
Volume 53	**Condition Assessment of High Voltage Insulation in Power System Equipment** R.E. James and Q. Su
Volume 55	**Local Energy: Distributed generation of heat and power** J. Wood
Volume 56	**Condition Monitoring of Rotating Electrical Machines** P. Tavner, L. Ran, J. Penman and H. Sedding
Volume 57	**The Control Techniques Drives and Controls Handbook, 2nd Edition** B. Drury
Volume 58	**Lightning Protection** V. Cooray (Editor)
Volume 59	**Ultracapacitor Applications** J.M. Miller
Volume 62	**Lightning Electromagnetics** V. Cooray
Volume 63	**Energy Storage for Power Systems, 2nd Edition** A. Ter-Gazarian
Volume 65	**Protection of Electricity Distribution Networks, 3rd Edition** J. Gers
Volume 66	**High Voltage Engineering Testing, 3rd Edition** H. Ryan (Editor)
Volume 67	**Multicore Simulation of Power System Transients** F.M. Uriate
Volume 68	**Distribution System Analysis and Automation** J. Gers
Volume 69	**The Lightning Flash, 2nd Edition** V. Cooray (Editor)
Volume 70	**Economic Evaluation of Projects in the Electricity Supply Industry, 3rd Edition** H. Khatib
Volume 72	**Control Circuits in Power Electronics: Practical issues in design and implementation** M. Castilla (Editor)
Volume 73	**Wide Area Monitoring, Protection and Control Systems: The enabler for smarter grids** A. Vaccaro and A. Zobaa (Editors)
Volume 74	**Power Electronic Converters and Systems: Frontiers and applications** A.M. Trzynadlowski (Editor)
Volume 75	**Power Distribution Automation** B. Das (Editor)
Volume 76	**Power System Stability: Modelling, analysis and control** B. Om P. Malik
Volume 78	**Numerical Analysis of Power System Transients and Dynamics** A. Ametani (Editor)
Volume 79	**Vehicle-to-Grid: Linking electric vehicles to the smart grid** J. Lu and J. Hossain (Editors)
Volume 81	**Cyber–Physical–Social Systems and Constructs in Electric Power Engineering** Siddharth Suryanarayanan, Robin Roche and Timothy M. Hansen (Editors)
Volume 82	**Periodic Control of Power Electronic Converters** F. Blaabjerg, K. Zhou, D. Wang and Y. Yang
Volume 86	**Advances in Power System Modelling, Control and Stability Analysis** F. Milano (Editor)
Volume 87	**Cogeneration: Technologies, Optimisation and Implementation** C.A. Frangopoulos (Editor)
Volume 88	**Smarter Energy: From smart metering to the smart grid** H. Sun, N. Hatziargyriou, H.V. Poor, L. Carpanini and M.A. Sánchez Fornié (Editors)
Volume 89	**Hydrogen Production, Separation and Purification for Energy** A. Basile, F. Dalena, J. Tong, T.N. Veziroğlu (Editors)
Volume 90	**Clean Energy Microgrids** S. Obara and J. Morel (Editors)
Volume 91	**Fuzzy Logic Control in Energy Systems with Design Applications in MATLAB®/Simulink®** İ.H. Altaş
Volume 92	**Power Quality in Future Electrical Power Systems** A.F. Zobaa and S.H.E.A. Aleem (Editors)
Volume 93	**Cogeneration and District Energy Systems: Modelling, analysis and optimization** M.A. Rosen and S. Koohi-Fayegh
Volume 94	**Introduction to the Smart Grid: Concepts, technologies and evolution** Salman K. Salman
Volume 95	**Communication, Control and Security Challenges for the Smart Grid** S.M. Muyeen and S. Rahman (Editors)
Volume 97	**Synchronized Phasor Measurements for Smart Grids** M.J.B. Reddy and D.K. Mohanta (Editors)
Volume 98	**Large Scale Grid Integration of Renewable Energy Sources** Antonio Moreno-Munoz (Editor)
Volume 100	**Modeling and Dynamic Behaviour of Hydropower Plants** N. Kishor and J. Fraile-Ardanuy (Editors)
Volume 101	**Methane and Hydrogen for Energy Storage** R. Carriveau and David S-K. Ting
Volume 108	**Fault Diagnosis of Induction Motors** Jawad Faiz, Vahid Ghorbanian and Gojko Joksimović
Volume 110	**High Voltage Power Network Construction** K. Harker
Volume 124	**Power Market Transformation** B. Murray
Volume 905	**Power system protection, 4 volumes**

Metaheuristic Optimization in Power Engineering

Jordan Radosavljević

The Institution of Engineering and Technology

Published by The Institution of Engineering and Technology, London, United Kingdom

The Institution of Engineering and Technology is registered as a Charity in England & Wales (no. 211014) and Scotland (no. SC038698).

First published 2018

The Institution of Engineering and Technology
Michael Faraday House
Six Hills Way, Stevenage
Herts, SG1 2AY, United Kingdom

www.theiet.org

British Library Cataloguing in Publication Data
A catalogue record for this product is available from the British Library

ISBN 978-1-78561-546-7 (hardback)
ISBN 978-1-78561-547-4 (PDF)

Typeset in India by MPS Limited

Contents

Preface xiii
Acknowledgements xv
Supplementary files xvi

1 Overview of metaheuristic optimization 1
1.1 Introduction 1
1.2 Description of metaheuristics 2
1.3 Principle of population-based metaheuristics 4
1.3.1 Genetic algorithm 6
1.3.2 Differential evolution 7
1.3.3 Evolutionary programing 7
1.3.4 Backtracking search optimization algorithm 8
1.3.5 Particle swarm optimization 9
1.3.6 Ant colony optimization 9
1.3.7 Artificial bee colony 9
1.3.8 Gravitational search algorithm 10
1.3.9 Wind-driven optimization 11
1.3.10 Colliding bodies optimization 12
1.3.11 Black hole algorithm 13
1.3.12 Gray wolf optimizer 14
1.3.13 Firefly algorithm 14
1.3.14 Cuckoo search algorithm 15
1.3.15 Moth swarm algorithm 15
1.3.16 Krill herd algorithm 16
1.3.17 Shuffled frog-leaping algorithm 17
1.3.18 Bacterial colony foraging optimization 17
1.3.19 Biogeography-based optimization 18
1.3.20 Teaching–learning-based optimization 19
1.3.21 League championship algorithm 19
1.3.22 Mine blast algorithm 20
1.3.23 Sine cosine algorithm 21
1.3.24 Harmony search 22
1.3.25 Imperialist competitive algorithm 23
1.3.26 Differential search algorithm 24
1.3.27 Glowworm swarm optimization 25
1.3.28 Spiral optimization algorithm 26
1.3.29 The Jaya algorithm 27
1.3.30 Creating a "new" algorithm 28

1.4 Criticism of metaheuristics 31
1.5 Educational software—*metahopt* 33
1.6 Conclusion 34
References 35

2 Overview of genetic algorithms 39
2.1 Introduction 39
2.2 Basic structure of the GA 40
2.3 Representation of individuals (encoding) 41
2.3.1 Binary encoding 42
2.3.2 Gray coding 43
2.3.3 Real-value encoding 43
2.4 Population size and initial population 44
2.5 Fitness function 44
2.5.1 Relative fitness 45
2.5.2 Linear scaling 46
2.6 Selection 46
2.6.1 Simple selection 48
2.6.2 Stochastic universal sampling 48
2.6.3 Linear ranking selection 49
2.6.4 Elitist selection 49
2.6.5 *k*-Tournament selection schemes 50
2.6.6 Simple tournament selection 50
2.7 Crossover 50
2.7.1 One-point crossover 51
2.7.2 Multipoint crossover 51
2.7.3 Uniform crossover 51
2.7.4 Shuffle crossover 52
2.7.5 Arithmetic crossover 52
2.7.6 Heuristic crossover 53
2.8 Mutation 53
2.9 GA control parameters 54
2.10 Multiobjective optimization using GA 55
2.11 Applications of GA to power system problems—literature overview 57
2.11.1 Optimal power flow 57
2.11.2 Optimal reactive power dispatch 60
2.11.3 Combined economic and emission dispatch 60
2.11.4 Optimal power flow in distribution networks 61
2.11.5 Optimal placement and sizing of distributed generation in distribution networks 63
2.11.6 Optimal energy and operation management of microgrids 64
2.11.7 Optimal coordination of directional overcurrent relays 65
2.11.8 Steady-state analysis of self-excited induction generator 66
2.12 Conclusion 66
References 67

3 Overview of particle swarm optimization 75
3.1 Introduction 75
3.2 Description of PSO 76
3.2.1 Parameters of PSO 79
3.2.2 General remarks about PSO 82
3.2.3 MATLAB® code of PSO 83
3.2.4 Example usage of PSO 85
3.3 PSO modifications 88
3.3.1 Population topology 88
3.3.2 Discrete binary PSO 89
3.3.3 Hybrid PSO 90
3.3.4 Adaptive PSO 90
3.4 Applications of PSO to power system problems—literature overview 94
3.4.1 Optimal power flow 94
3.4.2 Optimal reactive power dispatch 97
3.4.3 Economic dispatch 98
3.4.4 Optimal power flow in distribution networks 99
3.4.5 Optimal placement and sizing of distributed generation in distribution networks 100
3.4.6 Optimal energy and operation management of MGs 101
3.4.7 Optimal coordination of directional overcurrent relays 101
3.5 Conclusion 102
References 102

4 Overview of gravitational search algorithm 113
4.1 Introduction 113
4.2 Description of original GSA 115
4.2.1 Parameters of GSA 117
4.2.2 General remarks about GSA 118
4.2.3 MATLAB® code of GSA 120
4.2.4 Example usage of GSA 123
4.3 Binary gravitational search algorithm 125
4.4 Modified GSA 126
4.5 Opposition-based GSA 128
4.5.1 Current optimum opposition-based GSA 129
4.6 Adaptive gbest-guided GSA 131
4.6.1 Slow exploitation of GSA 131
4.6.2 Improving the exploitation of GSA 131
4.7 Self-adaptive GSA 133
4.8 Nondominated sorting GSA 135
4.8.1 Updating the external archive 136
4.8.2 Updating the list of moving agents 137
4.8.3 Updating the mass of moving agents 137
4.8.4 Updating the acceleration of agents 137
4.8.5 The use of mutation operator 138
4.8.6 Update and mutate the position of agents 138

4.9 Clustered-gravitational search algorithm 139
4.10 Hybrid PSO and GSA algorithm 140
4.11 Applications of GSA to power system problems—literature overview 145
4.11.1 Optimal power flow 145
4.11.2 Optimal reactive power dispatch 146
4.11.3 Economic dispatch using GSA 146
4.11.4 Optimal power flow in distribution networks 147
4.11.5 Optimal DG placement and sizing in distribution networks 147
4.11.6 Optimal energy and operation management of microgrids 148
4.11.7 Optimal coordination of overcurrent relays 149
4.12 Conclusion 149
References 150

5 Power-flow calculation 155
5.1 Introduction 155
5.2 Power-flow calculation in transmission networks 156
5.2.1 Power-flow equations 157
5.2.2 Bus classification 158
5.2.3 Solution methods 158
5.2.4 Power-flow software—*pfgui* 164
5.3 Power-flow calculation in distribution networks 165
5.3.1 Backward/forward sweep power-flow algorithm 170
5.3.2 Power-flow software—*pfdngui* 174
5.4 Conclusion 176
References 176

6 Optimal power flow in transmission networks 177
6.1 Introduction 177
6.2 Literature overview 178
6.3 Formulation of the OPF problem 185
6.3.1 Equality constraints 186
6.3.2 Inequality constraints 186
6.3.3 Objective function 188
6.3.4 Multiobjective function 191
6.3.5 Transient-stability-constrained OPF 192
6.4 Solution methodology for OPF problem 194
6.4.1 Overview of PSO 194
6.4.2 Application of PSO to the OPF problem 195
6.4.3 Overview of GSA 196
6.4.4 Application of GSA to the OPF problem 198
6.4.5 Overview of hybrid PSOGSA 199
6.4.6 Application of PSOGSA to the OPF problem 199

6.5 Simulation results 200
6.5.1 IEEE 30-bus test system 200
6.5.2 IEEE 118-bus test system 211
6.6 Solution software—*opfgui* 213
6.7 Conclusion 223
References 224

7 Optimal reactive power dispatch in transmission networks 235
7.1 Introduction 235
7.2 Problem formulation 236
7.3 ORPD using hybrid PSOGSA 238
7.3.1 Overview of PSOGSA 238
7.3.2 Application of PSOGSA to the ORPD problem 240
7.3.3 Simulation results of PSOGSA 241
7.4 ORPD using hybrid GSA–SQP algorithm 255
7.4.1 Application of hybrid GSA–SQP to the ORPD problem 256
7.4.2 Simulation results of hybrid GSA–SQP 257
7.5 Educational program package *ORPD* 257
7.6 Conclusion 263
References 263

8 Combined economic and emission dispatch 267
8.1 Introduction 267
8.2 Problem formulation 269
8.2.1 Fuel cost function 270
8.2.2 Emission function 270
8.2.3 Constraints 271
8.2.4 Slack generator calculation 272
8.3 Solution method 273
8.3.1 Overview of PSOGSA 273
8.3.2 PSOGSA implementation to the CEED problem 275
8.4 Simulation results 277
8.4.1 Test system 1 277
8.4.2 Test system 2 281
8.4.3 Test system 3 285
8.5 Educational software—*ceedgui* 286
8.6 Conclusion 292
References 293

9 Optimal power flow in distribution networks 297
9.1 Introduction 297
9.2 Deterministic optimal power flow 299
9.2.1 Objective function 300
9.2.2 Constraints 301

9.3 DG units modeling for OPF 302
9.3.1 Diesel generator 303
9.3.2 Fuel cell 304
9.3.3 Microturbine 304
9.3.4 Wind turbine 304
9.3.5 Photovoltaic 305
9.3.6 Mini hydropower plants 306
9.3.7 Electric grid 306
9.4 Solution methods 307
9.4.1 Genetic algorithm 307
9.4.2 Gravitational search algorithm 309
9.5 Probabilistic optimal power flow 311
9.5.1 Statistical characterization of the input random variables 311
9.5.2 Statistical evaluation of the output variables 313
9.5.3 Procedure for solving probabilistic OPF 314
9.6 Simulation results 316
9.6.1 Deterministic OPF analysis 318
9.6.2 Probabilistic OPF analysis 321
9.7 Solution software—*opfdngui* 324
9.8 Conclusion 327
References 332

10 Optimal Volt/Var control in distribution networks **337**
10.1 Introduction 337
10.2 Decomposition of the voltage-control problem 339
10.2.1 Seasonal control of voltage 340
10.3 Optimal Volt/Var control using metaheuristic optimization 350
10.3.1 Problem formulation 350
10.3.2 Solution method 352
10.3.3 Simulation results 355
10.4 Conclusion 359
References 360

11 Optimal placement and sizing of distributed generation in distribution networks **363**
11.1 Introduction 363
11.2 Preliminary locations of DG 369
11.3 Partial search of variants 372
11.3.1 Optimal DG placement by using partial search of variants 373
11.3.2 Optimal DG sizing by using partial search of variants 375
11.4 Genetic algorithm 377
11.4.1 Optimal DG placement and sizing by using GA 378
11.5 Simulation results 381

11.5.1 IEEE 31-bus system 381
11.5.2 Distribution network Zajecar 391
11.6 Educational program package *opsdg* 399
11.7 Conclusion 401
References 401

12 Optimal energy and operation management of microgrids 407
12.1 Introduction 407
12.2 Problem formulation of EOM 411
12.2.1 Objective function 412
12.2.2 Constraints 412
12.2.3 Distributed generation bids calculation 414
12.3 Solution method 416
12.3.1 Overview of PSO 417
12.3.2 Application of PSO to EOM 418
12.4 Probabilistic EOM of MG 419
12.4.1 Statistical characterization of the input random variables 420
12.4.2 Statistical evaluation of the output variables 421
12.4.3 Procedure for solving probabilistic EOM 422
12.5 Simulation results 424
12.5.1 Microgrid MG1 424
12.5.2 Microgrid MG2 430
12.5.3 MATLAB program *eom* used for deterministic EOM 437
12.6 Conclusion 444
References 444

13 Optimal coordination of directional overcurrent relays 449
13.1 Introduction 449
13.2 Problem formulation 450
13.2.1 Objective function 451
13.2.2 Limits of the settings 452
13.2.3 Limits of relay operation time 453
13.2.4 Coordination criteria 453
13.2.5 Modification of objective function for minimization of CTI 454
13.3 Solution method 455
13.3.1 Overview of GSA 455
13.3.2 Overview of SQP 457
13.3.3 Hybrid GSA–SQP algorithm 457
13.3.4 Implementation of hybrid GSA–SQP algorithm 458
13.4 Simulation results 459
13.4.1 Test system 1 459
13.4.2 Test system 2 461

13.4.3 Test system 3 463
13.4.4 Statistical evaluation of the results 466
13.5 Educational program package *ocdocr* 468
13.6 Conclusion 471
References 471

14 Steady-state analysis of self-excited induction generators **475**
14.1 Introduction 475
14.2 System configuration 477
14.3 Induction generator model 478
14.4 Steady-state equations of SEIG 479
14.5 Steady-state equations of parallel operated SEIGs 481
14.6 Solution method 484
14.6.1 Overview of genetic algorithm 485
14.6.2 Application of GA to SEIG 486
14.6.3 Application of GA to parallel operated SEIGs 488
14.7 Simulation results 490
14.7.1 Steady-state analysis of SEIG 490
14.7.2 Steady-state analysis of parallel operated SEIGs 494
14.8 Conclusion 501
References 502

Index **505**

Preface

Metaheuristic paradigm is a way to solve complex optimization problems by using simple mathematic rules and powerful computers. When we want to avoid optimization methods based on differential equations, metaheuristic optimization is a good alternative.

Generally, a metaheuristic procedure is designed to find, generate, or select a sufficiently good solution to an optimization problem, especially with incomplete or imperfect information or limited computation capacity. Metaheuristic optimization methods have become significant tools and often the only way to solve practical optimization problems. The basic requirement from these methods is to obtain the global solution or a solution close to the global optimum in a reasonable time. Modern and emerging power systems are an important application of such methods.

This book is a research monograph which deals with applications of different metaheuristic optimization methods to power system problems. The book is intended primarily for the researchers who are dealing with various aspects of power system optimization. The aim of the book is to explain the basic principles of metaheuristic optimization methods and their applications in power engineering. The first part of the book gives a brief description of selected metaheuristic optimization methods, whereas the second part covers the specific application of the metaheuristic optimization methods to solve various optimization problems in power systems. Each chapter contains a comprehensive review of recent literature. The organization of the book is as follows.

Chapter 1 gives a brief description of 30 metaheuristic population-based optimization methods, including well-established and recently developed algorithms. Chapter 2 deals with a detailed description of "classic" metaheuristic optimization method—genetic algorithm (GA). A literature overview of the GA's application in solving some of the most important problems in power systems is provided also. Chapter 3 discusses the basic concepts of particle swarm optimization (PSO) algorithm as the most general of all swarm intelligence algorithms. Some of the most popular applications of PSO related to power system problems are also noted. Chapter 4 deals with the gravitational search algorithm (GSA). The basic concept of GSA as well as their modified and improved versions is discussed.

Chapter 5 describes the classical methods for power flow computation in transmission and distribution networks. This chapter presents an introduction in the next chapters related to the optimal power flow in transmission and distribution networks, optimal reactive power dispatch, optimal sitting and sizing of distributed generation in distribution networks, etc.

Chapter 6 covers the applications of metaheuristic optimization methods in general, and especially the PSO, GSA, and hybrid PSO and GSA algorithms, for solving optimal power flow (OPF) in power systems. Chapter 7 discusses the applications of hybrid PSO and GSA algorithm and hybrid GSA and sequential quadratic programming (SQP) algorithm to the optimal reactive power dispatch (ORPD) problem. Chapter 8 deals with the applications of PSO, GSA, and hybrid PSO and GSA for solving the combined economic and emission dispatch (CEED) problem in power systems.

Chapter 9 deals with the applications of GA and GSA for the solution of the OPF in distribution networks with distributed generation (DG) units, including fuel cells, micro turbines, diesel generators, photovoltaic systems, and wind turbines. Chapter 10 discusses the applications of different metaheuristic optimization methods to the problem of optimal Volt/Var control in distribution networks considering the DG units. Chapter 11 deals with the determination of optimal placement and sizing of DG units in distribution networks using GA-based approach. Chapter 12 describes an efficient algorithm based on PSO for the energy and operation management (EOM) of microgrids (MGs), including different DG units and energy-storage devices. Owing to the stochastic nature of energy produced from renewable sources, as well as load uncertainties and market prices, a probabilistic approach in the EOM also is introduced.

Chapter 13 discusses the application of GSA and hybrid GSA-SQP algorithms to optimal coordination of directional overcurrent relays (DOCRs).

Chapter 14 describes how a metaheuristic optimization technique such as GA can be applied in solving a nonoptimization problem, i.e., in determining the steady-state performances of a SEIG for general operating conditions.

Chapters 1 and 3–13 contain description of the MATLAB® programs for the problems treated in these chapters. The programs have been developed to present the efficiency of different metaheuristic optimization methods in solving the power system problems. A download link is given for each program.

Jordan Radosavljević

Acknowledgements

I am thankful to editorial team of the IET, UK, with special mention to Senior Commissioning Editor Christoph von Friedeburg, and Editorial Assistant Olivia Wilkins for their encouragement to me to complete this book.

My thanks also go to my colleagues from the Electrical Engineering Department of the Faculty of Technical Sciences in Kosovska Mitrovica, with special mention to Professor Miroljub Jevtić for his continuous motivation in the writing and publication of the book and Miloš Milovanović for his help in preparing the figures.

Last but not the least, I want to thank my family – my wife Mirjana, daughter Jana, and son Bogdan – for their sustained support.

Supplementary files

Supplementary files to support this book can be obtained online from the IET Digital Library at http://digital-library.theiet.org.

From the homepage search for this book by title and once on the book's homepage click on the "supplementary material" tab.

For more information, email books@theiet.org.

Chapter 1

Overview of metaheuristic optimization

1.1 Introduction

The optimization can be defined as the process of finding the conditions that provide the minimum or maximum value of a chosen objective function. A constrained optimization problem can be mathematically described as follows [1]:

$$\text{Find } \mathbf{x} = \begin{bmatrix} x_1 \\ x_2 \\ \vdots \\ x_n \end{bmatrix} \quad \text{which} \quad \min/\max F(\mathbf{x}) \tag{1.1}$$

Subject to

$$\mathbf{g}(\mathbf{x}) = 0 \tag{1.2}$$

$$\mathbf{h}(\mathbf{x},) \leq 0 \tag{1.3}$$

$$\mathbf{x} \in \mathbf{X} \tag{1.4}$$

where F is objective function to be optimized (minimized or maximized), $\mathbf{x}$ is an n-dimensional vector of decision (control) variables, $\mathbf{g}(\mathbf{x})$ is a vector of equality constraints, $\mathbf{h}(\mathbf{x})$ is a vector of inequality constrains. Constraints (1.4) define the feasibility region of the problem decision variables, where $\mathbf{X}$ is the space of possible solutions.

Unconstrained optimization problems do not involve any constraints and can be defined by (1.1).

In general, the objective function F and constraints $\mathbf{g}$ and $\mathbf{h}$ can take different properties including nonlinearity, nonconvexity, and nondifferentiability, with both continuous and discrete control variables. Therefore, the optimization problem defined by (1.1)–(1.4) can be very complex.

There are two main approaches to solve the optimization problem: classical or exact optimization methods and heuristic optimization techniques. The exact algorithms are designed in such a way that it is guaranteed that they will find the optimal solution in a finite amount of time and have to prove that the obtained solution is optimal. On the other hand, heuristics do not have this guarantee, and therefore generally return solutions that are worse than optimal or with guarantee that the solution they find is within a certain percentage of the optimal solution. However, a lot of problems of interesting practical combinatorial optimization are not easily tackled by exact methods because exact algorithms must exhaustively examine every single solution in the solution space. Moreover, developing an efficient exact method is a nontrivial task, even for relatively easy problems [2].

Exact optimization methods, such as the Augmented Lagrangian, Newton–Raphson, sequential quadratic programing, dynamic and integer programing, and interior point method, are based on an estimation of the global optimum. Classical methods such as linear programing and nonlinear programing are efficient approaches that can be used to solve special cases of optimization problem in power system applications [1]. However, due to difficulties of differentiability, nonlinearity, and nonconvexity, these methods failed to provide the global optimum and only reached the local one. Moreover, these methods exhibit some limitations depending on the type of problem, e.g., when the objective function is not available in algebraic form. It becomes essential to develop optimization methods able to overcome these drawbacks and handle such difficulties [3]. With this goal in mind, in past decades, many heuristic and metaheuristic methods have been developed to solve the complex constrained optimization problems.

1.2 Description of metaheuristics

The word heuristics comes from the Greek word "heurisko" which means "to find," "to know," or "to guide an investigation." It implies that heuristic algorithms are actually algorithms created by experimentation in order to obtain a satisfactory solution. An important feature of heuristic algorithms is that they can solve the problems of high complexity, at least with approximate accuracy. Heuristics is a set of rules based on experience which are used in solving a problem. The basic advantage of heuristic methods is their speed, which allows obtaining satisfactory solutions for large-scale practical optimization problems. The disadvantage of heuristic methods is that there is no guarantee of optimality of the solution obtained. However, for complex optimization problems, the primary goal is to get a solution, regardless of its quality. When some solution exists, various techniques can be applied to improve its quality. In fact, this is the basic idea for building metaheuristic optimization methods [4].

Classical heuristic methods were mainly developed to solve some specific, individual problems using the familiar features of a given problem in solving it. On the contrary, the metaheuristic optimization methods consist of general set of rules that can be applied to solve a variety of optimization problems. Many metaheuristic optimization methods have been developed by mimic of some well-known processes, primarily in biology, physics, society, and nature in general.

The recent trend in the literature is to name all stochastic algorithms with randomization and local search as "metaheuristic." The randomization process generates arbitrary solutions, which explore the search space and are responsible to achieve global solution. The local search is responsible to determine convergence and focus on achieving good solutions in a specific region [5].

Recently, a number of authors adopt the definition of Sorensen and Glover [2] for the term metaheuristic. The definition is as follows:

> *A metaheuristic is a high-level problem-independent algorithmic framework that provides a set of guidelines or strategies to develop heuristic optimization algorithms. The term is also used to refer to a problem-specific implementation of a heuristic optimization algorithm according to the guidelines expressed in such a framework.*

According to this definition, a metaheuristic is not an algorithm, i.e., it is not a sequence of actions that needs to follow such a strict guide. Rather, it is a consistent set of ideas, concepts, and operators that can be used to design heuristic optimization algorithms [2].

A metaheuristic algorithm should be equipped with two major characteristics to ensure finding global optimum. These two main characteristics are exploration and exploitation. The exploration is the ability of expanding search space, whereas the exploitation is the ability of finding the optima around a good solution. The main differences between the existing metaheuristics concern the particular way in which they try to achieve a balance between exploration and exploitation. The metaheuristic optimization methods can be classified by different criteria. A fundamental classification of metaheuristic optimization methods is based on the number of solution in one iteration. According to this criteria, the metaheuristic methods are classified on [6]:

1. Single-solution-based metaheuristics, also called trajectory methods, such as simulated annealing (SA), tabu search (TS), greedy randomized adaptive search procedure, variable neighborhood search, guided local search, and iterated local search.
2. Population-based metaheuristics, such as genetic algorithm (GA), particle swarm optimization (PSO), gravitational search algorithm (GSA), and many others.

The single-solution-based metaheuristic methods are more exploitation oriented, whereas population-based metaheuristic methods are more exploration oriented.

Another classifications of metaheuristics can be found in the literature. The nature-inspired metaheuristics are broadly classified into evolutionary algorithms, swarm intelligence, physics-based algorithms, bio-inspired algorithms, and others [5]. A classification of the population-based metaheuristic optimization algorithms is presented in Table 1.1. Numerous "modified," "adapted," "improved," "opposition-based," and "multiobjective" versions of these methods can be found in the literature.

Table 1.1 Classification of population-based metaheuristic algorithms

Type/Inspiration	Algorithm	Reference	Year
Evolutionary algorithms	Genetic algorithm (GA)	[7]	1975
	Differential evolution (DE)	[8]	1997
	Evolutionary programing (EP)	[9,10]	1966
	Backtracking search optimization algorithm (BSA)	[11]	2013
Swarm intelligence	Particle swarm optimization (PSO)	[12]	1995
	Ant colony optimization (ACO)	[13]	1992
	Bee colony optimization-based algorithms	[14,15]	2007
Physical algorithms	Gravitational search algorithm (GSA)	[16]	2009
	Wind-driven optimization (WDO)	[17]	2013
	Colliding bodies optimization (CBO)	[18]	2014
	Black hole (BH)	[19]	2013
Bio-inspired algorithms	Gray wolf optimizer (GWO)	[20]	2014
	Firefly algorithm (FA)	[21]	2008
	Cuckoo search (CS)	[22]	2009
	Moth swarm algorithm (MSA)	[23]	2016
	Krill herd (KH) algorithm	[24]	2012
	Shuffled frog-leaping algorithm (SFLA)	[25]	2006
	Bacterial colony foraging optimization (BCFO)	[26]	2014
	Biogeography-based optimization (BBO)	[27]	2008
Other algorithms	Teaching–learning-based optimization (TLBO)	[28]	2011
	League championship algorithm (LCA)	[29]	2014
	Mine blast algorithm (MBA)	[30]	2013
	Sine cosine algorithm (SCA)	[31]	2016
	Harmony search (HS)	[32]	2001
	Imperialist competitive algorithm (ICA)	[33]	2007
	Differential search algorithm (DSA)	[34]	2012
	Glowworm swarm optimization (GSO)	[35]	2009
	Spiral optimization algorithm (SOA)	[36]	2011
	The Jaya algorithm (Jaya)	[37]	2016

1.3 Principle of population-based metaheuristics

In the population-based metaheuristic optimization algorithms, the population is defined by a set of individuals (agents) which represent potential solutions of the optimization problem. The number of agents (N) is named as the size of the population. In general, an agent can be represented as vector whose elements are the values of the control variables of the optimization problem. The number of control variables (n) is the search space dimension of the optimization problem.

The essence of metaheuristic methods is iterative correction of the solution, i.e., generating a new population by applying algorithmic operators with stochastic search mechanism on agents from the current population. The way in which the algorithmic operators are defined constitutes the originality (if is) of a particular metaheuristic optimization method. The efficiency and performance of metaheuristic optimization methods are dependent on the proper setting of the corresponding algorithmic parameters.

The main performances of metaheuristics are fast search of large solution spaces, ability to find global solutions, and avoiding local optimum. Their main advantage compared to the classical (exact) optimization methods is that they are not limited with requirements for differentiability, nonconvexity, and continuity of the objective function or types of control variables. Moreover, these methods can be used for practical optimization problems taking into account various types of objective function and constraints.

In recent years, various population-based metaheuristic optimization methods listed in Table 1.1 have been suggested for solving the different engineering problems.

The basic elements of metaheuristic optimization methods can be defined as follows:

Agent, $\mathbf{x}(t)$: It is a candidate solution represented by an n-dimensional vector, where n is the number of control variables. At time (iteration) t, the ith agent $\mathbf{x}_i(t)$ can be described as $\mathbf{x}_i(t) = \left[x_i^1(t), \ldots, x_i^d(t), \ldots, x_i^n(t)\right]$, where $x_i^d(t)$ is the position of the ith agent with respect to the dth dimension, i.e., the value of the dth control variable in the ith candidate solution at iteration t.

Population, $\mathbf{POP}(t)$: It is a set of N agents at time (iteration) t, i.e., $\mathbf{POP}(t) = [\mathbf{x}_1(t), \ldots, \mathbf{x}_N(t)]^T$, or in expanded form:

$$\mathbf{POP}(t) = \begin{bmatrix} x_1^1(t), & x_1^2(t), & \ldots, & x_1^d(t), & \ldots, & x_1^n(t) \\ x_2^1(t), & x_2^2(t), & \ldots, & x_2^d(t), & \ldots, & x_2^n(t) \\ & & & \vdots & & \\ x_i^1(t), & x_i^2(t), & \ldots, & x_i^d(t), & \ldots, & x_i^n(t) \\ & & & \vdots & & \\ x_N^1(t), & x_N^2(t), & \ldots, & x_N^d(t), & \ldots, & x_N^n(t) \end{bmatrix} \tag{1.5}$$

Space of possible solutions, $\mathbf{X}$: It is an n-dimensional solution space which is defined by lower and upper limits of control variables.

Fitness is a direct metric of the performance of the individual population member (agent). The fitness of each agent of the population is calculated from the value of the function being optimized.

General structure of the metaheuristic optimization methods can be represented as follows:

Initialization

1. Defining the objective function $F(\mathbf{x}_i)$ and the space of possible solutions $\mathbf{X}$;
2. Generate initial population of N agents: $\mathbf{POP}(1) = [\mathbf{x}_1(1), \ldots, \mathbf{x}_i(1), \ldots, \mathbf{x}_N(1)]^T \subseteq \mathbf{X}$

Usually, the initial positions of each agent are randomly selected between minimum and maximum values of the control variables.

Set the iteration counter: $t = 1$

Iterative procedure

3. Calculate the fitness value $F(\mathbf{x}_i(t))$ for each agent $\mathbf{x}_i(t)$, $i = 1,\ldots,N$ in the current population $\mathbf{POP}(t)$.
4. Generate new population $\mathbf{POP}(t+1) = [\mathbf{x}_1(t+1),\ldots,\mathbf{x}_i(t+1),\ldots, \mathbf{x}_N(t+1)]^T \subseteq \mathbf{X}$ by applying the algorithmic operators on search agents from the current population $\mathbf{POP}(t)$.
5. Repeat the iterative procedure until the stop criteria is reached.
6. The optimal solution $\mathbf{x}^*$ is determined.

1.3.1 Genetic algorithm

The GA as an optimization method was created by John Holland, who elaborated and realized an idea how to transform the characteristics of natural evolution into a computer program [7]. Through a series of GA operations, a new population is obtained and its individuals are created by the individuals from the previous population according to the natural evolution principles: the selection, crossover, and mutation. The basic steps of the GA are as follows:

1. Representation of individuals. All data (variables) that make an individual are written in a string. A string is composed of substrings. Each substring represents a binary encoded variable on which the process of optimization is carried out.
2. Initialization. Individuals with random strings are generated that set up the initial population.
3. Fitness function calculation. It is used to rate the quality of an individual and it represents an equivalent of the function that should be optimized, that is, objective function.
4. Selection. During the selection process, the individuals that will participate in the reproduction (parents) are selected. The point of the selection is to store and transfer good individuals to the next generation.
5. Crossover. The way in which coded column parts (substrings) are crossed over actually makes a GA. Crossover is an exchange process of column parts between two individuals, that is, "parents." One or two new individuals engender by the crossover, that is, a "child." The possibility of inheriting the first parent's characteristics by a child is introduced during this process.
6. Mutation. Mutation is a way to give a new piece of information to an individual. Mutation represents an accidental bit variation of an individual, generally with a constant probability for each bit within a population.
7. Ending conditions. The GA is an iterative process which ends when a maximum number of generations is achieved or when another criterion is fulfilled, such as a minimum offset from the best fitness value and medium fitness value of all individuals in a current population. Otherwise, return to step 3.

1.3.2 Differential evolution

Differential evolution (DE) algorithm was developed by Storn and Price [8]. This is a population-based algorithm which uses operators similar as GA: mutation, crossover, and selection. DE differs from GA in a mutation scheme that makes DE self-adaptive and in the selection process. In DE, all the solutions have the same chance of being selected as parents. DE employs a greedy selection process: the better one of new solution and its parent wins the competition providing significant advantage of converging performance over GAs.

The main steps of the DE algorithm can be stated as follows:

1. Initialization. DE starts with a randomly chosen population of N individuals in n-dimensional search space. Each individual is represented as a search vector of n elements.
2. Mutation. DE generates new vectors by adding the weighted difference between two population vectors to a third vector. A mutant vector is generated according to:

$$\mathbf{v}_i(t+1) = \mathbf{x}_{r_1}(t) + F \cdot (\mathbf{x}_{r_2}(t) - \mathbf{x}_{r_3}(t)) \tag{1.6}$$

   where indexes r_1, r_2, r_3 are randomly chosen integers from set $\{1,2,\ldots,N\}$; F is a real number between 0 and 2 which controls the amplification of the differential variation.
3. Crossover. The mutated vector's variables are then mixed with the variables of another predetermined vector, the target vector, to yield the so-called trial vector. The crossover is performed on each of the n variables:

$$u_i^d(t+1) = \begin{cases} v_i^d(t+1) & \text{if rand}(0,1) < \text{CR} \\ x_i^d(t) & \text{else} \end{cases} \tag{1.7}$$

4. Selection. If the trial vector yields a lower cost function value than the target vector, the trial vector replaces the target vector in the following generation. Each population vector has to serve once as the target vector so that N competitions take place in one generation.
5. Repeat steps 2–4 until the maximum number of iterations is reached.
6. Return best solution.

1.3.3 Evolutionary programing

Evolutionary programing (EP) is a metaheuristic population-based technique developed by Fogel *et al.* [9]. Starting from a randomly generated initial population, during each iteration the EP produce a new population through the use of a mutation operator, competition, and selection. The mutation operator makes a new individual (potential solution) by perturbing each component of an existing individual by a random amount. Hence, the updating equation is as follows: (control variable) of ith individual is updated as:

$$x_i^d(t+1) = x_i^d(t) + N\left[0, \left(\sigma_i^d\right)^2\right] \tag{1.8}$$

where x_i^d is the dth component (control variable) in ith individual $\mathbf{x}_i = [x_i^1, \ldots, x_i^d, \ldots, x_i^n]$; $N[0, (\sigma_i^d)^2]$ is a Gaussian random number with a mean of zero and a standard deviation of σ_i^d. The σ_i^d is calculated as [10]

$$\sigma_i^d = \left(x_{i,\max}^d - x_{i,\min}^d\right) \cdot \frac{fit_{\max} - fit_i}{fit_{\max} - a^t} \tag{1.9}$$

where fit_i is the fitness of individual i; $fit_{\max}$ is the maximum fitness within the current population; $x_{i,\min}^d$ and $x_{i,\max}^d$ are the lower and upper limits of the dth control variable; a is a positive number constant slightly less than unity; and t is the current iteration number.

The quality of each of the individuals is measured by their fitness, which is related to the objective function of the problem. Through the use of a competition scheme, the individuals in each population compete with each other. The winning individuals form a resultant population, which is regarded as the next generation. For optimization to occur, the competition scheme must be such that the more optimal solutions have a greater chance of survival than the poorer solutions. Through this the population evolves toward the global optimal point. The EP technique is iterative and the process is terminated by a stopping criteria. The criteria widely used is stopped after a specified number of iterations.

1.3.4 Backtracking search optimization algorithm

The backtracking search optimization algorithm (BSA) is a stochastic search algorithm developed by Civicioglu [11]. The author of BSA was motivated by studies that attempt to develop simpler and more effective search algorithm with as few control parameters as possible. The general structure of BSA can be explained by dividing its functions into five segments: initialization, selection-I, mutation, crossover, and selection-II.

BSA initializes the population **P** by random selection between lower and upper limits of the control variables. In selection-I stage, BSA determines the historical population **oldP** in order to calculate the search direction. In the mutation process of BSA, a trial population **Mutant** is formed as a function of **P** and **oldP**. Because the historical population is used in the calculation of the search-direction matrix, BSA generates a trial population, taking partial advantage of its experiences from previous generations. BSA's crossover process generates crossover population **T** as the final form of trial population. The initial value of the trial population is **Mutant**, as obtained in the mutation process. Crossover process contains two steps. The first step calculates a binary integer-valued matrix (*map*) of size $N \cdot n$ that indicates the individuals of **T** to be manipulated by using the relevant individuals of **P**. In selection-II stage, the individual $\mathbf{T}_i$ that have better fitness values than the corresponding individual $\mathbf{P}_i$ are used to update the $\mathbf{P}_i$ based on a greedy selection. If the best individual of **P** ($\mathbf{P}_{best}$) has a better fitness value than the global minimum value obtained so far by BSA, the global minimizer is updated to be $\mathbf{P}_{best}$, and the global minimum value is updated to be the fitness value of $\mathbf{P}_{best}$. Detailed equation of BSA can be found in [11].

1.3.5 Particle swarm optimization

The PSO algorithm was developed by Kennedy and Eberhart [12] and is based on simulation of bird flocking in two-dimensional space. It uses a number of particles (candidate solutions) which fly around in the search space to find best solution. Meanwhile, the particles all look at the best particle (best solution) in their paths. In other words, particles consider their own best solutions (pbest) as well as the best solution found so far (gbest). Each particle tries to modify its position using the following information: the current position, the current velocity, the distance between the current position and pbest, and the distance between the current position and gbest.

In each iteration, the velocities of particles are calculated using the following equation:

$$\mathbf{v}_i(t+1) = w \cdot \mathbf{v}_i(t) + C_1 \cdot r_1 \cdot (\mathbf{pbest}_i(t) - \mathbf{x}_i(t)) + C_2 \cdot r_2 \cdot (\mathbf{gbest}(t) - \mathbf{x}_i(t)) \tag{1.10}$$

After updating the velocities, the positions of particles can be calculated as

$$\mathbf{x}_i(t+1) = \mathbf{x}_i(t) + \mathbf{v}_i(t+1) \tag{1.11}$$

where $\mathbf{v}_i(t)$ is velocity of particle i at iteration t, w is a weighting function, C_1 and C_2 are positive constants, r_1 and r_2 are uniformly distributed random numbers in [0, 1], $\mathbf{x}_i(t)$ is the current position of particle i at iteration t, $\mathbf{pbest}_i$ is the individual best of particle i at iteration t, and **gbest** is the best solution found so far. The first part of (1.10) provides exploration ability for PSO. The second and third parts represent private thinking and collaboration of particles, respectively.

1.3.6 Ant colony optimization

Dorigo [13] developed the ant colony optimization (ACO) algorithm inspired by ants' behavior in determining the optimal path from the nest to the food source. Initially, ants wander randomly for food in the surrounding regions of nest. An ant's movement is observed by the neighboring ants with the pheromone intensity it lays down while searching for food. Once a food source is found, the pheromone intensity of the path increases due to the movement of ant from source to nest and other ants instead of searching at random, they follow the trail. With the progress in time, the pheromone intensity starts to evaporate and reduce its attraction. The amount of time taken for an ant to travel to food source and back to the nest is directly proportional to the quantity of pheromone evaporation. In addition, evaporation of the pheromone has the advantage of allowing the algorithm to avoiding the convergence toward a local optimum solution. So with time an optimal shortest path is achieved to maintain the high pheromone intensity [5]. The ACO algorithm was successfully applied to a large number of combinatorial optimization problems.

1.3.7 Artificial bee colony

Artificial bee colony (ABC) algorithm was inspired by the intelligent behavior of bees and developed by Karaboga and Basturk [15]. In the ABC algorithm, there are three types of bees: employed bees, onlookers, and scouts. The position of a food source represents a possible solution to the optimization problem, and the nectar amount of a food source corresponds to the quality of the solution represented by that food source.

In the employed bee phase, the employed bees are responsible for searching for new food sources in the neighborhood of current food sources and calculating the nectar amount of these sources. The location of the new food sources is determined as follows:

$$\mathbf{x}_i(t+1) = \mathbf{x}_i(t) + \phi \cdot (\mathbf{x}_i(t) - \mathbf{x}_k(t)) \quad i = 1, 2, \ldots, N; \;\; k \in \{1, \ldots, N\} \tag{1.12}$$

where $\mathbf{x}_k(t)$ is a randomly chosen solution different from $\mathbf{x}_i(t)$, and $\mathbf{x}_i(t+1)$ is the new solution (food source). ϕ is a uniform random number between $[-1, 1]$.

If the new food source has equal or better nectar than the old source, it is replaced with the old one in the memory. Otherwise, the old one is retained in the memory.

In the onlooker bee phase, onlooker bees select a food source for themselves with a probabilistic calculation method by using the fitness values of the food sources. The onlookers search for better food source in the neighborhood of current food source by using (1.12). If the new nectar amount is better than the nectar amount of the old resource, then the old resource is abandoned. Otherwise, the abandonment counter of that food resource is increased by one. This process is repeated until all onlookers are distributed among the food sources.

If the nectar amount of a food source has been exhausted or the profitability of the food source decreases under a certain level, the employed bee associated with that food source becomes a scout. This scout starts searching a new food source randomly without any guidance in the search space. This abandoning and scouting mechanism assist the algorithm to escape local optimums.

1.3.8 Gravitational search algorithm

The GSA is a metaheuristic optimization algorithm developed by Rashedi *et al.* [16]. In GSA, the search agents are a collection of masses which interact with each other based on the Newtonian gravity and the laws of motion. The position of the mass corresponds to the solution of the problem, and its gravitational and inertial masses are determined using a fitness function. In other words, each mass presents a solution. The algorithm is navigated by properly adjusting the gravitational and inertial masses.

After evaluating the current population fitness, the mass of each agent is calculated as follows:

$$M_i(t) = \frac{m_i(t)}{\sum_{j=1}^{N} m_j(t)} \tag{1.13}$$

where

$$m_i(t) = \frac{fit_i(t) - worst(t)}{best(t) - worst(t)} \tag{1.14}$$

where $fit_i(t)$ represent the fitness value of the agent i at iteration t. $best(t)$ and $worst(t)$ are the best and worst fitness of all agents, respectively. According to Newton gravitation theory, the total force that acts on the ith agent at iteration t is specified as follows:

$$\mathbf{F}_i(t) = \sum_{j \in Kbest, j \neq i} r \cdot G(t) \frac{M_j(t) \cdot M_i(t)}{R_{i,j}(t) + \varepsilon} \left(\mathbf{x}_j(t) - \mathbf{x}_i(t)\right) \tag{1.15}$$

where r is a random number within the interval [0, 1]. $G(t)$ is gravitational constant at iteration t, $M_i(t)$ and $M_j(t)$ are masses of agents i and j, ε is a small constant, and $R_{ij}(t)$ is the Euclidian distance between the two agents i and j, $Kbest$ is the set of first K agents with the best fitness value and biggest mass.

According to the law of motion, the acceleration of the ith agent, at iteration t is given by the following equation:

$$\mathbf{a}_i(t) = \frac{\mathbf{F}_i(t)}{M_i(t)} \tag{1.16}$$

The velocity and position of an agent are updated as follows:

$$\mathbf{v}_i(t+1) = r_i \cdot \mathbf{v}_i(t) + \mathbf{a}_i(t) \tag{1.17}$$

$$\mathbf{x}_i(t+1) = \mathbf{x}_i(t) + \mathbf{v}_i(t+1) \tag{1.18}$$

where r_i is a uniform random variable in the interval [0, 1].

The gravitational constant $G(t)$ in (1.15) is a function of the initial value G_0 and time t:

$$G(t) = G_0 \cdot \exp\left(-\alpha \cdot \frac{t}{t_{\max}}\right) \tag{1.19}$$

The parameters of maximum iteration $t_{\max}$, population size N, initial gravitational constant G_0, and constant α control the performance of GSA.

1.3.9 Wind-driven optimization

The wind-driven optimization (WDO) was developed by Bayraktar *et al.* [17]. The core equations of WDO are inspired by the wind in the Earth's atmosphere where the motion of an infinitesimally small air parcel is analyzed. To find the velocity and position displacement of the air parcel, the Newton's second law of motion is used. There are four major forces that can either cause the wind to move in a certain direction or deflect it from its existing path. These forces are the pressure gradient force, the friction force, the gravitational force, and the Coriolis force. The physical equations that govern each of these forces and detailed descriptions are given in [17]. The sum of these forces can be inserted into Newton's second law of motion and then the velocity and position displacement of each air parcel can be computed. The parcel's velocity is calculated using the following equation:

$$\mathbf{v}_{new} = (1-\alpha)\mathbf{v}_{cur} - g\mathbf{x}_{cur} + \left(RT\left|\frac{1}{i}-1\right|\left(\mathbf{x}_{opt}-\mathbf{x}_{cur}\right)\right) + \left(\frac{c\mathbf{v}_{cur}^{other\ dim}}{\mathrm{i}}\right) \tag{1.20}$$

where $\mathbf{v}_{cur}$ is the current iteration velocity, $\mathbf{x}_{cur}$ is the current position of the air parcel in the search space, $\mathbf{x}_{opt}$ is the optimum position of the air parcel's in the search space at the current iteration, i represents the ranking among all air parcels (the best solution $\mathbf{x}_{opt}$ has the lowest pressure with rank 1), $\mathbf{v}_{cur}^{other\ dim}$ is the velocity in one of the other dimensions, α is the friction coefficient, g is the gravitational constant, R is the universal gas constant, T is the temperature, and c is a constant that represents the rotation of the Earth. The coefficients α, g, RT, and c are the algorithm parameters that must be specified prior to starting an optimization.

At each iteration, the velocity and the position of all parcels need to be updated. Once the new velocity is calculated according (1.21), the position can be updated as follows:

$$\mathbf{x}_{new} = \mathbf{x}_{cur} + \mathbf{v}_{new} \tag{1.21}$$

where $\mathbf{x}_{new}$ is the new position of the air parcel in the search space for the next iteration.

1.3.10 Colliding bodies optimization

The colliding bodies optimization (CBO) algorithm was proposed by Kaveh and Mahdavi [18]. The CBO algorithm is based on one-dimensional collisions between bodies, with each search agent (potential solution) being considered as an object or body with mass. After a collision of two moving bodies having specified masses and velocities, these bodies are separated with new velocities. This collision causes the agents to move toward better positions in the search space.

The massed objects are composed of two main equal groups, i.e., stationary and moving objects, where the moving objects move to follow stationary objects and a collision occurs between pairs of objects. This is done for two purposes: (i) to improve the positions of moving objects and (ii) to push stationary objects toward better positions. After the collision, new positions of colliding bodies are updated based on new velocity by using the collision laws.

The CBO algorithm can be described in following steps:

1. Initialization. The initial position of each colliding body (CB) is determined with random initialization in the search space.
2. Determine the magnitude of the body mass for each CB based on the objective function value of the agents in the current population.

3. Perform the arrangement of the CBs objective function values in ascending order, and the sorted CBs are equally dividing into two groups: (i) the lower half of CBs (stationary CBs) and (ii) the upper half of CBs (moving CBs).
4. Calculate the velocity of each moving CB and the velocity of each stationary CB after the collision.
5. Update the position of each moving CB and the position of each stationary CB using the calculated velocities after the collision.
6. Repeat steps 2–5 until a termination criterion is satisfied, that is maximum iteration number.

1.3.11 Black hole algorithm

Hatamlou [19] developed a population-based optimization algorithm named the black hole (BH) algorithm inspired by BH phenomenon. A BH in space is formed when a massive star collapses. The gravitational force of the BH is too high that even the light cannot escape from it. The sphere-shaped boundary of a BH in space is known as the event horizon. Nothing can escape from within the event horizon because nothing can go faster than light. The radius of the event horizon is termed as the Schwarzschild radius:

$$R = \frac{2GM}{c^2} \tag{1.22}$$

where G is the gravitational constant, M is the mass of the BH, and c is the speed of light.

In the BH algorithm, a population consists of stars. Each star presents a candidate for a solution of the optimization problem. The best candidate among all the candidates at each iteration is selected as a BH and all the other candidates form the normal stars. The creation of the BH is not random and it is one of the real candidates of the population. Then, all the candidates are moved toward the BH based on their current location and a random number. The main steps in the BH algorithm are as follows:

1. Initialization. Randomly generate population of N candidate solutions—the stars in n search space.
2. Evaluate the fitness value for each star of the population.
3. Select the best star that has the best fitness value as the BH.
4. Change the location of each star according to following equation:

   $$\mathbf{x}_i(t+1) = \mathbf{x}_i(t) + r \cdot (\mathbf{x}_{BH} - \mathbf{x}_i(t)) \quad i = 1, 2, \ldots, N \tag{1.23}$$

   where $\mathbf{x}_i(t)$ and $\mathbf{x}_i(t+1)$ are the locations of the ith star at the iteration t and $t+1$, respectively; $\mathbf{x}_{BH}$ is the location of the BH in the search space; r is a random number in the interval [0,1]; N is the population size, i.e., the number of stars (candidate solutions).
5. If a star reaches a location with lower cost than the BH, exchange their locations.

6. Calculate the radius of the event horizon using the following equation:

$$R = \frac{f_{BH}}{\sum_{i=1}^{N} f_i} \tag{1.24}$$

where f_{BH} is the fitness value of the BH and f_i is the fitness value of the ith star.
7. If the distance between a star and the BH is less than R, that star is collapsed and a new star is created and distributed randomly in the search space.
8. Repeat steps 2–7 until the maximum iteration number is reached.
9. Take the BH in last iteration as optimal solution of the problem.

1.3.12 Gray wolf optimizer

The gray wolf optimizer (GWO) algorithm is based on mimics of the leadership hierarchy and hunting mechanism of gray wolves in nature and developed by Mirjalili *et al.* [20]. In the hierarchy of GWO, four types of members (search agents) can be considered: alpha (α), beta (β), delta (δ), and omega (ω). The dominance gradually decreases from α wolves to ω wolves.

The mathematical model of hunting mechanism of gray wolves consists of the three steps, such as searching for prey, encircling prey, and attacking prey:

$$\mathbf{D} = \left|\mathbf{C} \cdot \mathbf{x}_p(t) - \mathbf{x}(\mathbf{t})\right| \tag{1.25}$$

$$\mathbf{x}(t+1) = \mathbf{x}_p(t) - \mathbf{A} \cdot \mathbf{D} \tag{1.26}$$

$$\mathbf{A} = 2\mathbf{a} \cdot \mathbf{r}_1 - \mathbf{a} \tag{1.27}$$

$$\mathbf{C} = 2\mathbf{r}_2 \tag{1.28}$$

$$\mathbf{x}(t+1) = \frac{\mathbf{x}_1(t) + \mathbf{x}_2(t) + \mathbf{x}_3(t)}{3} \tag{1.29}$$

where t is the current iteration, $\mathbf{A}$ and $\mathbf{C}$ are coefficient vectors, $\mathbf{x}_p$ is the position vector of prey, $\mathbf{x}$ is the position vector of a gray wolf (α, β, and δ), $\mathbf{a}$ is vector whose components have linearly decreased from 2 to 0 over the course of iterations, $\mathbf{r}_1$ and $\mathbf{r}_2$ are random vectors in [0,1]. The search agents ω update their positions according to the position of three best search agents. The gray wolves finish the hunt by attacking the prey when it stops moving. This can be mathematically modeled by decreasing the value of vector $\mathbf{a}$ from 2 to 0 with iterations.

1.3.13 Firefly algorithm

The firefly algorithm was inspired by the flashing light of fireflies and developed by Yang [21]. All the fireflies are considered unisexual and their attraction is directly proportional to the intensity of their flash. Therefore, if a firefly particle had the choice of moving toward either of two fireflies, it will be more attracted toward the firefly with higher brightness and moves in that direction. If there are no fireflies nearby, the firefly will move in a random direction. The brightness of flash is associated with the fitness function. As a firefly's attractiveness is proportional to the light intensity seen by adjacent fireflies, the attractiveness β of a firefly can be defined as a function of the Cartesian distance r between the fireflies:

$$\beta = \beta_0 \exp\left(-\gamma r^2\right) \tag{1.30}$$

where β_0 is the attractiveness at $r = 0$ and γ is the absorption coefficient. The movement of a firefly i is attracted to another more attractive (brighter) firefly j is determined by

$$\mathbf{x}_i(t+1) = \mathbf{x}_i(t) + \beta_0 \exp\left(-\gamma r_{ij}^2\right)\left(\mathbf{x}_j(t) - \mathbf{x}_i(t)\right) + \alpha\varepsilon \tag{1.31}$$

where α is the randomization parameter and ε is the vector of random numbers taken from Gaussian distribution. At the end of all generations, the firefly with the highest brightness, i.e., the best fitness value is concluded as the optimal solution to the problem.

1.3.14 Cuckoo search algorithm

Cuckoo search algorithm is based on the obligate brood parasitic behavior of some cuckoo species in combination with the Lévy flight behavior of some birds and fruit flies [22]. This algorithm uses three idealized rules: (i) each cuckoo lays one egg at a time and dumps it in a randomly chosen nest; (ii) the best nests with high quality of eggs (fitness) will carry over to the next generations; and (iii) the number of available host nests is fixed, and the egg laid by a cuckoo is discovered by the host bird with a probability $p_a \in [0, 1]$. In this case, the host bird can either throw the egg away or abandon the nest, and build a completely new nest.

A new cuckoo (solution) by Lévy flights is generated as follows:

$$\mathbf{x}_i(t+1) = \mathbf{x}_i(t) + \alpha \cdot \text{Lévy}(\lambda) \tag{1.32}$$

where $\alpha > 0$ is the step size which should be related to the scales of the problem of interest. The Lévy flight essentially provides a random walk while the random step length is drawn from a Lévy distribution which has an infinite variance with an infinite mean. Here the steps essentially form a random walk process with a power-law step-length distribution with a heavy tail. Some of the new solutions should be generated by Lévy walk around the best solution obtained so far, this will speed up the local search. However, a substantial fraction of the new solutions should be generated by far field randomization and whose locations should be far enough from the current best solution, this will make sure the system will not be trapped in a local optimum.

1.3.15 Moth swarm algorithm

In the moth swarm algorithm (MSA), the possible solution of optimization problem is represented by position of light source, and the fitness/quality of this solution is considered as luminescence intensity of the light source [23]. These assumptions have been used to approximate the characteristics of the proposed algorithm. In addition, the proposed moth swarm is considered to consist of three groups of moths, which are as follows:

Pathfinders, whose main task is to discriminate the best positions as a light sources to guide the movement of the main swarm (i.e., light the way);
Prospectors, that tends to wander into a random spiral path within the neighborhood of the light sources, which have been marked by the pathfinders;
Onlookers is a group of moths that drift directly toward the best global solution (moonlight), which has been obtained by prospectors.

At any iteration, each moth (search agent) is incorporated into the optimization problem to find the luminescence intensity of its corresponding light source (fitness). The best fitnesses in the swarm are considered as the positions of the pathfinders, and guidance for the next update iteration. Hence, the second and third best groups take the name of the prospectors and onlookers, respectively. The equations for updating the positions of moths are given in [23].

1.3.16 Krill herd algorithm

Krill herd (KH) algorithm was developed by Gandomi and Alavi [24]. The KH algorithm is based on the simulation of the herding behavior of krill individuals. The minimum distances of each individual krill from food and from highest density of the herd are considered as the objective function for the krill movement.

As it is well known, a metaheuristic algorithm must have balanced two main characteristics, namely, the exploration (or random search) and exploitation (or local search) to achieve highest performance in solving optimization problem. The KH algorithm involves three movements and follows the search directions to enhance the objective function value. These movements are

1. movement induced by the presence of other individuals
2. foraging activity
3. random diffusion.

Regular KH algorithm may be expressed by Lagrangian model in an n-dimensional search space as follows:

$$\frac{d\mathbf{x}_i}{dt} = \mathbf{N}_i + \mathbf{F}_i + \mathbf{D}_i \tag{1.33}$$

where $\mathbf{N}_i$ is the motion induced by other krill individuals, $\mathbf{F}_i$ is the foraging motion, and $\mathbf{D}_i$ is the physical diffusion of the ith krill individuals.

For more precise modeling of the krill behavior, two adaptive genetic operators such as crossover and mutation are added to the KH algorithm.

Generally, the KH algorithm can be introduced by the following steps [24]:

1. Data structures: Define the simple bounds, determination of algorithm parameter(s), etc.
2. Initialization: Randomly create the initial population in the search space.

3. Fitness evaluation: Evaluation of each krill individual according to its position.
4. Motion calculation:
   i. Motion induced by the presence of other individuals,
   ii. Foraging motion,
   iii. Physical diffusion,
5. Implement the genetic operators (crossover and mutation).
6. Updating: Updating the krill individual position in the search space using the following equation:

$$\mathbf{x}_i(t+\Delta t) = \mathbf{x}_i(t) + \Delta t\frac{d\mathbf{x}_i}{dt} \tag{1.34}$$

   where parameter Δt depends on the search space and it can be obtained from the following formula:

$$\Delta t = C_t \sum_{j=1}^{n} \left(UB_j - LB_j\right) \tag{1.35}$$

   where n is the total number of variables, and UB_j and LB_j are upper and lower bounds of the jth variables ($j = 1,2,\ldots,n$), respectively. C_t is a constant number between [0,2].
7. Repeating: Go to step III until the stop criteria is reached.
8. End

1.3.17 Shuffled frog-leaping algorithm

Shuffled frog-leaping algorithm (SFLA) was proposed by Eusuff *et al.* [25] for solving combinatorial optimization problems. The SFLA mimics the nature of frogs in the memeplexes. Each frog represents a feasible solution to an optimization problem. The algorithm contains elements of local search and global information exchange. The SFLA consists of a set of interacting virtual population of frogs partitioned into different memeplexes. The virtual frogs act as hosts or carriers of memes where a meme is a unit of cultural evolution. The algorithm performs simultaneously an independent local search in each memeplex. The local search is completed using a PSO-like method adapted for discrete problems but emphasizing a local search. To ensure global exploration, the virtual frogs are periodically shuffled and reorganized into new memplexes in a technique similar to that used in the shuffled complex evolution algorithm. In addition, to provide the opportunity for random generation of improved information, random virtual frogs are generated and substituted in the population [25].

1.3.18 Bacterial colony foraging optimization

Bacterial foraging optimization algorithm is inspired by an activity called "chemotaxis" exhibited by bacterial foraging behaviors [26]. Motile bacteria such as *Escherichia coli* and Salmonella propel themselves by rotation of the flagella.

To move forward, the flagella rotates counterclockwise and the organism "swims" (or "runs"), while a clockwise rotation of the flagellum causes the bacterium to randomly "tumble" itself in a new direction and swim again. Alternation between "swim" and "tumble" enable the bacterium to search for nutrients in random directions. Swimming is more frequent as the bacterium approaches a nutrient gradient. Tumbling, hence direction changes, is more frequent as the bacterium moves away from some food to search for more. Basically, bacterial chemotaxis is a complex combination of swimming and tumbling that keeps bacteria in places of higher concentrations of nutrients. Chemotaxis can also be considered as the optimal foraging decision making capabilities of bacteria.

The original bacterial colony foraging optimization (BCFO) system consists of four principal mechanisms, namely, chemotaxis, swarming, reproduction, and elimination-dispersal. The main idea of BCFO proposed in [26] is to develop an adaptive and cooperative life-cycle model by combining bacterial chemotaxis, cell-to-cell communication, and self-adaptive searching strategies. The BCFO is a more bacterially realistic model that the bacteria split and die dynamically throughout the foraging process and the population size varies as the algorithm runs. The cell-to-cell communication enables the bacteria tumble toward better directions in the chemotactic steps to speed up convergence. With the self-adaptive searching strategy, each bacterium can maintain an appropriate balance between exploration and exploitation [26].

1.3.19 Biogeography-based optimization

Biogeography-based optimization (BBO) algorithm developed by Simon [27] is strongly related with the equilibrium theory of island biogeography. The BBO is based on the mathematical equations that govern the geographical distribution of biological organisms. Biogeography describes how a species migrates from one island to another, how new species arises, and how it becomes extinct. A habitat is defined as an island (area) that is geographically isolated from other islands. In BBO, the term "island" is used descriptively rather than literally. That is, an island is any habitat that is geographically isolated from other habitats. Therefore, the BBO use the more generic term "habitat" rather than "island." Geographical areas that are well suited as residences for biological species are said to have a high habitat suitability index (HSI). HSI is influenced by factors such as rainfall, diversity of vegetation, diversity of topographic features, land area, temperature, and so on. The variables that characterize habitability are called suitability index variables (SIVs). SIVs can be considered as the independent variables of the habitat and HSI calculation is carried out using these variables. Habitats with a high HSI tend to have a large number of species, while those with a low HSI have fewer numbers of species. The migration of some species from a habitat to an exterior habitat is known as emigration process and an entry of some species into one habitat from an outside habitat is known as immigration process. Habitats with high HSI have low species immigration rate because they are nearly saturated with species and are more static in their species distribution compared to low HSI habitats. By the same token high HSI habitats have higher emigration rate. The species on high HSI islands have more opportunities to emigrate to neighboring habitats and to share their characteristics with local habitats. Habitats with a low HSI have a high species immigration rate because of their sparse populations.

BBO algorithm maintains a set of candidate solutions called habitats, and each habitat feature is measured by a SIV. The HSI is used as a measure of how good a solution is, which is analogous to fitness in other population-based optimization algorithms. The greater the total number of species on the habitat, which corresponds to a high HSI, the better the solution it contains. The number of species present on the habitat is determined by a balance between the rate at which the new species arrive and the rate at which the old species become extinct on the habitat. In BBO, each individual has its own immigration rate (λ) and emigration rate (μ). These parameters are affected by the number of species (S) in a habitat and are used to probabilistically share information between habitats. Habitats with smaller populations are more vulnerable to extinction (i.e., the immigration rate is high). But as more species inhabit the habitat, the immigration rate reduces and the emigration rate increases. In BBO, good solutions (i.e., habitats with many species) tend to share their features with poor solutions (i.e., habitats with few species), and poor solutions accept a lot of new features from good solutions. Maximum immigration rate (I) occurs when island is empty and decreases as more species are added and maximum emigration rate (E) occurs when all possible species S_{max} are present on the habitat [6].

1.3.20 Teaching–learning-based optimization

Teaching–learning-based optimization (TLBO) is a metaheuristic optimization method proposed by Rao *et al.* [28]. This optimization algorithm does not require any algorithm-specific parameters, except population size and maximum number of iterations. Like other population-based algorithms, TLBO starts with a randomly generated population of candidate solutions. Then, the process of TLBO is divided into two parts, namely, the "Teacher Phase" and the "Learner Phase." In the "Teacher Phase," a teacher improves the mean level of learners. The knowledge of a class increases depending upon a good teacher because he/she brings the level of his/her learners to his/her level of knowledge. However, in actual life, this is not always the case because the level of learners depends on other factors like their aptitudes and their efforts and commitment to learn. Thus, a teacher can only increase the mean level of his/her learners. In the "Learner Phase," the learners improve their knowledge by interacting with other learners, i.e., between themselves. A learner i interacts with another learner j randomly selected. A learner learns something new, i.e., increases his knowledge if the second learner has more knowledge than him. Detailed explanations of TLBO are given in [28].

1.3.21 League championship algorithm

League championship algorithm was developed by Kashan [29], as a new metaheuristic algorithm to solve continuous optimization problems. Like other population-based metaheuristic optimization algorithms, league championship algorithm (LCA) works with a population of solutions to find the optimal solution. Each team (individual) in the league (population) represents a feasible solution to the problem being solved. These teams compete in an artificial league for several weeks (iterations). Based on the league schedule at each week, teams play in pairs (i.e., team i plays against team j) and the outcome is determined in terms of win or loss based on each team playing strength (corresponding to the fitness value) resultant from a particular team formation (solution). In the recovery period, keeping track of the previous week events, each team devises the required changes in its formation to set up a new formation (a new solution is generated) for the next week contest and the championship goes on for a number of seasons (stopping criterion) [38].

Like other evolutionary algorithms, the LCA works with a population of individuals. Hence, in the initialization step, a league (population) of N (the league size) teams (solutions) is generated and their playing strengths (fitness values) are evaluated. If a function of n variables is considered, hence each team is composed of n players corresponding to the number of variables. In this step, the teams' best formations take the initialization values. After that comes the competition step. Based on the league schedule, the said teams compete in pairs for $S \times (N - 1)$ weeks where S is the number of seasons and a week is noted as t. After each competition or game between team i and team j, based on the playing strength of each team, the outcome is given in terms of win or loss; there is no tie. In the recovery step, each team elaborate a new formation taking into account the team's current best formation and previous week events. In LCA, the selection is a greedy one. It replaces the current best formation with a more productive team formation with a better playing strength. In other words, if the new formation is the fittest one (i.e., the new solution is the best solution obtained so far for the ith member of the population), hereafter consider the new formation as the team's current best formation. The algorithm stops after the stopping criterion is met [29,38].

1.3.22 Mine blast algorithm

Mine blast algorithm was developed by Sadollah *et al.* [30]. The idea of the proposed algorithm is based on the observation of a mine bomb explosion, in which the thrown pieces of shrapnel collide with other mine bombs near the explosion area resulting in their explosion. To understand this situation, consider a mine field where the aim is to clear the mines. Hence, the goal is to find the mines, while importantly to find the one with the most explosive effect located at the optimal point $\mathbf{x}^*$ which can cause the most casualties (min or max $f(\mathbf{x})$ per $\mathbf{x}^*$).

The mine bombs of different sizes and explosive powers are planted under the ground. When a mine bomb is exploded, it spreads many shrapnel pieces and the casualties ($f(\mathbf{x})$) caused by each piece of shrapnel are calculated. A high value for casualties per piece of shrapnel in an area may indicate the existence of other mines which may or may not have higher explosive power. Each shrapnel piece has definite directions and distances to collide with other mine bombs which may lead to the explosion of other mines due to the collision. The collision of shrapnel pieces with other mines may lead us to discover the most explosive mine. The casualties caused by the explosion of a mine bomb are considered as the fitness of the objective function at the mine bomb's location. The domain (mine field) solution may be divided into infinite grid where there is one mine bomb in each portion of the grid [30].

1.3.23 Sine cosine algorithm

Recently, Mirjalili [31] proposed a novel population-based optimization algorithm called sine cosine algorithm (SCA). The particularity of the SCA can be outlined as follows [31,39]:

1. The SCA creates multiple initial random candidate solutions and requires them to fluctuate outwards or toward the best solution based on sine and cosine functions.
2. Exploration phase is performed when the sine and cosine functions return a value greater than 1 or less than -1.
3. The exploitation phase is performed when sine and cosine functions return value between 1 and -1.
4. A specified random and adaptive variables are integrated within the algorithm to balance between exploration and exploitation during search process. The structure of the standard mechanism search of the SCA can be summarized as follows [31,39].

Phase 1: Initialization. Like many population-based optimization techniques, the SCA starts the optimization search process with a random solution.

Phase 2: Evaluate and update solution. This random solution is evaluated repeatedly by a specified objective function and improved by a set of rules for exploration and exploitation stages. These two equations were updated based on a switching parameter and are expressed as follows:

$$x_i^d(t+1) = \begin{cases} x_i^d(t) + r_1 \cdot \sin(r_2) \cdot \left| r_3 \cdot p_i^d(t) - x_i^d(t) \right|, & r_4 < 0.5 \\ x_i^d(t) + r_1 \cdot \cos(r_2) \cdot \left| r_3 \cdot p_i^d(t) - x_i^d(t) \right|, & r_4 \geq 0.5 \end{cases} \tag{1.36}$$

where $x_i^d(t)$ is the position of the solution i in dth dimension at iteration t; $p_i^d(t)$ is the destination point i in dth dimension at iteration t; r_1, r_2, r_3, and r_4 are the parameters of SCA.

As (1.36) shows, there are four main parameters in SCA: r_1, r_2, r_3, and r_4.

The parameter r_1 is designed to guide the next position's region, which may be between the solution and destination or outside it. In order to achieve balance between exploration and exploitation phase, this parameter is dynamically adjusted during search process using the following equation:

$$r_1 = a - a\frac{t}{t_{\max}} \tag{1.37}$$

with

$$a = 2 \cdot \left(1 - 0.5 \cdot r \cdot \sin\left(\frac{t}{t_{\max}}\right)\right) \tag{1.38}$$

where t is the current iteration, $t_{\max}$ is the maximum iterations number, and r is a random number in [0,1].

The parameter r_2 defines how far the movement should be toward or outwards the distinction

$$r_2 = 2 \cdot \pi \cdot r \tag{1.39}$$

where r is a random number in [0,1].

The parameter r_3 gives random weights for destination in order to stochastically emphasize ($r_3 > 1$) or de-emphasize ($r_3 < 1$) the effect of desalination in defining the distance

$$r_3 = 2 \cdot r \cdot \left|\sin\left(\frac{r}{wf}\right)\right| \tag{1.40}$$

where r is a random number in [0,1], and wf is a constant.

The parameter r_4 is a random number in [0,1] that switches the transition between the sine and cosine components in (1.37).

1.3.24 Harmony search

Harmony search (HS) is a music-based metaheuristic optimization algorithm proposed by Geem *et al.* [32]. HS is an example of a successful transformation of the qualitative process of improvisation into certain quantitative rules by idealization. The process of finding the perfect harmony in music is adapted to the rules of the optimization search algorithm, namely, the HS algorithm [40].

To clarify the HS in more detail, one idealized process of improvisation of an experienced musician will first be analyzed. In improvisation, the musician has three possible choices: (i) plays a famous piece of music (a series of pitches in harmony) exactly from his memory; (ii) plays something similar to a known piece (thus adjusting the pitch slightly); or (iii) composes new or random notes. In [32], the authors formalized these three options into a quantitative optimization process, where they became: usage of harmony memory, pitch adjustment, and randomization.

The usage of harmony memory ensures that the best harmonies will be carried over to the new harmony memory. This process is controlled by a parameter $r_{accept} \in [\,0,1]$, called harmony memory accepting rate.

The pitch adjustment is determined by a pitch bandwidth $\mathbf{b}_{range}$ and a pitch adjusting rate r_{pa}. The pitch bandwidth can be determined as follows:

$$\mathbf{b}_{range} = \frac{\mathbf{x}_{\max} - \mathbf{x}_{\min}}{N} \tag{1.41}$$

where $\mathbf{x}_{\max}$ and $\mathbf{x}_{\min}$ are the limits of control variables, and N is the total number of control variables. Though in music, pitch adjustment means to change the frequencies, it corresponds to generate a slightly different solution in the HS

$$\mathbf{x}_{new} = \mathbf{x}_{old} + \varepsilon \cdot \mathbf{b}_{range} \tag{1.42}$$

where $\mathbf{x}_{old}$ is the existing pitch (solution) from the harmony memory, and $\mathbf{x}_{new}$ is the new pitch after the pitch adjusting action; ε is a random number generator in the range of $[-1,1]$. Pitch adjustment is similar to the mutation operator in GAs. The pitch-adjusting rate (r_{pa}) is adopted to control the degree of the adjustment.

The third component of HS is the randomization, which is to increase the diversity of the solutions and probability to find the global optimal solution.

The three components in HS can be summarized by the following pseudo code [40]:

begin
    Define objective function, and control variables
    Generate initial harmonics (initial population)
    Define pitch adjusting rate (r_{pa}), pitch limits and bandwidth
    Define harmony memory accepting rate (r_{accept})
    while *($t < t_{max}$)*
    Generate new harmonics by accepting best harmonics
    Adjust pitch to get new harmonics (solutions)
        if *(rand > r_{accept}), choose an existing harmonic randomly*
        else if *(rand > r_{pa}), adjust the pitch randomly within limits*
        else *generate new harmonics via randomization*
        end if
    Accept the new harmonics (solutions) if better
    end while
    Find the current best solutions
end

1.3.25 Imperialist competitive algorithm

Social and political behavior of people served as an inspiration for the development of this algorithm [33]. The basic principle of the algorithm is based on the population of individuals called country, classified into two categories: imperialists and colonies. Countries that impose their control over less-developed countries are called imperialists. On the other hand, countries under the control of imperialists are colonies. In imperialist competitive algorithm (ICA), each country is a potential solution of the problem. Similarly as other population-based algorithms, ICA starts with an initial set of potential solution-initial population (countries in the world). Some of the best countries in the population are selected to be the imperialist states and the rest countries form the colonies of these imperialists. All the colonies of initial population are divided among the mentioned imperialists based on their power. The power of an empire which corresponds to the fitness value of the potential solution of the optimization problem is inversely proportional to its cost [33].

After dividing all colonies among imperialists, these colonies start moving toward their relevant imperialist country. The total power of an empire depends on both the power of the imperialist country and the power of its colonies. This fact is modeled by defining the total power of an empire as the power of imperialist country plus a percentage of mean power of its colonies.

After that, the imperialist competition begins among all the empires. Any empire that is not able to succeed in this competition and can't increase its power, or at least prevent decreasing its power, will be eliminated from the competition. During imperialistic competition, the power of powerful empires gradually increases, whereas the power of weaker ones decreases. Consequently, weak empires will lose their power and ultimately they will collapse. In this way, the ICA converge to a state in which only one empire remains in the world and all the other countries are colonies of that empire. In this world, all the colonies are in the same position and have the same cost as the imperialist [33].

1.3.26 Differential search algorithm

Differential search algorithm (DSA) was developed by Civicioglu [34]. This algorithm is based on simulation of the Brownian-like random-walk movement used by an organism to migrate.

Many species of living beings show seasonal migration behavior throughout the year. This is due to the fact that the capacity and efficiency of the food areas existing in the nature often vary due to the periodical climatic changes during the year. Therefore, migration behavior allows the living beings to move from an area where capacity and diversity of natural sources is reduced to a more efficient area. In the migration movement, the migrating species constitute a superorganism containing a large number of individuals. Then the superorganism starts to change its position by moving toward more fruitful areas. Movement of a superorganism can be described by a Brownian-like random-walk model [34].

The behavior of superorganisms has been modeled using a number of computational intelligence algorithms, such as PSO, CS, ACO, and ABC. Many species of predatory living beings, before moving or migrating to a site, control the fertility of this one. In other words, if a superorganism desires to move to a new site that can meet its needs, this superorganism settles in this new site at least for a period of time. However, if a more fertile area is found, the superorganism continues its migration [34].

In the DSA, it is assumed that a population composed of random solutions of the concerned problem corresponds to an artificial superorganism migrating. Throughout DSA, the artificial superorganism migrates to global minimum value of the problem. However, during this migration process, the artificial superorganism tests if some randomly selected positions or locations are suitable to stop over temporarily during the migration. If such a suitable location is found, the members of the artificial superorganism that made such a discovery immediately settle at the found location and then continue their migration from this location [34].

1.3.27 Glowworm swarm optimization

Glowworm swarm optimization (GSO) algorithm shares a few features with some well-known algorithms, such as ACO and PSO, but with several significant differences [35]. The agents in GSO are thought of as glowworms that carry a luminescence quantity called luciferin along with them. The glowworms encode the fitness of their current locations, evaluated using the objective function, into a luciferin value that they broadcast to their neighbors. The glowworm identifies its neighbors and computes its movements by exploiting an adaptive neighborhood, which is bounded above by its sensor range. Each glowworm selects, using a probabilistic mechanism, a neighbor that has a luciferin value higher than its own and moves toward it. These movements based only on local information and selective neighbor interactions enable the swarm of glowworms to partition into disjoint subgroups that converge on multiple optima of a given multimodal function [35]. The GSO algorithm can be summarized in following steps:

1. Define the optimization problem, i.e., the objective function and control variables;
2. Read GSO algorithm parameters;
3. Initialize initial luciferin value, local decision range r_o;
4. Initialize the glowworm within the limits of each variable;
5. Calculate objective function and find the luciferin value of all glowworms using following equation:

$$\ell_i(t+1) = (1-\rho)\ell_i(t) + \gamma J(\mathbf{x}_i(t+1)) \tag{1.43}$$

   where $\ell_i(t)$ is the luciferin level associated with the glowworm i at time (iteration) t; ρ is the luciferin decay constant ($0 < \rho < 1$); γ is the luciferin enhancement constant; and $J(\mathbf{x}_i(t))$ is the value of the objective function at agent i's location at time t;
6. Find the neighborhood glowworms having brighter glow and are in the local decision range;
7. Find the probability of glowworm moving toward a neighbor using equation:

$$p_{ij}(t) = \frac{\ell_j(t) - \ell_i(t)}{\sum_{k\in N_i(t)} \ell_k(t) - \ell_j(t)} \tag{1.44}$$

   where $j \in N_i(t), N_i(t) = \{j : d_{ij}(t) < r_d^i(t); \ell_i(t) < \ell_j(t)\}$ is the set of neighbors of glowworm i at time t, $d_{ij}(t)$ represents the Euclidian distance between glowworms i and j at time t, and $r_d^i(t)$ represents the variable neighborhood range associated with glowworm i at time t.
8. Update the glowworm movement using (1.45) and check the limits.

$$\mathbf{x}_i(t+1) = \mathbf{x}_i(t) + s\left(\frac{\mathbf{x}_j(t) - \mathbf{x}_i(t)}{\|\mathbf{x}_j(t) - \mathbf{x}_i(t)\|}\right) \tag{1.45}$$

   where $\mathbf{x}_i(t)$ is the location of glowworm (potential solution) i at time t; s (>0) is the step size.

9. Update the local decision range of all glowworms using (1.46):

$$r_d^i(t+1) = \min\{r_s, \max\{0, r_d^i(t) + \beta(n_t - |N_i(t)|)\}\} \tag{1.46}$$

where β is a constant parameter; n_t is a parameter used to control the number of neighbors; and r_s is the radial range of Luciferian sensor, $(0 < r_d^i \leq r_s)$.
10. Repeat the above steps 5–9 until maximum iterations are attained.
11. Display the results.

1.3.28 Spiral optimization algorithm

Spiral optimization algorithm, a metaheuristic optimization technique, was first proposed by Tamura and Yasuda [36]. SOA is a relatively simple search concept inspired by natural spiral phenomena. This algorithm utilizes logarithmic spirals generated by n-dimensional spiral models. The spiral model is a dynamical system whose state $\mathbf{x}(t) \in \Re^n (n \geq 2)$ converges to a center $\mathbf{x}^* \in \Re^n$ from an initial point $\mathbf{x}(0)$ with a logarithmic spiral trajectory [41]

$$\mathbf{x}(t+1) = \mathbf{x}^* + r\mathbf{R}(\theta)(\mathbf{x}(t) - \mathbf{x}^*) \quad t = 0, 1, 2, \ldots \tag{1.47}$$

where $r\mathbf{R}(\theta)(\mathbf{x}(t) - \mathbf{x}^*)$ is the spiral vector; $r > 0$ is the step rate of the distance between $\mathbf{x}(t)$ and $\mathbf{x}^*$ per t; $\theta \in (-\pi, \pi]$ is the rotation rate of $\mathbf{x}(t)$ around the center $\mathbf{x}^*$ per t, and $\mathbf{R}(\theta) \in \Re^{n\times n}$ is the composite rotation matrix generally defined by arbitrarily multiplying τ types of basic rotation matrices $\mathbf{R}_{i_\ell,j_\ell}(\theta) \in \Re^{n\times n}$, $(\ell = 1, ..., \tau)$ as follows:

$$\mathbf{R}(\theta) = (-1)^\beta \mathbf{R}_{i_1,j_1}(\theta) \times \cdots \times \mathbf{R}_{i_\tau,j_\tau}(\theta), \quad \beta \in \{0, 1\} \tag{1.48}$$

$$\mathbf{R}_{i_\ell,j_\ell} = \begin{array}{c} \\ \\ \\ i_\ell \\ \\ \\ \\ j_\ell \\ \\ \\ \\ \end{array} \begin{bmatrix} 1 & & & i_\ell & & & j_\ell & & \\ & \ddots & & & & & & & \\ & & 1 & & & & & & \\ & & & \cos\theta & & & -\sin\theta & & \\ & & & & 1 & & & & \\ & & & & & \ddots & & & \\ & & & & & & 1 & & \\ & & & \sin\theta & & & \cos\theta & & \\ & & & & & & & 1 & \\ & & & & & & & & \ddots \\ & & & & & & & & & 1 \end{bmatrix} \tag{1.49}$$

where $i_\ell, j_\ell \in \{1, \ldots, n\}$, $i_\ell < j_\ell$, $\ell = 1, \ldots, \tau$, and the blank elements indicate 0. n is the solution space dimension, that is number of elements of the vector $\mathbf{x}$, i.e., number of control variables of an optimization problem.

The SOA algorithm focuses on utilizing spiral trajectories that have both search diversification and intensification aspects. This algorithm uses N (≥ 2) (number of search points) of the spiral model (1.47) by replacing the fixed constant step rate r and the given constant center $\mathbf{x}^*$ with the tunable step rate $r(t)$ and the common center $\mathbf{x}^*(t)$ defined as the best solution until the tth iteration, respectively. The SOA algorithm to minimize an objective function is generally described as follows [41]:

Step 0: Preparation: Select the number of search points $N \geq 2$, the parameter θ, the rotation matrix $\mathbf{R}(\theta)$, the setting rule for the step rate $r(t)$, and the termination criterion (max iteration number).

Step 1: Initialization: Generate the initial search points $\mathbf{x}_i(0)$, $(i = 1, \ldots, m)$ and determine the center $\mathbf{x}^*(\mathbf{0}) = \mathbf{x}_{i_b}(0), i_b = \min\{\arg\min_{i=1,\ldots,m}\{f(\mathbf{x}_i(0))\}\}$, and then set $t = 0$.

Step 2: Update the step rate $r(t)$ according to the setting rule.

Step 3: Update the search points as follows:

$$\mathbf{x}_i(t+1) = \mathbf{x}^*(t) + r(t)\mathbf{R}(\theta)(\mathbf{x}_i(t) - \mathbf{x}^*(t)) \quad i = 1, \ldots, N \tag{1.50}$$

Step 4: Update the center as follows:

$$\mathbf{x}^*(t+1) = \begin{cases} \mathbf{x}_{i_b}(t+1)(\text{if } f(\mathbf{x}_{i_b}(t+1)) < f(\mathbf{x}^*(t))) \\ \mathbf{x}^*(t) \text{ otherwise} \end{cases} \tag{1.51}$$

where $i_b = \min\{\text{argmin}_{i=1,\ldots,m}\{f(\mathbf{x}_i(t+1))\}\}$

Step 5: Check the termination criterion. If the termination criterion is satisfied then terminate and output $\mathbf{x}^*(t)$. Otherwise, set $t = t + 1$ and return to Step 2.

1.3.29 The Jaya algorithm

Jaya is a population-based metaheuristic optimization algorithm proposed by Rao [37]. This algorithm is based on the concept that the solution obtained for a given problem should move toward the best solution and should avoid the worst solution. This algorithm requires only the common control parameters and does not require any algorithm-specific control parameters, except the two ordinary controlling parameters of population size (N) (i.e., the number of candidate solutions) and the total iterations. The basic Jaya algorithm has only one phase according to the aforementioned concept, making it a simple optimization technique [37].

The Jaya algorithm can be briefly described in the following steps:

Step 1: Define the objective function $f(\mathbf{x})$ to be optimized, and control variables in the vector $\mathbf{x}$ considered as a candidate solution; initialize the population size (N), number of control variables (n), limits of the control variables, and total iterations number (t_{max});

Step 2: Generate an initial random population of N candidate solutions (search agents) within the predefined limits of control variables;

Step 3: Calculate the value of the objective function for each candidate solution in the current population;

Step 4: Determine the best and worst solutions in the current population;

Step 5: Based on the best and worst solutions, update all candidate solutions as follows:

$$\begin{aligned} x_i^d(t+1) = x_i^d(t) + r_1^d(t)\left[x_{best}^d(t) - \left(\left|x_i^d(t)\right|\right)\right] \\ -r_2^d(t)\left[x_{worst}^d(t) - \left(\left|x_i^d(t)\right|\right)\right] \end{aligned} \tag{1.52}$$

where $x_i^d(t)$ is the value of dth control variable in ith candidate solution at iteration t; $r_1^d(t)$ and $r_2^d(t)$ are randomly generated numbers within the range of [0,1] for the dth control variable at iteration t; $x_{best}^d(t)$ is the best value of the dth control variable at iteration t; $x_{worst}^d(t)$ is the worst value of the dth control variable at iteration t;

Step 6: For each candidate solution, compare the objective function values for the previous $x_i^d(t)$ and updated $x_i^d(t+1)$ solution. Accept the updated solution if it is better than the previous solution. Otherwise, keep the previous solution;

Step 7: Repeat steps 3–6 until the maximum number of iterations is reached. Stop and report the optimal solution.

1.3.30 Creating a "new" algorithm

By analyzing the above-described population-based optimization algorithms, some general common rules can be established. The rules are as follows:

1. An initial population of N agents (potential solutions) is generated by random selection of values of control variables from the range of their predefined boundaries. Each agent is an n-dimensional vector whose elements are values of control variables.
2. Calculating the value of the objective function for each agent from the current population, determining the best and the worst value of the objective function, and the corresponding best (xbest) and worst (xworst) agents in the current population.
3. This step is essential in the process of defining the "new algorithm." Based on the best, xbest, worst, and xworst, a scheme is created to modify existing solutions from the current population to get new (better) solutions. Many authors with more or less success try to make this "scheme" by imitation of some occurrence in nature or society, and they represent this as a "new" optimization method. Basically, all updating schemes are based on modifying existing solutions according to their distance from the current best solution, plus a very important component that introduces a random variation (randomization) into the process of modifying (creating) a new solution.
4. The whole procedure described in points 2 and 3 is repeated until a maximum number of iterations (generations) is reached. The solution obtained in the last iteration is optimal solution of the optimization problem.

By following these rules, the author developed a "new" population-based optimization algorithm, named the JANA algorithm. The algorithm isn't inspired by any natural phenomena. Name of algorithm is the name of a dear person. The aim of the author was to show that everyone can suggest a "new algorithm" only if it applies simple general rules described above. It only needs to modify the updating scheme and maybe to "explain" it to some kind of occurrence in nature, for example, by the behavior of an exotic insect. The proposed JANA algorithm is a simple metaheuristic population-based optimization algorithm consisting of several simple steps.

Initialization

Step 1: Define the objective function $F(\mathbf{x})$; define the control variables in the vector $\mathbf{x}$ considered as a candidate solution; initialize the population size (N), number of control variables (n), limits of the control variables, and total iterations number ($t_{\max}$).

Step 2: Generate an initial random population of N candidate solutions (search agents) within the predefined limits of control variables.

Iterative procedure

Step 3: Calculate the fitness value $F(\mathbf{x}_i(t))$ for each agent $\mathbf{x}_i(t), \quad i = 1,\ldots,N$ in the current population, determine the best fitness value, and corresponding best agent $\mathbf{xbest}(t)$ in the current population.

Step 4: Update all candidate solutions as follows:

$$x_i^d(t+1) = x_i^d(t) + r_1 \cdot K_1 \left[x_i^d(t) - x_{best}^d(t)\right] + S \cdot r_2 \cdot K_2 \left[\left(x_i^d(t)\right)^2 - \left(x_{best}^d(t)\right)^2\right]$$
$$i = 1,\ldots,N; \quad d = 1,\ldots,n \tag{1.53}$$

where $x_i^d(t)$ is the value of dth control variable in ith candidate solution at iteration t; r_1 and r_2 are randomly generated numbers within the range of [0,1]; $x_{best}^d(t)$ is the best value of the dth control variable at iteration t; K_1 and K_2 are positive constants; S is a coefficient which value is determined in each iteration as follows: *if rand* > 0.5 *then* $S = 1$; *else* $S = -1$; *end.*

Step 5: Check the search space boundaries for all agents. The agents that go out of the search space are reinitialized randomly.

Step 6: Repeat the iterative procedure (steps 3–5) until the stop criteria is reached, i.e., $t = t_{\max}$.

Step 7: Report the best solution determined in the last iteration. Stop.

The MATLAB® code of the JANA algorithm is given as follows:

```
%Simple metaheuristic search algorithm (JANA)
%-----------------------------------------------
function []=JANA

%INITIALIZATION
%===============================================================
%Define the objective function in the function program Fobj.m
%Define the space of posible solutions in the function program bounds.m
%-----------------------------------------------
%Defining algirithm parameters:
 N=50;              %population size (number of particles N)
```

```
 tmax=200;           %maximum number of iterations
 type_opt='min';     %defining type of optimization ('min' or 'max')
 K1=1;               %constant
 K2=0.5;             %constant

%Defininig space of posible solution
[numbervar,lowerbound,upperbound]=bounds;

%Generating of a randomized initial population
if size(upperbound,2)==1
    X=rand(N,numbervar).*(upperbound-lowerbound)+lowerbound;
end
if size(upperbound,2)>1
    for i=1:numbervar
    high=upperbound(i);low=lowerbound(i);
    X(:,i)=rand(N,1).*(high-low)+low;
    end
end

%Create the best so far chart and average fitnesses chart
BestChart=[];

%ITERATIVE PROCEDURE
%==================================================================
for t=1:tmax
    %Checking allowable range
    for i=1:N
    %Agents that go out of the search space, are reinitialized randomly
    Tp=X(i,:)>upperbound;
    Tm=X(i,:)<lowerbound;
    X(i,:)=(X(i,:).*(~(Tp+Tm)))+((rand(1,numbervar).*(upperbound-
...lowerbound)+lowerbound).*(Tp+Tm));
    end

%Evaluation of the fitness value of each agent using function Fobj.m
for i=1:N
    x=X(i,:);
    fitness(i)=Fobj(x);
end
%Determination of the best fitness (Fbest) and corespoding best agent
%(xbest)
if type_opt=='min'
    [best best_X]=min(fitness);
elseif type_opt=='max'
    [best best_X]=max(fitness);
end
if t==1
    Fbest=best;
    xbest=X(best_X,:);
end
if type_opt=='min'
    if best<Fbest
        Fbest=best;
        xbest=X(best_X,:);
    end
elseif type_opt=='max'
    if best>Fbest
        Fbest=best;
        xbest=X(best_X,:);
    end
end
```

```
BestChart=[BestChart Fbest];

XBEST=ones(N,numbervar);
for k=1:N
   XBEST(k,:)=xbest;
end

%Updating the position (X) of each agent:
S=ones(N,numbervar);
for k=1:N
   for kk=1:numbervar
      if rand>0.5
         S(k,kk)=1;
      else
         S(k,kk)=-1;
      end
   end
end

X=X+K1*rand(N,numbervar).*(X-XBEST)+K2*S.*rand(N,numbervar).*((X.^2-
...XBEST.^2);

end   %end of iteration procedure

%RESULTS:
%=====================================================================
disp(' ');
%Best value of the objective function, Fbest
Fbest
%Optimal values of control variables
xbest

%Plotting convergence profile of algorithm:
plot(BestChart,'--k','LineWidth',2);
xlabel('\fontsize{11}\bf Iteration');ylabel('\fontsize{11}\bf Fobj');
legend('\fontsize{11}\bf JANA',1);grid on;

%FUNCTIONS
%=====================================================================
function [Fobj]=Fobj(x)
%Code of the objective function to be optimized
%Input: x - vector of control variavles
%Output: Fobj - value of objective function
%F17 standard benchmark function:
Fobj=(x(2)-(x(1)^2)*5.1/(4*(pi^2))+5/pi*x(1)-6)^2+...
     10*(1-1/(8*pi))*cos(x(1))+10;
return

function [numbervar,lowerbound,upperbound]=bounds
%Defining number of control variables,
%lower and upper limits of control variables
numbervar=2;  lowerbound(1:numbervar)=-5; upperbound(1:numbervar)=5;
return
```

1.4 Criticism of metaheuristics

As can be seen from Table 1.1, the population-based metaheuristic algorithms can be classified into several groups according to some similar characteristics. It is evident that there are very similar algorithms within one group in the sense that there is no significant difference in the structure of the algorithms. Out of the 29 methods discussed, even 50% (15 methods) were created in the last 6 years. This leads to two possible conclusions. First conclusion, the explosion of creativity and inspiration of the authors happened, and second perhaps more likely conclusion, is that there were an inflation of "novel" methods.

A detailed elaboration of this effect was conducted by Sorensen [2]. This author addressed a call for a more critical view on such methods, and attempted to answer the following questions:

- How and why were metaphors introduced to inspire the development of metaheuristic optimization methods?
- What are the main fallacies of most metaphor-based research on metaheuristics?
- Why is the field of metaheuristics so vulnerable to this type of "research"?
- How can these methods pass the peer-review test and why can they all present such good results, and as a rule always better than other methods?

The authors [2] have explained this on an imaginative way: a metaheuristic method (GAs, SA, TS, etc.) in the general sense is a cooking style, such as "French" or "Chinese" or "Cajun," and not a recipe such as "spaghetti carbonara 'a la Antonio Carluccio."

Further, Sorensen [2] gives some general remarks and guide to the authors who deal with metaheuristics as follows:

1. Any research on metaheuristics should be adequately framed in the general literature on metaheuristics and optimization in general, not just the literature on the specific method that is being developed or used. Adequately framing a method entails deconstructing it, showing which components it consists of, examining in which other metaheuristics these and similar components appear, and how these components were adapted to the specific problem that is being solved. For this to be at all possible, a new metaheuristic algorithm should be explained using the general optimization terminology (a solution should be called a "solution," for example), and not in the language of some obscure metaphor.
2. In general, all metaheuristic designs should return to a situation in which methods are developed based on insight into the structure of the problem. Especially, research in metaheuristics should be applauded if it yields insight into the reasons why specific methods work well on specific problems.
3. In the component-based view of metaheuristics, operators from one or a set of different metaheuristic frameworks can be combined into ever more powerful methods. Using a deconstruction process [42], it is possible to gain insight into the contribution of each component, which in turn allows the metaheuristic designer to remove the parts that are not essential to the functioning of the metaheuristic, and to draw important conclusions: (i) why the method works and (ii) which components are responsible for the core optimization power of the overall method.

4. It does not make sense to ask general questions such as "Is GA better than PSO?" as such questions would be akin to asking "Is the Chinese kitchen better than the French?" The answer to such questions can only be "it depends on the optimization problem."
5. Fair comparison of metaheuristic methods should be done by a rigorous statistical protocol.
6. Finally, "novel" metaheuristics based on new metaphors should be avoided if they cannot demonstrate a contribution to the field. To stress the point: renaming existing concepts does not count as a contribution. Even though methods may be called "novel" by their author, many present no new ideas, except for the occasional marginal variant of an already existing method. Moreover, these methods take up the space of truly innovative ideas and research, for example, in the analysis of existing heuristics. Because these methods invariably change the vocabulary, they are difficult to understand. Combined with the fact that the authors of these methods usually neglect to properly position "their" method in the metaheuristics literature, such methods present a loss of time and a step backward rather than forward.

1.5 Educational software—*metahopt*

The *metahopt* is an educational software tool developed in MATLAB GUI program environment to evaluate performance of different metaheuristic optimization methods using the standard benchmark functions. The *metahopt* can be freely downloaded from [43]. The program includes 23 benchmark functions and a number of metaheuristic optimization algorithms described above. The program generates the minimum of the chosen function and corresponding values of decision (control) variables, as well as the convergence profile of the selected optimization method, and statistical evaluation of the results. Using *metahopt*, the user can compare the performances of different optimization methods based on statistical evaluation of the results.

Figure 1.1 shows the main window of *metahopt* while it operates on a Windows-based computer. Shown is an example of the execution of the program in case minimization of the function F1 (Sphere Model) using PSO as the solution method.

As can be seen from Figure 1.1, the *metahopt* is organized in the following components:

1. Panel OPTIONS. It consists of two pop-up menus. In the first pop-up menu, the user performs selection of a benchmark function between 23 offered. Pop-up menu method enables to choose of a metaheuristic optimization method to minimize the function. The user can select one of many methods, such as PSO, GSA, PSOGSA, ABC, WDO, FFA, GWO, CS, MSA, BSA, TLBO, etc. When clicking on the optimization method, the panel to settings of the algorithm parameters will be opened. The user can use the default parameters by clicking on button Use default, or to enter the desired value for each of algorithm

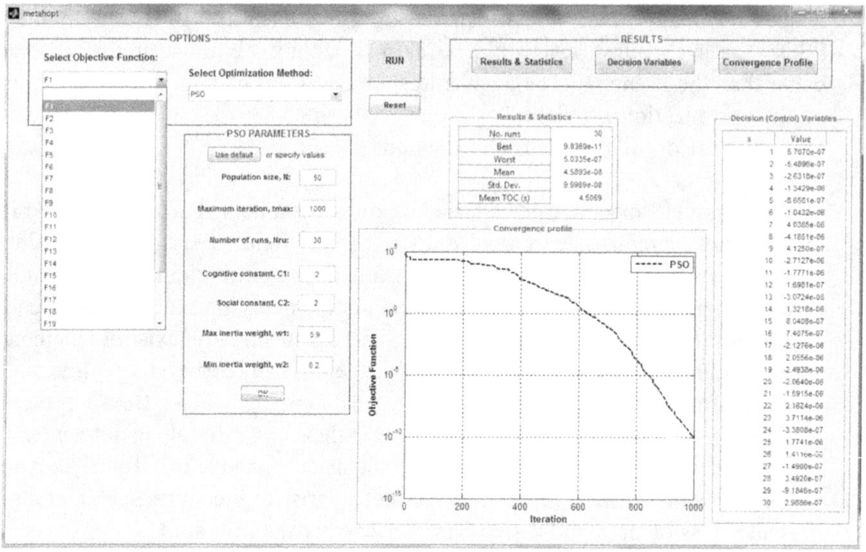

Figure 1.1 Main window of the metahopt

parameters. After selection of appropriate options on the panel OPTIONS, the user can start the program by clicking the button RUN. The button Function Details opens a pdf file with details related to benchmark functions.

2. Panel RESULTS. It manages the access to the different segments of the results obtained from the program. The best results, including objective function values in the last iteration, corresponding values of decision variables, and convergence profile of the optimization algorithm, are displayed immediately after execution of the program.

   Because of the stochastic nature of metaheuristic optimization methods, a very important segment of the program is statistical evaluation of the results. These statistical indicators are number of program execution, the minimum value of the objective function, the maximum value of the objective function, the mean value of the objective function, the standard deviation of the objective function, and the mean time of execution of the program. On the basis of these statistical indicators, a relevant comparison of different metaheuristic optimization methods can be achieved.

1.6 Conclusion

Metaheuristic optimization methods have become significant tools and often the only way of solving practical optimization problems. The basic requirement from these methods is to obtain the global solution or a solution close to the global optimum in a reasonable time.

Population-based metaheuristic optimization methods have several general features: (i) they are nature inspired; (ii) they have a stochastic dimension (involving random variables); (iii) they do not use the gradient or Hessian matrix of the objective function; and (iv) they have several parameters that need to be adjusted. These methods deal with a population of solutions.

Some of the general features that metaheuristic optimization methods should have are the following:

- They should be based on simple, universal, and easily understandable rules.
- Steps describing the metaheuristic method should be formulated with precise, preferably mathematical terms.
- All steps of the method should be in accordance with the rules that metaheuristics defined.
- The application of metaheuristics to a particular problem should ensure the global optimal solution or a solution close to the global optimum.
- The method should give equally good results for a wide range of examples from the same class, and not just for some selected test cases.

Lately there is an exponential growth of "novel" metaheuristic methods based on the imitation of some processes in nature and society. A more detailed analysis of these methods leads to the conclusion that most of them are actually only minor or higher modified versions of already established evolutionary (GA) or swarm intelligence (PSO)-based methods. Therefore, more attention should be paid to the analysis of existing well-established metaheuristics, rather than at all costs invent new and new methods.

References

[1] Leon-Aldaco S., Calleja H., Aguayo J. 'Metaheuristic optimization methods applied to power converters: a review'. *IEEE Transactions on Power Electronics*. 2015;**30**(12):6791–803

[2] Sorensen K. 'Metaheuristics—the metaphor exposed'. *International Transactions in Operational Research*. 2015;**22**:3–18

[3] Abou El Ela A.A., Abido M.A., Spea, S.R. 'Optimal power flow using differential evolution algorithm'. *Electrical Engineering*. 2009;**91**:69–78

[4] Varga Z.M. *Metaheurisitke*. Belgrade: University of Belgrade; 2008. p. 77

[5] Nanda S.J., Panda G. 'A survey on nature inspired metaheuristic algorithms for partitional clustering'. *Swarm and Evolutionary Computation*. 2014;**16**:1–8

[6] Boussaid I., Lepagnot J., Siarry P. 'A survey on optimization metaheuristics'. *Information Sciences*. 2013;**237**:82–117

[7] Holland J.H. *Adaptation in natural and artificial systems*. Cambridge: MIT Press; 1975

[8] Storn R., Price K. 'Differential evolution—a simple and efficient heuristic for global optimization over continuous spaces'. *Journal of Global Optimization*. 1997;**11**:341–59

[9] Fogel L.J., Owens A.J., Walsh M.J. *Artificial intelligence through simulated evolution*. New York: John Wiley & Sons; 1966. p. 160

[10] Yuryevich J., Wong K.P. 'Evolutionary programming based optimal power flow algorithm'. *IEEE Transactions on Power Systems*. 1999;**14**(4):1245–50

[11] Civicioglu P. 'Backtracking search optimization algorithm for numerical optimization problems'. *Applied Mathematics and Computation*. 2013;**219**:8121–44

[12] Kennedy J., Eberhart R.C. 'Particle swarm optimization'. *Proceedings of IEEE International Conference on Neural Networks*; Perth, Australia, 1995. IEEE; 1995. pp. 1942–8

[13] Dorigo M. *Optimization, learning and natural algorithms (Ph.D. thesis). Politecnico di Milano, Italy*, 1992

[14] Teodorović D., Lučić P., Marković G., Dell'Orco M. 'Bee colony optimization: principles and applications'. *Proceedings of 8th Seminar on Neural Network Applications in Electrical Engineering NEUREL 2006*; Belgrade, Serbia & Montenegro, Serbia, Sep 2006. IEEE; 2007. pp. 151–6

[15] Karaboga D., Basturk B. 'A powerful and efficient algorithm for numerical function optimization: artificial bee colony (ABC) algorithm'. *Journal of Global Optimization*. 2007;**39**:459–71

[16] Rashedi E., Nezamabadi-pour H., Saryazdi S. 'GSA: A gravitational search algorithm'. *Information Sciences*. 2009;**179**:2232–48

[17] Bayraktar Z., Komurcu M., Bosard J.A., Werner D.H. 'The wind driven optimization technique and its application in electromagnetic'. *IEEE Transactions on Antennas and Propagation*. 2013;**61**(5):2745–57

[18] Kaveh A., Mahdavai V.R. 'Colliding bodies optimization: a novel metaheuristic method'. *Computers and Structures*. 2014;**139**:18–27

[19] Hatamlou A. 'Black hole: a new heuristic optimization approach for data clustering'. *Information Sciences*. 2013;**222**:175–84

[20] Mirjalili S., Mirjalili S.M., Lewis A. 'Grey wolf optimizer'. *Advances in Engineering Software*. 2014;**69**:46–61

[21] Yang X.S. *Nature-inspired metaheuristic algorithms*. UK: Luniver Press; 2008

[22] Yang X.S. Deb S. 'Cuckoo search via Levy flights'. *Proceedings of World Congress on Nature and Biologically Inspired Computing*; Coimbatore, India, Dec 2009. IEEE; 2010. pp. 210–24

[23] Mohamed A.A., Mohamed Y.S., El-Gaafary A.A.M., Hemeida A.M. 'Optimal power flow using moth swarm algorithm'. *Electric Power System Research*. 2017;**142**:190–206

[24] Gandomi A.H., Alavi A.H. 'Krill herd: a new bio-inspired optimization algorithm'. *Communications in Nonlinear Science and Numerical Simulation*. 2012;**17**(12):4831–45

[25] Eusuff M., Lansey K., Pasha K. 'Shuffled frog-leaping algorithm: a memetic meta-heuristic for discrete optimization'. *Engineering Optimization*. 2006;**38**(2):129–54

[26] Chen H., Zhu Y., Hu K., Ma L. 'Bacterial colony foraging algorithm: combining chemotaxis, cell-to-cell communication, and self-adaptive strategy'. *Information Sciences*. 2014;**273**:73–100

[27] Simon D. 'Biogeography-based optimization'. *IEEE Transactions on Evolutionary Computation*. 2008;**12**:702–13

[28] Rao R.V., Savsani V.J., Vakharia D.P. 'Teaching-learning-based optimization: a novel method for constrained mechanical design optimization problems'. *Computer-Aided Design*. 2011;**43**(3):303–15

[29] Kashan A.H. 'League championship algorithm (LCA): an algorithm for global optimization inspired by sport championships'. *Applied Soft Computing*. 2014;**16**:171–200

[30] Sadollah A., Bahreininejad A., Eskandar H., Hamdi M. 'Mine blast algorithm: a new population based algorithm for solving constrained engineering optimization problems'. *Applied Soft Computing*. 2013;**13**:2592–612

[31] Mirjalili S. 'SCA: a sine cosine algorithm for solving optimization problems'. *Knowledge-Based Systems*. 2016;**96**:120–33

[32] Geem Z.W., Kim J.H., Loganathan G.V. 'A new heuristic optimization algorithm: harmony search'. *Simulation*. 2001;**76**:60–8

[33] Atashpaz-Gargari E., Lucas C. 'Imperialist competitive algorithm: an algorithm for optimization inspired by imperialistic competition'. *Proceedings of IEEE Congress on Evolutionary Computation*; Singapore, Singapore, Sep 2007. IEEE; 2008. pp. 4661–7

[34] Civicioglu P. 'Transforming geocentric Cartesian coordinates to geodetic coordinates by using differential search algorithm'. *Computers and Geosciences*. 2012;**46**:229–47

[35] Krishnand K.N., Ghose D. 'Glowworm swarm optimisation for simultaneous capture of multiple local optima of multimodal functions'. *Swarm Intelligence*. 2009;**3**(2):87–124

[36] Tamura K., Yasuda K. 'Primary study of spiral dynamics inspired optimization'. *IEEJ Transactions on Electrical and Electronic Engineering*. 2011;**6**:98–100

[37] Rao R.V. 'Jaya: a simple and new optimization algorithm for solving constrained and unconstrained optimization problems'. *International Journal of Industrial Engineering Computations*. 2016;**7**:19–34

[38] Bouchekara H.R.E.H., Abido M.A., Mehasni R. 'Optimal power flow using the league championship algorithm: a case study of the Algerian power system'. *Energy Conversion and Management*. 2014;**87**:58–70

[39] Mahdad B., Srairi K. 'A new interactive sine cosine algorithm for loading margin stability improvement under contingency'. *Electrical Engineering*. 2017; doi: 10.1007/s00202-017-0539-x

[40] Yang X.S. Harmony Search as a Metaheuristic Algorithm. in: *Music-Inspired Harmony Search Algorithm: Theory and Applications* (Editor Z. W. Geem), Studies in Computational Intelligence. Berlin: Springer; 2009;**191**:1–14

[41] Tamura K., Yasuda K. 'The spiral optimization algorithm: convergence conditions and settings'. *IEEE Transactions on Systems, Man, and Cybernetics: Systems*. 2017; doi: 10.1109/TSMC.2017.2695577

[42] Watson J.P., Howe A.E., Darrell W.L. 'Deconstructing Nowicki and Smutnicki's i-TSAB tabu search algorithm for the job-shop scheduling problem'. *Computers & Operations Research*. 2006;**33**(9):2623–44

[43] Radosavljević J. *Metahopt—Matlab GUI based software for evaluation of metaheuristic optimization methods*. 2017. Available at https://www.researchgate.net/profile/Jordan_Radosavljevic2/contributions

Chapter 2

Overview of genetic algorithms

2.1 Introduction

Genetic algorithms (GAs) belong to the group of metaheuristic optimization methods. They are based on mimicking the process of evolution in nature. Evolution refers to constant adaptation of living beings to varying conditions in the environment. Individuals with the largest ability to adapt have the best chances to survive. There is an ongoing ruthless fight for survival in nature resulting in the survival of the fittest and perishing of the weakest individuals. In order for a species to survive during evolution, it must adapt to the surrounding conditions and the environment for these constantly changes. Each succeeding generation of any species must retain the good properties of the preceding generation while improving and altering them so that the quality of individuals in the population is continuously enhanced [1].

In the process of evolution, the fittest individuals dominate over the weaker ones and produce offspring. An individual's ability to survive is determined by the genetic material it carries, or its genes. It means that genes of dominant individuals are passed from one generation to another, whereas genes of weaker individuals are gradually lost. In addition to this occurrence, when reproduction takes place in nature, genes are recombined thus causing differences amongst individuals of the same species, as well as similarities of individuals to their parents. If only genes are observed, it may be said that each new generation of individuals has a different gene order, whereby some individuals are better and some are worse compared with those from the preceding generation. In GAs, the change of genes occurring at reproduction is called crossover. In addition to crossover, there is another phenomenon that can be observed, but to a significantly smaller extent. It includes an accidental change in the genetic material occurring under the influence of external factors. This phenomenon is called mutation. Crossover and mutation are denoted as genetic operators, whereas the process of choosing the fittest individuals in each generation is called selection [2].

In the 1970s, scientist John Holland came up with an idea how to suitably transform the aforementioned characteristics of natural evolution into a computer algorithm [3]. He commenced his research on an algorithm that manages strings of binary digits (0,1), where strings represent chromosomes. Holland's algorithm solved some rather difficult problems by employing simple coding and a reproduction mechanism. Just as it is the case in nature, this so-called simple GA was but a plain chromosome operator.

When he started developing these algorithms, they had no name. Owing to the fact that they originated from the science of genetics, they were called GAs. In the years that followed many a researcher began developing, improving, and applying GAs to solving various optimization problems.

There are two basic approaches to solving a problem by applying GAs: adapting the problem to the GA or adapting the GA to the particularities of the problem.

Nowadays, GAs may be characterized as a stochastic method of global search that mimics natural biological evolution. The basic idea underlying this method refers to selecting an initial population followed by making corrections to the currently best solution through a string of evolutionary generations of the population. The correction, that is, transforming the initial solutions and generating new ones is performed by implementing the basic genetic operators: selection, crossover, and mutation.

2.2 Basic structure of the GA

GAs operate with a population of individuals. Each individual constitutes a potential solution to a given optimization problem. An individual can be described as a set of variable conditions the values of which are being optimized. The quality of an individual is quantified over the value of the fitness function or the function of goodness. A population of individuals constitutes a set of solutions to the given optimization problem. One generation has a population with a certain number of individuals with better or worse fitness function values. The GA is a process carried out in sequences by iterations while applying the following three basic operators: selection, crossover, and mutation. At the end of each GA iteration, a new generation of individuals (solutions) is obtained. After a certain number of generations, the GA process is terminated once the preestablished condition for completion has been met. The best individual from the last generation constitutes a solution to the optimization problem, and it is typically quite close to the global optimum [4].

A single GA iteration is usually carried out in two phases. The process begins with a current population. Selection enables elimination of bad individuals (solutions) and survival of better ones (with better fitness values), thus creating an intermediate population (pairs of parents). Selection can be perceived as a process of paring couples—parents. The following phase includes the crossover and mutation operations. Crossover is a process during which new individuals—offspring—are created by exchanging parent traits (genes). It is followed by mutation, during which qualities of the newly created individuals are altered by random change of genes thus obtaining better individuals from generation to generation, which means that values of variable conditions approach the optimal values. The basic operations underlying the GA are the following:

- Representation (encoding) of individuals—potential solutions;
- Initialization—initial population;

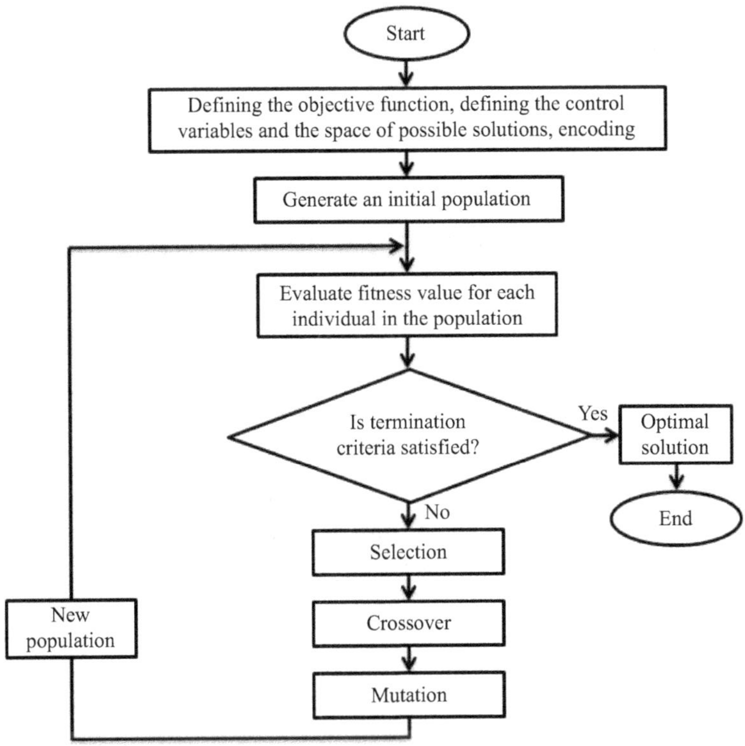

Figure 2.1 General flowchart of simple GA

- Defining a fitness function and evaluating fitness values of individuals;
- Selection;
- Crossover;
- Mutation;
- Defining control parameters.

The structure of a simple GA is shown in Figure 2.1.

2.3 Representation of individuals (encoding)

An individual constitutes a possible solution to a given optimization problem. Any individual may be represented as a string (sequence of data) in the form of a chromosome. Each piece of information in a string is a state variable in a given optimization problem. Since these are parts of a string, they may be referred to as substrings. The number of substrings and consequentially the string size, both depend on the number of variables being optimized. Generally, a string may be any kind of data structure describing an individual's characteristics. It is important for the GA that a string represents a possible solution to a given problem. Genetic operators should be defined for any data structure. That is why the manner in which presentation of data is carried out is of great significance for the efficiency of the GA. Therefore, the first decision to be made is how to represent variables in a string. Most commonly applied are binary and Gray encoding, although character encoding as well as direct real-value encoding are also used.

2.3.1 Binary encoding

In the majority of GAs, strings are represented binary (encoded) and their length is constant throughout the entire process. The practice has shown that the binary representation produces best results for most problems to which it may be applied. A string as a binary vector constitutes an encoded value $x \in [dg, gg]$. The length of the n binary number has an impact on the accuracy and it represents the number of bits, that is, of 1s and 0s in a single string (substring). It is possible to store 2^n various combinations of 0s and 1s in such a vector, that is, it is possible to store any number within the $[0, 2^n - 1]$ range. The binary vector $v(0) = [000\ldots0]$ represents the value $x = dg$, whereas the vector $v(2^n - 1) = [111\ldots1]$ represents the value $x = gg$. In general, if the binary number $b \in [0, 2^n - 1]$ is represented as the binary vector $v(b) = [B_{n-1}B_{n-2} \,\ldots\, B_1B_0]$, where $B_i = 0$ or 1, then

$$b = \sum_{i=0}^{n-1} B_i 2^i \tag{2.1}$$

is a binary number, and

$$x = dg + \frac{b}{(2^n - 1)}(gg - dg) \tag{2.2}$$

is the equivalent real number.

Decoding refers to the process of translating a binary number into a potential solution. In this instance, the potential solution includes any real number x in the $[dg, gg]$ interval. The binary vector $v(b)$ represents the value x calculated according to formula (2.2). Encoding, that is, calculating the binary number b for the specified real number x is carried out according to the following formula:

$$b = \frac{x - dg}{gg - dg}(2^n - 1) \tag{2.3}$$

Binary encoding is suitable for use owing to its simplicity. However, it has a major disadvantage: the Hamming distance between two successive numbers may be large, and in the worst case, it may be equal to the length of the binary sequence (string). The Hamming distance between two binary numbers is the number of bits in which they differ. For instance, the Hamming distance between the numbers $256_{10} = 100000000_2$ and $255_{10} = 011111111_2$ is 9 (all 9 bits are different). In other words, if at any one of its steps a GA found one good solution for $b = 011111111_2$, whereas the optimal solution is for $b = 100000000_2$, all nine bits must be changed. In order to overcome this disadvantage, Gray code is used to encode the numbers.

2.3.2 Gray coding

Consecutive numbers encoded in Gray code differ in only one bit. Therefore, the Hamming distance between the consecutive binary numbers (strings) is 1. The binary number is transformed into Gray code according to a specific algorithm. The algorithm of transformation of a binary number $b = b_m b_{m-1} \ldots b_1$ into *Gray* code $g = g_m g_{m-1} \ldots g_1$ and vice-versa is as follows:

$$g_m = b_m \tag{2.4}$$

$$g_k = b_k \oplus b_{k+1}, \quad k = 1, 2, \ldots, m-1 \tag{2.5}$$

$$b_k = \sum_{k=j}^{m} g_k (\text{mod } 2), \quad j = 1, 2, \ldots, m \tag{2.6}$$

Therefore, for instance, the numbers 7 and 8 are represented in the binary code as

$$7_{10} = 0111_2 \text{ и } 8_{10} = 1000_2, \tag{2.7}$$

whereas in Grey code, they are represented as

$$7_{10} = 0100_g \text{ и } 8_{10} = 1100_g. \tag{2.8}$$

2.3.3 Real-value encoding

Real-value encoding offers a number of advantages compared with binary encoding, particularly in instances of optimization of numerical functions. In this case, it is better to use floating point numbers, that is, the normalized exponential form composed of the mantissa m, the basis b, and the exponent e:

$$x = mb^e \tag{2.9}$$

As opposed to binary encoding, no particular coding mechanism needs to be embedded for such a view in a GA since according to the IEEE 854 standard, the chromosome itself is a real number only written as a floating point number.

Certain GAs allow specifying the dynamic range and resolution for each variable. For instance, if a variable has a range from 2.5 to 6.5 (the dynamic range is 4) and if a resolution of three decimal places is desired, then the dynamic range and the resolution require a string 12 bits long. In that case, the string containing only zeroes represents the value 2.5. A major advantage of representing variables in such a manner lies in the fact that the user is able to perceive population individuals as real-valued vectors rather than as bitstrings (sequences).

In theory, character (alphabetic) representation is also possible by using any alphanumeric characters. It means that an alphabet containing a number of characters or numbers is used rather than using 1s and 0s. However, most GAs use binary representation.

2.4 Population size and initial population

The population size refers to the number of individuals it consists of. It is usually constant, that is, it remains unaltered during the course of the GA run. The population size is a GA parameter. The larger the population, the more comprehensive the search space of possible solutions. However, it also means a higher computational time. The population size to be adopted depends primarily on the concrete problem being solved, followed by the size of the individual (string length), as well as the experience of the person performing the task. It seems that the population size tends to increase approximately linearly with the individual's string length. Nevertheless, the "optimal" population size, if it exists at all, primarily depends on the concrete problem. In general, the population size ranges between 20 and 200 individuals.

Most commonly, the initial population is generated by random selection of solutions (individuals) from the solution domain. The initial population can be uniform (all individuals are the same). It is also possible to embed into the initial population individuals as solutions obtained by another optimization method, thus increasing the GA convergence and speed.

2.5 Fitness function

The fitness function is equivalent to the objective function to be optimized. After an initial population has been created, a fitness function is computed for each individual, which is assigned a fitness value (goodness). A distinction should be made between the objective function and the fitness function. The objective function provides a measure of the given solution (set of variables) independent of other possible solutions. The fitness function transforms this measure into a rank of reproductive opportunities (survival), that is, the fitness value of an individual is defined with respect to other members of the current population. After decoding a chromosome, that is, transforming a genotype into a phenotype, each string is assigned a fitness value. A phenotype is used as an input in a fitness function. These fitness values are afterwards used for relative weighing (evaluation, measuring) of strings in a population. The higher the fitness value of an individual, the more likely it will survive and undergo crossover. The fitness function plays a key role in the selection process.

Major issues relating to defining and specifying a fitness function include the following:

- Dependence on whether the problem pertains to maximization or minimization;
- A fitness function may vary with the dynamics of GA runs;
- Assessing (evaluating) a fitness function may sometimes be only approximately computed;

- A fitness function should allocate rather various values to individuals so as to facilitate selection;
- Constraints must be encompassed;
- They should allow for a possibility to include various subobjective functions.

The fitness function is a black box for the GA. Internally, this may be achieved by a mathematical function, a simulation program, or a human expert who decides the quality of individuals. At the beginning of iterative search fitness values of individuals are usually assigned randomly. As the search progresses, certain values in all genes become dominant. Variations of fitness values decrease as the population converges. Change in the range of fitness values throughout the evolutionary process often leads to issues relating to premature convergence.

The simplest interpretation of the fitness function is the equivalent of the function to be optimized:

$$\text{fitness}(x) = f(x) \tag{2.10}$$

Computing a fitness function in the afore-specified manner is applicable in the instance of maximization of a function $f(x)$. In the instance of minimization, the above expression must be modified. As the function $f(x)$ must have no negative values since the fitness(x) would be negative (the probability of selection of an individual is in the majority of instances proportional to fitness, therefore a negative value of fitness is senseless), then in the instance of the minimization problem the fitness value should be computed as

$$\text{fitness}(x) = f(x) + M \tag{2.11}$$

M should be selected so that fitness is always a nonnegative number. However, this is not an adequate manner of determining the fitness function, which is apparent from the following example. If the function $f(x) = \sin(x)$ and $M = 1{,}000$, the fitness function for all the chromosomes will be approximately equal and its value will be approximately 1,000. This means that during the course of selection, all chromosomes will have a rather equal probability to be copied into the next population, that is, all of them are equal both good and bad, which is bad altogether!

As we have demonstrated, such a simplified manner of calculating fitness values most commonly fails to produce good results. Therefore, in all instances other than in simplest GAs, other forms of calculating a fitness function are resorted to.

2.5.1 Relative fitness

One of the ways is to use the so-called relative fitness. It is defined as [5]

$$F(c) = f(\Phi(c)) \tag{2.12}$$

where Φ is the function of the given problem, $f(.)$ converts the value of the objective function into a nonnegative number, and $F(.)$ is the resulting value of the fitness function.

In the largest number of instances, the fitness function value is proportional to the number of offspring produced by an individual into the next generation [6]. In such an instance, fitness of each individual, $F(c_i)$, is computed as

$$F(c_i) = \frac{\phi(c_i)}{\sum_{i=1}^{N} \phi(c_i)} \tag{2.13}$$

where N is the population size, and $\phi(c_i)$ is the value of the objective function corresponding to the individual i. Although such a definition of fitness ensures that each individual has a possibility to reproduce according to its corresponding fitness value, it fails to account for negative-objective function values.

2.5.2 Linear scaling

Linear scaling is also used for the purpose of defining a fitness function:

$$F(c) = \alpha \cdot f(c) + \beta \tag{2.14}$$

where α is a positively scaling factor for the maximization problem and a negatively scaling factor for the minimization problem. β ensures that the resulting fitness $F(\ )$ values are nonnegative. When linear scaling is used, the expected number of offspring is approximately proportional to performances of the corresponding individual. Owing to that, the fittest individuals may dominate the population in early generations thus leading to premature convergence toward local optima. In order to improve performance of a GA, the following conditions should be met when selecting a fitness function [5]:

- Individuals with similar genetic codes should have similar fitness values,
- The fitness function should be smooth,
- The fitness function should not have too many local maximum,
- The fitness function should not have a too isolated global maximum, etc.

2.6 Selection

Choosing individuals for reproduction is performed by selection. Individuals are selected according to their fitness values in such a manner that those with higher fitness values have a larger selection (survival) probability, whereas those with smaller fitness values are less likely to survive and contribute to creating a new generation. The selection procedure might also be carried out by sorting and selection of the best individuals (those with the largest fitness values). However, that would lead to premature convergence of the GA, that is, the optimization process would practically be completed during the several initial iterations. The problem lies in the fact that this procedure would lead to a loss of "good" genetic material that may be contained in bad individuals (those with small fitness values), which is why other selection mechanisms are applied. Selection schemes differ in the manner of selection of better individuals. According to the manner in which genetic material of better individuals is copied into the next iteration, selection schemes are classified into the following:

- Generation selection, wherein directly selected are better individuals whose genetic material is copied into the next iteration and
- Elimination selection, wherein selected are bad individuals for elimination, whilst better ones survive the selection process.

Depending on how better individuals are selected in generation selection, or how worse are selected in elimination selection, selection schemes are classified into:

- Proportional selections and
- Ranking selections.

It is noteworthy that a higher survival probability of a better individual in comparison with any other worse individual constitutes a common feature of all selection schemes. Proportionate selection schemes choose individuals with the probability proportional to its fitness value, that is, the selection probability depends on the ratio of the fitness value of an individual and the average fitness value of the population. Ranking selection scheme choose individuals with a probability depending on the position of the individual in the string of individuals sorted by their fitness values. Ranking selection schemes are classified into sorting and tournament selection schemes.

Ranking selection schemes include linear ranking selection and selection of the fittest, that is the (μ, λ) selection, and truncation selection. Tournament selection schemes are classified according to the number of individuals participating in the tournament. Here's a brief overview of the selection schemes and their corresponding subtypes.

Types of selection schemes:

1. **Proportional selection**
   - i. Simple selection
   - ii. Stochastic universal sampling
2. **Ranking selection**
   - i. Sorting selection
     - a. Linear ranking selection
     - b. Elitist selection
       - I. $(\mu + \lambda)$ selection
       - II. (μ, λ) selection
       - III. Truncation selection
   - ii. Tournament selection
     - a. k-tournament selection
     - b. Simple tournament selection

2.6.1 Simple selection

In simple selection, the probability of an individual being selected is proportional to its fitness value. Holland applied this selection scheme in a simple GA. Ordinal-based selection is often referred to as roulette-wheel selection. The roulette wheel analogy does not entirely correspond here since all numbers in the roulette wheel pockets take up equal wedges, whereas the imaginary area of a wedge an individual takes up is proportional to its fitness value, that is, to its probability of being selected.

The probability of an individual being selected, that is the area of its corresponding wedge, is determined based on the fitness value of the individual f_i and the total fitness value of the population Σf_i. The probability of selection of the ith individual $p(i)$ to undergo reproduction equals the quotient of its fitness value and the total fitness value of the population:

$$p(i) = \frac{f_i}{\sum_{i=1}^{N} f_i} \tag{2.15}$$

The population can accordingly be represented in the form of a roulette wheel, where larger wedges are assigned to those individuals with larger fitness values. The selection process is carried out by generating a random number r in the [0,1] interval. The individual i meeting the requirement $r \in [p(i-1), p(i)]$ is then selected and copied into the next population. The procedure is repeated as many times as there are individuals in the population. A single individual is selected at each iteration. The better the individual, the greater the probability it will be selected each time the wheel is spun. This allows for an individual to be copied into a new generation several times. This kind of selection logic may theoretically result, prior to crossover and mutation are performed, in a new generation being composed of a single individual in N copies (where N is the population size).

The issue of duplicate individuals occurring in a new generation has turned out to be a major drawback of this selection scheme. This issue may be resolved by applying various methods. The simplest one is implicit discarding of any such duplicate individual in the population. This is achieved by assigning zero to the fitness values of such individuals thereby preventing them to copy into the next generation.

2.6.2 Stochastic universal sampling

Stochastic universal sampling differs from ordinal-based proportionate selection in the manner in which individuals are selected. The selection probability is the same, and it is also computed according to expression (2.15). Simple proportional selection is carried out in N steps: a random number is generated N times, according to which it is then determined which individual is selected. Stochastic universal sampling employs a single step to select all the individuals: N equidistant pointers are generated by dividing the [0,1] interval into $N+1$ equal segments. The number of pointers indicating a given individual determines how many of its copies will be copied into the next population.

2.6.3 Linear ranking selection

In linear ranking selection, the selection probability of an individual is proportional to its rank in the population of individuals sorted according to their fitness values. The probability of selection in linear ranking selection is computed according to the following expression:

$$p(i) = \frac{i}{\sum_{i=1}^{N} i} = \frac{2i}{N(N+1)} \tag{2.16}$$

where the best individual is assigned the rank N and the worst one is assigned the rank 1.

The sum of selection probabilities of all individuals must be 1. A particularity of this selection scheme is that the probability of the ith individual to survive is i times higher than the probability of the worst individual assigned the rank 1.

2.6.4 Elitist selection

A predefined number of the best individuals are selected in elitist selection. There are three possible selection schemes: $(\mu+\lambda)$ selection, (μ, λ) selection, and truncation selection. The difference among the aforementioned selection schemes lies in the set of individuals from which the best ones are chosen. The best individuals may be sampled from the set consisting of parents only, parents and offspring together and offspring only. Parents are randomly sampled first, after which their offspring (with or without parents) are sorted and only a certain number of the best individuals are selected, where μ is the size of the parent population and λ is the size of the offspring population.

In $(\mu+\lambda)$ selection, μ parents are randomly sampled from the population. Crossover is performed to generate λ offspring. The μ fittest individuals are then selected into the next generation from the set of parents and offspring. The procedure is repeated until the new generation with N new individuals has been populated, that is, N/μ times or μ individuals are copied multiple times in order to populate the population.

In (μ, λ) selection, μ parents are also randomly selected from the population and then subjected to crossover so as to generate λ offspring. However, there are more offspring than parents $(\lambda \geq \mu)$ and the μ fittest offspring are selected into the next generation. The probability of selection of the ith individual for crossover is

$$p(i) = \begin{cases} 1/\mu, & 1 \leq i \leq \mu \\ 0, & \mu \leq i \leq N \end{cases} \tag{2.17}$$

In truncation selection n fittest individuals are sampled and copied N/n times. The probability of selection of the ith individual is $p(i) = 1$, for $i = 1,2,\ldots,n$, whereas for other individuals the probability is equal to zero.

2.6.5 k-Tournament selection schemes

In k-tournament selection ($k = 2,3,4,\ldots,N$) k individuals with equal probability are sampled and either the best or the worst individual is then selected. In generation tournament selection, k individuals are randomly sampled N times and the best individual amongst them is then selected and copied into the intermediate population.

In elimination tournament selection, k individuals are sampled M times and the worst one amongst them is eliminated. The parameter k is called the tournament size. According to the tournament size, tournament selection schemes are classified into binary tournament selection ($k = 2$), 3-tournament selection, 4-tournament selection, etc. Tournament selection does not require ranking of individuals, although the probability of selection is dependent on the individual's rank, as it is the instance in ranking selection schemes. In the tournament selection procedure, only the interrelation between k randomly sampled individuals is of significance. Fitness function values of individuals in each pair are used to determine which one is better/worse. Let the best individual be assigned the rank 1, and the worst the rank N. The best ranked individual may not be eliminated by means of any elimination tournament selection, since the remaining $k - 1$ randomly selected individuals are worse ranked than the best one. Retention of the best or several most fit individuals is called elitism. In elimination tournament selection schemes, elitism is inherently embedded, that is, the probability of the best individual being eliminated is equal to zero: $p(1) = 0$.

2.6.6 Simple tournament selection

Simple tournament selection is a specific form of binary tournament selection adapted to parallel runs. A one-step tournament is held to randomly select two pairs of individuals. The worse individual in each pair is selected for elimination. Crossover of the surviving individuals is performed to generate two offspring, which are then mutated and evaluated (their fitness value is determined). This pair of newly generated individuals compensates for the eliminated ones. In such a manner the GA with simple tournament selection performs both selection and reproduction in the same step.

Opting for the selection operator boils down to choosing the selection scheme most suitable for parallel runs. A number of authors agree that elimination tournament selection schemes are most suitable for parallel runs.

2.7 Crossover

Crossover is a term denoting the exchange (recombination) of genetic information during the course of reproduction. The crossover operation is what differs GAs from other optimization methods. Crossover is a process of exchange of bits (parts of strings) between two individuals—parents. It results in either two new individuals or one. There are several manners how parts (substrings) of individuals are selected for crossover such as one-point crossover, multipoint crossover, uniform crossover, shuffle crossover, arithmetic crossover, and heuristic crossover.

The crossover operator is applied to each pair of individuals with the given probability p_c. Applying a probability operator means that it will not be applied to certain individuals thus allowing them to be copied into the next generation unaltered. A random number r from the [0,1] interval is selected for each pair of individuals and crossover is performed if the mentioned number is smaller than the probability of crossover p_c.

2.7.1 One-point crossover

Traditional GAs employ one-point crossover. In this crossover operator, a single point on both parents' strings is selected at random. Crossover (exchange of bits) is carried out to the right of the selected crossover point. Whether a bit will remain unaltered after crossover depends on its position in a string and in the crossover scheme. There is no risk of full recombination since a portion of the bits, to the left of the crossover point, is copied to new individuals—offspring.

Example for one-point crossover:

Parent 1: [**1 1 0 0 1**| 0 1 0]
Parent 2: [0 0 1 0 0| **1 1 1**]
Offspring 1: [**1 1 0 0 1 1 1 1**]
Offspring 2: [0 0 1 0 0 0 1 0]

2.7.2 Multipoint crossover

In this crossover scheme, the number of points and crossover positions are selected at random for each individual crossover. For instance, in two-point crossover, offspring inherit genes located outside the crossover points. Bits between the crossover points are swapped, which is what makes offspring different from their parents.

Example for two-point crossover:

Parent 1: [**1 1 0**| 0 1 0| **1 0**]
Parent 2: [0 0 1| **0 0 1**| 1 1]
Offspring 1: [**1 1 0 0 0 1 1 0**]
Offspring 2: [0 0 1 0 1 0 1 1]

2.7.3 Uniform crossover

Uniform crossover refers to crossover with $b - 1$ crossover points (where b is the number of bits). The probability that an offspring will inherit one parent's gene is 0.5. It means that the inheritance probability is the same for both parents. If inheritance probabilities are different for certain genes, then crossover is referred to as n-uniform crossover. For instance, if $n = 0.3$, then there is a 30% probability that one bit will be inherited from one parent and a 70% probability that the other one will be inherited from the other parent. If the inheritance probability is different for certain genes, then a crossover mask is specified so as to define the inheritance probability for each individual gene.

Example for uniform crossover:

Parent 1: [**1** 1 **0** 0 1 **0** **1** 0]
Parent 2: [0 **0** 1 **0** **0** 1 1 **1**]
Offspring 1: [**1 0 0 0 0 0 1 1**]
Offspring 2: [0 1 1 0 1 1 1 0]

2.7.4 Shuffle crossover

Shuffle crossover is performed in three steps. Step one includes shuffling the bits in each parent. During step two, standard one-point or multipoint crossover is performed. Finally, in step three, the shuffled parent bits are unshuffled (the parents remain unaltered).

Example for shuffle crossover:

Select Shuffle Points
Parent 1: [1 1 1 0 1 0 0 1]
Parent 2: [1 0 0 0 1 0 1 1]

Shuffle Genes as Shuffle Points
Parent 1: [0 1 0 1 1 0 1 1]
Parent 2: [0 0 1 1 1 0 0 1]

Select 1-Point Crossover Point
Parent 1: [0 1 0 1 | 1 0 1 1]
Parent 2: [0 0 1 1 | 1 0 0 1]

Perform 1-Point Crossover Point
Offspring 1: [0 1 0 1 | 1 0 0 1]
Offspring 2: [0 0 1 1 | 1 0 1 1]

Select Unshuffled Points Same as Shuffled Points
Offspring 1: [0 1 0 1 1 0 0 1]
Offspring 2: [0 0 1 1 1 0 1 1]

Unshuffled the Genes in Offspring
Offspring 1: [1 1 0 0 1 0 0 1]
Offspring 2: [1 0 1 0 1 0 1 1]

2.7.5 Arithmetic crossover

Arithmetic crossover is suitable for optimization problems with constraints. Parents are first selected at random and then offspring gene values are calculated by using a function. The expression below shows two parent and two offspring individuals.

Parent 1: $[x_1\ x_2\ x_3\ x_4\ x_5\ x_6\ x_7\ x_8]$
Parent 1: $[y_1\ y_2\ y_3\ y_4\ y_5\ y_6\ y_7\ y_8]$
Offspring 1: $[p_1\ p_2\ p_3\ p_4\ p_5\ p_6\ p_7\ p_8]$
Offspring 2: $[p'_1\ p'_2\ p'_3\ p'_4\ p'_5\ p'_6\ p'_7\ p'_8]$

The gene values in position 1 for Offspring 1 and Offspring 2 are calculated in the following manner:

$$\begin{aligned} p_1 &= r \cdot x_1 + (1 - r) \cdot y_1 \\ p_1' &= (1 - r) \cdot x_1 + r \cdot y_1 \end{aligned} \tag{2.18}$$

where r is a random number in the interval [0,1], generated before each crossover operation is performed. If the value of the gene $p_1 < 0.5$, the gene in that position is assigned the value 0. Otherwise, it is assigned the value 1. The arithmetic operator ensures that each offspring gene is generated from the corresponding genes of both parents.

2.7.6 Heuristic crossover

Heuristic crossover employs parent fitness values to determine the search direction. The following formula is used to generate an offspring:

$$\begin{aligned} \mathit{Offspring}1 &= \mathit{BestParent} + r \cdot (\mathit{BestParent} - \mathit{WorstParent}) \\ \mathit{Offspring}2 &= \mathit{BestParent} \end{aligned} \tag{2.19}$$

where r is a random number in the [0,1] interval.

If the value of the random number r is such that one or more *Offspring*1 genes fall outside the threshold values, then such offspring must be discarded. For that reason, in heuristic crossover the parameter (n) is introduced and it denotes the number of attempts to generate a random number r, which results in feasible *Offspring*1 with the gene values within the threshold. If even after n attempts no feasible offspring is generated, then *Offspring*1 becomes *WorstParent*.

2.8 Mutation

Mutation refers to a random change of one or more genes (from 1 to 0, that is, from 0 to 1), which occurs in each generation with a certain predefined probability. The impact that mutation has on the quality of solutions is greatest over final generations when the majority of individuals is of similar quality. Mutation is a unary operator owing to the fact that is performed over a single individual. It constitutes a mechanism for avoiding local minima. For instance, if fitness values of all individuals in the population tend to the local minimum, only random search of the solution space may provide a better solution. It is sufficient that a single individual (generated by mutation) be better than the rest to have all individuals transferred to the space with a better solution over the next several generations. The role of mutation is also to restore the lost genetic material. If, for example, all individuals in a population have the same gene at a certain position in the string, it would be impossible to change it without mutation.

The mutation probability p_m is a GA parameter. If the mutation probability tends to one, then the algorithm is converted into an algorithm for random search of the solution space. On the other hand, if the mutation probability tends to zero, the optimization process will most likely be terminated in the local optimum. The mutation probability depends on the population size and on the manner of representation of individuals. It is often held constant for the entire run of the GA process, although in many instances this approach will not produce optimal solutions. It may be varied during the GA run and if it is, it is usually increased. For instance, the mutation rate may begin at 0.001 and end at 0.1 or at a value obtained when the specified number of generations has been completed. Several mutation types may be employed:

- Simple mutation in which each bit in a string is flipped with the same probability p_m;

  [1 1 0 **0** 1 0 1 0] → [1 1 0 **1** 1 0 1 0]

- Scramble mutation refers to the type of mutation in which a random process is employed to select a chromosome string which will undergo mutation in addition to the first and second threshold (or sample), after which the genes are scrambled or generated at random (complete scramble mutation) or inverted (inversion scramble mutation).

  Example for complete scramble mutation: $[x_1\ \boldsymbol{x_2}\ \boldsymbol{x_3}\ \boldsymbol{x_4}\ \boldsymbol{x_5}\ x_6\ x_7\ x_8] \rightarrow [x_1\ \boldsymbol{x_4}\ \boldsymbol{x_2}\ \boldsymbol{x_5}\ \boldsymbol{x_3}\ x_6\ x_7\ x_8]$
  Example for inversion scramble mutation: $[x_1\ \boldsymbol{x_2}\ \boldsymbol{x_3}\ \boldsymbol{x_4}\ \boldsymbol{x_5}\ x_6\ x_7\ x_8] \rightarrow [x_1\ \boldsymbol{x_5}\ \boldsymbol{x_4}\ \boldsymbol{x_3}\ \boldsymbol{x_2}\ x_6\ x_7\ x_8]$

In other nonbinary coding types mutation can be achieved by a small shift of the gene value or by random selection of a new value from within the threshold, in which instance the mutation probability of an individual p_M is defined instead of the mutation probability of a single gene p_m. The relation between these two parameters is as follows:

$$p_M = 1 - (1 - p_m)^n \tag{2.20}$$

where n is the number of bits in an individual (chromosome). For instance, if $p_m = 0.01$ (one of the 100 bits will be changed) and let an individual be represented by a string of $n = 32$ bits, then $p_M = 0.275$. It means that 27 out of 100 individuals will be altered.

2.9 GA control parameters

The GA parameters include the population size, number of generations or iterations, crossover probability, and mutation probability. In elimination GAs the number of individuals for elimination is specified instead of the crossover probability.

The algorithm gives different results for different parameter values: a better or worse solution is achieved faster or slower. During GA runs these parameters may be either constant or variable. In general, the approach with variable parameters is better. Constant parameter values may decrease the efficiency of the GA. Altering the GA parameters may be a function of time or the number of iterations (which is most commonly the case), or it may be a function of the range of solution spread. If solutions are spread, the crossover probability should be increased whereas the mutation probability should be decreased. On the other hand, in the instance of uniform solutions, the mutation probability should be increased and the crossover should be decreased since if two identical solutions undergo crossover, no new solution will be produced.

The crossover and mutation probabilities mainly depend on the population size. In the majority of GAs, typical population size, crossover probability, and mutation probability values cover the following ranges: [20–200], [0.5–1.0], and [0.001–0.05], respectively.

Finding an optimal solution by implementing a GA is an iterative process, which is completed once a maximum number of generations has been reached or when another criterium has been met, such as e.g., minimal deviation of the best fitness value from the mean fitness value of a population in the current generation.

2.10 Multiobjective optimization using GA

A general multiobjective optimization problem (MOP) can be defined as follows:

$$MinF(x) = [f_1(x), f_2(x), \ldots, f_k(x)]^T \tag{2.21}$$

$$s.t.\, x \in S, \quad x = [x_1, x_2, \ldots, x_n]^T \tag{2.22}$$

where $[f_1(x), f_2(x), \ldots, f_k(x)]$ are k-objective functions, $[x_1, x_2, \ldots, x_n]$ are the n optimization parameters, and $S \in R^n$ is the solution or parameter space.

Unlike of single-objective optimization, there may not exist one solution which is best with respect to all objectives. In a typical MOP, there exists a set of solutions, known as Pareto-optimal solutions or nondominated solutions, which are superior to the rest of solutions in the search space when all objectives are considered but are inferior to other solutions in the space in one or more objectives [7].

In solving an MOP, two conceptually distinct types of problem difficulty can be identified: search and decision-making. The first aspect refers to the optimization process in which the feasible set is sampled for Pareto-optimal solutions. The second aspect addresses the problem of selecting a suitable compromise solution from the Pareto-optimal set.

Since GA works with a population of potential solutions, a set of Pareto-optimal solutions can be obtained using GA. It has been proven that nondominated sorting genetic algorithm (NSGA) [7] and NSGA-II [8] are intelligent optimization algorithms able to balance the trade-offs between conflicting objectives.

NSGA proposed in [7] differs from simple GA in relation of the selection operator, whereas the crossover and mutation operators remain as usual. Figure 2.2 shows a flowchart of this algorithm.

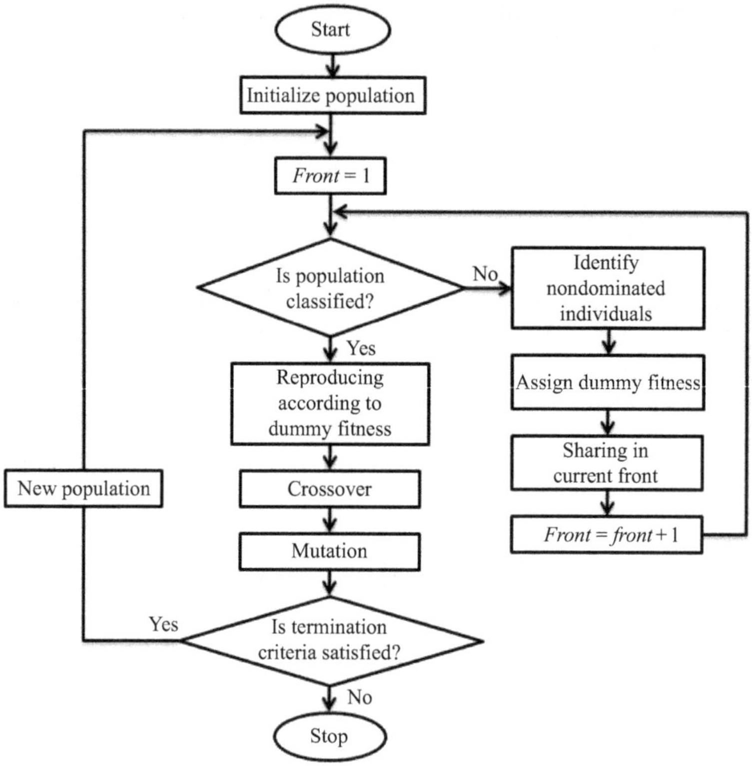

Figure 2.2 Flowchart of NSGA

Before the selection is performed, the population is ranked on the basis of an individual's nondomination. The nondominated individuals present in the population are first identified from the current population. Then, all these individuals are assumed to constitute the first nondominated front in the population and assigned a large dummy fitness value. The same fitness value is assigned to give an equal reproductive potential to all these nondominated individuals. To maintain diversity in the population, these classified individuals are then shared with their dummy fitness values. Sharing is achieved by performing selection operation using degraded fitness values that are obtained by dividing the original fitness value of an individual by a quantity proportional to the number of individuals around it. This causes multiple optimal points to co-exist in the population. After sharing, these nondominated individuals are ignored temporarily to process the rest of the population in the same way to identify individuals for the second nondominated front. These nondominated points are then assigned a new dummy fitness value that is kept smaller than the minimum shared dummy fitness of the previous front. This process is continued until the entire population is classified into several fronts. The population is then reproduced according to the dummy fitness values. A stochastic remainder proportional selection is used in this process [7].

The efficiency of NSGA lies in the way multiple objectives are reduced to a dummy fitness function using a nondominated sorting procedure. Another aspect of NSGA is that practically any number of objectives can be solved. Both minimization and maximization problems can also be handled by this algorithm.

The main shortcomings of the NSGA are as follows [8]:

1. The NSGA has a computational complexity of $O(MN^3)$, where M is the number of objectives and N is the population size. This makes NSGA computationally expensive for large population size.
2. Lack of elitism, which is the possibility of loss of good solutions once they are found.
3. The need for specifying a sharing parameter.

To overcome these difficulties, the authors proposed an improved version of NSGA, called NSGA-II [8]. In NSGA-II, a fast nondominated sorting approach with $O(MN^2)$, computational complexity is presented. Also, a selection operator is introduced that creates a mating pool by combining the parent and offspring populations and selecting the best (with respect to fitness and spread) solutions.

Moreover, the authors [8] modify the definition of dominance in order to solve constrained multiobjective problems efficiently. Simulation results on difficult test problems shows that the NSGA-II has excellent performances in solving both unconstrained and constrained MOPs.

2.11 Applications of GA to power system problems—literature overview

The GA and its various modified versions are applied for solving optimization problems in various areas. This chapter provides a literature overview of the application of GAs in solving some of the most important problems in power systems, such as optimal power flow (OPF), optimal reactive power dispatch (ORPD), economic dispatch (ED), optimal location and sizing of distributed generation (DG) in distribution networks, optimal coordination of overcurrent relays, optimal energy management of microgrids (MGs), etc.

2.11.1 Optimal power flow

A power flow study is a steady-state analysis whose target is to determine the voltages, currents, and active and reactive power flows in a system under a given load conditions. Power flow analysis is fundamental to the study of power systems. Some of the system variables, such as generator active power outputs, generator voltages, transformer tap settings, reactive power output of shunt Var compensators, etc., can be controlled independently, and thus have a direct impact on the power flows, that is on the steady state of the power system. These variables are defined as control variables. In principle, a desired steady state of the power system can be realized under different combinations of the control variables. This brings to the OPF problem.

The OPF is one of the most important tools for achieving the economic and secure operation of the power system. The OPF problem solution aims to optimize a chosen objective function through optimal adjustment of the power system control variables while at the same time satisfying various operating constraints. The objective function can take different forms. Classical objective function for the OPF is minimization of total generating cost. However, many other objective functions are possible, such as minimization of system losses, voltage profile improvement, voltage stability enhancement, etc. In its most general formulation, the OPF is a nonlinear, nonconvex, large-scale, static optimization problem with both continuous and discrete control variables. In recent two decades, many authors suggest GAs for solving the OPF problem.

One of the first applications of GA on the OPF problem is described in [9]. The authors are proposed an improved GA with the dynamical hierarchy of the coding system. This improvement means that the GA has a high potential in solving the OPF for large-scale systems under both normal and contingency conditions.

Bakirtzis *et al.* [10] proposed an enhanced GA which incorporated problem-specific operators to solve larger OPF problems. The approach is similar to the one in [9] with the following improvements: switchable shunt devices and transformer taps are modeled as discrete control variables. In GA the control variables are self-constrained. Dependent variables (state variables), such as slack bus power, load bus voltages, generator reactive power outputs, and transmission line loadings are restricted in [9,10] by adding them as the quadratic penalty terms on the objective function. The authors of [11] are discussed about the impact of these penalty terms on the accuracy and convergence of GA in solving the OPF problem.

The authors in [12] have proposed two-stage methodology for solving the OPF problem using GA. The first part employs the GA to obtain a feasible solution subject to desired load convergence, while the other part employs GA to obtain the optimal solution.

A GA with the representation of the control variables in their natural form was successfully applied in [13] for solving the OPF which aims to security enhancement of the power system.

In [14] the authors have used the GA in solving the OPF problem with the voltage angles at generator-buses as control variables and a specific procedure for the nonrandomized initialization of control variables, such as the generator active power outputs and the generator voltage magnitudes.

A real-coded mixed-integer GA is proposed in [15] for solving the OPF problems with considering transmission security and bus-voltage constraints under normal and contingent states. Each individual in this algorithm is a mixture of continuous and discrete control variables. Two arithmetic crossover and mutation schemes are proposed in [14] to deal with the continuous/discrete control variables, respectively.

An improved GA with an orthogonal-array-based crossover operator has been proposed in [16] for solving multicontingency transient stability constrained OPF problems. Based on simulation results the authors have concluded that the improved GA has better performances than simple GA in solving the OPF problems in terms of variance of solution and the convergence speed.

A GA-based approach for solving the OPF problem considering flexible alternating current (AC) transmission system (FACTS) devices is proposed in [17]. The FACTS devices are used to eliminate line over loads in the power system following single-line outages. First step of this approach is determination of the optimal locations of thyristor-controlled series capacitors (TCSCs) based on an index called the single contingency sensitivity index. Next step is determination of the TCSC parameters by solving the OPF problem using GA.

To overcome the high CPU time execution and improve the qualities of the solution in practical large-scale OPF problems, the authors [18] proposed an efficient parallel GA for the solution of large-scale OPF with consideration shunt FACTS devices under severe loading conditions. The OPF problem is decomposed in two sub problems: (i) the active power dispatch to minimize the total generation cost and (ii) the reactive power dispatch to adjust the voltage source and reactive power compensation within their specified constraints limits to reduce voltage deviation and the thermal transmission line. These two subproblems are solved separately by efficient parallel GA.

Authors in [19] proposed enhanced GA in combination with a new decoupled quadratic load flow routine to solve the multiobjective OPF problem. The multiobjective OPF problem is defined as simultaneous optimization of three conflicting objectives: total generation cost, active power losses, and voltage stability index.

An adapted GA with adjusting population size is used in [20] to solve OPF with different objectives that reflect generation costs minimization and voltage profile improvement.

NSGA II approach is proposed in [21,22] for solving the voltage stability constrained optimal power flow (VSC-OPF).

In [23,24], OPF problem of purely AC power systems is extended to a OPF problem of integrated AC–DC power system by taking into consideration power transfer control characteristics of HVDC links. Then, the OPF of integrated AC–DC power systems is solved by GA.

Reference [25] presents an alternative approach to OPF problem using line flow-based formulation and decoupled approach. The OPF solution was obtained using GA toolbox available in MATLAB®.

To overcome the excessive execution time, the authors [26] proposed an efficient evolutionary algorithm (EEA)-based OPF using the concept of incremental power flow model. The proposed approach was implemented using enhanced GA. Based on simulation results, the authors concluded that the proposed EEA approach is approximately five times faster than any conventional evolutionary algorithm, with insignificant difference in the objective function value.

In recent years the authors pay attention to the OPF solution with the aim of transient stability enhancement. A successful application of parallel NSGA-II to solve the multiobjective OPF considering transient stability as an objective function is presented in [27].

The authors [28] prove excellent performances of a recently developed GA with new multiparent crossover in solving the OPF with different objective function and constraints.

The integration of renewable energy sources into power system is very challenging problem due to their intermittent nature. Reference [29] proposes a new strategy based on GA and two-point estimate method for solving the OPF problem taking into account uncertainties of output power of wind generators and photovoltaic generators, and load forecasts.

2.11.2 Optimal reactive power dispatch

The ORPD can be considered as a special case of the OPF problem. The ORPD problem solution aims to minimize a chosen objective function, such as power loss or voltage deviation, through optimal adjustment of the power system control variables, under specified active power outputs of all generators (except at the slack bus P_{Gslack}), while at the same time satisfying various operating constraints. Usually, the active power outputs of generators are obtained as the solution of the ED problem. The ED problem solution aims to minimize the cost of generating electric power. Mathematically, the ORPD problem can be formulated as a nonlinear optimization problem with constraints. Many authors proposed GA for solving the ORPD problem.

Reference [30] presents an adaptive GA for ORPD and voltage control of power systems. In the adaptive GA, the probabilities of crossover and mutation are varied depending on the fitness values of the solutions and the normalized fitness distances between the solutions in the evolution process to prevent premature convergence and refine the convergence performance of GAs.

An improved GA with integer/float mixed coding of control variables is proposed in [31] for solution of the ORPD problem in order to minimize the system active power loss.

The authors [32] used a binary-coded GA with tournament selection, two point crossover, and bit-wise mutation for solving the ORPD, including voltage stability limit.

Self-adaptive real-coded GA with the simulated binary crossover, the binary tournament selection, and polynomial mutation is used in [33] to solve the ORPD problem. The control variables of the ORPD problem are threefold: continuous (generator voltages), discrete (transformer tap ratios), and binary (Var sources).

The authors [34] concluded that the real-coded algorithm has better performance in solving the ORPD compared to evolutionary programing.

2.11.3 Combined economic and emission dispatch

The ED problem solution aims to minimize the cost of generation of electric power through optimal adjustment of the committed generating unit outputs, while at the same time satisfying all unit and system constraints. When the environmental concerns are combined with the ED, then the problem becomes a combined economic and emission dispatch (CEED) problem. This problem considers two objectives: minimization of the fuel cost and emission from the thermal power plants with both equality and inequality constraints. So, the CEED problem is a multiobjective mathematical problem in which conflicting objectives are optimized simultaneously. The environmental aspect adds complexity to the solution of the ED problem due to the nonlinear characteristics of the mathematical models used to represent emissions. In addition, the CEED problem can be complicated even further if nonsmooth and nonconvex fuel cost functions are used to model generators, such as valve point loading effects. All these considerations make the CEED problem a highly nonlinear and a multimodal optimization problem.

The authors [35] propose an improved GA with two fuzzy controllers based on some heuristics to adaptively adjust the crossover probability and mutation rate during the optimization process for solving the CEED problem.

The application of a GA with advanced crossover and mutation operators to solve the optimal power dispatch problem for a multinode auction market is proposed in [36].

The authors [37] prove the efficiency of GA for CEED problems in terms of total minimum generation cost, solution time, and convergence criteria.

GA with some specific operators like two-level crossovers, swap mutation, repair, and approximation operators is proposed in [38] to solve the unit commitment problem. The problem is solved in two phases. In the first phase, unit commitment is solved with prevailing constraints, without line flow constraint by GA. In the second phase, the violations in the lines are minimized for a committed schedule using GA-based OPF [38].

For solving the stochastic CEED problem in a hybrid generation system consisting of thermal/wind/solar units, the authors [39] proposed a multiobjective-controlled elitist NSGA-II procedure. Based on stochastic CEED model, including probability density functions of wind power and PV power, the multiobjective-controlled elitist NSGA-II derives a set of Pareto-optimal hybrid system configuration in terms of cost and emission with good diversity. Then, fuzzy cardinal priority ranking is used to obtain the best compromise solution [39].

2.11.4 Optimal power flow in distribution networks

Due to the growth of DG penetration, the nature of distribution network is altered from a passive network to an active one and a distribution company market which is operated by distribution system operator. Generally, if properly planned and controlled, the DG units may offer improved voltage profile and power losses reduction of the distribution network, better economics, and a reduced dependence on the local utility. Safe, reliable, and economic operation of the active distribution network can be achieved only by efficiently coordination of the DG units operation, voltage regulation [voltage regulators, under-load tap changer (ULTC)], and reactive power compensation (Var compensators). In fact, it is an OPF problem. The main objective of the OPF problem for a distribution network is to minimize the fuel cost of DG units, voltage profile improvement, Volt/Var coordination, and minimization of power losses, through optimal settings of the control variables while at the same time satisfying various distribution system operating constraints.

In recent two decades, many authors have proposed GA for solving the OPF in distribution networks.

One of the first applications of GA in optimization of the distribution network operation is given in [40]. The authors proposed a GA approach to minimize power loss at distribution system through controlling the tap of load tap changer, size of substation capacitor, local controller settings, and voltage amplitude of DGs.

The authors [41] proposed a strategy for Volt/Var control in distribution system based on day-ahead load forecast. The GA is used to minimize the power loss and improve the voltage profile for a whole day across the whole system through optimum dispatch schedules for on-load tap changer settings at substation and all shunt capacitor switching in the system.

Park *et al.* [42] presented a coordinated two-stage control method for reducing the operation numbers of ULTC and switching capacitors in distribution system. In the first stage the GA is used to finds the optimal dispatch schedule for the capacitors to reduce the power loss and to improve the voltage profile during a day. In the second stage each capacitor operates according to this dispatch schedule and the ULTC is controlled in real time to cover the difference between the actual and forecasted load levels [42].

An improved adaptive GA with adaptively changing of the searching step length and arithmetic mutation strategy is used in [43] for optimizing capacitor switching, in order to loss reduction in distribution system.

The GA is used in [44] to optimal Volt/Var control in distribution system with DG. Each control device, such as the load control transformer, step voltage regulator, shunt capacitor, shunt reactor, and static Var compensator, is cooperatively controlled for optimal voltage of each time section.

The authors [45] propose an adaptive hybrid GA with a local search based on a generalization of the branch-exchange procedure and dynamic tuning of the mutation operator for the optimal reconfiguration of distribution networks in order to power loss reduction.

The GA with Gray-encoding is proposed in [46] to optimize the wind generator voltages, reactive power of static compensators, and transformer taps for the real power loss reduction in a distribution system.

The authors in [47] propose a procedure for determining the optimal tap changer positions of off-voltage tap changing transformers in radial rural distribution networks in presence of distributed generators. The procedure is based on spatial network decomposition and as well as bringing down a branchy LV network to an equivalent line. Optimal voltage module values of PV nodes in the optimal seasonal control plan are determined by the application of GA.

A hybrid solution method based on integration of fuzzy rules and GA is proposed in [48] for optimal Volt/Var/total harmonic distortion (THD) control in distorted distribution systems serving nonlinear loads.

Similar problem related to optimal Volt/Var/THD control in distribution networks considering the PV inverters' reactive power as additional control variable is solved using a multiobjective GA [49].

In [50], the GA is used to finding the best solution for the problem of optimal Volt/Var/THD control in distribution network considering the uncertainties in load demand and wind-power generation.

Radosavljević *et al.* [51] proposed a GA-based approach for the solution of the OPF in distribution networks with DG units, including fuel cells, micro turbines, diesel generators, photovoltaic systems, and wind turbines. The OPF is formulated as a nonlinear MOP with equality and inequality constraints. Due to the stochastic nature of energy produced from renewable sources, i.e., wind turbines and photovoltaic systems, as well as load uncertainties, a probabilistic algorithm is introduced in the OPF analysis.

In [52], the GA is incorporated in a day-ahead coordinated dispatch method of reactive power in order to achieve OPF, minimize power losses, and minimize switching of capacitor banks in distribution network with DGs.

The authors [53] use MATLAB gamultiobj function to solve a MOP with the objective to minimize the voltage error at pilot buses, the reactive power deviation, and the voltage error at the generators.

2.11.5 Optimal placement and sizing of distributed generation in distribution networks

Generally, the distributed generators have positive technical implications on the operation of the distribution network, such as reduction of power loss, improving of voltage profile, and increasing the reliability and efficiency. Impact of DG on distribution network operation mostly depends on his location, size, and type. The task of determining the optimal location, size, and type of one or more distributed generators can be stated as a way to achieve the desired operational mode. In recent two decades, many authors prove excellent performances of GA in solving this nonlinear optimization problem with both continuous and discrete control variables.

Celli and Pilo [54] applied GA to find the optimal location of DGs on an existing medium voltage distribution network, considering all the technical constraints, like feeder capacity limits, feeder voltage profile, and three-phase short-circuit current in the network nodes.

In [55], the authors use the GA for finding the optimal types of DG and their corresponding locations and sizes in a distribution network in order to improving service reliability, reducing the customer interruption costs, and saving the power cost.

A method combining OPF and GA is proposed in [56] for finding the best combination of sites within a distribution network for connecting a predefined number of DGs.

The authors [57] showed that load models can significantly affect the optimal location and sizing of DGs in distribution networks. A multiobjective performance index (IMO) for the performance calculation of distribution systems for DG size and location planning with load model considers is defined. GA is used to find the best solution.

Based on simulation results, the authors [58] concluded that the GA method for optimal placement and sizing of the DGs in radial and networked distribution networks gives better results than an analytical approach.

The authors in [59,60] proposed a combined GA/particle swarm optimization for optimal location and sizing of DG in distribution systems. The problem is defined as multiobjective optimization with several indices such as network power losses, voltage deviation index, voltage stability index, and reliability index.

In [61], the authors used a MATLAB realization of GA for optimal allocation of different types of DGs in order to minimization of cumulative average daily active power losses.

The authors [62] evaluated the impact of micro DG on the overall process of low voltage network planning using GA.

Some of authors [63,64] devoted special attention on the problem of optimal DG placement in context of power quality improvement.

GA-based tabu search method is proposed in [65] to determination of optimal location of DG units as independent private sector in a distribution system.

The NSGA-II approach is proposed in [66] for solving the optimal sitting of DG units in power system as a probabilistic MOP with objectives such as total imposed costs, total network losses, customer outage costs as well as absorbed private investments.

The authors [67] considered optimal placement of DG as a tool to improving the distribution system reliability. The GA is used for solving the nonlinear optimization problem.

A combination of analytical and GA methods is proposed in [68] for determination of optimal location and size of DG units to minimize the distribution system power loss.

Wu *et al.* [69] proposed cloud-theory-adapted GA to determine the optimal location and the capacity of DG units in the distribution networks considering different aspects such as energy loss, voltage quality and stability, and line loadability.

2.11.6 Optimal energy and operation management of microgrids

MGs are local distribution networks comprising various DG units, controllable loads, and energy storage devices, which can operate either interconnected or isolated from the main distribution grid as a controlled entity. The MGs represent the core of smart distribution networks within modern power systems. Generally, if properly planned and controlled, MGs may offer improved power quality, better economics, and a reduced dependence on the local utility. The benefits can be achieved only through efficient coordinated operation and control of the generation units together with storage devices and controllable loads. This is a nonlinear optimization problem with constraints named optimal energy and operation management (EOM). In recent years, many authors have proposed GA-based approaches for solving the problem of optimal EOM in MGs.

Arabali *et al.* [70] used a GA together with a two-point estimate method to minimize cost and increase efficiency of a MG consisting of PV and wind generation, energy storage, and controllable heating, ventilation, and air conditioning load.

Mohamed and Koivo [71] applied GA for solving the optimal EOM problem which is considered as a nonlinear constrained multiobjective optimization, where the objective function include the costs of the emissions, NO_x, SO_2, and CO_2, start-up costs, as well as the operation and maintenance costs.

The NSGA-II is applied for optimal planning of an islanded MG system [72] and for optimal EOM of a grid connected MG [73] considering the uncertainties of wind speed, solar radiation, and load demand.

In [74], a matrix real-coded GA is proposed for load management in the MG under different operation policies.

2.11.7 Optimal coordination of directional overcurrent relays

Generally, a power system contains many sections, and each section must be protected against overcurrents. In order to ensure optimum supply reliability, it is important to obtain selectivity, i.e., to switch off the part of the power system where a fault occurs and to avoid switching off the parts of the system where no fault occurs. The goal in coordination of over-current relays is to find the appropriate time dial setting (TDS) and pick up current setting of relays, considering various constraints, such as coordination and boundary limits. The coordination of overcurrent relays has been formulated as an optimization problem and solved with different optimization methods.

Several authors have proposed GA to solve this problem. In [75], the authors applied simple GA for optimal overcurrent coordination in a radial MV distribution network with one supply point.

Yinhong *et al.* [76] proposed a constraint interval coding technique to enhance the efficiency and precision of GA in solving the problem of optimal relay coordination in the multiloop network considering the mixed protection system with inverse-time relays and definite-time relays.

Razavi *et al.* [77] introduced a powerful optimal coordination method based on GA with objective function developed to overcome the problems of miscoordination and continuous or discrete time setting multiplier or TDS. A similar approach has been used in [78,79].

Chabanloo *et al.* [80] proposed a GA-based method for optimal coordination of the overcurrent relays considering the dynamic model of overcurrent relays instead of the fixed characteristic curves. This approach is suitable for distribution networks with DGs which can cause some problems in coordination of protection devices, due to changes in short-circuit levels at different points.

In [81] the optimal coordination of directional overcurrent relays is modeled by attempting the practical issues, that is, full discrete variables for time–current curve, time multiplier setting, plug setting, and solved by an adopted GA with integer variables.

The authors [82] demonstrate the capability of GA to solve the problem of optimal coordination of directional overcurrent relays, considering more than one level of short-circuit current and nonstandardized inverse time curves.

Moravej *et al.* [83] used a NSGA-II to reduce the discrimination time of the primary and backup relays within optimal coordination of directional overcurrent relays. In this approach, the operation times of the primary and backup relays are minimized separately, using the concept of multiobjective optimization.

2.11.8 Steady-state analysis of self-excited induction generator

Better utilization of renewable energy may be achieved by developing small-scale autonomous power systems like wind and mini hydro power plants, which use the self-excited induction generators (SEIGs). Likewise, it is well known that if an appropriate capacitor bank is connected across an externally driven induction machine, an electromotive force is induced in the machine windings due to the excitation provided by the capacitor. The induced voltage and current would continue to rise, until the Var supplied by the capacitor is balanced by the Var demanded by the machine. This results in an equilibrium state being reached and the machine now operates as an SEIG at a voltage and frequency assigned by the value of the capacitance, speed of the prime mover, parameters of the machine, and the load.

The key issue that needs to be solved is the voltage and frequency regulation on the SEIG. The voltage and frequency variables are nonlinearly dependent on the speeds, magnetizing reactances, excitation capacitances, and loads. Any regulation strategy requires an efficient procedure for computing of the frequency and magnetizing reactances, that is, the steady-state condition of SEIGs.

Joshi, Sandhu *et al.* [84–88] have contributed a significant amount of literature regarding the use of GA in performance analysis of SEIGs. In these approaches, the GA is employed to find the generated frequency and magnetizing reactance of a three-phase SEIG which supplies a three-phase balanced load.

Radosavljević *et al.* [89] proposed a general procedure for a steady-state analysis of a three-phase SEIG supplying balanced/unbalanced loads. This approach has been adapted [90] for a case when SEIGs operated in parallel. The symmetrical component theory is used for the transformation of a complex three-phase generator capacitance–load system to a simple equivalent passive circuit. The GA is applied for the determination of unknown variables by minimizing the impedance module of the equivalent circuit.

2.12 Conclusion

GA is a very powerful method for solving practical, complex, nonlinear optimization problems. The advantages of GA making it so efficient and popular to use are as follows:

- The objective function to be optimized is entirely optional, that is, there are no special requirements such as continuity, differentiability, etc.;
- They are applicable to a large number of problems of various natures;
- The algorithm structure offers great opportunities in terms of upgrade and increase in the algorithm efficiency;

- Reliability of the results obtained may be increased by simple repetition of the process;
- The result obtained constitutes a set of solutions, rather than a single solution. If a GA fails to find a global optimum, then it provides a solution that may be accepted as feasible;
- They solve any problem that may be presented as optimization-related, regardless of whether the state variables are real numbers, bits, or characters;
- They are easily applied to multidimensional problems;
- Availability of program support. There are ready-made software packages that may be applied to solving concrete problems. For instance, the global optimum toolbox in the MATLAB software package.

The disadvantages of GAs are as follows:

- They need to be adjusted to specified constraints;
- It is often necessary to adjust the problem to the algorithm;
- Parameters exert a considerable impact on efficiency. There is no universal rule for parameter setting;
- Convergence is slower than in other numerical methods. Due to a large number of operations they perform GAs are slow;

It is impossible to obtain a solution that is 100% reliable.

GA has found application in practically all areas of the power engineering for solving very complex problems.

References

[1] Beasley D., Bull D.R., Martin R.R. 'An overview of genetic algorithms: Part 1, Fundamentals'. *University Computing*. 1993;**15**(2):58–69

[2] Golub M. *Genetski algorithm* [online]. 2004. Available from http://www.zemris.fer.hr/~golub/ga/ga_skripta1.pdf [Accessed 20 Apr 2017]

[3] Holland J.H. *Adaptation in natural and artificial systems*. Cambridge: MIT Press; 1975

[4] Lee K.Y., El-Sharkawi M.A. *Modern heuristic optimization techniques with applications to power systems*. New York: IEEE Power Engineering Society (02TP160); 2002. p. 221

[5] Varga Z.M. *Metaheuristik*. Belgrade: University of Belgrade; 2008. p. 77

[6] Goldberg D.E. *Genetic algorithm in search, optimization, and machine learning*. Boston: Addison Wesley Publishing Company; 1989

[7] Srinivas N., Deb K. 'Multiobjective optimization using nondominated sorting in genetic algorithms'. *Evolutionary Computation*. 1995;**2**(3):221–48

[8] Deb K., Pratap A., Agarwal S., Meyarivan T. 'A fast and elitist multi-objective genetic algorithm: NSGA-II'. *IEEE Transactions on Evolutionary Computation*. 2002;**6**(2):182–97

[9] Lai L.L., Ma J.T., Yokoyama R., Zhao M. 'Improved genetic algorithms for optimal power flow under both normal and contingent operation states'.

International Journal of Electrical Power & Energy Systems. 1997;**19**(5):287–92

[10] Bakirtzis A., Biskas P., Zoumas C., Petridis V. 'Optimal power flow by enhanced genetic algorithm'. *IEEE Transactions on Power Systems*. 2002;**17**(2):229–36

[11] Sood Y., Padhy N., Gupta H. 'Discussion of optimal power flow by enhanced genetic algorithm'. *IEEE Transactions on Power Systems*. 2003;**18**(3):1219

[12] Osman M.S., Abo-Sinna M.A., Mousa A.A. 'A solution to the optimal power flow using genetic algorithm'. *Applied Mathematics and Computation*. 2004;**155**:391–405

[13] Devaraj D., Yegnanarayana B. 'Genetic-algorithm-based optimal power flow for security enhancement'. *IEE Proceedings—Generation, Transmission and Distribution*. 2005;**152**(6):899–905

[14] Todorovski M., Rajicic D. 'An initialization procedure in solving optimal power flow by genetic algorithm'. *IEEE Transactions on Power Systems*. 2006;**21**(2):480–7

[15] Gaing Z.L., Chang R.F. 'Security-constrained optimal power flow by mixed-integer genetic algorithm with arithmetic operators'. *Proceedings of IEEE Power Engineering Society General Meeting*; Montreal, Canada, June 2006. IEEE; 2006. pp. 1–8

[16] Chan K.Y., Ling S.H., Chan K.W., Iu H.H.C., Pong G.T.Y. 'Solving multi-contingency transient stability constrained optimal power flow problems with an improved GA'. *Proceedings of IEEE Congress on Evolutionary Computation (CEC 2007)*; Singapore, Sep 2007. pp. 2901–8

[17] Banu R., Devaraj D. 'Genetic algorithm approach for optimal power flow with FACTS devices'. *Proceedings of 4th International IEEE Conference Intelligent Systems*; Varna, Bulgaria, Sep 2008. IEEE; 2008. pp. 11–6

[18] Mahdad B., Bouktir T., Srairi K., Benbouzid M.E. 'Dynamic strategy based fast decomposed GA coordinated with FACTS devices to enhance the optimal power flow'. *Energy Conversion and Management*. 2010;**51**:1370–80

[19] Kumari M.S., Maheswarapu S. 'Enhanced genetic algorithm based computation technique for multi-objective optimal power flow solution'. *Electrical Power and Energy Systems*. 2010;**32**:736–42

[20] Attia A.F., Al-Turki Y.A., Abusorrah A.M. 'Optimal power flow using adapted genetic algorithm with adjusting population size'. *Electric Power Components and Systems*. 2012;**40**:1285–99

[21] Nithya C., Roselyn J.P., Devaraj D., Dash S.S. 'Voltage stability constrained optimal power flow using non-dominated sorting genetic algorithm-II (NSGA II)'. *Proceedings of the Third International Conference on Swarm, Evolutionary, and Memetic Computing*; Bhubaneswar, India, Dec 2012. pp. 793–801

[22] Panuganti S., Roselyn J.P., Devaraj D., Dash S.S. 'Voltage stability constrained optimal power flow using NSGA-II'. *Computational Water, Energy, and Environmental Engineering*. 2013;**2**:1–8

[23] Kılıç U., Ayan K. 'Optimal power flow solution of two-terminal HVDC systems using genetic algorithm'. *Electrical Engineering*. 2014;**96**(1): 65–77

[24] Kılıç U., Ayan K. 'Transient stability constrained optimal power flow solution of ac-dc systems using genetic algorithm'. *Proceedings of 3rd International Conference on Electric Power and Energy Conversion Systems*; Istanbul, Turkey, Oct 2013. IEEE; 2014. pp. 1–6

[25] Balasubramanian S., Gokhale R.V., Sekar A. 'A new AC optimal power flow formulation and solution using genetic algorithm based on P–Q decomposition'. *Proceedings of North American Power Symposium (NAPS)*; Charlotte, USA, Oct 2015. IEEE; 2015. pp. 1–5

[26] Reddy S.S., Bijwe P.R., Abhyankar A.R. 'Faster evolutionary algorithm based optimal power flow using incremental variables'. *Electrical Power and Energy Systems*. 2014;**54**:198–210

[27] Ye C-J., Huang M-X. 'Multi-objective optimal power flow considering transient stability based on parallel NSGA-II'. *IEEE Transaction on Power Systems*. 2015;**30**(2):857–66

[28] Bouchekara H.R.E.H., Chaib A.E., Abido M.A. 'Optimal power flow using GA with a new multi-parent crossover considering: prohibited zones, valve-point effect, multi-fuels and emission'. *Electrical Engineering*. 2016; doi: 10.1007/s00202-016-0488-9

[29] Reddy S.S. 'Optimal power flow with renewable energy resources including storage'. *Electrical Engineering*. 2017;**99**(2):685–95

[30] Wu Q.H., Cao Y.J., Wen J.Y. 'Optimal reactive power dispatch using an adaptive genetic algorithm'. *Electrical Power and Energy Systems*. 1998;**20**(8):563–9

[31] Zhang H., Zhang L., Meng F. 'Reactive power optimization based on genetic algorithm'. *Proceedings of International Conference on Power System Technology POWERCON '98*; Beijing, China, Aug 1998. IEEE; 1998. pp. 1448–53

[32] Durairaj S., Kannan P.S., Devaraj D. 'Application of genetic algorithm to optimal reactive power dispatch including voltage stability constraint'. *Journal of Energy & Environment*. 2005;**4**:63–73

[33] Subbaraj P., Rajnarayanan P.N. 'Optimal reactive power dispatch using self-adaptive real coded genetic algorithm'. *Electric Power Systems Research*. 2009;**79**:374–81

[34] Kumar S.K.N., Renuga P. 'Reactive power planning using real GA comparison with evolutionary programming'. *International Journal of Recent Trends in Engineering*. 2009;**1**(3):124–8

[35] Song Y.H., Wang G.S., Wang P.Y., Johns A.T. 'Environmental/economic dispatch using fuzzy logic controlled genetic algorithms'. *IEE Proceedings—Generation, Transmission, and Distribution*. 1997;**144**(4):377–82

[36] Numnonda T., Annakkage U.D. 'Optimal power dispatch in multinode electricity market using genetic algorithm'. *Electric Power Systems Research*. 1999;**49**:211–20

[37] Raglend I.J., Veeravalli S., Sailaja K., Sudheera B., Kothari D.P. 'Comparison of AI techniques to solve combined economic emission dispatch problem with line flow constraints'. *Electrical Power and Energy Systems.* 2010;**32**:592–8

[38] Kumar V.S., Mohan M.R. 'Solution to security constrained unit commitment problem using genetic algorithm'. *Electrical Power and Energy Systems.* 2010;**32**:117–25

[39] Abul'Wafa A.R. 'Optimization of economic/emission load dispatch for hybrid generating systems using controlled Elitist NSGA-II'. *Electric Power Systems Research.* 2013;**105**:142–51

[40] Niknam T., Ranjbar A.M., Shirani A.R. 'Impact of distributed generation on Volt/Var control in distribution networks'. *Proceedings of IEEE Power Tech Conference; Bologna, Italy*, Jun 2003. IEEE; 2004. pp. 1–7

[41] Hu Z., Wang X., Chen H., Taylor G.A. 'Volt/VAr control in distribution systems using a time-interval based approach'. *IEE Proceedings—Generation, Transmission, and Distribution.* 2003;**150**(5):548–54

[42] Park J-y., Nam S-r., Park J-k. 'Control of a ULTC considering the dispatch schedule of capacitors in a distribution system'. *IEEE Transactions on Power Systems.* 2007;**22**(2):755–61

[43] Zhang D., Fu Z., Zhang L. 'Joint optimization for power loss reduction in distribution systems'. *IEEE Transactions on Power Systems.* 2008;**23**(1):161–9

[44] Senjyu T., Miyazato Y., Yona A., Urasaki N., Funabashi T. 'Optimal distribution voltage control and coordination with distributed generation'. *IEEE Transactions on Power Delivery.* 2008;**23**(2):1236–42

[45] Queiroz L.M.O., Lyra C. 'Adaptive hybrid genetic algorithm for technical loss reduction in distribution networks under variable demands'. *IEEE Transactions on Power Systems.* 2009;**24**(1):445–53

[46] Hong Y-Y., Luo Y-F. 'Optimal VAR control considering wind farms using probabilistic load-flow and Gray-based genetic algorithms'. *IEEE Transactions on Power Delivery.* 2009;**24**(3):1441–9

[47] Radosavljević J., Jevtić M., Klimenta D. 'Optimal seasonal voltage control in rural distribution networks with distributed generators'. *Journal of Electrical Engineering.* 2010;**61**(6):321–31

[48] Ulinuha A., Masoum M.A.S., Islam S. 'Hybrid genetic-fuzzy algorithm for volt/var/total harmonic distortion control of distribution systems with high penetration of non-linear loads'. *IET Generation, Transmission & Distribution.* 2011;**5**(4):425–39

[49] Jashfar S., Esmaeili S. 'Volt/var/THD control in distribution networks considering reactive power capability of solar energy conversion'. *Electrical Power and Energy Systems.* 2014;**60**:221–33

[50] Jashfar S., Hosseini-Biyouki M.M., Esmaeili S. 'A stochastic programming to Volt/Var/Total harmonic distortion control in distribution networks including wind turbines'. *Electric Power Components and Systems.* 2015;**43**(7):733–46

[51] Radosavljević J., Jevtić M., Klimenta D., Arsić N. 'Optimal power flow for distribution networks with distributed generation'. *Serbian Journal of Electrical Engineering*. 2015;**12**(2):145–70

[52] Zhang L., Tang W., Liang J., Cong P., Cai Y. 'Coordinated day-ahead reactive power dispatch in distribution network based on real power forecast errors'. *IEEE Transactions on Power Systems*. 2016;**31**(3):2472–80

[53] Castro J.R., Saad M., Lefebvre S., Asber D., Lenoir L. 'Optimal voltage control in distribution network in the presence of DGs'. *Electrical Power and Energy Systems*. 2016;**78**:239–47

[54] Celli G., Pilo F. 'Optimal distributed generation allocation in MV distribution networks'. *Proceedings of 22nd IEEE Power Engineering Society International Conference on Power Industry Computer Applications*; Sydney, Australia, May 2001. IEEE; 2002. pp. 81–6

[55] Teng J-H., Luor T-S., Liu Y-H. 'Strategic distributed generator placements for service reliability improvements'. *Proceedings of IEEE Power Engineering Society Summer Meeting*; Chicago, USA, Jul 2002. IEEE; 2005. pp. 719–24

[56] Harrison G.P., Piccolo A., Siano P., Wallace A.R. 'Distributed generation capacity evaluation using combined genetic algorithm and OPF'. *International Journal of Emerging Electric Power Systems*. 2007;**8**(2):1–13

[57] Singh D., Singh D., Verma K.S. 'Multiobjective optimization for dg planning with load models'. *IEEE Transactions on Power Systems*. 2009;**24**(1):427–36

[58] Singh R.K., Goswami S.K. 'Optimum siting and sizing of distributed generations in radial and networked systems'. *Electric Power Components and Systems*. 2009;**37**:127–45

[59] Moradi M.H., Abedini M. 'A combination of genetic algorithm and particle swarm optimization for optimal DG location and sizing in distribution systems'. *Electrical Power and Energy Systems*. 2012;**34**:66–74

[60] Bohre A.K., Agnihotri G., Dubey M. 'Optimal sizing and sitting of DG with load models using soft computing techniques in practical distribution system'. *IET Generation, Transmission & Distribution*. 2016;**10**(11):2606–21

[61] Prenc R., Škrlec D., Komen V. 'Distributed generation allocation based on average daily load and power production curves'. *Electrical Power and Energy Systems*. 2013;**53**:612–22

[62] Mendoza J.E., López M.E., Fingerhuth S.C., Pena H.E., Salinas C.A. 'Low voltage distribution planning considering micro distributed generation'. *Electric Power Systems Research*. 2013;**103**:233–40

[63] Biswas S., Goswami S.K., Chatterjee A. 'Optimal distributed generation placement in shunt capacitor compensated distribution systems considering voltage sag and harmonics distortions'. *IET Generation, Transmission & Distribution*. 2014;**8**(5):783–97

[64] Savić A., Đurišić Ž. 'Optimal sizing and location of SVC devices for improvement of voltage profile in distribution network with dispersed photovoltaic and wind power plants'. *Applied Energy*. 2014;**134**:114–24

[65] Mohammadi M., Nafar M. 'Optimal placement of multitypes DG as independent private sector under pool/hybrid power market using GA-based Tabu Search method'. *Electrical Power and Energy Systems*. 2013;**51**:43–53
[66] Dehghanian P., Hosseini S.H., Moeini-Aghtaie M., Arabali A. 'Optimal siting of DG units in power systems from a probabilistic multi-objective optimization perspective'. *Electrical Power and Energy Systems*. 2013;**51**:14–26
[67] Dezaki H.H., Agheli A., Vahidi B., Abyaneh H.A. 'Optimized placement of connecting the distributed generations work stand alone to improve the distribution systems reliability'. *Journal of Electrical Engineering*. 2013;**64**(2):76–83
[68] Vatani M., Alkaran D.S., Sanjari M.J., Gharehpetian G.B. 'Multiple distributed generation units allocation in distribution network for loss reduction based on a combination of analytical and genetic algorithm methods'. *IET Generation, Transmission & Distribution*. 2016;**10**(1):66–72
[69] Wu C., Lou Y., Lou P., Xiao H. 'DG location and capacity optimization considering several objectives with cloud theory adapted GA'. *International Transactions on Electrical Energy Systems*. 2014;**24**:1076–88
[70] Arabali A., Ghofrani M., Etezadi-Amoli M., Fadali M.S., Baghzouz Y. 'Genetic-algorithm-based optimization approach for energy management'. *IEEE Transactions on Power Delivery*. 2013;**28**(1):162–70
[71] Mohamed F.A., Koivo H.N. 'Online management genetic algorithms of microgrid for residential application'. *Energy Conversion and Management*. 2012;**64**:562–8
[72] Guo L., Liu W., Jiao B., Hong B., Wang C. 'Multi-objective stochastic optimal planning method for stand-alone microgrid system'. *IET Generation, Transmission & Distribution*. 2014;**8**(7):1263–73
[73] Kuznetsova E., Ruiz C., Li Y-F., Zio E. 'Analysis of robust optimization for decentralized microgrid energy management under uncertainty'. *Electrical Power and Energy Systems*. 2015;**64**:815–32
[74] Chen C., Duan S., Cai T., Liu B., Hu G. 'Smart energy management system for optimal microgrid economic operation'. *IET Renewable Power Generation*. 2011;**5**(3):258–67
[75] So C.W., Li K.K., Lai K.T., Fung K.Y. 'Application of genetic algorithm for overcurrent relay coordination'. *Proceedings of the 6th International Conference on Developments in Power System Protection*; Nottingham, UK, Mar 1997. London: IET; 2002. pp. 66–9
[76] Yinhong L., Dongyuan S., Xianzhong D. 'A study on optimal coordination of definite-time relay based on genetic algorithm'. *Proceedings of Transmission and Distribution Conference and Exposition*; Dallas, USA, Sep 2003. IEEE; 2004. pp. 279–84
[77] Razavi F., Abyaneh H.A., Al-Dabbagh M., Mohammadi R., Torkaman H. 'A new comprehensive genetic algorithm method for optimal overcurrent relays coordination'. *Electric Power Systems Research*. 2008;**78**:713–20

[78] Mohammadi R., Abyaneh H.A., Razavi F., Al-Dabbogah M., Sadeghi S.H.H. 'Optimal relays coordination efficient method in interconnected power systems'. *Journal of Electrical Engineering*. 2010;**61**(2):75–83

[79] Adelnia F., Moravej Z., Farzinfar M. 'A new formulation for coordination of directional overcurrent relays in interconnected networks'. *International Transactions on Electrical Energy Systems*. 2015;**25**:120–37

[80] Chabanloo R.M., Abyaneh H.A., Agheli A., Rastegar H. 'Overcurrent relays coordination considering transient behaviour of fault current limiter and distributed generation in distribution power network'. *IET Generation, Transmission & Distribution*. 2011;**5**(9):903–11

[81] Azari A., Akhbari M. 'Optimal coordination of directional overcurrent relays in distribution systems based on network splitting'. *International Transactions on Electrical Energy Systems*. 2015;**25**:2310–24

[82] Salazar C.A.C., Enríquez A.C., Schaeffer S.E. 'Directional overcurrent relay coordination considering non-standardized time curves'. *Electric Power Systems Research*. 2015;**122**:42–9

[83] Moravej Z., Adelnia F., Abbasi F. 'Optimal coordination of directional overcurrent relays using NSGA-II'. *Electric Power Systems Research*. 2015;**119**:228–36

[84] Joshi D., Sandhu K.S., Soni M.K. 'Performance analysis of three-phase self-excited induction generator using genetic algorithm'. *Electric Power Components and Systems*. 2006;**34**:461–70

[85] Joshi D., Sandhu K.S. 'Voltage control of compensated self-excited induction generator using genetic algorithm'. *Proceedings of the 2nd International Conference on Renewable Energy Sources*; Corfu, Greece, Oct 2008. pp. 143–9

[86] Joshi D., Sandhu K.S. 'Excitation control of self excited induction generator using genetic algorithm and artificial neural network'. *International Journal of Mathematical Models and Methods in Applied Sciences*. 2009;**3**(1):68–75

[87] Joshi D., Sandhu K.S., Soni M.K. 'Voltage control of self-excited induction generator using genetic algorithm'. *Turkish Journal of Electrical Engineering & Computer Sciences*. 2009;**17**(1):87–97

[88] Vadhera S., Sandhu K.S. 'Constant voltage operation of self excited induction generator using optimization tools'. *International Journal of Energy and Environment*. 2008;**2**(4):191–8

[89] Radosavljević J., Klimenta D., Jevtić M. 'A genetic algorithm-based approach for a general steady-state analysis of three-phase self-excited induction generator'. *Revue roumaine des sciences techniques Série Électrotechnique et Énergétique*. 2012;**57**(1):10–9

[90] Radosavljević J., Klimenta D., Jevtić M. 'Steady-state analysis of parallel-operated self-excited induction generators supplying an unbalanced load'. *Journal of Electrical Engineering*. 2012;**63**(4):213–23

Chapter 3
Overview of particle swarm optimization

3.1 Introduction

Swarm intelligence can be defined as the collective intelligent behavior of natural or artificial systems. This behavior is characteristically for large groups of individuals acting in a specific area and communicates with each other. This communication can be direct, as is the case with the bees, or indirectly through the environment, thus serving the ants in search of food.

In these communications between individuals, the transferred information may contain a certain error, so there exists a probability that individuals will find a better location and inform the adjacent entities that will further propagate the information.

It is this interaction between the individuals, through which they receive information about neighboring favorable locations and their combination with their knowledge of the best location that has so far been found, that leads the individuals in finding new, more better locations.

Intelligent agents are simple organisms whose behavior is not difficult to implement as a computer program. The challenge in applying the swarm intelligence algorithms is in the correct selection of natural models and adapting of the problem which should be solved to the characteristics of the algorithm.

Particle swarm algorithm (PSO) is the most general of all swarm intelligence algorithms. The task of the algorithm is finding the global optimum in a multidimensional search space. Kennedy and Eberhart [1] developed PSO based on the analogy of swarm of bird and fish school.

The PSO algorithm is very popular because its simplicity of implementation and a relatively small number of parameters. Despite the simplicity of the algorithm, it showed very good results in solving various optimization problems in almost all areas of science and engineering. It is not limited with requirements for differentiability, nonconvexity, and continuity of the objective function or types of control variables. The PSO can be used for practical optimization problems taking into account various types of objective function and constraints. This is confirmed by the large number of research papers that use the PSO. An analysis of IEEE Xplore and Google Scholar citations and publications showed that the PSO is exponentially growing by year [2].

Some of the most popular applications of PSO are related to power system problems, such as optimal operation, control, and planning.

3.2 Description of PSO

PSO is basically developed through simulation of bird flocking in two-dimensional space. Bird flocking optimizes a certain objective function. PSO uses a number of particles (search agents) that fly around in the search space to find best solution. Meanwhile, the particles all look at the best particle (best solution) in their paths. In other words, particles consider their own best solutions as well as the best solution found so far.

The basic elements of the PSO method are briefly stated and defined as follows [3]:

Particle, $\mathbf{x}(t)$: It is a candidate solution represented by an n-dimensional vector, where n is the number of control variables. At time (iteration) t, the ith particle $\mathbf{x}_i(t)$ can be described as $\mathbf{x}_i(t) = \left[x_i^1(t), \ldots, x_i^d(t), \ldots, x_i^n(t)\right]$, where $x_i^d(t)$ is the position of the ith particle with respect to the dth dimension, i.e., the value of the dth control variable in the ith candidate solution.

Population, $\mathbf{POP}(t)$: It is a set of N particles at time (iteration) t, i.e., $\mathbf{POP}(t) = [\mathbf{x}_1(t), \ldots, \mathbf{x}_N(t)]^T$, or in expanded form:

$$\mathbf{POP}(t) = \begin{bmatrix} x_1^1(t), & x_1^2(t), & \ldots, & x_1^d(t), & \ldots, & x_1^n(t) \\ x_2^1(t), & x_2^2(t), & \ldots, & x_2^d(t), & \ldots, & x_2^n(t) \\ & & & \vdots & & \\ x_i^1(t), & x_i^2(t), & \ldots, & x_i^d(t), & \ldots, & x_i^n(t) \\ & & & \vdots & & \\ x_N^1(t), & x_N^2(t), & \ldots, & x_N^d(t), & \ldots, & x_N^n(t) \end{bmatrix} \tag{3.1}$$

Swarm: It is an apparently disorganized population of moving particles that tend to cluster together while each particle seems to be moving in a random direction.

Particle velocity, $\mathbf{v}(t)$: It is the velocity of the moving particles represented by an n-dimensional vector. At time (iteration) t, the ith particle velocity $\mathbf{v}_i(t)$ can be described as $\mathbf{v}_i(t) = \left[v_i^1(t), \ldots, v_i^d(t), \ldots, v_i^n(t)\right]$, where $v_i^d(t)$ is the velocity component of the ith particle with respect to the dth dimension.

Individual best, $\mathbf{pbest}_i(t)$: It is individual best. As a particle moves through the search space, it compares its fitness value at the current position to the best fitness value it has ever attained at any time up to the current time. The best position that is associated with the best fitness encountered so far is called the individual best, $\mathbf{pbest}(t)$. For the ith particle, individual best can be expressed as $\mathbf{pbest}_i(t) = \left[pbest_i^1(t), \ldots, pbest_i^d(t), \ldots, pbest_i^n(t)\right]$.

Global best, $\mathbf{gbest}(t)$: It is the best position among all individual best positions achieved so far. Hence, global best can be expressed as $\mathbf{gbest}(t) = \left[gbest^1(t), \ldots, gbest^d(t), \ldots, gbest^n(t)\right]$.

Fitness is a direct metric of the performance of the individual population member (agent). The fitness of each agent of the population is calculated from the value of the function being optimized.

Stopping criteria: These are the conditions under which the search process will terminate. It is often stated that the search will terminate if the number of iterations t reaches the maximum allowable number t_{max}.

In PSO, each agent tries to modify its position using the following information: the current position, the current velocity, the distance between the current position and **pbest**, and the distance between the current position and **gbest** [4–6]. This modification can be represented by the concept of velocity. Velocity of each agent can be modified by the following equation:

$$\mathbf{v}_i(t+1) = w \cdot \mathbf{v}_i(t) + C_1 \cdot r_1 \cdot (\mathbf{pbest}_i(t) - \mathbf{x}_i(t)) + C_2 \cdot r_2 \cdot (\mathbf{gbest}(t) - \mathbf{x}_i(t)) \tag{3.2}$$

where C_1 and C_2 are positive constants defined as acceleration coefficients, r_1 and r_2 are uniformly distributed random numbers in [0, 1], and w is an inertia weight. This w plays the role of balancing the global search and local search [4].

After updating the velocity, the current position of particle (searching point in the solution space) can be modified by the following equation:

$$\mathbf{x}_i(t+1) = \mathbf{x}_i(t) + \mathbf{v}_i(t+1) \tag{3.3}$$

The first term of (3.2), $w \cdot \mathbf{v}_i(t)$, provides exploration ability for PSO. For initial stages of the search process, large w to enhance the global exploration is recommended, while, for last stages, the weight factor is reduced for better local search.

The second term of (3.2), $C_1 \cdot r_1 \cdot (\mathbf{pbest}_i(t) - \mathbf{x}_i(t))$, represents the cognitive part of PSO where the particle changes its velocity based on its own thinking and memory.

The third term of (3.2), $C_2 \cdot r_2 \cdot (\mathbf{gbest}(t) - \mathbf{x}_i(t))$, represents the social part of PSO where the particle changes its velocity based on the social–psychological adaptation of knowledge.

Figure 3.1 shows a concept of modification of searching point by PSO, and Figure 3.2 shows a searching concept with agents in a solution space. Each agent changes its current position using the integration of vectors as shown in Figure 3.1.

Figure 3.3 shows the general flowchart of PSO. The general flowchart of PSO can be described as follows:

Step 1: Generation of initial condition of each particle. Initial searching points ($\mathbf{x}_i$) and velocities ($\mathbf{v}_i$) of each particle are usually generated randomly within the allowable range. The current searching point is set to **pbest** for each particle. The best evaluated value of **pbest** is set to **gbest** and the agent number with the best value is stored.

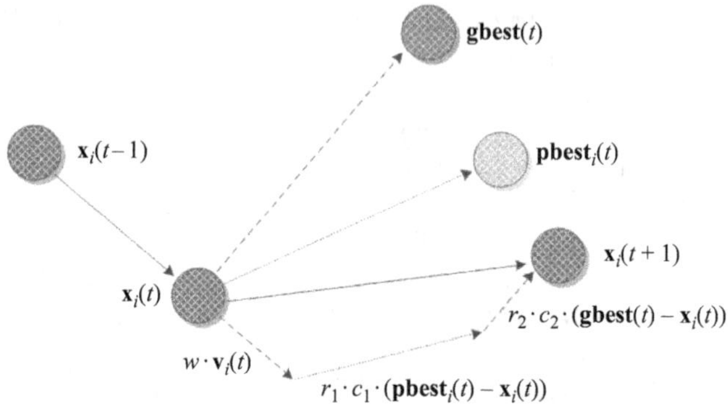

Figure 3.1 Concept of modification of a searching point by PSO

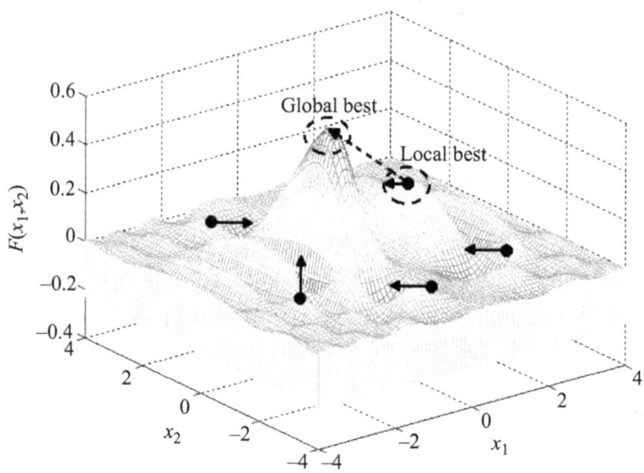

Figure 3.2 Searching concept of PSO with particles in a solution space

Step 2: Evaluation of searching point of each particle. The objective function value is calculated for each particle. If the value is better than the current **pbest** of the particle, the **pbest** value is replaced by the current value. If the best value of **pbest** is better than the current **gbest**, **gbest** is replaced by the best value and the particle number with the best value is stored.

Step 3: Update the velocity of each particle using (3.2). Update the position of each agent using (3.3).

Step 4: Checking the stopping criteria. The search will terminate if the number of iteration t reaches the predetermined maximum iteration number, t_{max}. Otherwise, go to Step 2.

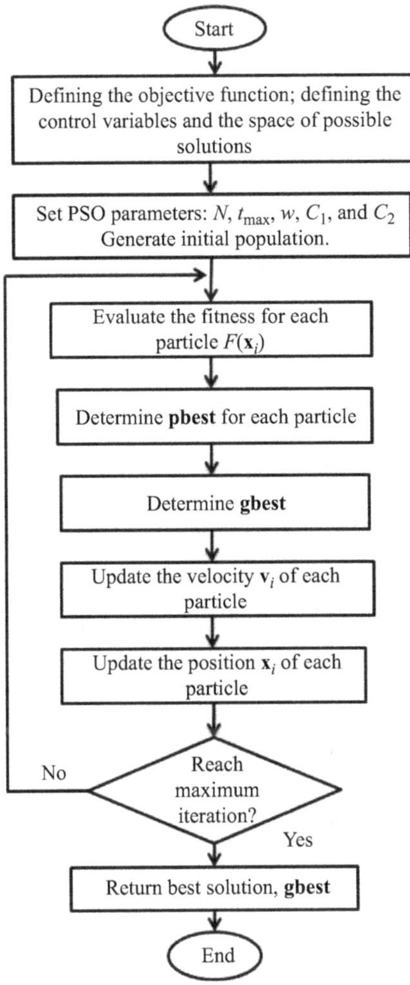

Figure 3.3 General flowchart of PSO

3.2.1 Parameters of PSO

The total number of iterations t_{max}, population size N, weight w, and constants C_1 and C_2 control the performance of PSO. A primary drawback of standard PSO is premature convergence when the parameters are not chosen correctly, especially while handling problems with many local optima [7].

There are several general rules for setting PSO parameters [8]:

1. Fixed parameters of PSO. This is the most commonly used approach in PSO literature. In this case, parameter values are either set as widely accepted values from literature or set by user based on strategy trial-and-error.
2. Dynamic or adaptive parameters of PSO.

3. Finding optimal PSO parameters using a heuristic algorithm. The problem of this strategy is that it adds significantly to the complexity of the problem.
4. Using parameter-free PSO variants which do not require any parameter setting process.

3.2.1.1 Population size and initial population

The population size refers to the number of particles. As in other metaheuristic optimization algorithms, the population size affects performance of PSO. The population size is usually constant, that is, it remains unchanged during the run the PSO. There is no general rule in literature for selection of population size. The larger the population, the more comprehensive the search space of possible solutions. However, it also means a higher computational time. The population size to be adopted depends primarily on the concrete problem being solved, followed by the dimension of particle (number of control variables), as well as the experience of the user performing the task. It seems that the population size tends to increase approximately linearly with the dimension of problem. Nevertheless, the "optimal" population size, if it exists at all, primarily depends on the concrete problem.

Most commonly, the initial population is generated by random selection of solutions (particles) from the solution domain. The initial population can be uniform (all individuals are the same). It is also possible to embed into the initial population individuals as solutions obtained by another optimization method. For example, nonlinear simplex method [9] and plowing method [10] have been used for PSO initialization. In this way the PSO convergence and speed is increased.

3.2.1.2 Maximum velocity

According to (3.2), the velocity update equations are stochastic. As a consequence, the velocities may become too high so that the particles become uncontrolled and exceed search space. Therefore, velocities should be bounded so that each component of velocity, v_i^d, is kept within the range $[-v_i^{d,\max}, v_i^{d,\max}]$. The selection of the parameter $v_i^{d,\max}$ must be carefully performed since it affects the balance between exploration and exploitation [2].

If it is too high, the particles move unpredictably and are swiftly attracted to global best (gbest) without enough exploration of search space. So, they may exceed search space, and the risk of getting trapped in local optima also increases. On the other hand, if it is too small, the movement of particles gets excessively bounded, the computational overhead increases, and the algorithm may be unable to converge [8]. To ensure uniform velocity through all dimensions, Abido [3] proposed $v_i^{d,\max}$ as follows:

$$v_i^{d,\max} = \frac{x_i^{d,\max} - x_i^{d,\min}}{N_I} \tag{3.4}$$

where N_I is the number of intervals in the dth dimension selected by user, $x_i^{d,\max}$ and $x_i^{d,\min}$ are the maximum and minimum values of particle i in dth dimension.

3.2.1.3 Inertia weight

The inertia weight w in (3.2) plays the role of balancing the global search (exploration) and local search (exploitation) of PSO. It can be a positive constant or even a positive linear or nonlinear function of time. The authors [4] experimented with different values of w in the range [0, 1.4] and concluded that the faster convergence is obtained by setting w between [0.8, 1.2]. In addition, in literature, some different approaches for setting inertia weight, such as fuzzy adaptive, chaotic, exponential, Gaussian, parallel, simulated annealing inertia weight, can be found [8]. However, based on a comparative study taking into account both simplicity and efficiency, the authors [11] concluded that the linearly decreasing inertia weight is the most appropriate method for setting inertia weight.

The weighting function given by (3.5), with $w_{max} = 0.9$ and $w_{min} = 0.4$, is widely used in literature:

$$w = w_{min} + \frac{(t_{max} - t)(w_{max} - w_{min})}{(t_{max} - 1)} \tag{3.5}$$

where t_{max} is the total number of iterations, and t is the current iteration.

3.2.1.4 Acceleration coefficients

Cognitive acceleration coefficient C_1 and social acceleration coefficient C_2 in (3.2) affect the maximum step size of a particle in an iteration. C_1 regulates the maximum step size of a particle in the direction of the pbest, while C_2 regulates the maximum step size in the direction of the gbest [12].

If the value of these coefficients is too high, the particles move abruptly and the risk of getting trapped in false optima increases. On the contrary, if their value is too low, the particles move too slowly, computational effort increases significantly and it is likely that the algorithm could not converge. On the other hand, the relative value of these two acceleration coefficients is critical and affects algorithm's behavior. When the value of C_1 increases, it enhances particles' attraction toward **pbest** and decreases their attraction toward **gbest**. Also, increasing C_2 in relation to C_1 increases attraction of particles toward **gbest**. In literature, setting $C_1 = C_2 = 2$ has been proposed as a generally acceptable setting for most of problems and is widely used in practical applications of PSO. But normally, by implementing some trial-and-error, more efficient computational behavior in comparison to the abovementioned generally accepted values can be achieved [8].

3.2.1.5 Constriction coefficients

There are other ways in the literature to define the acceleration coefficients. One such way is the constriction factor approach, given in [2,6,13]. The constriction factor approach utilizes the eigen-value analysis and controls the system behavior so that the system does not diverge in a real value region and can search different regions efficiently. The velocity of the constriction factor approach can be expressed as follows [2,6]:

$$\mathbf{v}_i(t+1) = \chi[\mathbf{v}_i(t) + C_1 \cdot r_1 \cdot (\mathbf{pbest}_i(t) - \mathbf{x}_i(t)) + C_2 \cdot r_2 \cdot (\mathbf{gbest}(t) - \mathbf{x}_i(t))] \tag{3.6}$$

$$\chi = \frac{2}{\varphi - 2 + \sqrt{\varphi^2 - 4\varphi}} \tag{3.7}$$

where $\varphi = C_1 + C_2$, $\varphi > 4$.

In the construction method, φ is commonly set to 4.1, $C_1 = C_2$, and the constant multiplier χ is approximately 0.73. This results in the previous velocity being multiplied by 0.73 and each of the two terms $(\mathbf{pbest}_i(t) - \mathbf{x}_i(t))$ and $(\mathbf{gbest}(t) - \mathbf{x}_i(t))$ being multiplied by a random number limited by $0.73 \cdot 2.05 \approx 1.496$.

Obviously, a PSO with constriction is algebraically equivalent to a PSO with inertia. Equations (3.2) and (3.6) can be transformed into one another via the mapping $w \leftrightarrow \chi$ and $C_i \leftrightarrow \chi C_i$. Hence, the optimal settings suggested in the constriction method correspond to $w = 0.73$ and $C_1 = C_2 = 1.496$ for PSO with inertia [2].

3.2.2 General remarks about PSO

PSO is a metaheuristic optimization method used for searching nonlinear multidimensional search spaces. The features of the PSO are as follows [6,8,14,55]:

- The algorithm is based on researches about swarms such as flock of birds and fish schooling.
- It is based on a simple concept, and its coding is very easy.
- It has less parameters to be tuned by user, but the efficiency of PSO is highly dependent on proper parameter settings.
- It provides fast convergence, high accuracy and requires less computational burden.
- The direction of a particle in the search space is calculated using only two best positions, $\mathbf{pbest}_i$ and $\mathbf{gbest}$. PSO uses a kind of memory for updating the velocity (due to $\mathbf{pbest}_i$ and $\mathbf{gbest}$). All the particles tend to converge to the best solution quickly.
- In PSO, the fitness values of particles are not important in the updating procedure, and updating is performed without considering the distance between solutions.
- Its behavior is not highly affected by increase in dimensionality.
- It is efficient in solving practical optimization problems taking into account various types of objective function and constraints.
- Because of its simplicity, PSO is very suitable for hybridization with other optimization methods.
- The initialization of particles in PSO is done randomly. If the particles initialized are located in a local space, then the chances of getting trapped in local minima are increased.
- The speed of search depends on the separation of particles.

3.2.3 MATLAB® code of PSO

```
%Particle Swarm Optimization (PSO)
%-------------------------------------------------------------------
function [] = PSO

%INITIALIZATION
%===================================================================
%Defining the objective function in the function program Fobj.m
%Defining the space of possible solutions in the function program bounds.m
%-------------------------------------------------------------------
%Defining parameters of PSO:
%-------------------------------------------------------------------
 N=25;                %population size (number of particles N)
 tmax=100;            %maximum number of iterations
 C1=2;                %cognitive constant C1
 C2=2;                %cocial constant C2
 wmax=0.9;            %max inertia weight in w, wmax
 wmin=0.4;            %min inertia weight in w, wmin

%Defining type of optimization (min or max)
%-------------------------------------------------------------------
%Type of optimization: 'min' or 'max'
type_opt='min';

%Generating of the initial population. Initial positions of particles are
%randomly selected between min and max values of the control variables
%-------------------------------------------------------------------
[numbervar,lowerbound,upperbound]=bounds;
if size(upperbound,2)==1
   X=rand(N,numbervar).*(upperbound-lowerbound)+lowerbound;
end
if size(upperbound,2)>1
    for i=1:numbervar
    high=upperbound(i);low=lowerbound(i);
    X(:,i)=rand(N,1).*(high-low)+low;
    end
end

%Create the best so far fitness chart
%-------------------------------------------------------------------
BestChart=[];

%Initialize velocity of agents
V=zeros(N,numbervar);


%ITERATIVE PROCEDURE
%===================================================================
for t=1:tmax

for i=1:N
%Agents that go out of the search space, are reinitialized randomly.
 Tp=X(i,:,t)>upperbound;
 Tm=X(i,:,t)<lowerbound;
 X(i,:,t)=(X(i,:,t).*(~(Tp+Tm)))+((rand(1,numbervar).*(upperbound-...
lowerbound)+lowerbound).*(Tp+Tm));

 x=X(i,:,t);  %position of particle i at iteration t
```

```
% Calculate the fitness value for each particle
%-----------------------------------------------------
fitness(i,t)=Fobj(x);

%Determination of pbest (individual best) and gbest (global best)
%-----------------------------------------------------
switch type_opt
    case 'min'
        [pbest_fitness,index]=min(fitness(i,:));
        pbest=X(i,:,index);
        if(i==1 && t==1)
            gbest=pbest;
            gbest_fitness=pbest_fitness;
        elseif fitness(i,t)<gbest_fitness
            gbest_fitness=fitness(i,t);
            gbest=X(i,:,t);
        end
    case 'max'
        [pbest_fitness,index]=max(fitness(i,:));
        pbest=X(i,:,index);

        if(i==1 && t==1)
            gbest=pbest;
            gbest_fitness=pbest_fitness;
        elseif fitness(i,t)>gbest_fitness
            gbest_fitness=fitness(i,t);
            gbest=X(i,:,t);
        end
    otherwise
        error('fitness_type mismatch')
end

%Updating the inertia weight, w
%-----------------------------------------------------
w(t)=wmin+((tmax - t)*(wmax - wmin))/(tmax-1);

%Updating the velocity (V) of each particle
%-----------------------------------------------------
V(i,:,(t+1))=w(t)*V(i,:,t)+C1*rand(1,numbervar).*(pbest-X(i,:,t))...
            +C2*rand(1,numbervar).*(gbest-X(i,:,t));

%Updating the position (X) of each particle
%-----------------------------------------------------
X(i,:,(t+1))= X(i,:,t) + V(i,:,(t+1));
end

%Best solution in each iteration is stored in the vector GlobBestCos:
GlobBestCos(t)= gbest_fitness;

%Display the results in each iteration
%-----------------------------------------------------
%fprintf('PSO| Iter:%3d -->  Fbest: %9.5f\n',t,gbest_fitness);
%-----------------------------------------------------
BestChart=[BestChart gbest_fitness];

end  %end of iteration  procedure
%-----------------------------------------------------
```

```
%RESULTS
%=========================================================================
disp(' ');
%Best value of the objective function, Fbest
Fbest=GlobBestCos(tmax)

%Optimal values of control variables, xbest
xbest=gbest

%Plotting convergence profile of PSO:
plot(BestChart,'--k','LineWidth',2);
xlabel('\fontsize{12}\bf Iteration');ylabel('\fontsize{11}\bf Fobj');
legend('\fontsize{12}\bf PSO',1);grid on;

%Functions
%=========================================================================
function [Fobj]=Fobj(x)
%Code of the objective function to be optimized
%Input: x - vector of control variables
%Output: Fobj - value of objective function
%- - - - - - - - - - - - - - - - - - - - - - - - - - - - - - - - - - - - -
Fobj=1.8.^(-1.5*sqrt(x(1)^2+x(2)^2)).*sin((x(1)-0.5).^2).*cos(0.5*...
     (x(2)-0.5).^2);
return

function [numbervar,lowerbound,upperbound]=bounds
%Defining number of control variables,
%lower and upper limits of control variables
%- - - - - - - - - - - - - - - - - - - - - - - - - - - - - - - - - - - - -
numbervar=2;  lowerbound(1:numbervar)=-4; upperbound(1:numbervar)=4;
return
%=========================================================================
```

3.2.4 Example usage of PSO

To show how GSA works, the program is applied to find minimum and maximum of the function given as follows:

$$F(x_1,x_2) = 1.8^{-1.5\sqrt{x_1^2+x_2^2}} \cdot \sin\left((x_1-0.5)^2\right) \cdot \cos\left(0.5(x_2-0.5)^2\right) \tag{3.8}$$

within the range $-4 \leq x_1,\ x_2 \leq 4$.

Graph of this function is shown in Figure 3.4.

The results of PSO obtained in the case of minimization are as follows:

```
>> PSO
 Fbest =
   -0.2328
xbest =
   -1.6232    0.0723
```

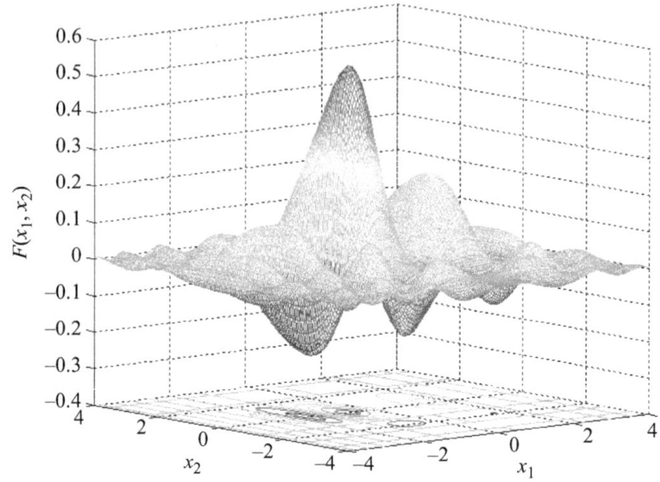

Figure 3.4 Graph of the function F

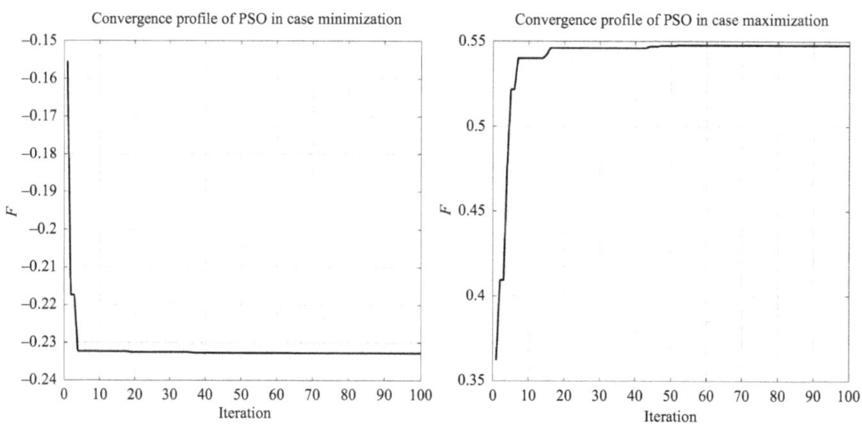

Figure 3.5 Convergence of PSO

The results of PSO obtained in the case of maximization are as follows:

```
>> PSO
 Fbest =
   0.5477
xbest =
   -0.5894    0.0340
```

Figure 3.5 shows the convergence profile of PSO in both cases, minimization and maximization of the function F.

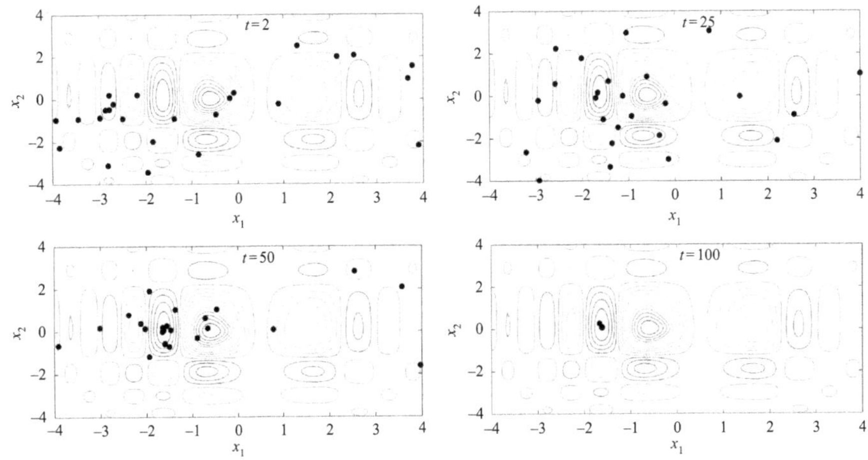

Figure 3.6 Positions of particles in the solution space for case minimization of the F

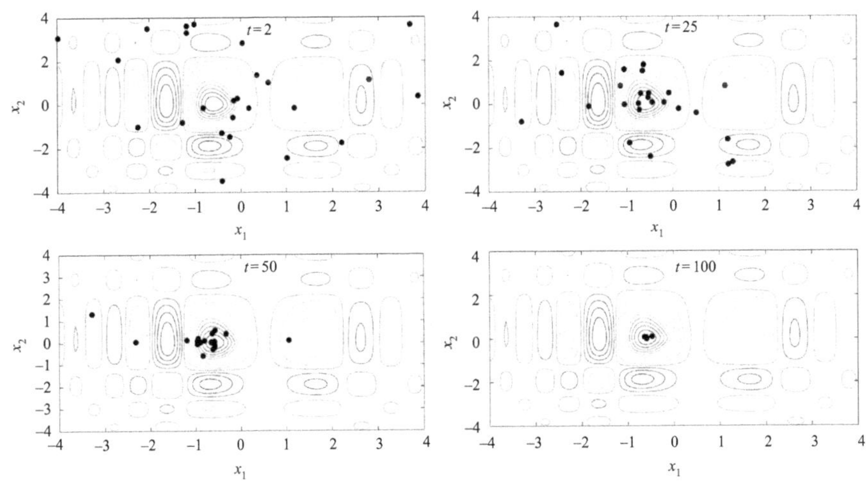

Figure 3.7 Positions of particles in the solution space for case maximization of the F

The positions of particles in the search space for iteration 2, 15, 50, and 100 in finding the minimum and maximum of the function *F* are shown in Figures 3.6 and 3.7, respectively. As can be seen from these figures, all particles move to the global optimum position—best fitness value.

3.3 PSO modifications

In the last two decades, numerous modifications of the standard PSO algorithm have appeared in the literature. These PSO variants can be divided in several groups depending on the type of population topology, the type of variables, the type of problems, the way of defining parameters, the population dynamics, the number of swarms and relation between swarms, the PSO work independently or in combination with other metaheuristic algorithm, etc.

3.3.1 Population topology

Population topology defines the way of transferring information between particles. The PSO algorithm can be described generally as a population of vectors whose trajectories oscillate around a region which is defined by each individual's previous best success and the success of some other particle. Various methods have been used to identify "some other particle" to influence the individual [15]. The most commonly used topologies are *gbest*, *lbest*, and von Neumann topology.

- In *gbest* or completely connected topology [Figure 3.8(a)], each particle is connected to all other particles. Information on the best solution is global and at all times known to all particles. The trajectory of each particle's search is influenced by the **gbest** of the entire population. While it is conceptualized as a fully connected graph, in practice it only meant that the program needed to keep track of the best function result that had been found, and the index of the particle that found it [1].
- *lbest* or ring lattice topology, in which each particle is connected with two particles creating a closed circuit as shown in Figure 3.8(b). The *lbest* topology allows each particle to be influenced by two neighbors of the population. In the *lbest* variant of PSO, particles have information only of their own and their nearest array neighbor' bests (**lbest**), rather than that of the entire group. Namely, in (3.2), **gbest** is replaced by **lbest** in the *lbest* model of PSO [16]. This topology has the advantage of allowing parallel search, as subpopulations could converge in diverse regions of the search space [2].

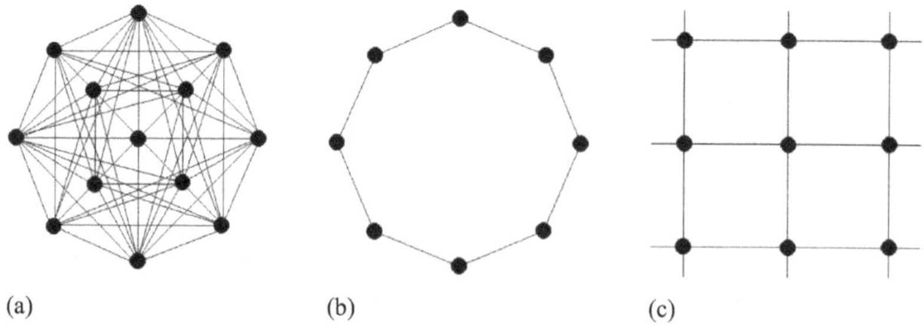

Figure 3.8 Topologies: (a) gbest, (b) pbest, and (c) von Neumann

- *von Neumann* topology in which each particle has four neighbors. This topology defined neighborhoods on a grid and possesses some of the parallelism of *lbest*. Thus, the graph is more densely connected than *lbest*, but less densely than *gbest*.

Abovementioned topologies of PSO are static. However, several authors suggest that adaptive topology might be beneficial. These adaptive topologies are defined as dynamic topologies. Some of these approaches are as follows:

- Combination of the *lbest* and *gbest* topology by introducing a variable neighborhood operator [17].
- Using a weighted Euclidean distance in identifying the interaction partner for a particle [18].
- Creating random subpopulations of n size and occasionally randomized all the connections [19]. In this way the authors introduced a dynamic multiswarm PSO.
- Arranging the particles in a dynamic hierarchy, with each particle being influenced by its own previous success and that of the particle directly above it [20].
- Dividing the population in subpopulations, each maintaining its own order and structure [13].

3.3.2 Discrete binary PSO

The standard PSO works with real numbers and continuous optimization problems. However, many of practical engineering problems are formulated as combinatorial optimization problems with both continuous and discrete decision variables [6]. As any problem, discrete or continuous, can be expressed in a binary notation, the authors of original PSO have developed a discrete binary version of PSO [21].

In the proposed binary PSO, the velocity is used as a probability threshold to determine whether dth component of particle i (x_i^d) should be evaluated as a zero or a one. One of the functions accomplishing this feature is the sigmoid function, which usually utilized with neural networks [6]:

$$\mathrm{sig}\left(v_i^d\right) = \frac{1}{1 + e^{-v_i^d}} \tag{3.9}$$

The disposition of particle should be adjusted for success of the particle and the group. Therefore, the updating formula each v_i^d should be consists of the distance between the current position of particle and the individual best position (**pbest**$_i$) and the distance between the current position of particle and the global best position of all particles in the group (**gbest**). Namely, similar to the original continuous version, the updating formula for binary version of PSO can be described as follows [6,21]:

$$v_i^d(t+1) = v_i^d(t) + r_1 \cdot \left(pbest_i^d(t) - x_i^d(t)\right) + r_2 \cdot \left(gbest^d(t) - x_i^d(t)\right) \tag{3.10}$$

$$\rho_i(t+1) < \mathrm{sig}\left(v_i^d(t+1)\right) \text{ then } x_i^d(t+1) = 1; \quad \text{else } x_i^d(t+1) = 0 \tag{3.11}$$

where r_1 and r_2 are positive uniformly distributed random numbers with predefined upper limits, and ρ_i is a uniformly distributed random number in [0, 1].

In the binary version of PSO, the limits of random numbers r_1 and r_2 are often set so that the two rand limits sum to 4. The second and third terms of (3.10) can be weighted like the standard continuous version of PSO. v_i^d can be limited so that $\mathrm{sig}(v_i^d)$ does not approach too closely to 0.0 or 1.0. This ensures that there is always some chance of a bit flipping. A constant parameter $V_{\max}$ is often set in [−4.0, 4.0]. The entire algorithm of the binary version of PSO is almost the same as that of the standard continuous version except the above decision equations [6].

3.3.3 Hybrid PSO

The search process in PSO is very much depending on pbest and gbest. Therefore, the searching area may be limited by pbest and gbest. This leads to increase in the computation time to find the effective area in the solution space. In order to overcome this weakness of PSO, the author [22] proposed a hybrid PSO (HPSO) method. The HPSO method utilizes the basic structure of PSO and the tournament selection method used in evolutionary programing such as of genetic algorithm (GA). By introduction of the tournament selection mechanism in PSO algorithm, effect of pbest and gbest is gradually eliminated by the selection. This can result in the broader search area. The tournament selection is performed as follows. The fitness of each particle, based on its current position, is compared to the fitness of k other particles and scores a point for each with worst fitness. The population is the sorted using this score with the particles having the highest scores appearing at the head of the population. The fitness of the individual best positions for the particles are not considered during the scoring or sorting of the particle. Particle positions with low evaluation values are replaced by those with high evaluation values. This selection method is applied before the manipulation of the population with PSO-updating operators [22]. It should be noted that the moved particles will still contain their pbest information which will affect their next position. In this way, both intensive search in a current effective area and dependence on the past high evaluation position are realized at same time in HPSO. The number of highly evaluated particles is increased, while the number of lowly evaluated particles is reduced in each iteration. While the difference between HPSO and PSO is fairly minor, the addition of this selection method should provide HPSO with a more exploitative search mechanism that should find better optima more consistently than PSO [22]. A general flowchart of HPSO is shown in Figure 3.9. The exchange rate at the selection method is added as a new optimization parameter of PSO [6].

3.3.4 Adaptive PSO

As mentioned before, the PSO parameters such as inertia weight and acceleration coefficients have crucial impact on efficiency and convergence of PSO. In order to achieve better dynamic balance between global and local search ability of PSO, some authors introduced adaptive inertia weight, adaptive acceleration coefficients, or both adaptive inertia weight and acceleration coefficients.

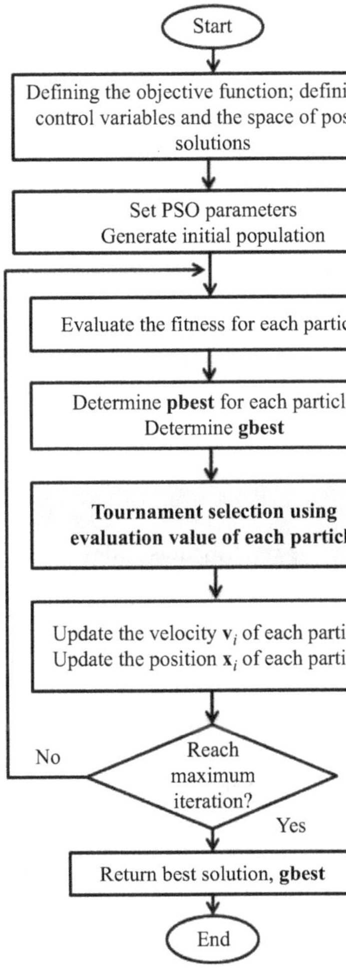

Figure 3.9 General flowchart of HPSO

3.3.4.1 PSO with adaptive inertia weight

The authors [23] proposed a fuzzy adaptive PSO (FPSO) by using a fuzzy inertia weight control and a fuzzy location updated control. The FPSO utilize a fuzzy adaptive process for inertia weight w:

$$w' = \mu_1 w, \quad \mu_1 = \begin{cases} 1, & t \leq \dfrac{t_{\max}}{5} \\ e^{-\frac{(t-20)^2}{\sigma^2}}, & t > \dfrac{t_{\max}}{5} \end{cases} \tag{3.12}$$

and a fuzzy parameter to control the particle position change:

$$\mu_2 = \begin{cases} 1, & t \le T \\ \dfrac{1}{1 + \left(\dfrac{t-30}{a}\right)^2}, & t > T \end{cases} \tag{3.13}$$

In (3.12), the membership function is Gauss distribution, t is the current iteration, $t_{\max}$ is the maximum iteration number, and w is the linear decreasing inertia weight (3.5). In formula (3.13), a is a constant, generally obtained in the interval [5,10], and T is a given threshold related to iterations maximum number $t_{\max}$.

So, the new updating equations for velocity and position of particles are as follows:

$$\mathbf{v}_i(t+1) = w' \cdot \mathbf{v}_i(t) + C_1 \cdot r_1 \cdot (\mathbf{pbest}_i(t) - \mathbf{x}_i(t)) + C_2 \cdot r_2 \cdot (\mathbf{gbest}(t) - \mathbf{x}_i(t)) \tag{3.14}$$

$$\mathbf{x}_i(t+1) = \eta_2 \cdot \mathbf{x}_i(t) + \mathbf{v}_i(t+1) \tag{3.15}$$

The general steps of FPSO can be described as follows:

Step 1: Initialize the velocity and position of all particles.
Step 2: Evaluate the fitness for each particle and determine the **pbest** and **gbest**.
Step 3: Using the fuzzy process on the inertia weight and the location update parameter, update the velocity and position of particles according to (3.14) and (3.15).
Step 4: Repeat steps 2 and 3 until the stop criteria is reached.
Step 5: Return best solution. Stop.

3.3.4.2 PSO with adaptive acceleration coefficients

In order to enhance the global search in the early part of the optimization and to encourage the particles to converge toward the global optima at the end of the search, the authors [24] proposed time-varying acceleration coefficients for the PSO. In this approach, the acceleration coefficients are changed with the time in such a manner that the cognitive component of (3.2) is reduced, while the social component is increased as the search proceeds. A large cognitive component and small social component at the beginning allow particles to move around the search space, instead of moving toward the population best. On the other hand, a small cognitive component and a large social component in the later stages of the optimization allow the particles to converge to the global optimum. Therefore, the time-varying acceleration coefficients are proposed as follows [24]:

$$C_1 = C_{1i} + (C_{1i} - C_{1f}) \cdot \frac{t}{t_{\max}} \tag{3.16a}$$

$$C_2 = C_{2i} + (C_{2i} - C_{2f}) \cdot \frac{t}{t_{\max}} \tag{3.16b}$$

where C_{1i}, C_{1f}, C_{2i}, and C_{2f} are initial and final values of cognitive and social acceleration coefficients, respectively. The authors [24] recommended settings for these values are as follows: $C_{1i} = 2.5$, $C_{1f} = 0.5$, $C_{2i} = 0.5$, and $C_{2f} = 2.5$.

3.3.4.3 PSO with adaptive inertia weight and acceleration coefficients

A significant improvement in the performance of the PSO was achieved by the authors [25]. They introduced an adaptive PSO (APSO), which adapts inertia weight and acceleration coefficients. This algorithm can perform a global search over the entire search space with faster convergence speed than standard PSO. In APSO, first, a real-time evolutionary state estimation is performed to identify one of the following four defined evolutionary states, including exploration, exploitation, convergence, and jumping out in each generation. Then, the automatic control of inertia weight and acceleration coefficients is conducted. An elitist learning strategy is performed when the evolutionary state is classified as convergence state.

The evolutionary state estimation in APSO takes into account the population distribution information and particle fitness in every generation. This procedure contain following steps:

1. Calculate the mean distance of each particle i to all the other particles using Euclidean distance:

$$d_i = \frac{1}{N-1} \sum_{j=1, j \neq i}^{N} \sqrt{\sum_{d=1}^{n} \left(x_i^d - x_j^d \right)} \tag{3.17}$$

   where N is the population size (number of particles), and n is the dimension of problem (number of decision variables).
2. Compute an evolutionary factor f as follows:

$$f = \frac{d_g - d_{\min}}{d_{\max} - d_{\min}} \tag{3.18}$$

   where d_g is the d_i for the global best particle (gbest), $d_{\min}$ and $d_{\max}$ are the minimum and maximum distances obtained by comparing all d_is.
3. Classify f into four different sets using fuzzy classification. The sets represent the states of exploration, exploitation, convergence, and jumping out.

Next segment is adaptation of the inertia weight and acceleration coefficients.

The inertia weight adaptation is performing according to parameter f using a sigmoid mapping:

$$w(f) = \frac{1}{1 + 1.5e^{-2.6f}} \in [0.4,\ 0.9] \ \forall f \in [0,\ 1] \tag{3.19}$$

The acceleration coefficients adaptation is conducted according to the previously defined evolutionary states in the following four strategies:

Strategy 1: If the algorithm is in exploration state, C_1 is increased while C_1 is decreased.
Strategy 2: If the algorithm is in exploitation state, C_1 is increased slightly while C_1 is decreased slightly.
Strategy 3: If the algorithm is in convergence state, C_1 and C_1 are increased slightly.
Strategy 4: If the algorithm is in jumping out state, C_1 is decreased while C_1 is increased.

A general flowchart of the APSO is shown in Figure 3.10.

Based on the benchmark test, the authors [25] concluded that APSO achieved an improvement in all conditions of optimal globality, accuracy of solution, reliability, and computation time, both in unimodal and multimodal problems.

3.4 Applications of PSO to power system problems—literature overview

In the last two decades, PSO is applied in various scientific areas and engineering disciplines. During this period, thousands of papers were published presenting successful applications of the PSO on different optimization problems. Based on analysis of IEEE Xplore and Google Scholar citations and publications, the authors [2] showed that the number of publications with PSO applications is exponentially growing by year.

A high number of these publications are related to application of the PSO and its modifications in solving the power system problems [26]. Specifically, the PSO is successfully applied in solving the problems related to optimal planning, control, and operation of power systems, distribution systems and microgrids (MGs).

The following is a literature overview of the applications of the PSO to some specific problems in power systems, including optimal power flow (OPF), optimal reactive power dispatch (ORPD), economic dispatch (ED), optimal Volt/Var control in distribution networks (DNs), optimal placement and sizing of distributed generation, optimal energy management of MGs, and optimal coordination of overcurrent relays.

3.4.1 Optimal power flow

First application of PSO for solving the OPF problem is described in [3]. This paper is one of the most cited papers related to applications of metaheuristic optimization methods in solving the OPF problem.

The authors [27] applied of an improved PSO (IPSO) to solve the OPF problems. This improvement is based on incorporating a biology concept "passive congregation" in the standard PSO.

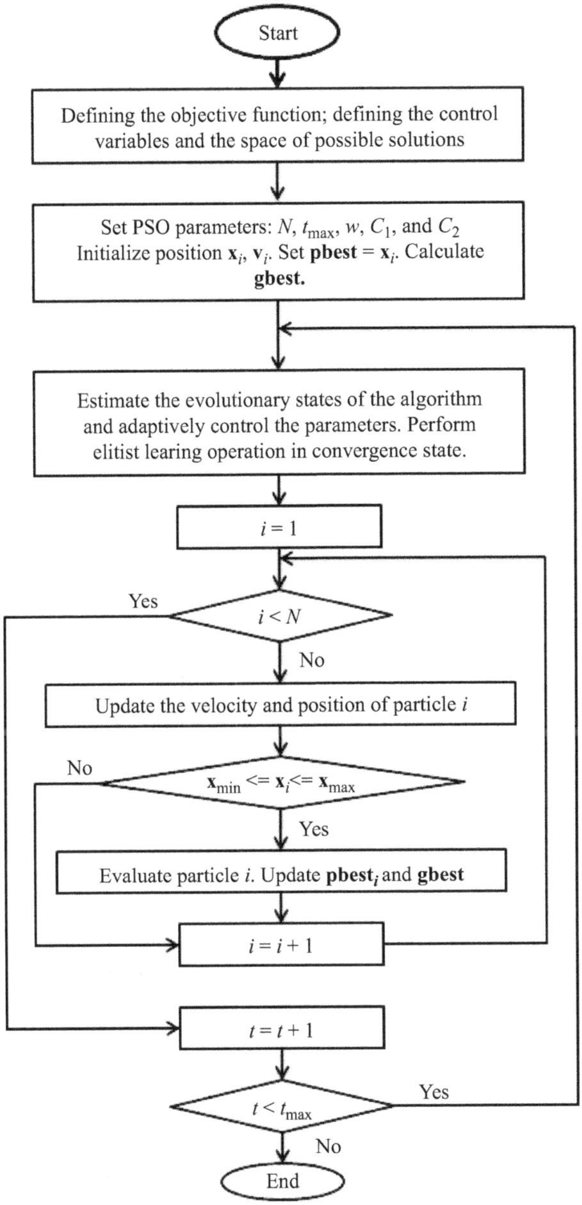

Figure 3.10 Flowchart of APSO

Zhao *et al.* [28] used standard PSO for OPF problem with nonstationary multistage assignment penalty function.

Wang *et al.* [29] proposed a modified PSO algorithm and its application on the OPF problem. In the modified PSO algorithm, particles not only studies from itself and the best one but also from other individuals. In this way, the opportunity to find the global optimum is increased and the influence of the initial position of the particles is decreased.

In order to solve the constrained OPF with both continuous and discrete control variables, the author [30] proposed an efficient mixed-integer PSO with two mutation schemes.

Esmin *et al.* [31] expanded the standard PSO to a HPSO by introducing the arithmetic mutation from the GA field. The HPSO is applied in [31] to solve the OPF problem with objective to power loss minimization.

Dutta and Sinha [32] used PSO to solve the OPF problem which is decoupled into two subproblems, namely, the ED and the reactive power dispatch.

The authors [33] proved the feasibility and robustness of PSO in solving a challenging OPF problem, namely the transient-stability constrained OPF problem.

Yumbla *et al.* [34] used the PSO algorithm with reconstruction operators to solve the OPF problem taking into account both operating security constraints and system capacity requirements. In this approach, the reconstruction operators and an external penalty are adopted to handle the constraints of the OPF problem. The use of reconstruction operators allows to increase the suitable particles in the searching space.

Abido [35] proposed a multiobjective version of the conventional PSO technique by redefining of global best and local best individuals in multiobjective optimization domain. The proposed multiobjective PSO is applied to OPF problem.

Liu *et al.* [36] introduced a multistart technique to overcome premature convergence of the PSO algorithm in solving the OPF problem.

In [37], fuzzy logic is applied in combination with PSO algorithm for solving OPF to enhancement of the power system security.

Niknam *et al.* [38] proposed an IPSO with a chaos algorithm to compute the inertia weight and a self-adaptive method to compute acceleration coefficients for the multiobjective OPF problem. In this approach, a fuzzy decision-based mechanism is used to select the best compromise solution of Pareto set obtained by the IPSO.

The authors [39] incorporated one new index named, iteration best, in the equation for speed updating of standard PSO. By adding this new index, the OPF solution quality is improved.

Kumar and Chaturvedi [40] employed the integration of fuzzy system and PSO algorithm to solve the OPF problems. The simulation results show the effectiveness and robustness of proposed PSO-fuzzy approach.

Similar applications of the standard PSO to solving the OPF problem is described in [41–43], and a comparison of the performance of PSO along with evolutionary programing, and GA in solving the multiobjective OPF problem is given in [44].

Chang *et al.* [45] proposed an evolutionary PSO approach for the OPF problem considering up-spinning reserves, down-spinning reserves and the operational constraints of the generation unit in a power system with both thermal generators and wind turbine generators.

3.4.2 Optimal reactive power dispatch

The ORPD can be considered as a special case of the OPF problem for a fixed economic active power dispatch. The ORPD problem solution aims to optimize a chosen objective function through optimal adjustment of the reactive power control variables such as generator voltages, transformer tap settings, and other sources of reactive power such as shunt Var compensators. The ORPD provides minimization of system power losses, improvement of voltage profile, system security, power transfer capability, and efficiency. Similarly to the OPF, the ORPD problem can be mathematically formulated as a nonlinear optimization problem with constraints. Some authors suggest PSO for the ORPD problem.

Still in 2000, Yoshida *et al.* [46] used a PSO for reactive power and voltage control considering voltage security assessment. This problem is considered as a mixed-integer nonlinear optimization problem with continuous and discrete control variables.

Coath *et al.* [47] applied standard PSO to solve the reactive power and voltage control problem in power systems, considering wind turbine generators as one of the power generation sources.

In 2005, several papers appeared which suggests the application of the PSO for solving the ORPD problem. The authors [48] integrate the multiagent system and the PSO algorithm in a MAPSO optimization method to solve the reactive power dispatch problem. Reference [49] presents an IPSO algorithm for ORPD and voltage control of power systems. The IPSO algorithm uses more particles' information to control the mutation operation and adaptive strategy for choosing parameters. Zhang and Liu [50] successfully applied PSO method to the ORPD problem in a practical power system in China.

Vlachogiannis and Lee [51] applied the parallel vector evaluated PSO for reactive power control formulated as a multiobjective problem.

In [52], the authors proved the efficiency of the APSO for practical problems such as the power system reactive power optimization.

Pouya and Lesani [53] applied the θ-PSO, as an extension of the standard PSO algorithm, to solve the problem of reactive power market.

Mahadevan and Kannan [54] compared performances of standard PSO and comprehensive learning PSO (CLPSO) in solving the ORPD problem and concluded that the CLPSO method can avoid the shortcoming such as the premature convergence of PSO method and can discover higher quality solution for the ORPD problems.

Badar *et al.* [55] used PSO with dynamic weight, and Singh *et al.* [56] applied PSO with an aging leader and challengers for the solution of ORPD problems. The authors demonstrate the effectiveness and robustness of these modified versions of standard PSO for solving the ORPD problems.

Recently, the evolutionary PSO method that combines features of classical optimization methods and PSO algorithm is proposed in [57] to solve the ORPD problem with discrete control variables such as in-phase transformer taps.

3.4.3 Economic dispatch

The ED problem solution aims to minimize the cost of generation of electric power through optimal adjustment of the committed generating unit outputs, while at the same time satisfying all unit and system constraints. When the environmental concerns are combined with the ED, then the problem becomes a combined economic and emission dispatch (CEED) problem.

The authors [26] presented a survey of PSO applications in solving the ED problems until 2007.

Park *et al.* [58,59] proposed the PSO for solving the ED problems with nonsmooth cost functions. In these papers, a modified PSO mechanism is suggested to deal with the equality and inequality constraints in the ED problems.

Pancholi and Swarup [60,61] applied the PSO algorithm to solve the ED problem with security constraints related to line flows and voltage constraints. Based on simulation results, the authors conclude that the PSO approach is computationally faster than linear programing, quadratic programing, and GA in solving the ED problems.

Kumar *et al.* [62] used PSO for solving the CEED problems. The authors concluded that the PSO gives better results compared with GAs.

Gaing [63] successfully applied PSO for solving the ED problem in power system considering practical operating constraints of generator such as such as ramp rate limits, prohibited operating zone, and nonsmooth cost functions. Same author applied the PSO to solve the constrained dynamic ED problem [64].

The authors [65] employed PSO algorithm to solve a bid-based dynamic ED model with various constraints such as ramp rates, transmission line capacity, and emission constraints to maximize the social profit in a competitive electricity market.

The application of modified PSO called fitness distance ratio PSO algorithm is proposed in [66] to determine the optimal power dispatch of the independent power producers in deregulated environment, with considering transient stability limits of the power producers and physical transmission line limits with contingency.

To solve nonconvex ED problems, the authors [67] introduced a split-up in the convergence behavior of the classical PSO and integrated a simple local search procedure in this modified PSO.

In addition to the OPF and ORPD problems, Abido made a significant contribution in application of PSO in solving the ED problem [68], whereas Park *et al.* continued their researches presented in [58,59] by proposing an IPSO for solving ED problems with nonconvex cost functions [69].

Raglend *et al.* [70] presented a comparative study of the performances of PSO along with GA, evolutionary programing, and differential evolution in solving the CEED problem on the IEEE 30-bus system and 15-units system.

A multiobjective PSO with a fuzzy membership function is used in [71], and a modified adaptive θ-PSO with a fuzzy clustering technique is applied in [72] to solve a highly nonlinear and nonconvex CEED problem.

Ghorbani *et al.* [73] used PSO algorithm with smart inertia factor, in which inertia coefficients are controlled with respect to cost function in each population, to solve the nonconvex ED problem.

Gholami *et al.* [74] investigated the influence of plug-in electric vehicles and renewable energy sources on the CEED results in a power system. The PSO algorithm is applied to solve the CEED nonlinear problem.

Recently, Hosseinnezhad *et al.* [75] proposed species-based quantum PSO for ED problem considering transmission network losses and practical constraints of plants, whereas Plathottam and Salehfar [76] formulated an ED problem for a power system area with high penetration of renewable generation and energy storage.

3.4.4 Optimal power flow in distribution networks

Main objective of the OPF problem for a DN is to minimize the fuel cost of DG units, voltage profile improvement, Volt/Var coordination, and minimization of power losses though optimal settings of the control variables while at the same time satisfying various distribution system operating constraints. The OPF analysis in DNs implies the inclusion of different types of DG units, taking into account the uncertainties of the input variables, the multiobjective optimization and choice of an efficient method for solving the OPF problem. Some authors have shown that the PSO is the one of suitable method for these problems.

Niknam [77] used standard PSO and Niknam *et al.* [78] used a FPSO to determine active and reactive power generated by DG units, main substation, capacitors, and also tap of tap-changer transformers in DNs with objective to minimize the energy cost, considering the physical and technical constraints. In the continuation of the research, Niknam *et al.* [79] proposed a multiobjective fuzzy adaptive chaotic PSO for the multiobjective daily optimal operation management of DNs with the aim to minimize the total electrical energy losses, the total electrical energy cost, and the total pollutant emission produced by sources.

Madureira and Lopes [80] used evolutionary PSO for optimal Volt/Var control in DN with large integration of DG units and MGs.

In [81], the optimal operation of the DN with renewable energy sources is considered as a mixed integer, nonlinear, and multiobjective problem, and a binary PSO and fuzzy set theory are proposed for solving this problem.

Cheng and Chen [82] applied a dynamically adaptive multiobjective PSO algorithm, and Chen *et al.* [83] used a fuzzy adaptive HPSO to solve the problem of ORPD and voltage control of DNs with DG units.

A modified PSO for discrete control variables is used in [84] to solve the problem of daily coordination between capacitors and on-load tap changers with objective to minimize the total energy loss and the voltage deviations in a DN.

Recently, Meirinhos *et al.* [85] proposed an innovative approach for voltage control at the MV level of DN, which is based on a preventive day-ahead analysis that uses data from load and renewable energy sources forecasting tools to establish a plan for operation of the different distributed energy resources for the next day. The approach is formulated as a multitemporal OPF solved by a variant of the evolutionary PSO algorithm.

3.4.5 Optimal placement and sizing of distributed generation in distribution networks

Impact of DG on DN operation most depends on its location, size, and type. Determining optimal location and size of a DG unit aims to maximize the positive effects it has on the DN operation, such as reduction of the power loss, improvement in the voltage profile, network reliability increasing, etc. This is a nonlinear combinatorial optimization problem with both discrete and continuous control variables.

The authors [86] combine PSO with load flow algorithm to solve the problem of optimal locating and sizing of the DG units in a single step, i.e., finding the best combination of location and size simultaneously.

In [87,88], PSO is applied for optimal sitting and sizing of DG to voltage profile improvement, loss reduction, and total harmonic distortion (THD) reduction in DNs.

Prommee and Ongsakul proposed an APSO [89] and an improved reinitialized social structure PSO [90] for solving optimal multiple DG placement in a DN in order to minimize the real power losses.

The authors [91] introduced a hybrid method, which employs discrete PSO and OPF to achieve optimal site and size of DG units.

PSO was used in [92] to determine the optimal size and location of different types of DGs by minimizing the system power loss considering the exact loss formula as objective function.

Siano and Mokryani [93] proposed a hybrid optimization method for optimal allocation of wind turbines in radial DNs. This method used PSO to select the optimal sizes of wind turbines, while the market-based OPF is used to determine the optimal number of wind turbines at each candidate bus.

Ameli *et al.* [94] applied multiobjective PSO to find the optimal location and size of DGs as well as their optimal-generated electricity prices in a competitive market, whereas Zeinalzadeh *et al.* [95] proposed a new application of multiobjective PSO with the aim of determining optimal location and size of DGs and shunt capacitor banks simultaneously with considering load uncertainty in distribution systems.

Reference [96] propose a long-term scheduling for optimal allocation and sizing of different types of DG units in the DNs in order to minimize power losses. PSO is used to solve the optimization problem.

Nasri *et al.* [97] introduced an optimal DG planning methodology based on PSO algorithm in order to minimize the system energy losses. The planning problem is formulated for dispatchable and nondispatchable DG units in different scenarios.

An improved multiobjective PSO with a preference strategy has been proposed and successfully applied to determine the optimal siting and sizing of DG units into the distribution system [98].

Bohre *et al.* [99] introduced a novel multiobjective function for optimal sizing and placement of DGs considering different indices such as active and reactive power loss indices, voltage deviation index, reliability index, and shift factor indices. PSO was used to optimize the multiobjective function.

Recently, Kumawat *et al.* [100] proposed a PSO-based method for determining the optimal size and location of DGs in DN to minimize annual energy loss, considering time-varying characteristics of electrical load demand.

3.4.6 Optimal energy and operation management of MGs

MGs are local DNs comprising various distributed generation units, controllable loads, and energy storage devices, which can operate either interconnected or isolated from the main distribution grid as a controlled entity.

Moghaddam *et al.* [101] proposed and implemented an expert multiobjective adaptive modified PSO algorithm for optimal operation of a typical MG with renewable energy sources accompanied by a back-up Micro-Turbine/Fuel Cell/Battery hybrid power source to level the power mismatch or to store the surplus of energy when it's needed. The problem is formulated as a nonlinear constraint multiobjective optimization problem to minimize the total operating cost and the net emission simultaneously.

Chung *et al.* [102] developed an effective control parameter-tuning method based on PSO in order to coordinate multiple MG generators for both grid-connected and autonomous modes.

Al-Saedi *et al.* [103] used PSO for real-time self-tuning of the power control parameters in a controller scheme for control the flow of the active and reactive power between the main grid and the MG composed of DG units.

An IPSO algorithm combined with Monte-Carlo simulation is used in [104] to solve the dynamic ED of a MG system with both renewable and nonrenewable energy sources.

Radosavljević *et al.* presented an efficient algorithm based on PSO for energy and operation management of a MG including different DG units and energy storage devices. Owing to the stochastic nature of energy produced from renewable sources, i.e., wind turbines and photovoltaic systems, as well as load uncertainties and market prices, a probabilistic approach in the energy and operation management is introduced in [105].

In [106], the optimal power dispatch problem of interconnected MGs with minimum operating cost is solved by using PSO and compared by Monte-Carlo simulation method considering power and load uncertainties.

Recently, Mohan *et al.* [107] proposed a stochastic weight tradeoff PSO based backward–forward sweep OPF method to obtain the online optimal schedules of distributed energy resources in MG considering renewable energy, grid power trade, and demand response.

3.4.7 Optimal coordination of directional overcurrent relays

Generally, a power system contains many sections, and each section should be protected against overcurrents. In order to ensure optimum supply reliability, it is important to obtain selectivity, i.e., to switch off the part of the power system where a fault occurs and to avoid switching off the parts of the system where no fault occurs. The goal in coordination of over-current relays is to find the appropriate time dial setting and pick up current setting of relays, considering various constraints, such as coordination and boundary limits. The coordination of overcurrent relays has been formulated as an optimization problem and solved with different optimization methods.

Several authors applied PSO to solve this problem.

In [108], a modified PSO algorithm is used to calculate the optimal relay settings. The modification of PSO algorithm is consistently using the interior point method to obtain the initial feasible solutions.

Mansour *et al.* [109] proposed two modifications of the standard PSO algorithm to solve the problem of optimal coordination of directional overcurrent relays. The first modification introduce the repair algorithm that gives the PSO algorithm the capability of tackling the coordination constraints imposed on the relays, while searching for an optimal setting. Another modification introduce the simplex method for initializing PSO, instead the random initialization.

In [110], the author modified PSO based on (i) the split-up of the cognitive part into the best and not-best components, (ii) personal best position exchanging, (iii) new velocity limiter. The modified PSO is used for optimal coordination of distance and directional over-current relays in power systems with series capacitors integrated into the transmission lines.

Alam *et al.* [111] investigated the performances of different metaheuristic optimization methods, including PSO, for optimal coordination of directional overcurrent relays on several power systems of different sizes.

3.5 Conclusion

PSO is a simple but very efficient optimization technique based on simulation of swarm of bird in two-dimensional space. A certain objective function is optimized through an iterative procedure of updating the swarm position by sharing information between particles, i.e., potential solutions of the problem. Basic version of PSO is developed for continuous optimization problems. However, numerous modified and adapted versions of the basic PSO allow solving problems with discrete variables. PSO is very suitable for hybridization with other metaheuristic algorithms or with classic optimization methods. This makes it very efficient and widely applicable in solving complex optimization problems in all areas of science and engineering. Over the past 20 years, many authors have demonstrated the excellent performance of PSO in solving various practical optimization problems in the field of power engineering, such as OPF, optimal Volt/Var control, optimal placement and sizing of capacitors and DG units, etc.

References

[1] Kennedy J., Eberhart R.C. 'Particle swarm optimization'. *Proceedings of IEEE International Conference on Neural Networks*; Perth, Australia, 1995. IEEE; 1995. pp. 1942–48

[2] Poli R., Kennedy J., Blackwell T. 'Particle swarm optimization an overview'. *Swarm Intelligence.* 2007;**1**:33–57

[3] Abido M.A. 'Optimal power flow using particle swarm optimization'. *Electrical Power and Energy Systems*. 2002;**24**:563–71

[4] Shi Y., Eberhart R.C. 'A modified particle swarm optimizer'. *Proceedings of IEEE International Conference on Evolutionary Computation*; Anchorage, Alaska, May 1998. IEEE; 1998. pp. 69–73

[5] Mirjalili S., Hashim S.Z.M. 'A new hybrid PSOGSA algorithm for function optimization'. *Proceedings of International Conference on Computer and Information Application*; Tianjin, China, Dec 2010. IEEE; 2010. pp. 374–77

[6] Lee K.Y., El-Sharkawi M.A. *Modern heuristic optimization techniques with applications to power systems*. IEEE Power Engineering Society (02TP160); 2002. p. 221

[7] Frank S., Stepanovice I., Rebennack S. 'Optimal power flow: a bibliographic survey II: non-deterministic and hybrid methods'. *Energy Systems*. 2012:**3**(3):259–89

[8] Jordehi A.R., Jasni J. 'Parameter selection on particle swarm optimization: a survey'. *Journal of Experimental & Theoretical Artificial Intelligence*. 2013:**25**(4):527–42

[9] Parsopoulos K.E., Vrahatis M.N., Grmela A., Matorkis N. 'Initializing the particle swarm optimizer using the nonlinear simplex method'. In *Advances in Intelligent Systems, Fuzzy Systems, Evolutionary Computation*. Interlaken: WSEAS Press; 2002. pp. 216–21

[10] Norouzzadeh M.S., Ahmadzadeh M.R., Palhang, M. 'Plowing PSO: a novel approach to effectively initializing particle swarm optimization'. *Proceedings of 3rd IEEE International Congress on Computer Science and Information Technology*; Chengdu, China, Jul 2010. IEEE; 2010. pp. 705–9

[11] Han W., Yang P., Ren H., Sun J. 'Comparison study of several kinds of inertia weight for PSO'. *Proceedings of the IEEE International Conference on Progress in Informatics and Computing*; Shanghai, China, Dec 2010. IEEE; 2011. pp. 280–84

[12] Nanchian S., Majumdar A., Pal B.C. 'Three-phase state estimation using hybrid particle swarm optimization'. *IEEE Transactions on Smart Grid.* 2017;**8**(3):1035–45

[13] Clerc M. *Particle swarm optimization*. London: ISTE; 2006. p. 243

[14] Rashedi E., Nezamabadi-pour H., Saryazdi S. 'GSA: a gravitational search algorithm'. *Information Sciences.* 2009;**179**:2232–48

[15] Kennedy J., Mendes R. 'Population structure and particle swarm performance'. *Proceedings of the IEEE Congress on Evolutionary Computation*; Honolulu, HI, USA, May 2002. IEEE; 2002. pp. 1671–76

[16] Kennedy J., Eberhart R.C. *Swarm intelligence*. San Francisco: Morgan Kaufmann Publishers; 2001. p. 497

[17] Suganthan P.N. 'Particle swarm optimiser with neighbourhood operator'. *Proceedings of the IEEE Congress on Evolutionary Computation*; Washington, DC, USA, Jul 1999. IEEE; 2002. pp. 1958–62

[18] Peram T., Veeramachaneni K., Mohan C. 'Fitness-distance ratio based particle swarm optimization'. *Proceedings of the IEEE Swarm Intelligence Symposium*; Indianapolis, IN, USA, Apr 2003. IEEE; 2003. pp. 174–81
[19] Liang J.J., Suganthan P.N. 'Dynamic multi-swarm particle swarm optimizer with a novel constraint-handling mechanism'. *Proceedings of the IEEE Congress on Evolutionary Computation*; Vancouver, BC, Canada, Jul 2006. IEEE; 2006. pp. 9–16
[20] Janson S., Middendorf M. 'A hierarchical particle swarm optimizer and its adaptive variant'. *IEEE Transactions on System Man and Cybernetics B*. 2005;**35**(6):1272–82
[21] Kennedy J., Eberhart R.C. 'A discrete binary version of the particle swarm algorithm'. *Proceedings of the 1997 IEEE International Conference on Systems, Man, and Cybernetics*; Orlando, FL, USA, Oct 1997. IEEE; 2002. pp. 4104–09
[22] Angeline P.J. 'Using selection to improve particle swarm optimization'. *Proceedings of the 1998 IEEE International Conference on Evolutionary Computation Proceedings*, 1998. IEEE World Congress on Computational Intelligence; Anchorage, AK, USA, May 1998. IEEE; 2002. pp. 84–89
[23] Liu C., Ouyang C., Ouyang C. 'An adaptive fuzzy weight PSO algorithm'. *Proceedings of 2010 Fourth International Conference on Genetic and Evolutionary Computing*; Shenzhen, China, Dec 2010. IEEE; 2011. pp. 8–10
[24] Ratnaweera A., Halgamuge S.K., Watson H.C. 'Self-organizing hierarchical particle swarm optimizer with time-varying acceleration coefficients'. *IEEE Transactions on Evolutionary Computation*. 2004;**8**(3):240–55
[25] Zhan Z.H., Zhang J., Li Y., Chung H.S.H. 'Adaptive particle swarm optimization'. *IEEE Transactions on Systems, Man, and Cybernetics—Part B: Cybernetics*. 2009;**39**(6):1362–81
[26] Yang B., Chen Y., Zhao Z. 'Survey on applications of particle swarm optimization in electric power systems'. *Proceedings of IEEE International Conference on Control and Automation*; Guangzhou, China, Jun 2007. IEEE; 2007. pp. 481–86
[27] He S., Wen J.Y., Prempain E., Wu Q.H., Fitch J., Mann S. 'An improved particle swarm optimization for optimal power flow'. *Proceedings of 2004 International Conference on Power System Technology*; Singapore, Singapore, Nov 2004. IEEE; 2005. pp. 1633–37
[28] Zhao B., Guo C., Cao Y. 'Improved particle swam optimization algorithm for OPF problems'. *Proceedings of IEEE PES Power Systems Conference and Exposition*; New York, NY, USA, Oct 2004. IEEE; 2005. pp. 233–38
[29] Wang C.R., Yuan H.J., Huang Z.Q., Zhang J.W., Sun C.J. 'A modified particle swarm optimization algorithm and its application in optimal power flow problem'. *Proceedings of the Fourth International Conference on Machine Learning and Cybernetics*; Guangzhou, China, Aug 2005. IEEE; 2005. pp. 2885–89
[30] Gaing Z.L. 'Constrained optimal power flow by mixed-integer particle swarm optimization'. *2005 IEEE Power Engineering Society General Meeting*; San Francisco, CA, USA, Jun 2005. IEEE; 2005. pp. 1–8

[31] Esmin A.A.A., Lambert-Tores G., de Souza A.C.Z. 'A hybrid particle swarm optimization applied to loss power minimization'. *IEEE Transactions on Power Systems*. 2005;**20**(2):859–66

[32] Dutta P., Sinha A.K. 'Voltage stability constrained multi-objective optimal power flow using particle swarm optimization'. *1st International Conference on Industrial and Information Systems*; Peradeniya, Sri Lanka, Aug 2006. IEEE; 2007. pp. 161–66

[33] Mo N., Zou Z., Chan K., Pong T. 'Transient stability constrained optimal power flow using particle swarm optimisation'. *IET Generation, Transmission & Distribution*. 2007;**1**(3):476–83

[34] Yumbla P., Ramirez J., Coello C. 'Optimal power flow subject to security constraints solved with a particle swarm optimizer'. *IEEE Transactions on Power Systems*. 2008;**23**(1):33–40

[35] Abido M.A. 'Multiobjective particle swarm optimization for optimal power flow problem'. *Proceedings of 12th International Middle-East Power System Conference*; Aswan, Egypt, Mar 2008. IEEE; 2008. pp. 392–96

[36] Liu W., Li M., Wang X. 'An improved particle swarm optimization algorithm for optimal power flow'. *Proceedings of IEEE 6th International Power Electronics and Motion Control Conference*; Wuhan, China, May 2009. IEEE; 2009. pp. 2448–50

[37] Swaroopan N.M.J, Somasundaram P. 'Optimal power flow for security enhancement using fuzzified particle swarm optimization'. *Proceedings of International Conference on Sustainable Energy and Intelligent Systems*; Chennai, India, Jul 2011. IEEE; 2012. pp. 474–79

[38] Niknam T., Narimani M.R., Aghaei J., Azizipanah-Abarghooee R. 'Improved particle swarm optimisation for multi-objective optimal power flow considering the cost, loss, emission and voltage stability index'. *IET Generation, Transmission & Distribution*. 2012;**6**(6):515–27

[39] Ghanizadeh T., Babaei E. 'Optimal power flow using iteration particle swarm optimization'. *Proceedings of IEEE 5th India International Conference on Power Electronics*; Delhi, India, Dec 2012. IEEE; 2013. pp. 1–6

[40] Kumar S., Chaturvedi D.K. 'Optimal power flow solution using fuzzy evolutionary and swarm optimization'. *Electrical Power and Energy Systems*. 2013;**47**:416–23

[41] Turkay B.E., Cabadag R.I. 'Optimal power flow solution using particle swarm optimization algorithm'. *EuroCon 2013*; Zagreb, Croatia, Jul 2013. IEEE; 2013. pp. 1418–24

[42] Choudhury D., Patra S. 'Multi objective optimal power flow using particle swarm optimization technique'. *International Conference on Signal Processing, Communication, Power and Embedded System*; Odisha, India, Oct 2016. IEEE; 2016. pp. 1261–66

[43] Reddy S.S., Rathnam S. 'Optimal power flow using glowworm swarm optimization'. *Electrical Power and Energy Systems*. 2016;**80**:128–39

[44] Kahouradze S., Mahmoudi A., Mokhlis H.B. 'A comparative study of multi-objective optimal power flow based on particle swarm, evolutionary

programming, and genetic algorithm'. *Electrical Engineering.* 2015; **97**(1):1–12

[45] Chang Y.C., Lee T.Y., Chen C.L., Jan R.M. 'Optimal power flow of a wind-thermal generation system'. *Electrical Power and Energy Systems.* 2014;**55**:312–20

[46] Yoshida H., Kawata K., Fukuyama Y., Takayama S., Nakanishi Y. 'A particle swarm optimization for reactive power and voltage control considering voltage security assessment'. *IEEE Transactions on Power Systems.* 2000; **15**(4):1232–39

[47] Coath G., Al-Dabbagh M., Halgamuge S.K. 'Particle swarm optimisation for reactive power and voltage control with grid-integrated wind farms'. *Proceedings of 2004 IEEE Power Engineering Society General Meeting*; Denver, CO, USA, Jun 2004. IEEE; 2005. pp. 1–6

[48] Zhao B., Guo C.X., Cao Y.J. 'A multiagent-based particle swarm optimization approach for optimal reactive power dispatch'. *IEEE Transactions on Power Systems.* 2005;**20**(2):1070–78

[49] Zhao B., Guo C.X., Cao Y.J. 'An improved particle swarm optimization algorithm for optimal reactive power dispatch'. *IEEE Power Engineering Society General Meeting*; San Francisco, CA, USA, Jun 2005. IEEE; 2005. pp. 1–8

[50] Zhang W., Liu Y. 'Reactive power optimization based on PSO in a practical power system'. *IEEE Power Engineering Society General Meeting*; Denver, CO, USA, Jun 2004. IEEE; 2005. pp. 1–5

[51] Vlachogiannis J.G., Lee K.Y. 'Reactive power control based on particle swarm multi-objective optimization'. *Proceedings of the 13th International Conference on Intelligent Systems Application to Power Systems*; Arlington, VA, USA, Nov 2005. IEEE; 2006. pp. 494–98

[52] Li D., Gao L., Lu S., Ma J., Li Y. 'Adaptive particle swarm optimization algorithm for power system reactive power optimization'. *Proceedings of the 2007 American Control Conference*; New York, USA, Jul 2007. IEEE; 2007. pp. 4733–37

[53] Pouya K., Lesani H. 'An angle-based PSO approach for reactive power management problem'. *Proceedings of IEEE/PES Power Systems Conference and Exposition*; Seattle, WA, USA, Mar 2009. IEEE; 2009. pp. 1–6

[54] Mahadevan K., Kannan P.S. 'Comprehensive learning particle swarm optimization for reactive power dispatch'. *Applied Soft Computing.* 2010;**10**:641–52

[55] Badar A.Q.H., Umre B.S., Junghare A.S. 'Reactive power control using dynamic particle swarm optimization for real power loss minimization'. *Electrical Power and Energy Systems.* 2012;**41**:133–36

[56] Singh R.P., Mukherjee V., Ghoshal S.P. 'Optimal reactive power dispatch by particle swarm optimization with an aging leader and challengers'. *Applied Soft Computing.* 2015;**29**: 298–309

[57] Ferreira E.C., Neto M.S.I., Asada E.N. 'Metaheuristic strategies for solving the optimal reactive power dispatch with discrete variables'. *Proceedings of*

12th IEEE International Conference on Industry Applications; Curitiba, Brazil, Nov 2016. IEEE; 2017. pp. 1–6

[58] Park J.B., Lee K.S., Shin J.R, Lee K.Y. 'Economic load dispatch for non-smooth cost functions using particle swarm optimization'. *Proceedings of 2003 IEEE Power Engineering Society General Meeting*; Toronto, Canada, Jul 2003. IEEE; 2004. pp. 938–43

[59] Park J.B., Lee K.S., Shin J.R., Lee K.Y. 'A particle swarm optimization for economic dispatch with nonsmooth cost functions'. *IEEE Transactions on Power Systems*. 2005;**20**(1):34–42

[60] Pancholi R.K. 'Particle swarm optimization for economic dispatch with line flow and voltage constraints'. *Proceedings of Conference on Convergent Technologies for Asia-Pacific Region*; Bangalore, India, Oct 2003. IEEE; 2004. pp. 450–55

[61] Pancholi R.K., Swarup K.S. 'Particle swarm optimization for security constrained economic dispatch'. *Proceedings of International Conference on Intelligent Sensing and Information Processing*; Chennai, India, Jan 2004. IEEE; 2004. pp. 7–12

[62] Kumar A.I.S., Dhanushcodi K., Kumar J.J., Paul C.K.C. 'Particle swarm optimization solution to emission and economic dispatch problem'. *Proceedings of TENCON 2003. Conference on Convergent Technologies for the Asia-Pacific Region*; Bangalore, India, Oct 2003. IEEE; 2004. pp. 435–39

[63] Gaing Z.L. 'Particle swarm optimization to solving the economic dispatch considering the generator constraints'. *IEEE Transactions on Power Systems*. 2003;**18**(3):1187–95

[64] Gaing Z.L. 'Constrained dynamic economic dispatch solution using particle swarm optimization'. *Proceedings of IEEE Power Engineering Society General Meeting*; Denver, CO, USA, Jun 2004. IEEE; 2005. pp. 1–6

[65] Zhao B., Guo C., Cao Y. 'Dynamic economic dispatch in electricity market using particle swarm optimization algorithm'. *Proceedings of the 5lh World Congress on Intelligent Control and Automation*; Hangzhou, China, Jun 2004. IEEE; 2004. pp. 5050–54

[66] Anitha M., Subramanian S., Gnanadass R., Ajjarapu V. 'Transient stability constrained optimal power dispatch with linear ramping model'. *Proceedings of IEEE Power and Energy Society General Meeting—Conversion and Delivery of Electrical Energy in the 21st Century*; Pittsburgh, PA, USA, Jul 2008. IEEE; 2008. pp. 1–7

[67] Selvakumar A.I., Thanushkodi K. 'A new particle swarm optimization solution to nonconvex economic dispatch problems'. *IEEE Transactions on Power Systems*. 2007;**22**(1):42–51

[68] Abido M.A. 'Multiobjective particle swarm optimization for environmental/economic dispatch problem'. *Electric Power Systems Research*. 2009;**79**:1105–13

[69] Park J.B., Jeong Y.W., Shin J.R., Lee K.Y. 'An improved particle swarm optimization for nonconvex economic dispatch problems'. *IEEE Transactions on Power Systems*. 2010;**25**(1):156–66

[70] Raglend I.J., Veeravalli S., Sailaja K., Sudheera B., Kothari D.P. 'Comparison of AI techniques to solve combined economic emission dispatch problem with line flow constraints'. *Electrical Power and Energy Systems*. 2010;**32**:592–98

[71] Hazra J., Sinha A.K. 'A multi-objective optimal power flow using particle swarm optimization'. *European Transactions on Electrical Power*. 2011;**21**:1028–45

[72] Niknam T., Doagou-Mojarrad H. 'Multiobjective economic/emission dispatch by multiobjective u-particle swarm optimisation'. *IET Generation, Transmission & Distribution*. 2012;**6**(5):363–77

[73] Ghorbani N., Vakili S., Babaei E., Sakhavati A. 'Particle swarm optimization with smart inertia factor for solving non-convex economic load dispatch problems'. *International Transactions on Electrical Energy Systems*. 2014;**24**:1120–33

[74] Gholami A., Ansari J., Jamei M., Kazemi A. 'Environmental/economic dispatch incorporating renewable energy sources and plug-in vehicles'. *IET Generation, Transmission & Distribution*. 2014;**8**(12):2183–98

[75] Hosseinnezhad V., Rafiee M., Ahmadian M., Ameli M.T. 'Species-based quantum particle swarm optimization for economic load dispatch'. *Electrical Power and Energy Systems*. 2014;**63**:311–22

[76] Plathottam S.J., Salehfar H. 'Unbiased economic dispatch in control areas with conventional and renewable generation sources'. *Electric Power Systems Research*. 2015;**119**:313–21

[77] Niknam T., 'An approach based on particle swarm optimization for optimal operation of distribution network considering distributed generators'. *Proceedings of 32nd Annual Conference on IEEE Industrial Electronics*; Paris, France, Nov 2006. IEEE; 2007. pp. 633–37

[78] Niknam T., Firouzi B.B., Ostadi A. 'A new fuzzy adaptive particle swarm optimization for daily Volt/Var control in distribution networks considering distributed generators'. *Applied Energy*. 2010;**87**:1919–28

[79] Niknam T., Meyimand H.Z., Mojarrad H.D. 'A novel multi-objective fuzzy adaptive chaotic PSO algorithm for optimal operation management of distribution network with regard to fuel cell power plants' *European Transactions on Electrical Power*. 2011;**21**:1954–83

[80] Madureira A.G., Lopes J.A.P. 'Coordinated voltage support in distribution networks with distributed generation and microgrids'. *IET Renewable Power Generation*. 2009;**3**(4):439–54

[81] Ebrahimi A., Ramezani M.R., Moieni M. 'Optimal operation of distribution network considering renewable energy sources by binary particle swarm optimization and fuzzy theory'. *Proceedings of CIRED 2012 Workshop Integration of Renewables into the Distribution Grid*; Lisbon, Portugal, May 2012. IEEE; 2012. pp. 1–4

[82] Cheng S., Chen M.Y. 'Multi-objective reactive power optimization strategy for distribution system with penetration of distributed generation'. *Electrical Power and Energy Systems*. 2014;**62**:221–28

[83] Chen S., Hu W., Su C., Zhang X, Chen Z. 'Optimal reactive power and voltage control in distribution networks with distributed generators by fuzzy adaptive hybrid particle swarm optimisation method'. *IET Generation, Transmission & Distribution*. 2015;**9**(11):1096–103

[84] Mokgonyana L., Zhang J., Zhang L., Xia X. 'Coordinated two-stage volt/var management in distribution networks'. *Electric Power Systems Research*. 2016;**141**:157–64

[85] Meirinhos J.L., Rua D.E., Carvalho L.M., Madureira A.G. 'Multi-temporal optimal power flow for voltage control in MV networks using distributed energy resources'. *Electric Power Systems Research*. 2017;**146**:25–32

[86] AlHajri M.F., AlRashidi M.R., El-Hawary M.E. 'Hybrid particle swarm optimization approach for optimal distribution generation sizing and allocation in distribution systems'. *Proceedings of Canadian Conference on Electrical and Computer Engineering*; Vancouver, BC, Canada, Apr 2007. IEEE; 2007. pp. 1290–93

[87] Alinejad-Beromi Y., Sedighizadeh M., Sadighi M. 'A particle swarm optimization for sitting and sizing of distributed generation in distribution network to improve voltage profile and reduce THD and losses'. *Proceedings of 43rd International Universities Power Engineering Conference*; Padova, Italy, Sep 2008. IEEE; 2008. pp. 1–5

[88] Amanifar O. 'Optimal distributed generation placement and sizing for loss and THD reduction and voltage profile improvement in distribution systems using particle swarm optimization and sensitivity analysis'. *Proceedings of 16th Conference on Electrical Power Distribution Networks*; Bandar Abbas, Iran, Apr 2011. IEEE; 2011. pp. 1

[89] Prommee W., Ongsakul W. 'Optimal multi-distributed generation placement by adaptive weight particle swarm optimization'. *Proceedings of International Conference on Control, Automation and Systems*; Seoul, Korea, Oct 2008. IEEE; 2008. pp. 1663–68

[90] Prommee W., Ongsakul W. 'Optimal multiple distributed generation placement in microgrid system by improved reinitialized social structures particle swarm optimization'. *European Transactions on Electrical Power*. 2011;**21**:489–504

[91] Gomez-Gonzalez M., Lopez A., Jurado F. 'Optimization of distributed generation systems using a new discrete PSO and OPF'. *Electric Power Systems Research*. 2012;**84**:174–80

[92] Kansal S., Kumar V., Tyagi B. 'Optimal placement of different type of DG sources in distribution networks'. *Electrical Power and Energy Systems*. 2013;**53**:752–60

[93] Siano P., Mokryani G. 'Assessing wind turbines placement in a distribution market environment by using particle swarm optimization'. *IEEE Transactions on Power Systems*. 2013;**28**(4):3852–64

[94] Ameli A., Bahrabi S., Khazaeli F., Haghifam M.R. 'A multiobjective particle swarm optimization for sizing and placement of DGs from DG owner's and distribution company's viewpoints'. *IEEE Transactions on Power Delivery*. 2014;**29**(4):1831–40

[95] Zeinalzadeh A., Mohammadi Y., Moradi M.H. 'Optimal multi objective placement and sizing of multiple DGs and shunt capacitor banks simultaneously considering load uncertainty via MOPSO approach'. *Electrical Power and Energy Systems*. 2015;**67**:336–49

[96] Karimyan P., Gharehpetian., Abedi M., Gavili A. 'Long term scheduling for optimal allocation and sizing of DG unit considering load variations and DG type'. *Electrical Power and Energy Systems*. 2014;**54**:277–87

[97] Nasri A., Golshan M.E.H., Najed S.M.S. 'Optimal planning of dispatchable and non-dispatchable distributed generation units for minimizing distribution system's energy loss using particle swarm optimization'. *International Transactions on Electrical Energy Systems*. 2014;**24**:504–19

[98] Cheng S., Chen M.Y., Fleming P.J. 'Improved multi-objective particle swarm optimization with preference strategy for optimal DG integration into the distribution system'. *Neurocomputing*. 2015;**148**:23–9

[99] Bohre A.K., Agnihotri G., Dubey M. 'Optimal sizing and sitting of DG with load models using soft computing techniques in practical distribution system'. *IET Generation, Transmission & Distribution*. 2016;**10**(11): 2606–21

[100] Kumawat M., Gupta N., Jain N., Bansal R.C. 'Swarm-intelligence-based optimal planning of distributed generators in distribution network for minimizing energy loss'. *Electric Power Components and Systems*, 2017;**0**(0):1–12

[101] Moghaddam A.A., Seifi A., Niknam T, Pahlavani M.R.A. 'Multi-objective operation management of a renewable MG (micro-grid) with back-up micro-turbine/fuel cell/battery hybrid power source'. *Energy*. 2011;**36**:6490–507

[102] Chung I.Y., Liu W., Cartes D.A., Moon S.I. 'Control parameter optimization for multiple distributed generators in a microgrid using particle swarm optimization'. *European Transactions on Electrical Power*. 2011;**21**:1200–16

[103] Al-Saedi W., Lachowicz S.W., Habibi D., Bass O. 'Power flow control in grid-connected microgrid operation using particle swarm optimization under variable load conditions'. *Electrical Power and Energy Systems*. 2013;**49**:76–85

[104] Wu H., Liu X., Ding M. 'Dynamic economic dispatch of a microgrid: mathematical models and solution algorithm'. *Electrical Power and Energy Systems*. 2014;**63**:336–46

[105] Radosavljević J., Jevtić M., Klimenta D. 'Energy and operation management of a microgrid using particle swarm optimization'. *Engineering Optimization*. 2016;**48**(5):811–30

[106] Nikmehr N., Ravadanegh S.N. 'A study on optimal power sharing in interconnected microgrids under uncertainty'. *International Transactions on Electrical Energy Systems*. 2016;**26**:208–32

[107] Mohan V., Singh J.G., Ongsakul W., Suresh M.P.R. 'Performance enhancement of online energy scheduling in a radial utility distribution microgrid'. *Electrical Power and Energy Systems*. 2016;**79**:98–107

[108] Zeineldin H.H., El-Saadany E.F., Salama M.M.A. 'Optimal coordination of overcurrent relays using a modified particle swarm optimization'. *Electric Power Systems Research*. 2006;**76**:988–95

[109] Mansour M.M., Mekhamer S.F., El-Kharbawe N.E.S. 'A modified particle swarm optimizer for the coordination of directional overcurrent relays'. *IEEE Transactions on Power Delivery*. 2007;**22**(3):1400–10

[110] Moravej Z., Jazaeri M., Gholamzadeh M. 'Optimal coordination of distance and over-current relays in series compensated systems based on MAPSO'. *Energy Conversion and Management*. 2012;**56**:140–51

[111] Alam M.N., Das B., Pant V. 'A comparative study of metaheuristic optimization approaches for directional overcurrent relays coordination'. *Electric Power Systems Research*. 2015;**128**:39–52

Chapter 4

Overview of gravitational search algorithm

4.1 Introduction

Gravitational search algorithm (GSA) belongs to the nature-inspired metaheuristic optimization methods. A metaheuristic optimization method consists of a generalized set of rules that can be applied to solve a variety of optimization problems. Many metaheuristic optimization methods have been developed on the model of some well-known processes in nature. For example, well-known genetic algorithm is based on mimicking of the process of evolution in biology; simulated annealing emulates the physical process of annealing, etc.

Metaheuristic optimization methods are the population-based stochastic search techniques. The population is defined by a set of individuals (agents) which represent potential solutions of the optimization problem. The number of agents (N) is named as the size of the population. In general, an agent can be represented as vector whose elements are the values of the control variables of the optimization problem. The number of control variables (n) is the search space dimension of the optimization problem.

The essence of metaheuristic methods is iterative correction of the solution, i.e., generating a new population by applying algorithmic operators with stochastic search mechanism on agents from the current population. The way in which the algorithmic operators are defined constitutes the essence of a particular metaheuristic optimization method. The efficiency and performance of metaheuristic optimization methods are dependent on the proper setting of the corresponding algorithmic parameters.

The main performances of metaheuristics are fast search of large solution spaces, ability to find global solutions and avoiding local optimum. Their main advantage compared to the classical (deterministic) optimization methods is that they are not limited with requirements for differentiability, nonconvexity and continuity of the objective function or types of control variables. Moreover, these methods can be used for practical optimization problems taking into account various types of objective function and constraints.

In recent years, various population-based metaheuristic optimization methods have been suggested for solving the different engineering problems.

The basic elements of metaheuristic optimization methods can be defined as follows:

Agent, $\mathbf{x}(t)$: It is a candidate solution represented by an n-dimensional vector, where n is the number of control variables. At time (iteration) t, the ith agent $\mathbf{x}_i(t)$ can be described as $\mathbf{x}_i(t) = \left[x_i^1(t), \ldots, x_i^d(t), \ldots, x_i^n(t)\right]$, where $x_i^d(t)$ is the position of the ith agent with respect to the dth dimension, i.e., the value of the dth control variable in the ith candidate solution.

Population, $\mathbf{POP}(t)$: It is a set of N agents at time (iteration) t, i.e., $\mathbf{POP}(t) = [\mathbf{x}_1(t), \ldots, \mathbf{x}_N(t)]^T$, or in expanded form:

$$\mathbf{POP}(t) = \begin{bmatrix} x_1^1(t), & x_1^2(t), & \ldots, & x_1^d(t), & \ldots, & x_1^n(t) \\ x_2^1(t), & x_2^2(t), & \ldots, & x_2^d(t), & \ldots, & x_2^n(t) \\ & & & \vdots & & \\ x_i^1(t), & x_i^2(t), & \ldots, & x_i^d(t), & \ldots, & x_i^n(t) \\ & & & \vdots & & \\ x_N^1(t), & x_N^2(t), & \ldots, & x_N^d(t), & \ldots, & x_N^n(t) \end{bmatrix} \tag{4.1}$$

Space of possible solutions, $\mathbf{X}$: It is an n-dimensional solution space which is defined by lower and upper limits of control variables.

Fitness is a direct metric of the performance of the individual population member (agent). The fitness of each agent of the population is calculated from the value of the function being optimized.

General structure of metaheuristic optimization methods can be represented as follows:

Initialization

1. Defining the objective function $F(\mathbf{x}_i)$ and the space of possible solutions $\mathbf{X}$.
2. Generate initial population of N agents: $\mathbf{POP}(1) = [\mathbf{x}_1(1), \mathbf{x}_2(1), \ldots, \mathbf{x}_N(1)]^T \subseteq \mathbf{X}$.

   The initial positions of each agent are randomly selected between minimum and maximum values of the control variables.

   Set the iteration counter: $t = 1$

Iterative procedure

3. Calculate the fitness value $F(\mathbf{x}_i(t))$ for each agent $\mathbf{x}_i(t)$, $i = 1, \ldots, N$ in the current population $\mathbf{POP}(t)$.
4. Generate new population $\mathbf{POP}(t+1) = [\mathbf{x}_1(t+1), \mathbf{x}_2(t+1), \ldots, \mathbf{x}_N(t+1)]^T \subseteq \mathbf{X}$ by applying the algorithmic operators on search agents from the current population $\mathbf{POP}(t)$.
5. Repeat the iterative procedure until the stop criteria is reached.
6. The optimal solution $\mathbf{x}^*$ is determined.

End

4.2 Description of original GSA

Rashedi *et al.* [1] developed an efficient population-based metaheuristic optimization algorithm, named GSA, which is inspired by the Newton's laws of gravity and motion for interaction between masses. GSA has been verified as high quality performance in solving different optimization problems in literature.

In GSA, the search agents are a collection of masses which interact with each other based on the Newtonian gravity and the laws of motion. In this algorithm, agents are considered as objects and their performances are measured by their masses. All these objects attract each other by the gravity force, and this force causes a global movement of all objects toward the objects with heavier masses. The position of the mass corresponds to the solution of the problem, and its gravitational and inertial masses are determined using a fitness function. In other words, each mass presents a solution. The algorithm is navigated by properly adjusting the gravitational and inertial masses. By lapse of time, the masses will be attracted by the heaviest mass which it represents an optimum solution in the search space. The interaction of agents in the GSA is illustrated in Figure 4.1.

The GSA could be considered as an isolated system of masses. It is like a small artificial world of masses obeying the Newtonian laws of gravitation and motion [1].

Law of Gravity: Each particle attracts every other particle, and the gravitational force between two particles is directly proportional to the product of their masses and inversely proportional to the distance between them, R. The authors [1] used R instead of R^2, based on experimentation where R provide better results than R^2 in all experimental cases.

Law of Motion: The current velocity of any mass is equal to the sum of the fraction of its previous velocity of mass and the variation in the velocity. Variation in the velocity or acceleration of any mass is equal to the force acted on the system divided by mass of inertia.

In a system with N agents (masses), the position and velocity of the ith agent is defined by

$$\mathbf{x}_i = \left(x_i^1, \ldots, x_i^d, \ldots, x_i^n\right) \quad \text{for} \quad i = 1, 2, \ldots, N \tag{4.2}$$

$$\mathbf{v}_i = \left(v_i^1, \ldots, v_i^d, \ldots, v_i^n\right) \quad \text{for} \ i = 1, 2, \ldots, N \tag{4.3}$$

where n is the search space dimension of the problem, x_i^d defines the position of the ith agent in the dth dimension, and v_i^d is the velocity of the ith agent in the dth dimension.

After evaluating the current population fitness, the mass of each agent is calculated as follows:

$$M_i(t) = \frac{m_i(t)}{\sum_{j=1}^{N} m_j(t)} \tag{4.4}$$

$$m_i(t) = \frac{fit_i(t) - worst(t)}{best(t) - worst(t)} \tag{4.5}$$

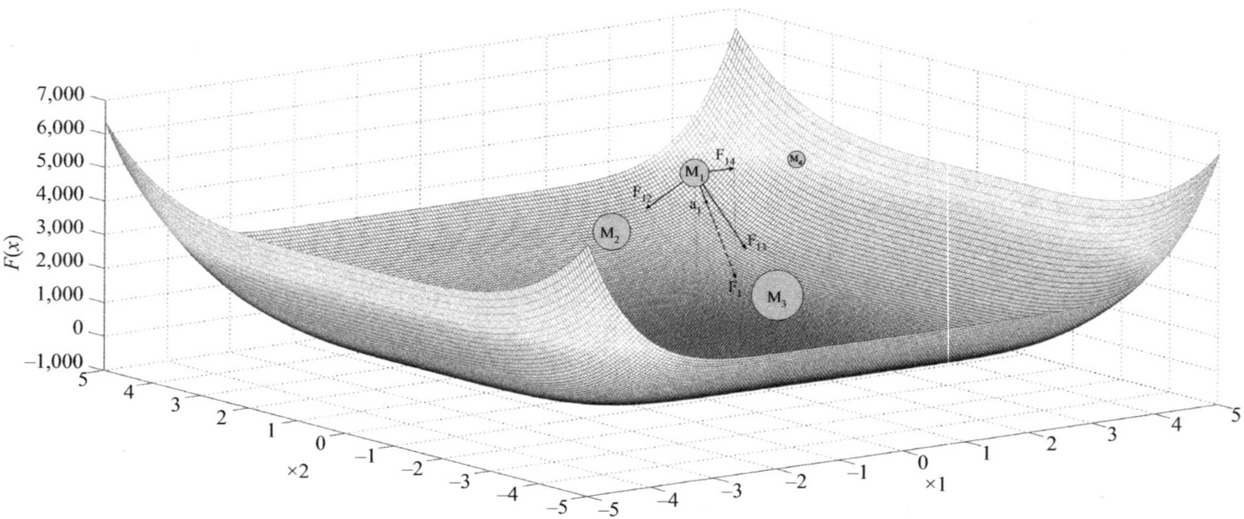

Figure 4.1 Interaction of agents in GSA

where $fit_i(t)$ represent the fitness value of the agent i at iteration t. $best(t)$ and $worst(t)$ are the best and worst fitness of all agents, respectively, and defined as follows (for a minimization problem):

$$best(t) = \min_{j \in \{1,\dots,N\}} fit_j(t) \tag{4.6}$$

$$worst(t) = \max_{j \in \{1,\dots,N\}} fit_j(t) \tag{4.7}$$

For a maximization problem, *best* and *worst* are defined by (4.8) and (4.9), respectively:

$$best(t) = \max_{j \in \{1,\dots,N\}} fit_j(t) \tag{4.8}$$

$$worst(t) = \min_{j \in \{1,\dots,N\}} fit_j(t) \tag{4.9}$$

According to Newton gravitation theory, the total force that acts on the ith agent at iteration t is specified as follows:

$$\mathbf{F}_i(t) = \sum_{j \in Kbest, j \neq i} r \cdot G(t) \frac{M_j(t) \cdot M_i(t)}{R_{i,j}(t) + \varepsilon} \left(\mathbf{x}_j(t) - \mathbf{x}_i(t)\right) \tag{4.10}$$

where r is a random number within the interval [0,1]; $G(t)$ is gravitational constant at iteration t; $M_i(t)$ and $M_j(t)$ are masses of agents i and j; ε is a small constant; $R_{ij}(t)$ is the Euclidian distance between the two agents i and j, $R_{ij}(t) = \|\mathbf{x}_i(t), \mathbf{x}_i(t)\|_2$; *Kbest* is the set of first K agents with the best fitness value and biggest mass.

According to the law of motion, the acceleration of the ith agent, at iteration t, is given by the following equation:

$$\mathbf{a}_i(t) = \frac{\mathbf{F}_i(t)}{M_i(t)} \tag{4.11}$$

The searching strategy on this notion can be defined to find the next velocity and position of an agent. Next velocity of an agent is defined as a function of its current velocity added to its current acceleration. Hence, the velocity and position of an agent are updated as follows:

$$\mathbf{v}_i(t+1) = r_i \cdot \mathbf{v}_i(t) + \mathbf{a}_i(t) \tag{4.12}$$

$$\mathbf{x}_i(t+1) = \mathbf{x}_i(t) + \mathbf{v}_i(t+1) \tag{4.13}$$

where r_i is a uniform random variable in the interval [0,1]. This random number is utilized to give a randomized characteristic to the search.

4.2.1 Parameters of GSA

It must be pointed out that the gravitational constant $G(t)$ in (4.10) is important in determining the performance of GSA. It is initialized at the beginning and will be reduced with time to control the search accuracy. In other words, the gravitational constant is a function of the initial value G_0 and time (iteration) t:

$$G(t) = G_0 \cdot e^{-\alpha(t/t_{\max})} \tag{4.14}$$

The parameters of maximum iteration $t_{\max}$, population size N, initial gravitational constant G_0, and constant α control the performance of GSA. According to (4.14), the value of G is large at the beginning of the algorithm to increase the exploration ability. On the other hand, large value for G means high power attraction between the masses that cause more movement and complexity. As time passes, G should be decreased to find the optima around a good solution. Therefore, the value of G deeply affects the performance of algorithm so that it should be controlled carefully. Equation (4.14) is an exponential function that is known as decay curve. Decay rate is $\alpha/t_{\max}$. Decay rate determines the rate of decreasing of G and subsequently the velocity and movement rate of agents. G_0 is the initial value. Therefore, appropriate setting of G_0 and alpha is important [2].

Parameter *Kbest* in (4.10) is used to perform a good compromise between exploration and exploitation. In order to avoid trapping in a local optimum, the algorithm must use the exploration at beginning. By lapse of iterations, exploration must fade out and exploitation must fade in. *Kbest* determines the number of agents that attract others and must be controlled to maintain appropriate balance between the exploration and exploitation. *Kbest* is calculated based on a function of time that is large in early iterations, but it is decreased with time. At the early iterations, almost all agents attract the others to prevent GSA from trapping in local optima. This behavior guarantees the exploration in the search process of GSA. In the next iterations, the value of *Kbest* should be decreased to emphasize the exploitation instead of exploration and improve the convergence rate. The large value for *Kbest* decreases the convergence rate and gives the algorithm the chance to explore the search space more and prevent from trapping in local optima. But, with increasing *Kbest*, the complexity and computational time is increased. Therefore, (4.15) is proposed in [2] to control the value of *Kbest* during the progress of the algorithm.

$$Kbest = K_0 \left[\frac{t}{t_{\max}}(1-p) + 1 \right] \tag{4.15}$$

where t is the current iteration, $t_{\max}$ is the maximum number of iterations and p is the fractional of total agents in the last iteration. The authors [2] recommended $0.2 < K_0 \leq 1$. By reducing K_0, the complexity of GSA is reduced.

4.2.2 General remarks about GSA

Figure 4.2 shows the general flowchart of GSA.

To see how the GSA is efficient, some remarks are noted [1]:

- Since each agent could observe the performance of the others, the gravitational force is an information-transferring tool.

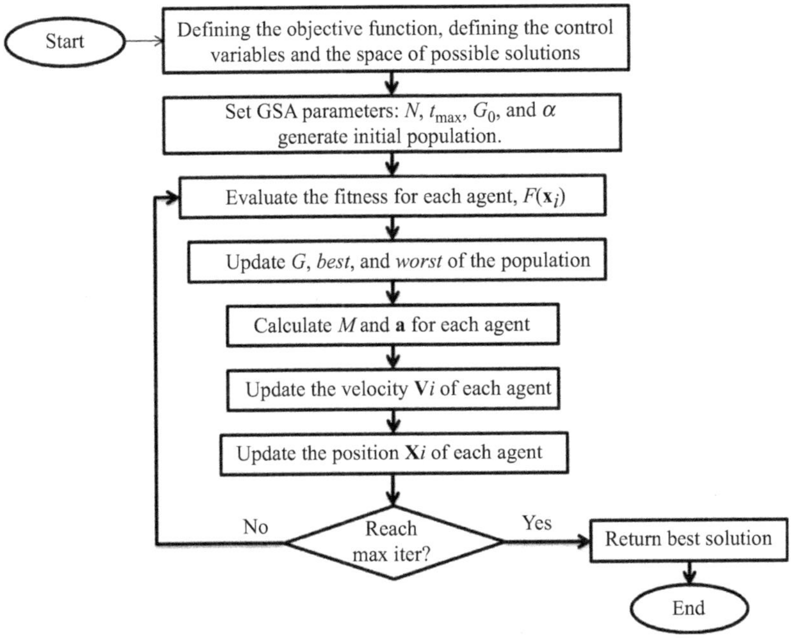

Figure 4.2 General flowchart of GSA

- Due to the force that acts on an agent from its neighborhood agents, it can see space around itself.
- A heavy mass has a large effective attraction radius and hence a great intensity of attraction. Therefore, agents with a higher performance have a greater gravitational mass. As a result, the agents tend to move toward the best agent.
- The inertia mass is against the motion and make the mass movement slow. Hence, agents with heavy inertia mass move slowly and hence search the space more locally. So, it can be considered as an adaptive learning rate.
- Gravitational constant adjusts the accuracy of the search, so it decreases with time.
- GSA is a memory-less algorithm. However, it works efficiently like the algorithms with memory. Experimental results show the good convergence rate of the GSA.
- A bigger inertia mass provides a slower motion of agents in the search space and hence a more precise search. Conversely, a bigger gravitational mass causes a higher attraction of agents. This permits a faster convergence.

4.2.3 MATLAB® code of GSA

Based on the above equations and general flowchart in Figure 4.2, a MATLAB code of GSA was developed. The MATLAB code of GSA is given as follows:

```
%Gravitational Search Algorithm (GSA)
%- - - - - - - - - - - - - - - - - - - - - - - - - - - - - - - - - - - -
function []=GSA

%INITIALIZATION
%======================================================================
%Defining the objective function in the function program Fobj.m
%Defining the space of possible solutions in the function program bounds.m
%- - - - - - - - - - - - - - - - - - - - - - - - - - - - - - - - - - - -
%Defining parameters of GSA:
%- - - - - - - - - - - - - - - - - - - - - - - - - - - - - - - - - - - -
N=25;            %population size (number of agents)
tmax=100;        %maximum number of iterations
G0=100;          %parameter of gravitational constant
alpha=20;        %parameter of gravitational constant
p=2;             %fractional of total agents in the last iteration
K0=1;            %parameter of Kbest

%Defining type of optimization: 'min' or 'max'
%- - - - - - - - - - - - - - - - - - - - - - - - - - - - - - - - - - - -
type_opt='min';

%Generating of the initial population. Initial positions of agents are
%randomly selected between min and max values of the control variables
%- - - - - - - - - - - - - - - - - - - - - - - - - - - - - - - - - - - -
[numbervar,lowerbound,upperbound]=bounds;
if numbervar==1
   X=rand(N,numbervar).*(upperbound-lowerbound)+lowerbound;
end
if numbervar>1
   for i=1:numbervar
   X(:,i)=rand(N,1).*(upperbound(i)-lowerbound(i))+lowerbound(i);
   end
end

%Create the best so far fitness chart
%- - - - - - - - - - - - - - - - - - - - - - - - - - - - - - - - - - - -
BestChart=[];

%Initialize velocity of agents
V=zeros(N,numbervar);

%ITERATIVE PROCEDURE
%======================================================================
for t=1:tmax

for i=1:N
%Agents that go out of the search space, are reinitialized randomly .
Tp=X(i,:)>upperbound;
Tm=X(i,:)<lowerbound;
X(i,:)=(X(i,:).*(~(Tp+Tm)))+((rand(1,numbervar).*(upperbound-lowerbound)...
      +lowerbound).*(Tp+Tm));
end
```

```
%Evaluation of the fitness value of each agent using function Fobj.m
%-------------------------------------------------------
for i=1:N
    x=X(i,:);
    fitness(i)=Fobj(x);
end

%Determination of the best fitness (Fbest) and corresponding best agent
%(xbest)
%-------------------------------------------------------
if type_opt=='min'
    [best best_X]=min(fitness);
elseif type_opt=='max'
    [best best_X]=max(fitness);
end
if t==1
    Fbest=best;
    xbest=X(best_X,:);
end
if type_opt=='min'
    if best<Fbest
        Fbest=best;
        xbest=X(best_X,:);
    end
elseif type_opt=='max'
    if best>Fbest
        Fbest=best;
        xbest=X(best_X,:);
    end
end
BestChart=[BestChart Fbest];

%Calculation of the gravitational constant (G)
%-------------------------------------------------------
G=G0*exp(-alpha*t/tmax);

%Calculation of the mass (M)
%-------------------------------------------------------
Fmax=max(fitness); Fmin=min(fitness); Fmean=mean(fitness);
[i N]=size(fitness);
if Fmax==Fmin
    M=ones(N,1);
else
    if type_opt=='min'
       best=Fmin;worst=Fmax;
    else
       best=Fmax;worst=Fmin;
    end
        M=(fitness-worst)./(best-worst);
end
M=M./sum(M);

%Calculation of the acceleration (a)
%-------------------------------------------------------
Kbest=K0*((1-p)/tmax*t+1);
Kbest=round(N*Kbest);
[Ms ds]=sort(M,'descend');
for i=1:N
    E(i,:)=zeros(1,numbervar);
```

```
    for ii=1:Kbest
        j=ds(ii);
        if j~=i
            R=norm(X(i,:)-X(j,:),2); %Euclidian distance.
            for k=1:numbervar
                E(i,k)=E(i,k)+rand*(M(j))*((X(j,k)-X(i,k))/(R+eps));
            end
        end
    end
end
a=E.*G;
%Updating the velocity (V) of each agent
%- - - - - - - - - - - - - - - - - - - - - - - - - - - - - - - - - - - - - -
V=rand(N,numbervar).*V+a;

%Updating the position (X) of each agent
%- - - - - - - - - - - - - - - - - - - - - - - - - - - - - - - - - - - - - -
X=X+V;

%Display the results in each iteration
%- - - - - - - - - - - - - - - - - - - - - - - - - - - - - - - - - - - - - -
fprintf('GSA| Iter:%3d - -> Fbest: %9.5f\n',t,Fbest);
%- - - - - - - - - - - - - - - - - - - - - - - - - - - - - - - - - - - - - -
end %end of iteration procedure

%RESULTS:
%==========================================================================
%Best value of the objective function, Fbest
Fbest

%Optimal values of control variables, xbest
xbest

%Plotting convergence profile of GSA:
%- - - - - - - - - - - - - - - - - - - - - - - - - - - - - - - - - - - - - -
plot(BestChart,'- -k','LineWidth',2);
xlabel('\fontsize{11}\bf Iteration');ylabel('\fontsize{11}\bf Fobj');
legend('\fontsize{11}\bf GSA',1);grid on;

%Functions
%==========================================================================
function [Fobj]=Fobj(x)
%Code of the objective function to be optimized
%Input: x - vector of control variables
%Output: Fobj - value of objective function
%- - - - - - - - - - - - - - - - - - - - - - - - - - - - - - - - - - - - - -
Fobj=3*(1-x(1)).^2.*exp(-(x(1).^2+(x(2)+1).^2))...
    -10*(x(1)/5-x(1).^3-x(2).^5).*exp(-(x(1).^2+x(2).^2))-
    (exp(-((x(1)+1).^2+x(2).^2)))/3;
return

function [numbervar,lowerbound,upperbound]=bounds
%Defining number of control variables,
%lower and upper limits of control variables
%- - - - - - - - - - - - - - - - - - - - - - - - - - - - - - - - - - - - - -
numbervar=2; lowerbound(1:numbervar)=-3; upperbound(1:numbervar)=3;
return
%==========================================================================
```

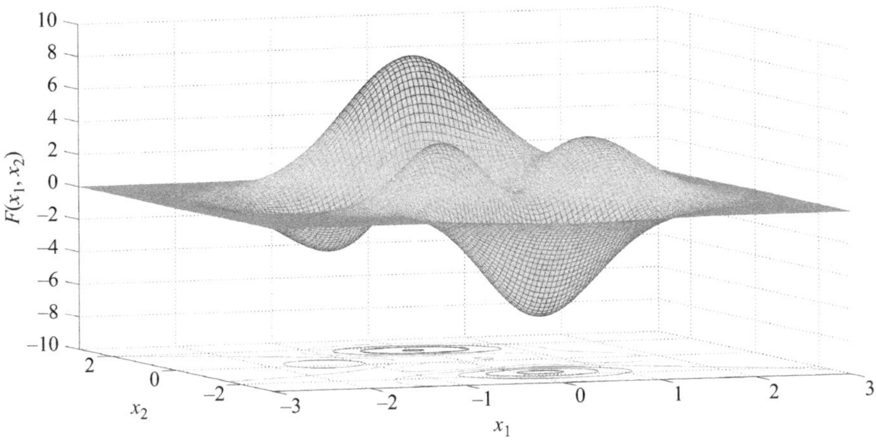

Figure 4.3 Graph of the function F

4.2.4 Example usage of GSA

To show how GSA works, the program is applied to find minimum and maximum of the function given as follows:

$$F(x_1,x_2) = 3(1-x_1)^2 \cdot e^{-\left(x_1^2+(x_2+1)^2\right)} - 10\left(\frac{x_1}{5} - x_1^3 - x_2^5\right) \cdot e^{-\left(x_1^2+x_2^2\right)} - \frac{e^{-\left((x_1+1)^2+x_2^2\right)}}{3} \tag{4.16}$$

within the range $-3 \le x_1,\ x_2 \le 3$

Graph of this function is shown in Figure 4.3.

The results of GSA obtained in the case of minimization are as follows:

```
>> GSA
  Fbest =
     -6.5511
xbest =
     0.2283   -1.6255
```

The results of GSA obtained in the case of maximization are as follows:

```
>> GSA
  Fbest =
     8.1062
xbest =
     -0.0093   1.5814
```

Figure 4.4 shows the convergence profile of GSA in both cases, minimization and maximization of the function *F*.

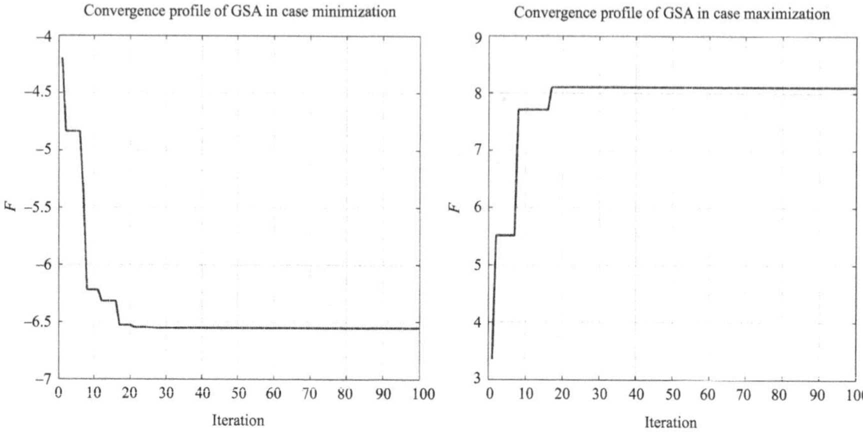

Figure 4.4 Convergence of GSA

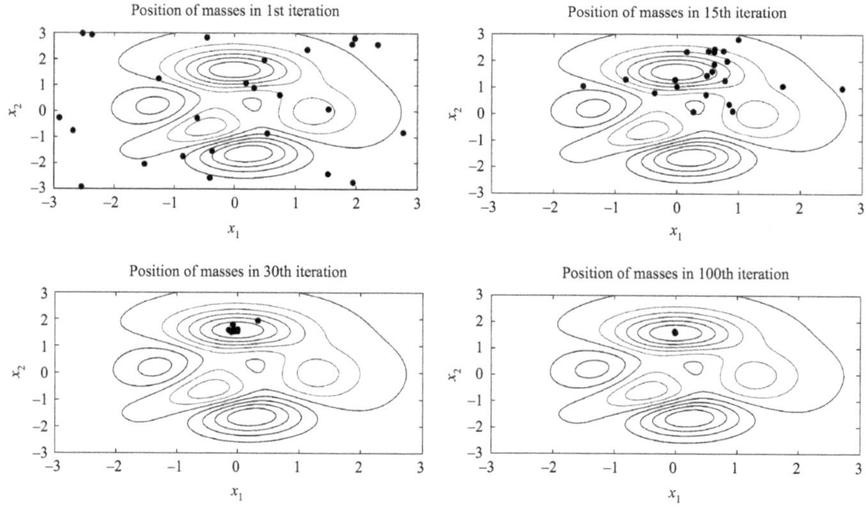

Figure 4.5. Positions of search agents in the solution space for case maximization of the F

The positions of search agents (mass) for iteration 1, 15, 30, and 100 in finding the maximum of the function F are shown in Figure 4.5. As can be seen from this figure, all masses move to the global maximum position—best fitness value.

4.3 Binary gravitational search algorithm

The authors of original GSA [1] have introduced a binary version of the algorithm [3]. Optimization problems, defined in the real space, may be considered in the binary space, too. The solution is to display real digits with some bits in the binary mode. The binary search space is considered as a hypercube in which an agent may move to nearer and farther corners of the hypercube by flipping various numbers of the bits.

In developing a binary version of GSA, some basic concepts of GSA are modified. In discrete binary environment, every dimension can take only 0 or 1. Moving through a dimension means that the corresponding variable value changes from 0 to 1 or vice versa. In order to introduce a binary mode for the gravitational algorithm, the updating procedure of the force, acceleration, and velocity may be considered similar to the original (continuous) algorithm [Eqs. (4.10)–(4.13)]. The main difference between continuous and binary gravitational search algorithm (BGSA) is that in the binary algorithm, the position updating means a switching between "0" and "1" values. This switching should be done according to the mass velocity. The idea of the author [3] is to update the position in a manner that the current bit value is changed with a probability that is calculated according to the mass velocity. In other words, BGSA updates the velocity based on (4.12) and considers the new position to be either 1 or 0 with the given probability.

Before defining a transfer function to map the velocity to the probability of position updating, it will be useful to remind some basic concepts of GSA [1,3].

- A large absolute value of the velocity shows that the current position of the mass is not proper and a great movement is required to reach the optimum position.
- A small absolute value of the velocity indicates that the current position of the mass is close to the optimum position, and there is a small distance reaching to the optimum position. Then, after finding the optimal solution, the velocity becomes zero.

Based on the above concepts of the GSA, in the implementation of the BGSA, the following concepts should be taken into account [3]:

- A large absolute value of velocity must provide a high probability of changing the position of the mass respect to its previous position (from 1 to 0 or vice versa).
- A small absolute value of the velocity must provide a small probability of changing the position. In other words, a zero value of the velocity represents that the mass position is good and must not be changed.

Based on the abovementioned concepts, a proper probability function should be defined such that for a small $|v_i^d|$, the probability of changing x_i^d must be near zero and for a large $|v_i^d|$, the probability of x_i^d movement must be high. The authors [3] proposed the function $S(v_i^d)$ to transfer v_i^d into a probability function:

$$S(v_i^d(t)) = |\tanh(v_i^d(t))| \tag{4.17}$$

The function $S(v_i^d)$ is bounded within interval [0,1] and increases with increasing $|v_i^d|$. Once $S(v_i^d)$ is calculated, the agents will move according to the rule explained in (4.18):

$$\begin{aligned} &\text{if } rand < S(v_i^d(t+1)) \text{ then } x_i^d(t+1) = \text{complement}(x_i^d(t)) \\ &\text{else } x_i^d(t+1) = x_i^d(t) \end{aligned} \tag{4.18}$$

To achieve a good convergence rate, the velocity should be limited, $|v_i^d| < v_{\max}$, and the distance R computed based on Hamming distance. The Hamming distance between two binary numbers is the number of bits in which they differ.

In the BGSA, the gravitational constant is considered as a linear decreasing function $G(t) = G_0(1 - t/t_{\max})$. Furthermore, parameter *Kbest* in (4.10) is initialized to N (total number of agents) and is decreased linearly to 1.

4.4 Modified GSA

The random number r_i in (4.12) causes that the velocity of any agent is a stochastic variable. So, every agent moves along an uncontrolled trajectory in the search space. In order to control the oscillations of trajectory of agents, the authors [4] proposed a modification of GSA by introducing a reasonable interval to clamp the agent's velocity:

$$-V_{\max} \le v_i^d \le V_{\max} \tag{4.19}$$

where $V_{\max}$ is a maximum velocity, which defines the maximum change one agent can undergo in its positional coordinates during an iteration.

The value of $V_{\max}$ is problem-dependent parameter. If the value of this parameter is too large, then the agents may move erratically, going beyond a good solution; on the other hand, if is too small, then the agent's movement is limited and the optimal solution may not be reached [4].

A population-based heuristic algorithm, such as GSA, should be equipped with two major characteristics to ensure finding global optimum. These two main characteristics are exploration and exploitation. The exploration is the ability of expanding search space, where the exploitation is the ability of finding the optima around a good solution. In premier iterations, a heuristic search algorithm explores the search space to find new solutions. To avoid trapping in a local optimum, the algorithm must use the exploration in the first few iterations. By lapse of iterations, exploration fades out and exploitation fades in, so the algorithm tunes itself in semioptimal points. To have a high performance search, an essential key is a suitable trade-off between exploration and exploitation. In the modified GSA, a time-varying maximum velocity is proposed to improve the search performance and controlling the balance between exploration and exploitation. The equation for time-varying maximum velocity is given as

$$V_{\max} = V_{\max 0}\left[1 - \left(\frac{t}{t_{\max}}\right)^{h}\right] \tag{4.20}$$

where h is a positive constant, and $V_{\max 0}$ is the initial velocity limit. The authors [4] suggest choosing h by trial-and-error and $V_{\max 0}$ as a fraction β of the distance between the bounds of the search space for each control variable:

$$V_{\max 0} = \beta(x_{\max} - x_{\min}), \quad 0 < \beta \leq 1 \tag{4.21}$$

In initial stage of the modified GSA, a relatively large value of $V_{\max}$ enables an initial global search. The gradual reduction of $V_{\max}$ over the iterations decreases the amplitude of agent's oscillations and narrows down the algorithm exploration. It encourages the agents to concentrate the search effort around the best solutions found so far and converge to the global optima in the latter part of the optimization [4]. Figure 4.6 shows a general flowchart of the modified GSA.

The above-described modification of the GSA can be incorporated in the MATLAB code of the original GSA given in Section 4.2.3, as follows:

```
.
.
.
%Updating the velocity (V) of each agent
%- - - - - - - - - - - - - - - - - - - - - - - - - - - - - - - - -
V=rand(N,numbervar).*V+a;
%=============================================================
%Determine the max velocity Vmax:
beta=0.5;
h=1;
Vmax0=beta*(upperbound-lowerbound);
Vmax=Vmax0*(1-(t/tmax)^h);
for kk=1:N
    for k=1:numbervar
        if V(kk,k)<-Vmax(k)
            V(kk,k)=-Vmax(k);
        elseif V(kk,k)>Vmax(k)
            V(kk,k)=Vmax(k);
        end
    end
end
%=============================================================
%Updating the position (X) of each agent
%- - - - - - - - - - - - - - - - - - - - - - - - - - - - - - - - -
X=X+V;
.
.
.
```

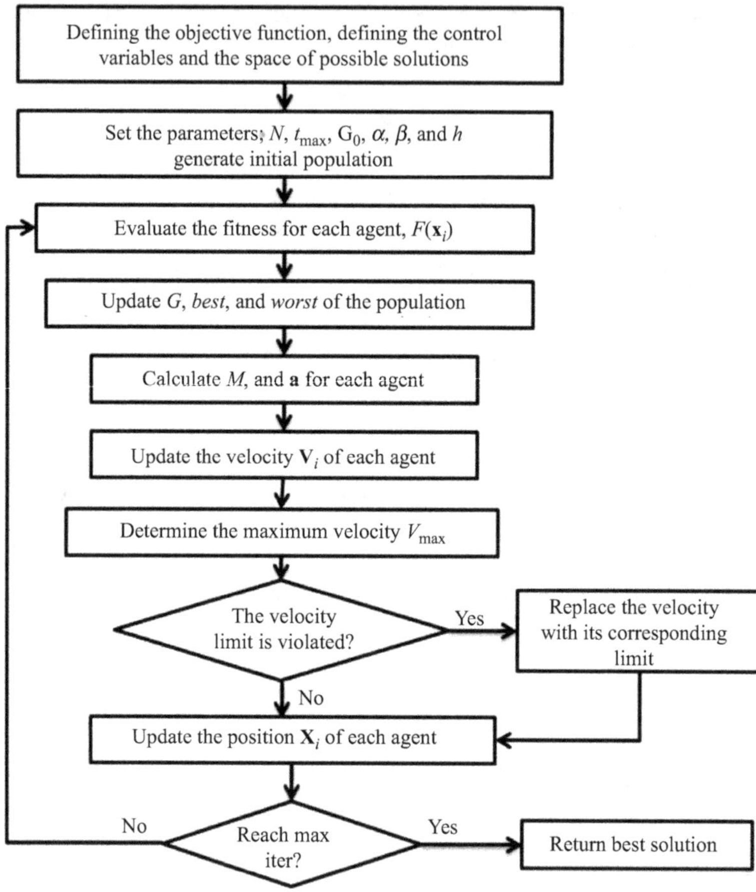

Figure 4.6 General flowchart of modified GSA

4.5 Opposition-based GSA

The introduction of the opposition solutions, i.e., the opposite population in a metaheuristic optimization method can increase its ability to find a global optimal solution. In general, the opposite number of a variable can be defined as its reflected value through the center of its domain [5,6].

For example, if $\mathbf{x}_i = \left(x_i^1, \ldots, x_i^d, \ldots, x_i^n\right)$ is a point in n-dimensional search space, where $x_i^1, \ldots, x_i^d, \ldots, x_i^n$ are real numbers and $x_i^d \in \left[a_i^d, b_i^d\right]$, $d = 1, \ldots, n$, and $i = 1, \ldots, N$, then the opposite point of $\mathbf{x}_i$ is $\breve{\mathbf{x}}_i = \left(\breve{x}_i^1, \ldots, \breve{x}_i^d, \ldots, \breve{x}_i^n\right)$, where

$$\breve{x}_i^d = a_i^d + b_i^d - x_i^d \tag{4.22}$$

where $d = 1, \ldots, n$ and $i = 1, \ldots, N$.

Now, by employing the opposite point definition, the opposition-based gravitational search algorithm (OGSA) can be defined [7,8].

Let $\mathbf{x}_i = \left(x_i^1, \ldots, x_i^d, \ldots, x_i^n\right)$ be a point in n-dimensional search space, i.e., a potential solution of the optimization problem. Assume $f = (.)$ is a fitness function which is used to evaluate the potential solution's fitness. According to the definition of the opposite point, $\breve{\mathbf{x}}_i = \left(\breve{x}_i^1, \ldots, \breve{x}_i^d, \ldots, \breve{x}_i^n\right)$ is the opposite of $\mathbf{x}_i = \left(x_i^1, \ldots, x_i^d, \ldots, x_i^n\right)$. Now, if $f\left(\breve{\mathbf{x}}_i\right) \leq f(\mathbf{x}_i)$, for a minimization problem, then the point $\mathbf{x}_i$ can be replaced with $\breve{\mathbf{x}}_i$; otherwise, the algorithm continue with $\mathbf{x}_i$. Hence, the point and its opposite point are evaluated simultaneously in order to continue with the fitter one.

The strategy of opposition-based learning is incorporated in original GSA. First step is generating an initial population, and second step is updating agents' positions and producing a new population. In the OGSA, the original GSA is chosen as a parent algorithm, and opposition-based ideas are embedded in it with an intention to exhibit accelerated convergence profile.

The steps of the OGSA are as follows:

Step 1: Search space identification. Initialize GSA parameters like N, T, G_0, α, and J_r.
Step 2: Initialization: Generate random population of N agents and form the population POP.
Step 3: Calculate the fitness value for each member of the POP.
Step 4: Update the $G(t)$, $best(t)$, $worst(t)$, and $M_i(t)$ for $i = 1,2, \ldots, N$.
Step 5: Calculation of the total force in different directions.
Step 6: Calculation of acceleration and velocity of each agent.
Step 7: Update each agent's position.
Step 8: Opposition-based generation jumping

$$\text{if } rand < J_r (J_r \text{ is jumping rate})$$

Generate the opposite population OPOP using (4.22)
Select N fittest agents from set {POP, OPOP} as current population POP.

Step 9: Repeat steps 3–8 until the stop criteria is reached.
Step 10: Return best solution. Stop.

4.5.1 Current optimum opposition-based GSA

The lack of application of the opposition of the population at metaheuristics optimization methods is to increase the computation time, because the population size is increased from N to $2N$. This problem can be avoided by using the current optimum opposition instead of the opposite population. The opposite point using the current optimum is in the neighborhood of the global optimum during the process of evolution, especially in later stages [6,9]. The opposition of the current optimal solution is defined as follows:

Let $\mathbf{x}_i = \left(x_i^1, \ldots, x_i^d, \ldots, x_i^n\right)$ be a point in n-dimensional search space, $\mathbf{x}_{best} = \left(x_{best}^1, \ldots, x_{best}^d, \ldots, x_{best}^n\right)$ be the best solution in the current population, and

$x_i^d, x_{best}^d \in [a^d, b^d]$, $d = 1,\ldots,n$. The current optimum opposite point of $\mathbf{x}_i$ is denoted by $\breve{\mathbf{x}}_i^{coo} = (\breve{x}_i^{1,coo},\ldots,\breve{x}_i^{d,coo},\ldots,\breve{x}_i^{n,coo})$, where

$$\breve{x}_i^{d,coo} = 2x_{best}^d - x_i^d \tag{4.23}$$

where $d = 1,\ldots,n$ and $i = 1,\ldots,N$.

Here $\mathbf{x}_{best}$ is the center of opposition.

It is possible that the opposite solution $\breve{x}_i^{d,coo}$, $1 \le d \le n$, jumps out of the box constraint $[a^d, b^d]$. In this case, the opposite solution is assigned to a value as follows:

$$\breve{x}_i^{d,coo} = \begin{cases} a^d & \text{if } \breve{x}_i^{d,coo} < a^d \\ b^d & \text{if } \breve{x}_i^{d,coo} > b^d \end{cases} \tag{4.24}$$

Now, a current optimum opposition-based gravitational search algorithm (COOGSA) can be described. The COOGSA is based on modification of the original GSA by introducing the opposition of the current optimal solution in each iteration of the algorithm. The original GSA is chosen as a parent algorithm, and the opposition of the current optimum is used in the stage of updating agents' positions to improve the convergence and increase probability to find global optimum.

Let $\mathbf{x}_{best}(t) = (x_{best}^1(t),\ldots,x_{best}^d(t),\ldots,x_{best}^n(t))$ be the best solution in the current iteration t. The position of the ith agent in the dth dimension is updated based on the movement probability $P_i(t)$ as follows:

$$x_i^d(t+1) = \begin{cases} \breve{x}_i^{d,coo} & \text{if } \quad rand_i < P_i \\ x_i^d & \qquad \text{otherwise} \end{cases} \tag{4.25}$$

where $rand_i$ is a random number in [0,1] and $\breve{x}_i^{d,coo}$ is the current optimum opposite point of x_i^d when x_{best}^d is the center of opposition. The movement probability P_i makes the fittest candidate solutions to move more slowly. It can be calculated as follows [6]:

$$P_i(t) = \frac{best(t) - fit_i(t)}{best(t) - worst(t)} \tag{4.26}$$

The different steps of the proposed COOGSA are as follows:

Step 1: Search space identification. Initialize GSA parameters like N, T, G_0, and α.
Step 2: Initialization: Generate random population of N agents and form the population **POP**.
Step 3: Calculate the fitness value for each member of the **POP**.
Step 4: Update the $G(t)$, $best(t)$, $worst(t)$, and $M_i(t)$ for $i = 1,2,\ldots,N$.
Step 5: Calculation of the total force in different directions.
Step 6: Calculation of acceleration and velocity of each agent.
Step 7: Update each agent's position.
Step 8: Calculation of the opposition of the current optimum solution using (4.23) and (4.24).

Step 9: Calculation of the movement probability using (4.26).
Step 10: Update each agent's position using (4.25).
Step 11: Repeat steps 3–10 until the stop criteria is reached.
Step 12: Return best solution. Stop.

4.6 Adaptive gbest-guided GSA

One of the major weaknesses of the original GSA is slow exploitation that might results in degraded performance when dealing with real engineering problems. To overcome this problem, the authors [10] proposed an adaptive gbest-guided GSA. In this approach, the best mass is archived and utilized to accelerate the exploitation phase of original GSA.

As well known, a population-based metaheuristic optimization algorithm should be equipped with two major characteristics to ensure finding global optimum. These two main characteristics are exploration and exploitation. The exploration is the ability of expanding search space, where the exploitation is the ability of finding the optima around a good solution.

4.6.1 Slow exploitation of GSA

In GSA, the gravitational constant (G) defines the speed at which solutions change their location in solution space. According to (4.10), a high value of G results in high intensities of gravitational forces, resulting rapid movement in earlier iterations. However, G is progressively decreased according to (4.14), and this, combined with the slow movement of increasingly heavy agents, helps GSA during exploitation. So the exploitation phase coincides with less intensity of attractive force and slow movement. Unfortunately, heavy masses with slow movement and less intensity of attractive force significantly degrade the speed of convergence as well. Therefore, it seems that GSA suffers from slow search speed in the exploitation phase originating from these factors. It is worth mentioning that GSA has no memory for saving the best solution obtained so far, so the best solution might be lost as the best mass is attracted away by other less-fit masses. All these problems motivate the authors [10] to develop the solution discussed in the following section.

4.6.2 Improving the exploitation of GSA

The basic idea of the proposed method is to save and use the location of the best mass to speed up the exploitation phase. There are two benefits in this method: accelerating the movement of particles toward the location of the best mass, which may help them to surpass it and be the best mass in the next iteration, and saving the best solution attained so far. Mathematically, the proposed method can be described as follows [10]:

$$\mathbf{v}_i(t+1) = r_i \cdot \mathbf{v}_i(t) + c_1' \cdot \mathbf{a}_i(t) + c_2' \cdot (\mathbf{gbest} - \mathbf{x}_i(t)) \tag{4.27}$$

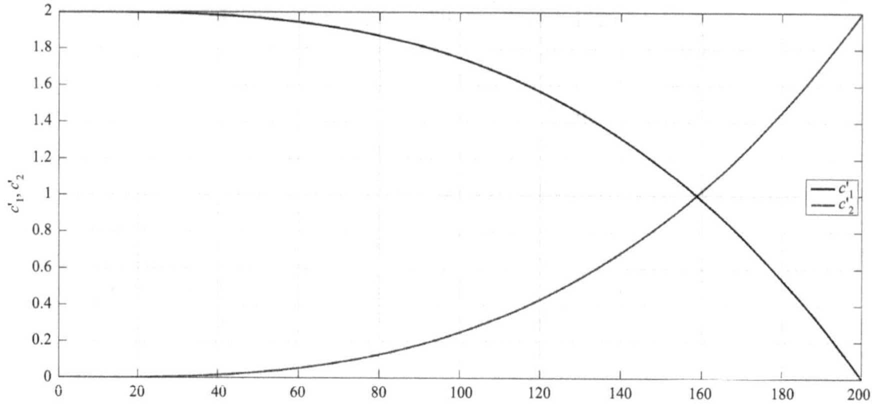

Figure 4.7 Adaptive coefficients c'_1 and c'_2

where $\mathbf{v}_i(t)$ is the velocity of agent i at iteration t, c'_1 and c'_2 are accelerating coefficients, r_i is a random number in [0,1], $\mathbf{a}_i(t)$ is the acceleration of agent i at iteration t, and **gbest** is the position of the best solution acquired so far.

The first component of (4.27), $r_i \cdot \mathbf{v}_i(t) + c'_1 \cdot \mathbf{a}_i(t)$, is the same as that of GSA, in which the exploration of the masses is emphasized. The second component, $c'_2 \cdot (\mathbf{gbest} - \mathbf{x}_i(t))$, is responsible for attracting masses toward the best masses obtained so far.

After calculating the velocity, the position of agent i can be updated as follows:

$$\mathbf{x}_i(t+1) = \mathbf{x}_i(t) + \mathbf{v}_i(t+1) \tag{4.28}$$

A problem may arise in that this method could affect the exploration phase as well, since it establishes a permanent element of velocity updating. In order to prevent the new updating velocity method from degrading the exploration ability, the authors [10] used adaptive values for c'_1 and c'_2 as follows:

$$c'_1 = -\frac{2t^3}{t_{max}^3} + 2 \tag{4.29}$$

$$c'_2 = \frac{2t^3}{t_{max}^3} \tag{4.30}$$

As can be seen from Figure 4.7, c'_1 adaptively decreases and c'_2 adaptively increases so that the masses tend to accelerate toward the best solution as the algorithm reaches the exploitation phase. Since there is no clear border between the exploration and exploitation phases in evolutionary algorithms, the adaptive method is the best option for allowing a gradual transition between these two phases. In addition, this adaptive approach emphasizes exploration in the first iterations and exploitation in the final iterations.

It is worth mentioning that the second part of (4.27), $c_2' \cdot (\mathbf{gbest} - \mathbf{x}_i(t))$, is quite similar to the social component of PSO, so the adaptive gbest-guided GSA could be also considered as a hybrid of GSA and PSO described in Section 4.10. A high value of c_1' biases toward GSA behavior, while a high value of c_2' emphasizes the social component of PSO in performing the search process. The adaptive method allows GSA to explore the search space and a PSO-like exploitation of the best solution discovered by GSA.

The general steps of the adaptive gbest-guided GSA are as follows:

Step 1: Search space identification. Initialize GSA parameters like N, T, G_0, α.
Step 2. Initialization: Generate random population of N agents and form the population **POP**.
Step 3: Calculate the fitness value for each agent in the **POP**.
Step 4: Update the $G(t)$, $best(t)$, $worst(t)$, and $M_i(t)$ for $i = 1,2,\ldots,N$.
Step 5: Calculation of the total force in different directions.
Step 6: Calculation of acceleration of each agent.
Step 7: Update the best solution acquired so far, **gbest**.
Step 8: Calculate the coefficients c_1' and c_2' using (4.29) and (4.30).
Step 9: Calculate the velocities of agents using (4.27).
Step 10: Update the positions of agents using (4.28).
Step 11: Repeat steps 3–10 until the stop criteria is reached.
Step 12: Return best solution, **gbest**. Stop.

Some remarks on the adaptive gbest-guided GSA and its advantages are the following [10]:

- The proposed method uses a memory (gbest) for saving the best solution obtained so far, in contrast to the original GSA, so it is accessible at any time and will not be lost.
- Each agent can observe the best solution (gbest) and move toward it, so masses are provided with a sort of social intelligence.
- The effect of gbest is emphasized in the exploitation phase by adapting c_1' and c_2'.
- The effect of gbest on agents is independent of their masses and so can be considered as an external force not subject to gravitational rules. This effectively prevents particles from gathering together and having extremely slow movement.
- The computational cost of this method is extremely low.

4.7 Self-adaptive GSA

The authors [11,12] used a self-adaptive mutation technique to improve the convergence characteristics of the original GSA. In this mutation technique, two methods are offered to modify the solutions. Each search agent according to a probability model chooses one of these methods. The probability model is based on the ability of each method to provide more optimal solutions. Using this mutation technique, the agents self-adaptively distinguish which of the proposed methods deserves to be employed for the problem in hand or for each stage of the optimization procedure.

Method 1: The GSA is a memory-less algorithm, i.e., the particles do not use the proper information found in previous iterations. In this method, the GSA save the best solution found so far, named **gbest**, and uses the updating process as follows:

$$\mathbf{x}_{i,\text{method1}}(t) = \mathbf{x}_i(t) + r_i(\mathbf{gbest}(t) - l \cdot \mathbf{Mean}(t)) \quad i = 1, \ldots, N_1 \tag{4.31}$$

where r_i is the random number in [0,1], **Mean** is the mean value of the population, N_1 is the number of agents which are selected updating Method 1, and l is a weighting factor.

Method 2: This updating method serves to improve the diversity of solutions, alleviate stagnation, and avoid being trapped in local optima. For each agent i, five particles are selected randomly as $m_1 \neq m_2 \neq m_3 \neq m_4 \neq m_5 \neq i$, and a trial solution is created as

$$\mathbf{x}_{i,\text{trial}}(t) = \mathbf{x}_{m_1}(t) + r_1(\mathbf{x}_{m_2}(t) - \mathbf{x}_{m_3}(t)) + r_2(\mathbf{x}_{m_4}(t) - \mathbf{x}_{m_5}(t)) \quad i = 1, \ldots, N_2 \tag{4.32}$$

where r_1 and r_2 are random numbers in [0,1], N_2 is the number of agents which are selected updating Method 2. Using the following scheme, a mutant solution is obtained:

$$x^d_{i,\text{method1}}(t) = \begin{cases} x^d_{i,\text{trial}}(t) & \text{if } (r \leq 0.5) \\ x^d_i(t) & \text{else} \end{cases} \tag{4.33}$$

where $x^d_i(t)$ and $x^d_{i,\text{trial}}(t)$ are the position of the ith agent with respect to the dth dimension in iteration t for the existing and trial agent i, respectively; r is the random number in [0,1]. Finally, between $\mathbf{x}_{i,\text{method2}}(t)$ and $\mathbf{x}_i(t)$, one with better fitness value is selected.

In the self-adaptive GSA, the probability of both aforementioned methods is considered as $prob_{\text{method}} = 0.5$, (method = 1, 2), and a parameter named accumulator is defined as $acum_{\text{method}} = 0$, (method = 1, 2).

In each iteration, the agents are sorted based on their fitness values, while $i = 1$ represents the agent with best fitness value and $i = N$ stands for the agent with the worst fitness value. Thereafter, a weight factor is allocated to each of them. The better solutions obtain larger weight factors as

$$w_i = \frac{\log(N - i + 1)}{\log(1) + \cdots + \log(N)}, \qquad i = 1, \ldots, N \tag{4.34}$$

where N is the population size.

The accumulator of each moving strategy is updated as

$$acum_{\text{method}} = acum_{\text{method}} + \frac{w_{ii}}{N_{\text{method}}}, \qquad ii = 1, \ldots, N_{\text{method}} \tag{4.35}$$

where N_{method} is the number of agents selecting methodth method and w_{ii} ($ii = 1,\ldots,N_{\text{method}}$) are the weight factors corresponding to them. The excitation probability is calculated as

$$prob_{\text{method}} = (1-\theta)prob_{\text{method}} + \theta\frac{acum_{\text{method}}}{t_{\max}}, \quad (\text{method} = 1,2) \tag{4.36}$$

where θ is a learning rate to control the learning speed in the self-adaptive GSA. It is assumed to be 0.142 in [11,12]. $t_{\max}$ is the maximum iteration number. Finally, the normalized probability values are calculated as follows:

$$prob_{\text{method}} = \frac{prob_{\text{method}}}{prob_1 + prob_2}, \quad (\text{method} = 1,2) \tag{4.37}$$

At each generation, each agent chooses the methodth mutation method by using the roulette wheel mechanism based on their probability values.

The self-adaptive GSA can be summarized in the following steps:

Step 1: Search space identification. Initialize GSA parameters like N, $t_{\max}$, G_0, α. Set $prob_{\text{method}} = 0.5$, ($\text{method} = 1, 2$), $acum_{\text{method}} = 0$ ($\text{method} = 1, 2$), $\theta = 0.142$.
Step 2. Initialization: Generate random population of N agents and form the population **POP**.
Step 3: Sort the agents based on their fitness values. Thereafter, determine **gbest**, the mean of the population, *worst*(t), and *best*(t).
Step 4: Update the $G(t)$, $best(t)$, $worst(t)$, and $M_i(t)$ for $i = 1,2,\ldots, N$.
Step 5: Update the overall force which applies on each agent.
Step 6: Calculate the acceleration and the velocity of each agent.
Step 7: Update the agents position using.
Step 8: For each agent, select one of the mutation method by the roulette wheel mechanism based on the $prob_{\text{method}}$.
Step 9: Apply the mutation approach. Find the new solution for each agent.
Step 10: Update the accumulator of both strategies.
Step 11: Update the $prob_{\text{method}}$ ($\text{method} = 1,2$).
Step 12: Go to step 3 until the stop criteria is reached (t reaches $t_{\max}$).
Step 13: Return best solution. Stop.

4.8 Nondominated sorting GSA

The authors [13] introduced the nondominated sorting gravitational search algorithm (NSGSA) for multiobjective optimization problems. The NSGSA use the concept of Pareto-optimality condition and nondominance.

The general iterative steps of NSGSA are as follows [13]:

Step 1: Search space identification; set the algorithm parameters: N, $t_{\max}$, G_0, α, N_{archive}, P_r, P_s, P_{elitism}, w_0, w_1, and β.
Step 2: Initialization: Generate random population of N agents.

Step 3: Evaluate fitness of each agent.
Step 4: Update external archive.
Step 5: Nondominated sorting.
Step 6: Update the list of moving agents.
Step 7: Update the mass of moving agents.
Step 8: Update agents' acceleration.
Step 9: Update agents' velocity.
Step 10: Update and mutate agents' positions.
Step 11: Repeat steps 3–10 until the maximum number of iteration is reached.
Step 12: Report Pareto-optimal solutions stored in archive. Stop.

4.8.1 Updating the external archive

The external archive is updated based on the Pareto-dominance criterion. If a member of the population is dominated by any member of the external archive, it will not be inserted to the archive. On the contrary, if it is nondominated with respect to all members of the archive or it dominates some members of the archive, it will be inserted to the archive and the dominated members will be removed. If the number of archived members exceeds the maximum archive length, the most crowded area must be determined to eliminate a member from that area. A spread indicator is introduced [13] to control the length of external archive. This indicator is based on the spread of the points within the Pareto front. To get a set of solutions that spans the whole Pareto-optimal area as homogeneously as possible, a crowding distance measure, $d_{c,i}$, is introduced as follows [13–15]:

$$d_{c,i} = \sqrt{\sum_{j=1}^{M} \left(d^{j}_{c,i}\right)^{2}} \quad (i = 1, 2, \ldots, N_{\text{archive}}) \tag{4.38}$$

where

$$d^{j}_{c,i} = fit^{j}_{i+1} - fit^{j}_{i-1} \quad (i = 1, 2, \ldots, N_{\text{archive}};\ \ j = 1, 2, \ldots, M) \tag{4.39}$$

where fit^{j}_{i+1} is the jth fitness of the point $i+1$, when the members of the archive are sorted versus the jth objective function. It is desired to have nearly equal crowding distances. To have nearly equal $d_{c,i}$, the deviation of crowding distance from their average, $\overline{d}_c$, should be minimized. For this reason, a spread indicator δ introduced:

$$\delta = \sum_{\substack{i=1 \\ i \notin E}}^{N_{\text{archive}}} \frac{\left|d_{c,i} - \overline{d}_c\right|}{(N_{\text{archive}} - M)\overline{d}_c} \tag{4.40}$$

where N_{archive} is archive length and E is the set of extreme points, the size of which is taken equal to M. To get a uniform spread of Pareto-archive spread indicator,

(δ) must be reduced. This has been done by reducing the difference between max $\{d_{c,i}\}$ and min$\{d_{c,i}\}$. In other words, the maximum $d_{c,i}$ must be decreased or the minimum $d_{c,i}$ must be increased. For this purpose, the two nearest neighbors in the objective space are determined and named as members 1 and 2. These two members are removed alternatively to calculate δ_1 and δ_2 corresponding to the removal of members 1 and 2, respectively. The member, the presence of which causes bigger δ, is then removed, unless it extends the Pareto front, the case of which the other member should be removed.

4.8.2 Updating the list of moving agents

For updating the list of moving agents for the next flight, M extreme points of the Pareto front (M single-objective optimal solutions of the external archive) are inserted to the list. Then, M points, located in the least crowded area, are inserted. These points are selected using the crowding distance assignment algorithm [15]. A percent of the archive members, P_{elitism}, is also selected randomly and inserted to the list of moving agents. Finally, the list length is limited to N (population size) by removing some of the worst particles, starting from the worst layer of the nondominated sorted agents. It should be noted that the initial velocity of the members, inserted from the external archive to the list of moving agents, must be set to zero.

4.8.3 Updating the mass of moving agents

In a multiobjective optimization problem, there are multiple objectives and a single fitness function must be defined corresponding to each solution. NSGSA utilizes the ranks of the layers, generated using the nondominated sorting algorithm to determine the fitness of each particle.

In NSGSA, the M members, imported from the extreme regions of the external archive, and the M members, imported from the least crowded area of this archive, are ranked as 1. The P_{elitism} percent of the archive members, chosen randomly and inserted to the list of moving particles, is ranked as 2. The members, remained from the last iteration, get ranks 3 and more, regarding the rank of the layers they belong to. The rank of each particle is then considered as its fitness and its mass is updated using (4.4). It should also be noted that the imported particles from the archive usually have bigger mass than the others due to their better fitness. Therefore, according to (4.11), their acceleration is less than the other particles. Moreover, as mentioned previously, the initial velocity of these new imported particles is set to zero. Therefore, these new particles move more slowly than the others and provide fine search around the Pareto front.

4.8.4 Updating the acceleration of agents

For updating the gravitational acceleration of agents, NSGSA uses the same Equations (4.4)–(4.11) as original GSA. The most important parameter of this equation is the gravitational constant, $G(t)$. In original GSA, $G(t)$ is suggested by (4.14), where α and G_0 are the function parameters. It is difficult to find a suitable constant G_0 for various optimization problems. In NSGSA [13], G_0 is proposed as

$$G_0 = \beta \max_{d \in \{1,\ldots,n\}} \left(\left|x_u^d - x_l^d\right|\right) \tag{4.41}$$

where β is a parameter of NSGSA. The quick decrease of $G(t)$ in (4.14) causes a fast decay of the exploration. To improve this problem, NSGSA utilizes a linear function of $G(t)$, defined as follows:

$$G(t) = G_0\left(1 - \frac{t}{t_{\max}}\right) \tag{4.42}$$

4.8.5 The use of mutation operator

In NSGSA, two mutation operators, called sign and reordering mutations, have been added to the original GSA to decrease the chance of trapping in local optima. Moreover, these two mutation operators promote and preserve diversity within the moving agents. The mutation operator in NSGSA works as follows: the velocity of agent is first mutated by the sign and then by the reordering mutations.

In sign mutation, to update the position of each agent, the sign of velocity vector changes temporally, with a predefined probability, P_s, as follows [13]:

$$v_i^{\prime d}(t+1) = s_i^d v_i^d(t+1) \tag{4.43}$$

where $d = 1,2,\ldots,n$; $i = 1,2,\ldots,N$

$$s_i^d = \begin{cases} -1 & rand < P_s \\ 1 & \text{otherwise} \end{cases} \tag{4.44}$$

$$\mathbf{x}_i(t+1) = \mathbf{x}_i(t) + \mathbf{v}_i^{\prime}(t+1) \tag{4.45}$$

where $v_i^{\prime d}(t+1)$ is the mutated velocity by the sign mutation operator, and *rand* is a uniform random number in [0,1].

In reordering mutation, some particles are chosen randomly to be mutated according to reordering mutation probability (P_r). Then, elements of the velocity vector are rearranged randomly.

4.8.6 Update and mutate the position of agents

NSGSA utilizes a combination of sign and reordering mutations to update the position of particles. Therefore, in NSGSA, the position of particles is updated as follows:

$$\mathbf{v}_i(t+1) = w(t) \cdot \mathbf{v}_i(t+1) + \mathbf{a}_i(t) \tag{4.46}$$

$$\mathbf{v}_i^{\prime}(t+1) = \textbf{Sign\_Mutate}(\mathbf{v}_i(t+1)) \tag{4.47}$$

$$\mathbf{v}_i''(t+1) = \mathbf{Reordering\_Mutate}(\mathbf{v}_i(t+1)) \tag{4.48}$$

$$\mathbf{x}_i(t+1) = \mathbf{x}_i(t) + \mathbf{v}_i''(t+1) \tag{4.49}$$

where $w(t)$ is the time varying inertia coefficient. In original GSA, this coefficient is taken a random number between 0 and 1. As in PSO, it is better to use a decreasing inertia weight than a random one to have proper exploration and exploitation in the first and the last iterations, respectively. NSGSA utilizes a time varying weighting coefficient as follows:

$$w(t) = w_0 - (w_0 - w_1)\frac{t}{t_{\max}} \tag{4.50}$$

4.9 Clustered-gravitational search algorithm

The authors of original GSA [1] and BGSA [3] proposed a novel version of GSA named Clustered-GSA (CGSA) [16] in order to reduce the computation complexity of the original GSA.

In CGSA, the number of agents is reduced during iterations using a clustering method. Clustering of the population is used to saving memory and reducing complexity of GSA.

In the CGSA, the equations for calculating mass, force and updating position and velocity of the agents are similar to the original version of GSA. But, in the CGSA, after a certain number of iterations, the population is clustered into the certain number of groups to reduce the population size. To do this, the whole objects are clustered into some groups, and each cluster is substituted with its center of mass. The new population for the next iteration would be these centers. For clustering, the k means algorithm is used.

The position of center of mass of cluster j in the dimension $d(x_{cmj}^d)$ is defined as follows [16]:

$$x_{cmj}^d(t) = \frac{1}{M_{jtot}}\sum_{i=1}^{Cnum_j} m_i x_i^d(t), \quad j = 1, 2, \ldots, kcluster \tag{4.51}$$

The velocity and mass of each cluster are the average velocity and average mass of the agents in that cluster, respectively. *kcluster* is the total number of clusters. $Cnum_j$ is the total number of agents in cluster j, and M_{jtot} is the sum of masses in cluster j.

The number of clusters (*kcluster*) and *Kbest* are decreased linearly in such way that *Kbest* is less than *kcluster* during the all iterations. In fact, the proposed algorithm starts with exploring well the search space and, as time passes, it switches to exploiting. This concept guides an algorithm to find a global optimum by controlling exploration and exploitation. The flowchart of the CGSA is shown in Figure 4.8.

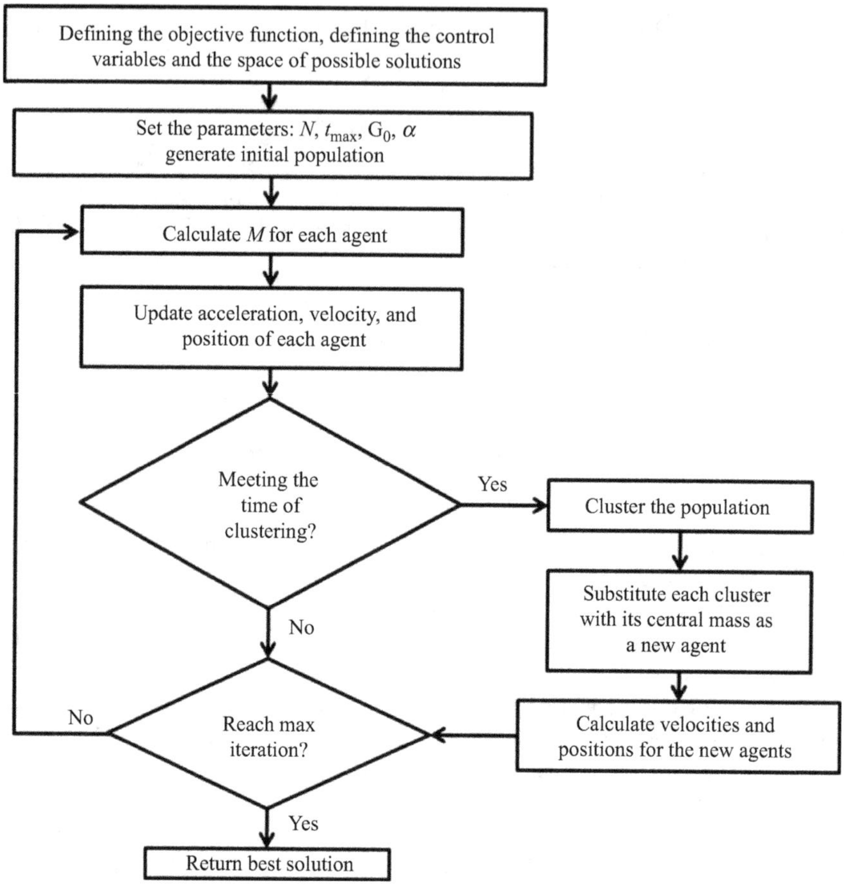

Figure 4.8 Flowchart of clustered-GSA

4.10 Hybrid PSO and GSA algorithm

A population-based heuristic algorithm should be equipped with two major characteristics to ensure finding global optimum. These two main characteristics are exploration and exploitation. The exploration is the ability of expanding search space, where the exploitation is the ability of finding the optima around a good solution. In premier iterations, a heuristic search algorithm explores the search space to find new solutions. To avoid trapping in a local optimum, the algorithm must use the exploration in the first few iterations. By lapse of iterations, exploration fades out and exploitation fades in, so the algorithm tunes itself in semioptimal points. To have a high performance search, an essential key is a suitable trade-off between exploration and exploitation. However, it often happens that with strengthening one's ability, the other will weaken and vice versa.

Because of the abovementioned points, the existing heuristic optimization algorithms are capable of solving finite set of problems. It has been proved that there is no algorithm, which can perform general enough to solve all optimization problems. Merging the optimization algorithms is a way to balance the overall exploration and exploitation ability. PSO is one of the most widely used evolutionary algorithms in hybrid methods due to its simplicity, convergence speed, and ability of searching global optimum.

The hybrid optimization algorithm of PSO and GSA was developed by Mirjalili and Hashim [17]. The basic idea of hybrid PSOGSA algorithm is to combine the ability for social thinking (**gbest**) in PSO with the local search capability of GSA. In order to combine these algorithms, (4.52) is proposed as follows:

$$\mathbf{v}_i(t+1) = r_1\mathbf{v}_i(t) + C_1 r_2 \mathbf{a}_i(t) + C_2 r_3 (\mathbf{gbest}(t) - \mathbf{x}_i(t)) \quad i = 1, 2, \ldots, N; \tag{4.52}$$

where $\mathbf{v}_i(t)$ is the velocity of the ith agent at iteration t; C_1 and C_2 are positive constants; r_1, r_2, and r_3 are uniformly distributed random numbers in [0,1]; $\mathbf{a}_i(t)$is the acceleration of the ith agent at iteration t; $\mathbf{gbest}(t)$ is the best solution so far at iteration t; and N is the size of population (the number of agents).

In each iteration, the positions of agents are updated as follows:

$$\mathbf{x}_i(t+1) = \mathbf{x}_i(t) + \mathbf{v}_i(t+1) \quad i = 1, 2, \ldots, N; \tag{4.53}$$

The different steps of the PSOGSA are the following:

1. Search space identification. Initialize PSOGSA parameters: N, t_{max}, G_0, α, C_1, and C_2;
2. Generate initial population. All agents are randomly initialized;
3. Fitness evaluation of the agents to update **gbest**;
4. Update the $G(t)$ using, $best(t)$, $worst(t)$, and $M_i(t)$ for the population;
5. Calculate forces using and accelerations for all agents;
6. Update velocity of all agents using (4.52);
7. Update position of all agents using (4.53);
8. Repeat steps 3–7 until the stop criteria is reached (usually max number of iterations);
9. Return best solution. Stop.

To see how PSOGSA is efficient, some remarks are noted [17]. In PSOGSA, the quality of solutions (fitness) is considered in the updating procedure. The agents near good solutions try to attract the other agents which are exploring the search space. When all agents are near a good solution, they move very slowly. In this case, the **gbest** helps them to exploit the global best. PSOGSA use a memory (**gbest**) to save the best solution found so far, so it is accessible anytime. Each agent can observe the best solution so far and tend toward it. With adjusting C_1 and C_2, the abilities of global search and local search can be balanced.

The MATLAB code of the hybrid PSOGSA algorithm is given below.

```
%Hybrid PSO and GSA (PSOGSA)
%------------------------------------------------------------------------
function []=PSOGSA

%INITIALIZATION
%========================================================================
%Defining the objective function in the function program Fobj.m
%Defining the space of possible solutions in the function program bounds.m
%------------------------------------------------------------------------
%Defining parameters of PSOGSA:
%------------------------------------------------------------------------
 N=25;           %population size (number of agents)
 tmax=100;       %maximum number of iterations
 C1=0.5;         %constant
 C2=1.5;         %constant
 G0=1;           %parameter of gravitational constant
 alpha=20;       %parameter of gravitational constant
 type_opt='min'; %type of optimization: 'min' or 'max'
%------------------------------------------------------------------------
%Generating of the initial population. Initial positions of agents are
%randomly selected between min and max values of the control variables
%------------------------------------------------------------------------
[numbervar,lowerbound,upperbound]=bounds;
if numbervar==1
    X=rand(N,numbervar).*(upperbound-lowerbound)+lowerbound;
end
if numbervar>1
    for i=1:numbervar
    X(:,i)=rand(N,1).*(upperbound(i)-lowerbound(i))+lowerbound(i);
    end
end
if type_opt=='min'
    Fbest=inf;
elseif type_opt=='max';
    Fbest=-inf;
end

%Initialize velocity (V), acceleration (a), mass (M), and force (F) of
%agents
V = randn(N,numbervar) ;
a=zeros(N,numbervar);
M(N)=0;
F=zeros(N,numbervar);
BestChart=[];    %Create the best so far fitness chart

%ITERATIVE PROCEDURE
%========================================================================
t = 0 ;                 %Iterations' counter
while ( t < tmax )
G=G0*exp(-alpha*t/tmax);  %Calculation of the gravitational constant (G)
t = t + 1;
F=zeros(N,numbervar);
M(N)=0;
a=zeros(N,numbervar);
for i=1:N
%Agents that go out of the search space, are reinitialized randomly .
    Tp=X(i,:)>upperbound;
```

```
    Tm=X(i,:)<lowerbound;
    X(i,:)=(X(i,:).*(~(Tp+Tm)))+((rand(1,numbervar).*...
            (upperbound-lowerbound)+lowerbound).*(Tp+Tm));
end

%Evaluation of the fitness value of each agent using function Fobj.m
%-----------------------------------------------------------
for i=1:N
    x=X(i,:);
    fitness(i)=Fobj(x);
end

%Determining the best and worst fitness in the population
%-----------------------------------------------------------
if type_opt=='min'
    [best best_X]=min(fitness);
    [worst worst_X]=max(fitness);
elseif type_opt=='max'
    [best best_X]=max(fitness);
    [worst worst_X]=min(fitness);
end

%Determination of the best fitness (Fbest) and corresponding best agent
%(xbest)
%-----------------------------------------------------------
if t==1
    Fbest=best;
    xbest=X(best_X,:);
end
if type_opt=='min'
    if best<Fbest
        Fbest=best;
        xbest=X(best_X,:);
    end
elseif type_opt=='max'
    if best>Fbest
        Fbest=best;
        xbest=X(best_X,:);
    end
end
BestChart=[BestChart Fbest];

%Calculating the masses for all agents
%-----------------------------------------------------------
M=(fitness-worst)./(best-worst);
M=M./sum(M);

%Calculating the forces for all agents
%-----------------------------------------------------------
for i=1:N
   for j=1:numbervar
      for k=1:N
        if(X(k,j)~=X(i,j))
        F(i,j)=F(i,j)+ rand*G*M(k)*M(i)*(X(k,j)-X(i,j))/abs(X(k,j)-...
        X(i,j));
        end
       end
   end
end
```

```
% Updating the velocities and accelerations of agents
%-------------------------------------------------------------
for i=1:N
    for j=1:numbervar
        if(M(i)~=0)
            a(i,j)=F(i,j)/M(i);
        end
    end
end

for i=1:N
   for j=1:numbervar
     V(i,j)=rand*V(i,j)+C1*rand*a(i,j)+C2*rand*(xbest(j)-X(i,j));
     %Eq. (4.52)
   end
end

%Updating the positions of agents
%-------------------------------------------------------------
X=X+V;   %Eq. (4.53)

end  % end of iteration procedure

%RESULTS
%=============================================================
%Best value of the objective function, Fbest
Fbest

%Optimal values of control variables, xbest
xbest

%Plot convergence characteristic of PSOGSA:
plot(BestChart,'-k','LineWidth',2);
xlabel('\fontsize{11}\bf Iteration');
ylabel('\fontsize{11}\bf Fobj');
legend('\fontsize{11}\bf PSOGSA',1);
return

%Functions
%=============================================================
function [Fobj]=Fobj(x)
%Code of the objective function to be optimized
%Input: x - vector of control variables
%Output: Fobj - value of objective function
%-------------------------------------------------------------
%Branian Function (F17 standard benchmark function)
Fobj=(x(2)-(x(1)^2)*5.1/(4*(pi^2))+5/pi*x(1)-6)^2+...
      10*(1-1/(8*pi))*cos(x(1))+10;
return

function [numbervar,lowerbound,upperbound]=bounds
%Defining number of control variables,
%lower and upper limits of control variables
%-------------------------------------------------------------
numbervar=2;
lowerbound(1:numbervar)=[-5 10];
upperbound(1:numbervar)=[0 15];
return
%=============================================================
```

4.11 Applications of GSA to power system problems—literature overview

Thanks to the excellent performance in finding the global optimum, the GSA has been applied in solving many problems in power engineering. Many authors have demonstrated the effectiveness of the GSA in solving different optimization problems, related to operation, control and planning of the power systems. This section gives an overview of use the GSA in solving some specific problems, such as optimal power flow (OPF), optimal reactive power dispatch (ORPD), economic dispatch (ED), optimal distributed generation (DG) placement and sizing, optimal coordination of overcurrent relays, etc.

4.11.1 Optimal power flow

The OPF problem solution aims to optimize a chosen objective function through optimal adjustment of the power system control variables while at the same time satisfying various operating constraints. The objective function can take different forms including minimization of total generating cost, minimization of active power losses, minimization of pollutant emissions, voltage profile improvement, voltage stability enhancement, etc. The set of control variables in the OPF problem consists of generator active power outputs, generator voltages, transformer tap settings, shunt Var compensations, etc. Mathematically, the OPF problem can be considered as a nonlinear, nonconvex, large-scale, static optimization problem with both continuous and discrete control variables. In recent years, many authors suggest GSA for solving the OPF problem.

One of the first applications of the GSA in solving the OPF problem is presented in [18]. The authors considered different objective functions such as the voltage stability enhancement in both nominal and contingency conditions, the fuel cost minimization including piecewise quadratic cost function, and cost function with valve point effect.

Bhattacharya and Roy [19] used GSA for solution of multiobjective OPF problems. The multiobjective OPF problem is formulated as simultaneous minimization of fuel cost, power loss and voltage deviation.

Sonmez *et al.* [20] applied GSA to solve the OPF considering flexible AC transmission systems (FACTS) devices. The authors conclude that the GSA has fast computing time and finds better results than other popular optimization techniques.

The authors [21] proposed an enhanced GSA for solving security constrained OPF. A new mutation operator is introduced in GSA for continuously generating of new masses throughout the solution space in order to enhance the exploration capability and avoiding being trapped in local minima. The solution approach consists of two levels: ED is solved to generate an initial solution, and security constrained OPF is solved to find global optimum.

The nondominated sorting multiobjective GSA is proposed and successfully applied in [14] for solving different multiobjective OPF problems. The authors used phase elitism nondominated sorting and crowding distance to locate and manage the Pareto-optimal solutions.

Radosavljević *et al.* [22,23] applied the hybrid PSOGSA algorithm for solving the OPF problem in power systems. Simulation results show that the hybridization of PSO and GSA algorithms provides effective and robust high-quality solution of the OPF problem.

4.11.2 Optimal reactive power dispatch

The ORPD can be considered as a special case of the OPF problem. The ORPD problem solution aims to minimize a chosen objective function, such as power loss or voltage deviation, through optimal adjustment of the power system control variables, such as generator voltages, transformer tap settings, shunt Var compensations, under specified active power outputs of all generators (except at the slack bus P_{Gslack}), while at the same time satisfying various operating constraints. Mathematically, the ORPD problem can be formulated as a nonlinear optimization problem with constraints. In recent years, several authors proposed GSA for solving the ORPD problem.

Similar to the case of the OPF problem, one of the first applications of GSA in solving the ORPD problem was proposed by Duman *et al.* [24,25]. Roy *et al.* [26] used GSA for solving multiobjective ORPD to simultaneously minimize the transmission losses and maximize the voltage stability margin. The efficiency of these approaches is proven through their application on the IEEE-30 bus, IEEE-57 bus, and IEEE-118 bus systems.

Niknam *et al.* [27] proposed opposition-based self-adaptive modified GSA for multiobjective ORPD and voltage control in power system operation. The proposed method was used to find the Pareto-optimal solutions for the multiobjective problem considering transmission losses, load voltage deviation, and voltage stability index.

Using the concept of opposition-based learning, the authors [7,8] introduced an OGSA and successfully applied this algorithm for the solution of ED problem [7] as well as for the solution of ORPD problem [8].

Radosavljević *et al.* used the hybrid PSOGSA algorithm [28] and proposed a hybridization of GSA with sequential quadratic programing method [29] to solve the ORPD problem.

4.11.3 Economic dispatch using GSA

The ED problem solution aims to minimize the cost of generation of electric power through optimal adjustment of the committed generating unit outputs, while at the same time satisfying all unit and system constraints. When the environmental concerns are combined with the ED, the problem becomes a combined economic and emission dispatch (CEED) problem. The environmental aspect adds complexity to the solution of the ED problem due to the nonlinear characteristics of the mathematical models used to represent emissions.

Duman *et al.* have contributed a significant amount of literature regarding to use of GSA in power system problems. In addition to the applications of GSA to OPF [18] and ORPD [24,25] problems, they used the GSA to solve ED problem considering valve-point effect [30], as well as for solution the CEED problems [31].

In [32], the CEED problem is solved by using GSA to minimize the fuel cost and emission of NO_x, considering both thermal generators and wind turbines.

In [33], quasioppositional GSA is proposed to solve nonlinear economic load dispatch optimization problems by considering valve-point loading effects.

Radosavljević [34] proposed and successfully applied the GSA for the CEED problem in power systems. Based on simulation results, the author concluded that the GSA provides effective and robust high-quality solution of the CEED problem.

Beigvand *et al.* [35] used GSA to solve the nonconvex combined heat and power ED problem. The effectiveness of the GSA in solving the problem is proven on several test systems with/without valve-point loading effect and network losses.

Hybrid PSOGSA algorithm provides effective, robust and high-quality solution of the CEED problem [36].

4.11.4 Optimal power flow in distribution networks

Due to the growth of DG penetration, the nature of distribution network is altered from a passive network to an active. Safe, reliable, and economic operation of the active distribution network can be achieved only by efficiently coordination of the DG units operation, voltage regulation (voltage regulators, Under-Load Tap Changer (ULTC) transformer), and reactive power compensation (Var compensators). In fact, it is an OPF problem.

In [37], researches employed the GSA to solve the OPF problem in distribution networks with DG units. In that article, the OPF is formulated as a nonlinear optimization problem considering the main operating constraints and characteristics of distribution network taking into account also the different DG units and load models.

Wang *et al.* [38] proposed an improved GSA for solving the optimal reactive power flow problem in distribution networks including the DGs with randomness power outputs. The authors used chaos optimization idea, inertia mass factor, memory, and group sharing mechanism to improve the original GSA in computation speed and convergence rate.

Black hole gravitational search algorithm, which is combined black hole algorithm with GSA, is proposed in [39] and applied to solve day-ahead reactive power optimization problem in distribution network.

4.11.5 Optimal DG placement and sizing in distribution networks

Impact of the DG units on distribution network operation mostly depends on his location, size, and type. In recent years, many researchers proposed GSA for solving this nonlinear optimization problem with both continuous and discrete control variables.

Usually, the main objective in optimal location and sizing of DG units are to minimize active power loss and improve voltage profile of distribution system. Mistry *et al.* [40] use GSA to achieve this objective.

The authors in [41,42] employed GSA along with PSO to find best location and size of multi-DG units in distribution system considering different aspects such as the total real power losses, the voltage profile, the megavolta ampere intake by the grid, the DG quantity and the greenhouse gases emission.

An improved GSA by introducing a chaotic gravitational constant is applied in [43] for determining the optimal placement and sizing of DG in a radial distribution system to minimize the total losses, total voltage harmonic distortion, and voltage deviation.

One more application of GSA to the optimal allocation and sizing of multi DG in distribution system for minimizing power loss is given in [44]. The authors concluded that the GSA has better performances in comparison to evolutionary programing.

Kundu and Chakrabarti [45] used GSA for optimal placement and sizing of DG units in order to improve nodal pricing and system voltage profiles.

Niknam *et al.* [46] proposed θ-self-adaptive GSA to achieve the optimal places for fuel cell power plants (FCPPs) in distribution networks and daily optimal active powers of distribution substation and FCPPs.

In addition to the implementation on the problem of optimal location and sizing of generators, the GSA is successfully applied in solving of the problems of optimal capacitor placement in radial distribution systems [47], as well as optimal placement of FACTS device in the power systems [48].

4.11.6 Optimal energy and operation management of microgrids

Microgrids (MGs) are local distribution networks comprising various DG units, controllable loads, and energy storage devices that can operate either interconnected or isolated from the main distribution grid as a controlled entity. The MGs represents the core of smart distribution networks within modern power systems. Generally, if properly planned and controlled, MGs may offer improved power quality, better economics and a reduced dependence on the local utility. The benefits can be achieved only through efficient coordinated operation and control of the generation units together with storage devices and controllable loads. This is a nonlinear optimization problem with constraints named optimal energy and operation management (EOM).

Niknam *et al.* [11] proposed a self-adaptive GSA for determining the optimal energy management of renewable MGs considering uncertainties in load demand, market prices, and the available electrical power of renewable DG units, such as wind turbines and photovoltaic systems.

Li *et al.* [49] applied an improved GSA based on opposition learning and elite strategy to solve a multiobjective EOM problem of MG. The multiobjective function consists of the fuel costs, operation and maintenance costs, and pollutant emissions of microsources.

Jamian *et al.* [50] used the GSA to optimize power outputs of the multiple DG units in autonomous distribution network in order to improve the voltage profile and reduce the power losses in the system.

BGSA is applied in [51] to optimize the operation scheduling of MG, which includes dynamic optimal unit commitment and power dispatch.

Authors in [52] presented a new application of the GSA for the optimal control of a wave energy conversion system under different operating conditions of the DC MG system.

4.11.7 Optimal coordination of overcurrent relays

Generally, a power system contains many sections, and each section must be protected against overcurrents. In order to ensure optimum supply reliability, it is important to obtain selectivity, i.e., to switch off the part of the power system where a fault occurs and to avoid switching off the parts of the system where no fault occurs. The goal in coordination of over-current relays is to find the appropriate time dial setting and pick up current setting of relays, considering various constraints, such as coordination and boundary limits. The coordination of over-current relays has been formulated as an optimization problem and solved with different optimization methods.

Recently, several authors proposed GSA for solving the problem of optimal relay coordination.

Tripathi *et al.* [53] proved that the GSA has better performances in solving the problem of optimal relay coordination in distribution systems with DG units than PSO algorithm.

In [54], GSA is implemented for optimal coordination of overcurrent relays for the protection of an offshore wind farm connected to onshore grid with VSC–HVDC.

Patel *et al.* [55] demonstrated a comparative study of protection coordination among various overcurrent relays using different optimization techniques, including GA, PSO, and GSA. Based on simulation studies on IEEE-34 bus benchmark distribution system with DG units, the authors concluded that the GSA gives the best results.

4.12 Conclusion

GSA is population-based metaheuristic optimization algorithm based on the law of gravity and the notion of mass interactions. In GSA, search agents are considered as objects and their performances are measured by their masses. All these objects attract each other by the gravity force, and this force causes a global movement of all objects toward the objects with heavier masses, i.e., toward the better solutions of the optimization problem.

Basic version of GSA is developed for continuous optimization problems. However, numerous modified versions of the basic GSA are adapted for solving continuous, discrete, and multiobjective optimization problems. Moreover, the GSA is suitable for hybridization with other metaheuristic algorithms and classical optimization methods which makes it very efficient and widely applicable in solving real optimization problems with different objective functions and control variables. Recently, many authors have applied GSA and its modified, adapted, and hybrid versions in solving complex optimization problems in the field of power engineering, such as OPF, ORPD, CEED, optimal coordination of overcurrent relays, etc. In addition to the GA and PSO, the GSA was established as one of most used metaheuristic optimization algorithm in all areas of engineering and science.

References

[1] Rashedi E., Nezamabadi-pour H., Saryazdi S. 'GSA: a gravitational search algorithm'. *Information Sciences*. 2009;**179**:2232–48

[2] Amoozegar M., Rashedi E. 'Parameter tuning of GSA using DOE'. *Proceedings of 4th International eConference on Computer and Knowledge Engineering*; Mashhad, Iran, Oct 2014. IEEE; 2014. pp. 1–6

[3] Rashedi E., Nezamabadi-pour H., Saryazdi S. 'BSGA: binary gravitational search algorithm'. *Natural Computing*. 2010;**9**:727–45

[4] Khajehzadeh M., Taha M.R., El-Shafie A., Eslami M. 'A modified gravitational search algorithm for slope stability analysis'. *Engineering Applications of Artificial Intelligence*. 2012;**25**(8):1589–97

[5] Tizhoosh H.R. 'Opposition-based learning: a new scheme for machine intelligence'. *Proceedings of International Conference on Computational Intelligence for Modelling, Control and Automation, and International Conference on Intelligent Agents, Web Technologies and Internet Commerce*; Vienna, Austria, Nov 2005. IEEE; 2006. pp. 695–701

[6] Seif Z., Amadi M.B. 'An opposition-based algorithm for function optimization'. *Engineering Applications of Artificial Intelligence*. 2015;**37**:293–306

[7] Shaw B., Mukherjee V., Ghoshal S.P. 'A novel opposition-based gravitational search algorithm for combined economic and emission dispatch problems of power systems'. *Electrical Power and Energy Systems*. 2012;**35**:21–33

[8] Shaw B., Mukherjee V., Ghoshal S.P. 'Solution of reactive power dispatch of power systems by an opposition-based gravitational search algorithm'. *Electrical Power and Energy Systems*. 2014;**55**:29–40

[9] Xu Q., Wang L., He B., Wang N. 'Modified opposition-based differential evolution for function optimization'. *Journal of Computational Information Systems*. 2011;**7**(5):1582–91

[10] Mirjalili S., Lewis A. 'Adaptive gbest-guided gravitational search algorithm'. *Neural Computation & Applications*. 2014;**25**(7):1569–84

[11] Niknam T., Golestaneh F., Malekpour A. 'Probabilistic energy and operation management of a microgrid containing wind/photovoltaic/fuel cell generation and energy storage devices based on point estimate method and self-adaptive gravitational search algorithm'. *Energy*. 2012;**43**:427–37

[12] Narimani M.R., Vehed A.A., Azizpanah-Abarghooee R. 'Enhanced gravitational search algorithm for multi-objective distribution feeder

reconfiguration considering reliability, loss and operational cost'. *IET Generation, Transmission & Distribution*. 2014;**8**(1):55–69

[13] Hadi N., Mahdi N., Patrick S. 'A multi objective gravitational search algorithm based on non-dominated sorting'. *International Journal of Swarm Intelligence Research*. 2012;**3**(3):32–49

[14] Bhowmik A.R., Chakraborty A.K. 'Solution of optimal power flow using nondominated sorting multi objective gravitational search algorithm'. *Electrical Power and Energy Systems*. 2014;**62**:323–34

[15] Deb K., Pratap A., Agarwal S., Meyarivan T. 'A fast and elitist multiobjective genetic algorithm: NSGA-II'. *IEEE Transactions on Evolutionary Computation*. 2002;**6**(2):182–97

[16] Shams M., Rashedi E., Hakimi A. 'Clustered-gravitational search algorithm and its application in parameters optimization of a low noise amplifier'. *Applied Mathematics and Computation*. 2015;**258**:436–53

[17] Mirjalili S., Hashim S.Z.M. 'A new hybrid PSOGSA algorithm for function optimization'. *Proceedings of International Conference on Computer and Information Application*; Tianjin, China, Dec 2010. IEEE; 2012. pp. 374–7

[18] Duman S., Guvenc U., Sonmez Y., Yorukeren N. 'Optimal power flow using gravitational search algorithm'. *Energy Conversion and Management*. 2012;**59**:86–95

[19] Bhattacharya A., Roy P.K. 'Solution of multi-objective optimal power flow using gravitational search algorithm'. *IET Generation, Transmission & Distribution*. 2012;**6**(8):751–63

[20] Sonmez Y., Duman S., Guvenc U., Yorukeren N. 'Optimal power flow incorporating FACTS devices using gravitational search algorithm'. *Proceedings of International Symposium on Innovations in Intelligent Systems and Applications*; Trabzon, Turkey, Jul 2012. IEEE; 2012. pp. 1–5

[21] Jahan M.S., Amjady N. 'Solution of large-scale security constrained optimal power flow by a new bi-level optimization approach based on enhanced gravitational search algorithm'. *IET Generation, Transmission & Distribution*. 2013;**7**(12):1481–91

[22] Radosavljević J., Klimenta D., Jevtić M., Arsić N. 'Optimal power flow using a hybrid optimization algorithm of particle swarm optimization and gravitational search algorithm'. *Electric Power Components and Systems*. 2015;**43**(17):1958–70

[23] Radosavljević J., Arsić N., Jevtić M. 'Optimal power flow using hybrid PSOGSA algorithm'. *Proceedings of 55th International Scientific Conference of Riga Technical University on Power and Electrical Engineering*; Riga, Latvia, Oct 2014. IEEE; 2014. pp. 136–40

[24] Duman S., Sonmez Y., Guvenc U., Yorukeren N. 'Application of gravitational search algorithm for optimal reactive power dispatch problem'. *Proceedings of International Symposium on INnovations in Intelligent SysTems and Applications*; Istanbul, Turkey, Jun 2011. IEEE; 2011. pp. 519–23

[25] Duman S., Sonmez Y., Guvenc U., Yorukeren N. 'Optimal reactive power dispatch using a gravitational search algorithm'. *IET Generation, Transmission & Distribution*. 2012;**6**(6):563–76
[26] Roy P.K., Mandal B., Bhattacharya K. 'Gravitational search algorithm based optimal reactive power dispatch for voltage stability enhancement'. *Electric Power Components and Systems*. 2012;**40**:965–76
[27] Niknam T., Narimani M.R., Azizipanah-Abarghooee R., Bahmani-Firouzi B. 'Multiobjective optimal reactive power dispatch and voltage control: a new opposition-based self-adaptive modified gravitational search algorithm'. *IEEE Systems Journal*. 2013;**7**(4):742–53
[28] Radosavljević J., Jevtić M., Milovanović M. 'A solution to the ORPD problem and critical analysis of the results'. *Electrical Engineering*. 2016; doi: 10.1007/s00202-016-0503-1
[29] Radosavljević J., Jevtić M. 'Solution of optimal reactive power dispatch by a hybrid GSA-SQP algorithm'. *Elektronika ir Elektrotechnika*. 2016;**22**(3):3–6
[30] Duman S., Guvenc U., Yorukeren N. 'Gravitational search algorithm for economic dispatch with valve-point effects'. *International Review of Electrical Engineering*. 2010;**5**(6):2890–5
[31] Guvenc U., Sonmez Y., Duman S., Yorukeren N. 'Combined economic and emission dispatch solution using gravitational search algorithm'. *Scientia Iranica Transactions D: Computer Science & Engineering and Electrical Engineering*. 2012;**19**(6):1754–62
[32] Mondal S., Bhattacharya A., Halder nee Dey S. 'Multi-objective economic emission load dispatch solution using gravitational search algorithm and considering wind power penetration'. *Electrical Power and Energy Systems*. 2013;**44**:282–92
[33] Bulbul S.M.A., Roy P.K. 'Quasi-oppositional gravitational search algorithm applied to complex economic load dispatch problem'. *Proceedings of 1st International Conference on Non Conventional Energy*; Kalyani, India, Jan 2014. IEEE; 2014. pp. 308–13
[34] Radosavljević J. 'Gravitational search algorithm for solving combined economic and emission dispatch'. *Proceedings of XIV International Symposium Infoteh-Jahorina*; Jahorina, Bosnia and Herzegovina, Mar 2015. pp. 148–53
[35] Beigvand S.D., Abdi H., Scala M.L. 'Combined heat and power economic dispatch problem using gravitational search algorithm'. *Electric Power Systems Research*. 2016;**133**:160–72
[36] Radosavljević J. 'A solution to the combined economic and emission dispatch using hybrid PSOGSA algorithm'. *Applied Artificial Intelligence*. 2016;**30**(5):445–74
[37] Radosavljević J., Jevtić M., Arsić N., Klimenta D. 'Optimal power flow for distribution networks using gravitational search algorithm'. *Electrical Engineering*. 2014;**96**:335–45
[38] Wang S., Wang N., Wang Q., Yang Y., Huang G. 'Reactive power optimization in distribution network by using improved gravitational search

algorithm'. *Proceedings of 5th International Conference on Electric Utility Deregulation and Restructuring and Power Technologies*; Changsha, China, Nov 2015. IEEE; 2015. pp. 178–82

[39] Pei H., Li Y., Liu K. 'A multi-object black hole gravitational search algorithm for day-ahead reactive optimization in distribution network'. *Proceedings of 2016 IEEE Chinese Guidance, Navigation and Control Conference*; Nanjing, China, Aug 2016. *IEEE*; 2016. pp. 901–6

[40] Mistry K., Bhavsar V., Roy R. 'GSA based optimal capacity and location determination of distributed generation in radial distribution system for loss minimization'. *Proceedings of 11th International Conference on Environment and Electrical Engineering; Venice*, Italy, May 2012. IEEE; 2012. pp. 1–6

[41] Tan W.S., Hassan M.Y., Rahman H.A., Abdullah P., Hussin F. 'Multi-distributed generation planning using hybrid particle swarm optimisation—gravitational search algorithm including voltage rise issue'. *IET Generation, Transmission & Distribution*. 2013;**7**(9):929–42

[42] Sharma D., Yadov N.K., Gunjan V.K., Bala A. 'Impact of distributed generation on voltage profile using different optimization techniques'. *Proceedings of International Conference on Control, Computing, Communication and Materials*; Allahabad, India, Oct 2016. IEEE; 2017. pp. 1–6

[43] Kadir A.F.A., Mohamed A., Shareef H., Ibrahim A.A., Khatib T., Elmenreich W. 'An improved gravitational search algorithm for optimal placement and sizing of renewable distributed generation units in a distribution system for power quality enhancement'. *Journal of Renewable and Sustainable Energy*. 2014;**6**(3):1–17

[44] Rasdi M.R.H.M., Musirin I., Hamid Z.A., Haris H.C.M. 'Gravitational search algorithm application in optimal allocation and sizing of multi distributed generation'. *Proceedings of 8th International Power Engineering and Optimization Conference*; Langkawi, Malaysia, Mar 2014. IEEE; 2014. pp. 364–8

[45] Kundu B., Chakrabarti A. 'OPF governed determination of location and size of distribution generators using gravitational search algorithm'. *Proceedings of 2nd International Conference on Control, Instrumentation, Energy & Communication*; Kolkata, India, Jan 2016. IEEE; 2016. pp. 446–50

[46] Niknam T., Bornapour M., Gheisari A. 'Combined heat, power and hydrogen production optimal planning of fuel cell power plants in distribution networks'. *Energy Conversion and Management*. 2013;**66**:11–25

[47] Shuaib Y.M., Kalavathi M.S., Rajan C.C.A. 'Optimal capacitor placement in radial distribution system using gravitational search algorithm'. *Electrical Power and Energy Systems*. 2015;**64**:384–97

[48] Packiasudha M., Suja S., Jerome J. 'A new cumulative gravitational search algorithm for optimal placement of FACT device to minimize system loss in the deregulated electrical power environment'. *Electrical Power and Energy Systems*. 2017;**84**:34–46

[49] Li P., Xu W., Zhou Z., Li R. 'Optimized operation of microgrid based on gravitational search algorithm'. *Proceedings of International Conference on Electrical Machines and Systems*; Busan, Korea, Oct 2013. IEEE; 2013. pp. 338–42

[50] Jamian J.J., Mustafa M.W., Mokhlis H., Baharudin M.A., Abdilahi A.M. 'Gravitational search algorithm for optimal distributed generation operation in autonomous network'. *Arabian Journal for Science and Engineering*. 2014;**39**:7183–8

[51] Li P., Zhou Z., Lin X., Yang X., Niu X. 'Dynamic optimal operation scheduling of microgrid using binary gravitational search algorithm'. *Proceedings of International Conference on Power System Technology*; Chengdu, China, Oct 2014. IEEE; 2014. pp. 3175–80

[52] Hasanien H.M., 'Gravitational search algorithm-based optimal control of Archimedes wave swing-based wave energy conversion system supplying a DC microgrid under uncertain dynamics'. *IET Renewable Power Generation*. 2017;**11**(6):763–70

[53] Tripathi J.M., Adhishree, Krishan R. 'Optimal coordination of overcurrent relays using gravitational search algorithm with DG penetration'. *Proceedings of 6th IEEE Power India International Conference*; Delhi, India, Dec 2014. IEEE; 2015. pp. 1–6

[54] Perveen R., Mohanty S.R., Kishor N. 'Optimal coordination of overcurrent relay for offshore wind farm integrated to onshore grid with VSC-HVDC'. *Proceedings of 6th IEEE Power India International Conference*; Delhi, India, Dec 2014. IEEE; 2015. pp. 1–6

[55] Patel A., Mondal S., Tiwari A.K., Choudhary N.K. 'Impact of distributed generation on coordination of overcurrent relays in a benchmark distribution system'. *1st IEEE International Conference on Power Electronics. Intelligent Control and Energy Systems*; Delhi, India, July 2016. IEEE; 2017. pp. 1–6

Chapter 5
Power-flow calculation

5.1 Introduction

The power system operation in a normal stationary state is based on the equation of the power balance. It means that the sum of power produced by the generators must at each moment be equal to the sum of the power of the consumers and the losses in the system. Without knowing the stationary state of the power system, any analysis is impossible related to system planning, system operation, and control actions in the system. In order to fully reconstruct the stationary state of the system, it is necessary to know the active and reactive powers in all branches of the network as well as complex voltages at all nodes of the network. For a given network configuration and power of loads, the parameters of stationary state of the power system can be determined by the power-flow calculation. Power-flow analysis is fundamental to the study of power systems.

The power-flow calculation forms the basic tool in the power system analysis. The power-flow calculation is used for different purposes including the power system planning, the power system control and monitoring, the contingency analysis, the state estimation, the dynamic simulations of power system (as basic or initial calculation), the power system optimization, etc. Actually, the power-flow calculation is the key step in solving optimization problems in power systems regardless to the optimization method.

In the power-flow study, the following data are known:

- network topology and parameters of system components,
- active and reactive powers of loads,
- active power outputs of generators,
- magnitude of voltage at generator buses, and
- magnitude and angle of voltage at slack bus.

The following data are determined in the power-flow calculation:

- magnitude and angle of voltage at load buses,
- angle of voltage at generator buses,
- active and reactive powers through all branches, and
- system power losses.

5.2 Power-flow calculation in transmission networks

In the power-flow calculation, the power system is assumed to be balanced. This allows a single-phase representation of the system. The single-phase scheme consists of single-phase equivalents of the power system components. The power-flow calculation is usually performed in the per unit (p.u.) system, so that all the system variables and parameters are expressed in the p.u. system. Transmission lines are modeled by equivalent π circuits with series admittances $\underline{y}_s = 1/\underline{z}_s = 1/(r_s + jx_s)$ and shunt capacitive admittances $\underline{y}_{sh} = jb_c$. Transformers with nominal turns ratio are modeled by series equivalent admittances $\underline{y}_t$ which correspond to equivalent impedances obtained by standard short-circuit tests $\underline{y}_t = 1/\underline{z}_t = 1/(r_t + jx_t)$. Transformers with off-nominal turns ratio (t) are represented by equivalent π circuit with admittances as follows:

first shunt admittance $\underline{y}_1 = \underline{y}_t \cdot (1 - t)/t^2$, series admittance $\underline{y}_2 = \underline{y}_t/t$, and second shunt admittance $\underline{y}_3 = \underline{y}_t \cdot (t - 1)/t$.

Generators are modeled by complex power injections at specified buses $\underline{S}_G = P_G + jQ_G$. Constant power loads are modeled as a specified quantity of active and reactive powers consumed at a bus $\underline{S}_L = P_L + jQ_L$. Shunt reactors and capacitors are modeled by corresponding admittance elements connected to ground. In this way, constant impedance loads can be modeled.

The first step in developing the mathematical model describing the power flow in the network is mathematical modeling of the network. Based on the single-line diagram of the power system, a bus admittance matrix can be formed. The bus admittance matrix $\underline{\mathbf{Y}}_{\text{bus}}$ represents mathematical model of the system in the power-flow study. The $\underline{\mathbf{Y}}_{\text{bus}}$ matrix is a $N \cdot N$ matrix (where N is the number of buses in the system) constructed from the admittances of the equivalent circuit elements making up the single-line diagram of the power system. There are some general and simple rules regarding the bus admittance matrix:

- The diagonal elements of the bus admittance matrix, $\underline{Y}_{ii}$, are determined as the sum of all the admittances terminating at bus i. Therefore, the admittance of elements connected between bus i and reference (ground) is added to the (i, i) entry of the admittance matrix. The admittance of elements connected between buses i and j is added to the (i, i) and (j, j) entries of the admittance matrix.
- Off-diagonal elements of the bus admittance matrix are negative of the sum of all connecting admittances between corresponding buses i and j. If there is no physical connection between the two buses (i and j), then $\underline{Y}_{ij} = 0$.
- The bus admittance matrix is symmetrical ($\underline{Y}_{ij} = \underline{Y}_{ji}$) if there are no phase-shifting transformers. It has weak diagonal dominance, and it is singular if there are no shunt branches to ground. The $\underline{\mathbf{Y}}_{\text{bus}}$ matrix is sparse with the degree of sparsity increasing with the network size.

5.2.1 Power-flow equations

The basic equation for power-flow analysis can be derived from the bus equations defining the relationships between network bus voltages and currents. Using the bus admittance matrix, the bus equations can be written as follows:

$$\begin{bmatrix} \underline{I}_1 \\ \underline{I}_2 \\ \cdots \\ \underline{I}_N \end{bmatrix} = \begin{bmatrix} \underline{Y}_{11} & \underline{Y}_{12} & \cdots & \underline{Y}_{1N} \\ \underline{Y}_{21} & \underline{Y}_{22} & \cdots & \underline{Y}_{2N} \\ \cdots & \cdots & \cdots & \cdots \\ \underline{Y}_{N1} & \underline{Y}_{N2} & \cdots & \underline{Y}_{NN} \end{bmatrix} \begin{bmatrix} \underline{V}_1 \\ \underline{V}_2 \\ \cdots \\ \underline{V}_N \end{bmatrix} \tag{5.1}$$

where N is the total number of buses, $\underline{Y}_{ij}$ are the elements of the bus admittance matrix, $\underline{V}_i$ are the bus voltages, and $\underline{I}_i$ are the currents injected at each bus, and $i = 1,2,\ldots,N$.

The effects of generators, loads, and other devices (e.g., reactive compensators) connected to the network buses are reflected in the bus current. Constant impedance loads are, however, included in the bus admittance matrix [1].

According to (5.1), the bus equation for the bus i can be written as

$$\underline{I}_i = \sum_{j=1}^{N} \underline{Y}_{ij}\underline{V}_j \tag{5.2}$$

The complex power injected at the bus i is expressed as follows:

$$\underline{S}_i = P_i + jQ_i = \underline{V}_i\underline{I}_i^* = \underline{V}_i \sum_{j=1}^{N} \underline{Y}_{ij}^*\underline{V}_j^* = V_i^2\underline{Y}_{ii}^* + \underline{V}_i \sum_{\substack{j=1 \\ j\neq i}}^{N} \underline{Y}_{ij}^*\underline{V}_j^* \tag{5.3}$$

On the other hand, it should be noted that the injected power at bus i is the difference between the power generation ($\underline{S}_{Gi}$) and the power consumption ($\underline{S}_{Li}$) at this bus:

$$\underline{S}_i = \underline{S}_{Gi} - \underline{S}_{Li} = (P_{Gi} - P_{Li}) + j(Q_{Gi} - Q_{Li}) = P_i + jQ_i \tag{5.4}$$

where P_i and Q_i are the active and reactive powers injected at bus i, respectively; P_{Gi} and Q_{Gi} are the active and reactive powers generated at bus i, respectively; P_{Li} and Q_{Li} are the active and reactive powers load at bus i,er respectively.

Expressing the complex elements of the bus admittance matrix in rectangular form ($\underline{Y}_{ij} = G_{ij} + jB_{ij}$), and the complex bus voltage in the polar form ($\underline{V}_i = V_i e^{j\theta}$), (5.3) becomes

$$\begin{aligned} \underline{S}_i &= V_i^2(G_{ii} - jB_{ii}) + V_i \sum_{j\in\alpha_i} V_j e^{j(\theta_i-\theta_j)}(G_{ij} - jB_{ij}) \\ &= V_i^2(G_{ii} - jB_{ii}) + V_i \sum_{j\in\alpha_i} V_j\left[\cos(\theta_i - \theta_j) + j\sin(\theta_i - \theta_j)\right](G_{ij} - jB_{ij}) \end{aligned} \tag{5.5}$$

Therefore, the active and reactive powers injected at bus i are as follows:

$$P_i = V_i^2 G_{ii} + V_i \sum_{\substack{j=1 \\ j \neq i}}^{N} V_j \left[G_{ij} \cos\left(\theta_i - \theta_j\right) + B_{ij} \sin\left(\theta_i - \theta_j\right) \right] \quad i = 1, 2, \ldots, N \tag{5.6}$$

$$Q_i = -V_i^2 B_{ii} + V_i \sum_{\substack{j=1 \\ j \neq i}}^{N} V_j \left[G_{ij} \sin\left(\theta_i - \theta_j\right) - B_{ij} \cos\left(\theta_i - \theta_j\right) \right] \quad i = 1, 2, \ldots, N \tag{5.7}$$

5.2.2 Bus classification

There are four variables that are associated with each bus: active power P, reactive power Q, voltage magnitude V, and voltage angle θ. In a system with N buses, there are $4N$ variables. Power-flow equations (5.6) and (5.7) form a system of $2N$ non-linear algebraic equations. So, $2N$ variables should be specified in order to make the system of equations solvable.

Accordingly, at each bus, two of the above four variables are specified and the remaining two are unknown variables. Depending upon which of the variables are specified, the system buses are classified:

- Load (PQ) bus: Active and reactive powers are specified. The unknown variables are the voltage magnitude and voltage angle. Load buses may contain generators with specified active and reactive power outputs. Generally, any bus with specified injected complex power can be considered as a load bus. The number of PQ buses is N_{PQ}.
- Voltage-controlled (PV) bus: Active power and voltage magnitude are specified. The unknown variables are the reactive power and voltage angle. The injected reactive power is limited depending on the characteristics of the individual source. A PV bus must have a variable source of reactive power such as generators, and reactive power compensators. The number of PV buses is N_{PV}.
- Slack (swing) bus: Voltage magnitude and phase angle are specified. The active and reactive powers are unknowns. Because the power losses in the system are not known a priori, at least one bus must have unspecified active and reactive powers. Thus, the slack bus is the only bus with known voltage. The bus selected as the slack bus must have a source of both real and reactive power, since the injected power at this bus must "swing" to take up the "slack" in the solution.

5.2.3 Solution methods

The power-flow calculation can be divided into two segments. In the first segment, the implicitly unknown variables such as the voltage angle of PV buses and the voltage magnitude and angle of PQ buses are determined by solving the nonlinear power-flow equations. In the second segment, after solving the implicitly unknowns, the explicitly unknowns such as the injected reactive power of PV buses and the active and reactive powers of slack bus are calculated directly from the corresponding (5.6) and (5.7). Finally, the power flows in each branch of the network as well as the power losses in the power system are calculated.

The solution of the nonlinear power-flow equations given by (5.3), i.e., (5.6) and (5.7) requires the use of iterative techniques for even the simplest power systems. A general procedure for calculating the power flow can be described in the following steps:

1. Specification of input data that implies defining the network topology (drawing the single-line diagram), and calculating the parameters network's elements in p.u. system.
2. Creation of the bus admittance matrix.
3. Bus classification.
4. Defining a convergence criteria.
5. Make an initial guess for the voltages (both magnitude and phase angle) at each bus in the system.
6. Substitute in the power-flow equations and determine the deviations from the solution.
7. Update the estimated voltages based on some commonly known numerical algorithms.
8. Repeat steps 6 and 7 until the convergence criteria is satisfied.

There are several methods for solving the power-flow equations, such as Gauss–Seidel, Newton–Raphson, and fast decoupled power-flow method.

5.2.3.1 Gauss–Seidel method

Equation (5.3) can be transformed as follows:

$$\frac{\underline{S}_i^*}{\underline{V}_i^*} = \frac{P_i - jQ_i}{\underline{V}_i^*} = \underline{Y}_{ii}\underline{V}_i + \sum_{\substack{j=1\\ j\neq i}}^{N} \underline{Y}_{ij}\underline{V}_j \tag{5.8}$$

From (5.8) the voltage $\underline{V}_i$ can be expressed as

$$\underline{V}_i = \frac{1}{\underline{Y}_{ii}}\left[\frac{\underline{S}_i^*}{\underline{V}_i^*} - \sum_{\substack{j=1\\ j\neq i}}^{N} \underline{Y}_{ij}\underline{V}_j\right] \tag{5.9}$$

Equation (5.9) is the essence of Gauss–Seidel method. The iterations begin with an initial guess of the magnitude and angle of voltages at all PQ buses, and of the voltage angle at all PV buses. For a PQ bus, active and reactive powers are known, and (5.9) is used to compute the voltage $\underline{V}_i$ at current iteration $(k+1)$ by using the best available voltages for all the buses, as follows:

$$\underline{V}_i^{(k+1)} = \frac{1}{\underline{Y}_{ii}} \left[\frac{\underline{S}_i^*}{\left(\underline{V}_i^{(k)}\right)^*} - \sum_{j=1}^{i-1} \underline{Y}_{ij} \underline{V}_j^{(k+1)} - \sum_{j=i+1}^{N} \underline{Y}_{ij} \underline{V}_j^{(k)} \right] \tag{5.10}$$

In case the bus i is a PV bus, the following procedure is applied [1]:

1. For the PV bus, the reactive power is not specified, so it must be calculated in each iteration:

$$Q_i = \mathrm{Im}\left\{ \underline{V}_i \sum_{j=1}^{N} \underline{Y}_{ij}^* \underline{V}_j^* \right\} \tag{5.11}$$

   where Q_i is calculated by using the best available values of bus voltages. If the Q_i is within the limits $Q_{i,\min}$ and $Q_{i,\max}$, it is used in (5.10) to compute the updated value of $\underline{V}_i$.

   Its real and imaginary components are multiplied by the ratio of the specified value of magnitude of the PV bus voltage to the magnitude of its updated value, thus complying with the magnitude constant. In other words, the magnitude of the voltage is forced to be the specified value and the angle of the voltage is computed using (5.10) and (5.11).
2. If Q_i computed by (5.11) exceeds either the minimum or maximum limit, it is set equal to the limit. The updated value of $\underline{V}_i$ is computed by treating the PV bus as a PQ bus.

The procedure is continued until the real and imaginary components of voltages at each bus computed by successive iterations converge to a specified tolerance.

5.2.3.2 Newton–Raphson method

Let us consider the following nonlinear equation:

$$f(x) = 0 \tag{5.12}$$

The Newton–Raphson solution method is based on sequential linearization of the nonlinear equation using Taylor's theorem. The solution procedure for this scalar nonlinear function can be described in the following steps:

1. Make an initial guess of x, $x^{(0)}$.
2. Represent $f(x)$ by a Taylor series about $x^{(k)}$

$$f(x) = f\left(x^{(k)}\right) + \left.\frac{df(x)}{dx}\right|_{x^{(k)}} \cdot \left(x - x^{(k)}\right) + \frac{1}{2}\left.\frac{d^2 f(x)}{dx^2}\right|_{x^{(k)}} \cdot \left(x - x^{(k)}\right)^2 + \text{higher order terms} \tag{5.13}$$

   where $k = 0,1,2,\ldots$

3. Approximate $f(x)$ by neglecting all terms except the first two:

$$f(x) = f\left(x^{(k)}\right) + \left.\frac{df(x)}{dx}\right|_{x^{(k)}} \cdot \left(x - x^{(k)}\right) = 0 \tag{5.14}$$

4. Use this linear approximation to determine $\Delta x^{(k)} = x - x^{(k)}$

$$\Delta x^{(k)} = -\left[\left.\frac{df(x)}{dx}\right|_{x^{(k)}}\right]^{-1} f\left(x^{(k)}\right) \tag{5.15}$$

5. Compute the value of x at iteration $(k+1)$:

$$x^{(k+1)} = x^{(k)} + \Delta x^{(k)} \tag{5.16}$$

6. The process is repeated until the error $\Delta f = \left|f\left(x^{(k+1)}\right) - f\left(x^{(k)}\right)\right|$ is less than a specified tolerance.

This procedure can be generalized for a set of nonlinear equations with N unknowns:

$$\begin{aligned} &f_1(x_1, x_2, \ldots, x_N) = 0 \\ &f_2(x_1, x_2, \ldots, x_N) = 0 \\ &\vdots \\ &f_N(x_1, x_2, \ldots, x_N) = 0 \end{aligned} \tag{5.17}$$

The set of equation (5.17) can be represented in matrix form:

$$\mathbf{f}(\mathbf{x}) = 0 \tag{5.18}$$

By analogy with the procedure for scalar equations, the Newton–Raphson iterative procedure for calculating the unknowns is as follows:

$$\Delta\mathbf{x}^{(k)} = -\left[\mathbf{J}\left(\mathbf{x}^{(k)}\right)\right]^{-1} \cdot \mathbf{f}\left(\mathbf{x}^{(k)}\right) \tag{5.19}$$

$$\mathbf{x}^{(k+1)} = \mathbf{x}^{(k)} + \Delta\mathbf{x}^{(k)} \tag{5.20}$$

where $\mathbf{x} = [x_1\ x_2\ \ldots\ x_N]^T$, and $\mathbf{f}(\mathbf{x}) = [f_1(\mathbf{x})\, f_2(\mathbf{x})\ \ldots\ f_N(\mathbf{x})]^T$. The superscript T denotes the transposition operator.

J is the Jacobian matrix, which is defined as follows:

$$\mathbf{J}\left(\mathbf{x}^{(k)}\right) = \left.\begin{bmatrix} \dfrac{\partial f_1(\mathbf{x})}{\partial x_1} & \dfrac{\partial f_1(\mathbf{x})}{\partial x_2} & \cdots & \dfrac{\partial f_1(\mathbf{x})}{\partial x_N} \\ \dfrac{\partial f_2(\mathbf{x})}{\partial x_1} & \dfrac{\partial f_2(\mathbf{x})}{\partial x_2} & \cdots & \dfrac{\partial f_2(\mathbf{x})}{\partial x_N} \\ \vdots & \vdots & \ddots & \vdots \\ \dfrac{\partial f_N(\mathbf{x})}{\partial x_1} & \dfrac{\partial f_N(\mathbf{x})}{\partial x_1} & \cdots & \dfrac{\partial f_N(\mathbf{x})}{\partial x_N} \end{bmatrix}\right|_{\mathbf{x}=\mathbf{x}^{(k)}} \tag{5.21}$$

For the power-flow problem, the nonlinear function $\mathbf{f}(\mathbf{x})$ takes the following form:

$$\mathbf{f}(\mathbf{x}) = \begin{bmatrix} \Delta\mathbf{P}(\mathbf{x}) \\ \Delta\mathbf{Q}(\mathbf{x}) \end{bmatrix} \tag{5.22}$$

where $\Delta\mathbf{P}(\mathbf{x})$ is a subvector consisting of nonlinear equations representing debalans of active powers in PQ and PV buses of the power system:

$$\begin{aligned} \Delta P_i(\mathbf{x}) = {} & P_{Gi}^{sp} - P_{Di}^{sp} - V_i^2 \cdot G_{ii} - V_i \cdot \sum_{j \in \alpha_i} V_j \cdot \left[G_{ij} \cdot \cos\left(\theta_i - \theta_j\right) \right. \\ & \left. + B_{ij} \cdot \sin\left(\theta_i - \theta_j\right) \right] = 0 \end{aligned} \tag{5.23}$$

$\Delta\mathbf{Q}(\mathbf{x})$ is a subvector consisting of nonlinear equations representing debalans of reactive powers in PQ buses of the power system:

$$\begin{aligned} \Delta Q_i(\mathbf{x}) = {} & Q_{Gi}^{sp} - Q_{Di}^{sp} + V_i^2 \cdot B_{ii} - V_i \cdot \sum_{j \in \alpha_i} V_j \cdot \left[G_{ij} \cdot \sin\left(\theta_i - \theta_j\right) \right. \\ & \left. - B_{ij} \cdot \cos\left(\theta_i - \theta_j\right) \right] = 0 \end{aligned} \tag{5.24}$$

and P_{Gi}^{sp}, Q_{Gi}^{sp} and P_{Di}^{sp}, Q_{Gi}^{sp} are specified generated and load powers, respectively; α_i is a set of bus indexes which are connected to bus i.

The vector $\mathbf{x}$ consists of

$$\mathbf{x} = \begin{bmatrix} \boldsymbol{\theta} \\ \mathbf{V} \end{bmatrix} \tag{5.25}$$

where $\boldsymbol{\theta}$ is a subvector of voltage angles at PQ and PV buses, and $\mathbf{V}$ is a subvector of voltage magnitudes at PQ buses.

The Jacobian matrix for the power-flow problem takes the following form:

$$\mathbf{J}(\mathbf{x}) = \begin{bmatrix} \mathbf{J}_{11} & \mathbf{J}_{12} \\ \mathbf{J}_{21} & \mathbf{J}_{22} \end{bmatrix} = \begin{bmatrix} \dfrac{\partial \Delta\mathbf{P}}{\partial \boldsymbol{\theta}} & \dfrac{\partial \Delta\mathbf{P}}{\partial \mathbf{V}} \\ \dfrac{\partial \Delta\mathbf{Q}}{\partial \boldsymbol{\theta}} & \dfrac{\partial \Delta\mathbf{Q}}{\partial \mathbf{V}} \end{bmatrix} \tag{5.26}$$

The elements of submatrix $\mathbf{J}_{11}$ are

$$\begin{aligned} \frac{\partial \Delta P_i}{\partial \theta_i} &= V_i \sum_{j \in \alpha_i} V_j \left[G_{ij} \sin\left(\theta_i - \theta_j\right) - B_{ij} \cos\left(\theta_i - \theta_j\right) \right] \\ \frac{\partial \Delta P_i}{\partial \theta_j} &= -V_i V_j \left[G_{ij} \sin\left(\theta_i - \theta_j\right) - B_{ij} \cos\left(\theta_i - \theta_j\right) \right] \qquad j \neq i \end{aligned} \tag{5.27}$$

The elements of submatrix $\mathbf{J}_{12}$ are

$$\begin{aligned} \frac{\partial \Delta P_i}{\partial V_i} &= -2 V_i G_{ii} - \sum_{j \in \alpha_i} V_j \left[G_{ij} \cos\left(\theta_i - \theta_j\right) + B_{ij} \sin\left(\theta_i - \theta_j\right) \right] \\ \frac{\partial \Delta P_i}{\partial V_j} &= -V_i \left[G_{ij} \cos\left(\theta_i - \theta_j\right) + B_{ij} \sin\left(\theta_i - \theta_j\right) \right] \qquad j \neq i \end{aligned} \tag{5.28}$$

The elements of submatrix $\mathbf{J}_{21}$ are

$$\begin{aligned}\frac{\partial \Delta Q_i}{\partial \theta_i} &= -\sum_{j \in \alpha_i} V_i V_j \left[G_{ij} \cos\left(\theta_i - \theta_j\right) + B_{ij} \sin\left(\theta_i - \theta_j\right)\right] \\ \frac{\partial \Delta Q_i}{\partial \theta_j} &= V_i V_j \left[G_{ij} \cos\left(\theta_i - \theta_j\right) + B_{ij} \sin\left(\theta_i - \theta_j\right)\right] \qquad j \neq i\end{aligned} \tag{5.29}$$

The elements of submatrix $\mathbf{J}_{22}$ are

$$\begin{aligned}\frac{\partial \Delta Q_i}{\partial V_i} &= 2V_i B_{ii} - \sum_{j \in \alpha_i} V_j \left[G_{ij} \sin\left(\theta_i - \theta_j\right) - B_{ij} \cos\left(\theta_i - \theta_j\right)\right] \\ \frac{\partial \Delta Q_i}{\partial V_j} &= -V_i \left[G_{ij} \sin\left(\theta_i - \theta_j\right) - B_{ij} \cos\left(\theta_i - \theta_j\right)\right] \qquad j \neq i\end{aligned} \tag{5.30}$$

Finally, according to (5.19) and (5.20), the application of the Newton–Raphson method to power-flow solution can be generalized as follows:

Set $k = 0$; make an initial guess of $\mathbf{x}$, $\mathbf{x}^{(k)} = [\boldsymbol{\theta}^{(k)}\ \mathbf{V}^{(k)}]^T$

While $\|\Delta\mathbf{P}(\mathbf{x}^{(k)})\| > \varepsilon_p$ and $\|\Delta\mathbf{Q}(\mathbf{x}^{(k)})\| > \varepsilon_q$ Do

$$\begin{bmatrix}\boldsymbol{\theta} \\ \mathbf{V}\end{bmatrix}^{(k+1)} = \begin{bmatrix}\boldsymbol{\theta} \\ \mathbf{V}\end{bmatrix}^{(k)} - \left(\left.\begin{bmatrix}\dfrac{\partial \Delta\mathbf{P}}{\partial \boldsymbol{\theta}} & \dfrac{\partial \Delta\mathbf{P}}{\partial \mathbf{V}} \\ \dfrac{\partial \Delta\mathbf{Q}}{\partial \boldsymbol{\theta}} & \dfrac{\partial \Delta\mathbf{Q}}{\partial \mathbf{V}}\end{bmatrix}\right|^{(k)}\right)^{-1} \cdot \begin{bmatrix}\Delta\mathbf{P} \\ \Delta\mathbf{Q}\end{bmatrix}^{(k)} \tag{5.31}$$

$k = k + 1$

End

5.2.3.3 Fast decoupled power-flow method—Stott method

The fast decoupled power-flow method was developed by Stott and Alsac [2]. This method is based on some approximations which simplify the power-flow problem. Equation (5.31) may be written as

$$\begin{bmatrix}\Delta\mathbf{P} \\ \Delta\mathbf{Q}\end{bmatrix} = -\begin{bmatrix}\dfrac{\partial \Delta\mathbf{P}}{\partial \boldsymbol{\theta}} & \dfrac{\partial \Delta\mathbf{P}}{\partial \mathbf{V}} \\ \dfrac{\partial \Delta\mathbf{Q}}{\partial \boldsymbol{\theta}} & \dfrac{\partial \Delta\mathbf{Q}}{\partial \mathbf{V}}\end{bmatrix} \begin{bmatrix}\Delta\boldsymbol{\theta} \\ \Delta\mathbf{V}\end{bmatrix} \tag{5.32}$$

Considering the physical weak coupling between P and V, and Q and θ the coupling submatrices $\partial\Delta\mathbf{P}/\partial\mathbf{V}$ and $\partial\Delta\mathbf{Q}/\partial\boldsymbol{\theta}$ can be neglected, transforming (5.32) in two separate equations:

$$\Delta\mathbf{P} = -\frac{\partial \Delta\mathbf{P}}{\partial \boldsymbol{\theta}} \Delta\boldsymbol{\theta} \tag{5.33}$$

$$\Delta\mathbf{Q} = -\frac{\partial \Delta\mathbf{Q}}{\partial \mathbf{V}} \Delta\mathbf{V} \tag{5.34}$$

Further simplifications can be made by recognizing that in practical power systems, the following approximations are valid: $\cos\left(\theta_i - \theta_j\right) \approx 1$; $G_{ij} \sin\left(\theta_i - \theta_j\right) \ll B_{ij}$; $Q_i \ll B_{ii} V_i^2$.

In addition, the following simplifications are used to calculate the elements of submatrices $\partial \Delta \mathbf{P} / \partial \boldsymbol{\theta}$ and $\partial \Delta \mathbf{Q} / \partial \mathbf{V}$: (i) the series resistance and shunt reactance of the network elements are neglected; (ii) all transformers have nominal transformation ratio, i.e., off-nominal ratio transformer taps are omitted; and (iii) the voltage magnitude at all buses is set to 1.0 p.u.

In this way, the submatrices $\partial \Delta \mathbf{P} / \partial \boldsymbol{\theta}$ and $\partial \Delta \mathbf{Q} / \partial \mathbf{V}$ are reduced to the constant matrices, and (5.33) and (5.34) become

$$\frac{\Delta \mathbf{P}}{\mathbf{V}} = -\mathbf{B}' \Delta \boldsymbol{\theta} \tag{5.35}$$

$$\frac{\Delta \mathbf{Q}}{\mathbf{V}} = -\mathbf{B}'' \Delta \mathbf{V} \tag{5.36}$$

where $\mathbf{B}'$ is a reduced susceptance matrix obtained from the network susceptance matrix $\mathbf{B}$ by removing the row and column which corresponds to the slack bus, $\mathbf{B}''$ is a reduced susceptance matrix obtained from the network susceptance matrix $\mathbf{B}$ by removing the rows and columns which corresponds to the slack bus and PV buses. It is clear that the network susceptance matrix $\mathbf{B}$ is the imaginary part of the bus admittance matrix $\underline{\mathbf{Y}}_{\text{bus}}$.

Finally, the fast decoupled (Stott–Alsac) power-flow method can be summarized as follows:

Set $k = 0$; make an initial guess of $\boldsymbol{\theta}$ and $\mathbf{V}$

While $\|\Delta \mathbf{P}(\mathbf{x}^{(k)})\| > \varepsilon_p$ and $\|\Delta \mathbf{Q}(\mathbf{x}^{(k)})\| > \varepsilon_q$ Do

$$\Delta \mathbf{P}^{(k)} = \Delta \mathbf{P}\left(\boldsymbol{\theta}^{(k)}, \mathbf{V}^{(k)}\right) \tag{5.37}$$

$$\boldsymbol{\theta}^{(k+1)} = \boldsymbol{\theta}^{(k)} - [\mathbf{B}']^{-1} \frac{\Delta \mathbf{P}^{(k)}}{\mathbf{V}^{(k)}} \quad \text{(for all buses except slack bus)} \tag{5.38}$$

$$\Delta \mathbf{Q}^{(k)} = \Delta \mathbf{Q}\left(\boldsymbol{\theta}^{(k+1)}, \mathbf{V}^{(k)}\right) \tag{5.39}$$

$$\mathbf{V}^{(k+1)} = \mathbf{V}^{(k)} - [\mathbf{B}'']^{-1} \frac{\Delta \mathbf{Q}^{(k)}}{\mathbf{V}^{(k)}} \quad \text{(for PQ buses)} \tag{5.40}$$

$k = k + 1$

End

5.2.4 Power-flow software—pfgui

The above power-flow computation methods were implemented through an educational MATLAB® GUI-based software package—*pfgui*. The program is available at [5].

The aim of this program is to encompass the main steps in solving the power-flow problem. These steps include: (i) selection of test system, and edit system data; (ii) selection of solution method; (iii) program execution; (iv) displaying the results; and (v) analysis of the results.

Figure 5.1 shows the main window of *pfgui* while it operates on a Windows-based computer. As can be seen from the figure, the *pfgui* is organized in following components:

1. Panel TEST SYSTEM. It consists of one pop-up menu and one button. Pop-up menu enables selection of a test system to be considered. It is available in 11 test systems; seven of which are standard IEEE test systems such as IEEE 9-bus, 14-bus, 30-bus, 39-bus, 57-bus, and 118-bus test systems. The button System Data opens the input file of the selected test system. The input file contains branch data and bus data of the selected test system. The data are given in matrix form. The user can change the data and save those modifications.
2. In the panel METHOD, the user selects the solution method. It is available as three standard power-flow methods: Newton–Raphson method, fast decoupled Stott method, and Gauss–Seidel method.
3. After selection of appropriate options on the panels TEST SYSTEM and METHOD, the user can start the program by clicking the button RUN power flow.
4. Panel RESULTS. It manages the access to the different segments of the power flow results generated from the program. The summary of results, such as the accuracy, number of iterations, total power of load, total power of generation, total power loss, maximum voltage, minimum voltage etc., are displayed immediately after execution of the program (Figure 5.1). Besides, the program generates other important results, also. When clicking the button Bus Voltages & Powers, the calculated values of bus voltages and bus powers will be displayed, as shown in Figure 5.2. The button Branch Power Flow enables to display the calculated branch power flows and power losses in the system (Figure 5.3). The system voltage profile is obtained by clicking the button Voltage Profile, as shown in Figure 5.4.

5.3 Power-flow calculation in distribution networks

Power-flow calculation is a function that creates a "image of the stationary state" of a distribution network at a given moment. On this "image," the powers and voltages at all nodes, power flow through all branches must be clearly seen, as well as power losses on all elements of the network. Power-flow calculation is one of the most important and most used functions in the analysis of distribution networks. The basic tool in solving different optimization problems in distribution networks, such as the OPF problem, Volt/Var control problem, optimal sitting and sizing of distributed generation (DG) units, etc., is the power-flow computation, regardless to optimization method that was used.

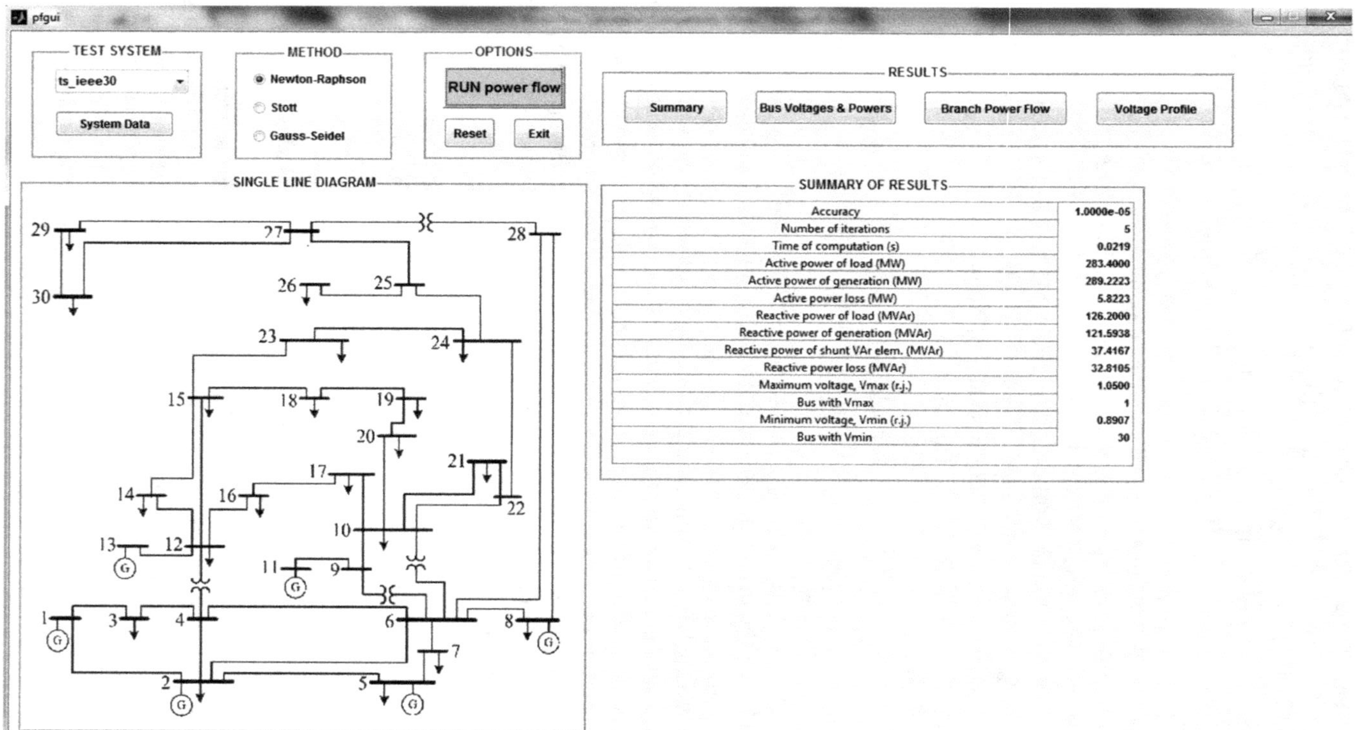

Figure 5.1 Main window of the pfgui program

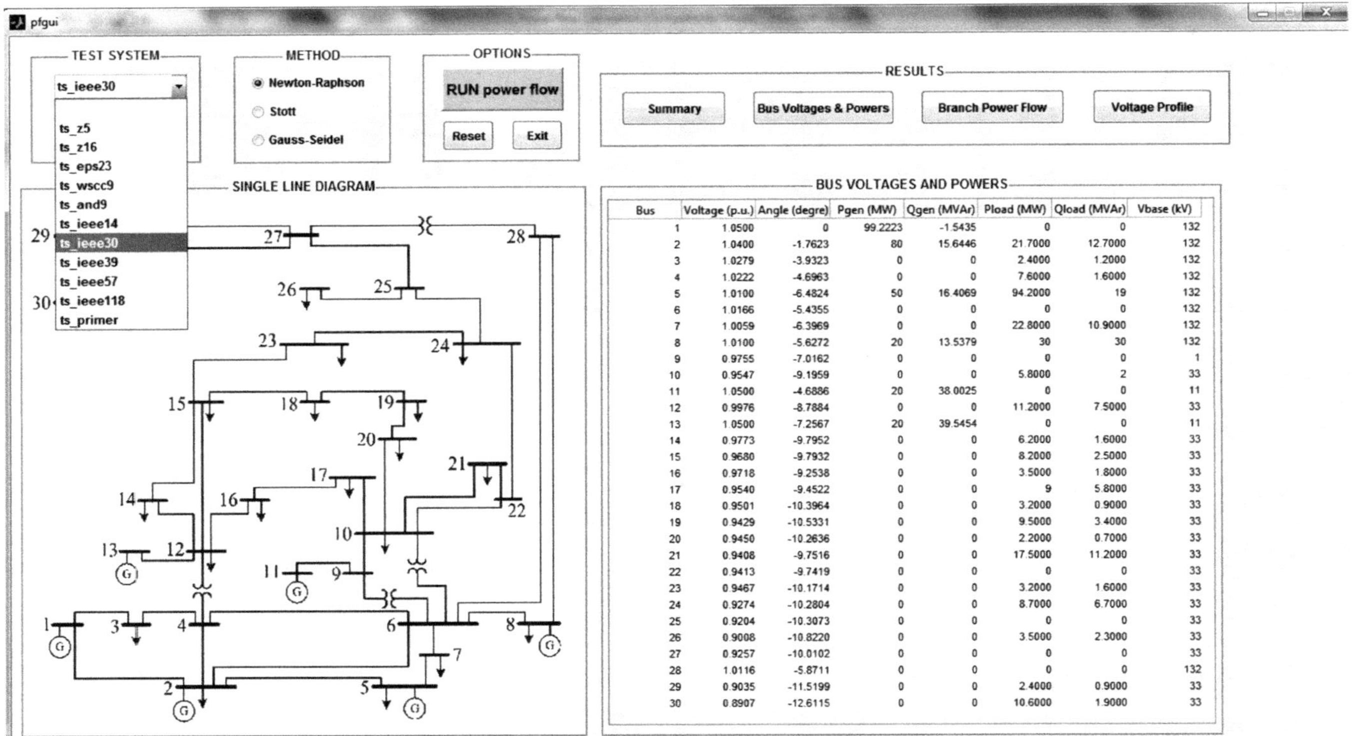

Bus	Voltage (p.u.)	Angle (degre)	Pgen (MW)	Qgen (MVAr)	Pload (MW)	Qload (MVAr)	Vbase (kV)
1	1.0500	0	99.2223	-1.5435	0	0	132
2	1.0400	-1.7623	80	15.6446	21.7000	12.7000	132
3	1.0279	-3.9323	0	0	2.4000	1.2000	132
4	1.0222	-4.6963	0	0	7.6000	1.6000	132
5	1.0100	-6.4824	50	16.4069	94.2000	19	132
6	1.0166	-5.4355	0	0	0	0	132
7	1.0059	-6.3969	0	0	22.8000	10.9000	132
8	1.0100	-5.6272	20	13.5379	30	30	132
9	0.9755	-7.0162	0	0	0	0	1
10	0.9547	-9.1959	0	0	5.8000	2	33
11	1.0500	-4.6886	20	38.0025	0	0	11
12	0.9976	-8.7884	0	0	11.2000	7.5000	33
13	1.0500	-7.2567	20	39.5454	0	0	11
14	0.9773	-9.7952	0	0	6.2000	1.6000	33
15	0.9680	-9.7932	0	0	8.2000	2.5000	33
16	0.9718	-9.2538	0	0	3.5000	1.8000	33
17	0.9540	-9.4522	0	0	9	5.8000	33
18	0.9501	-10.3964	0	0	3.2000	0.9000	33
19	0.9429	-10.5331	0	0	9.5000	3.4000	33
20	0.9450	-10.2636	0	0	2.2000	0.7000	33
21	0.9408	-9.7516	0	0	17.5000	11.2000	33
22	0.9413	-9.7419	0	0	0	0	33
23	0.9467	-10.1714	0	0	3.2000	1.6000	33
24	0.9274	-10.2804	0	0	8.7000	6.7000	33
25	0.9204	-10.3073	0	0	0	0	33
26	0.9008	-10.8220	0	0	3.5000	2.3000	33
27	0.9257	-10.0102	0	0	0	0	33
28	1.0116	-5.8711	0	0	0	0	132
29	0.9035	-11.5199	0	0	2.4000	0.9000	33
30	0.8907	-12.6115	0	0	10.6000	1.9000	33

Figure 5.2 Display of the pfgui results: bus voltages and powers

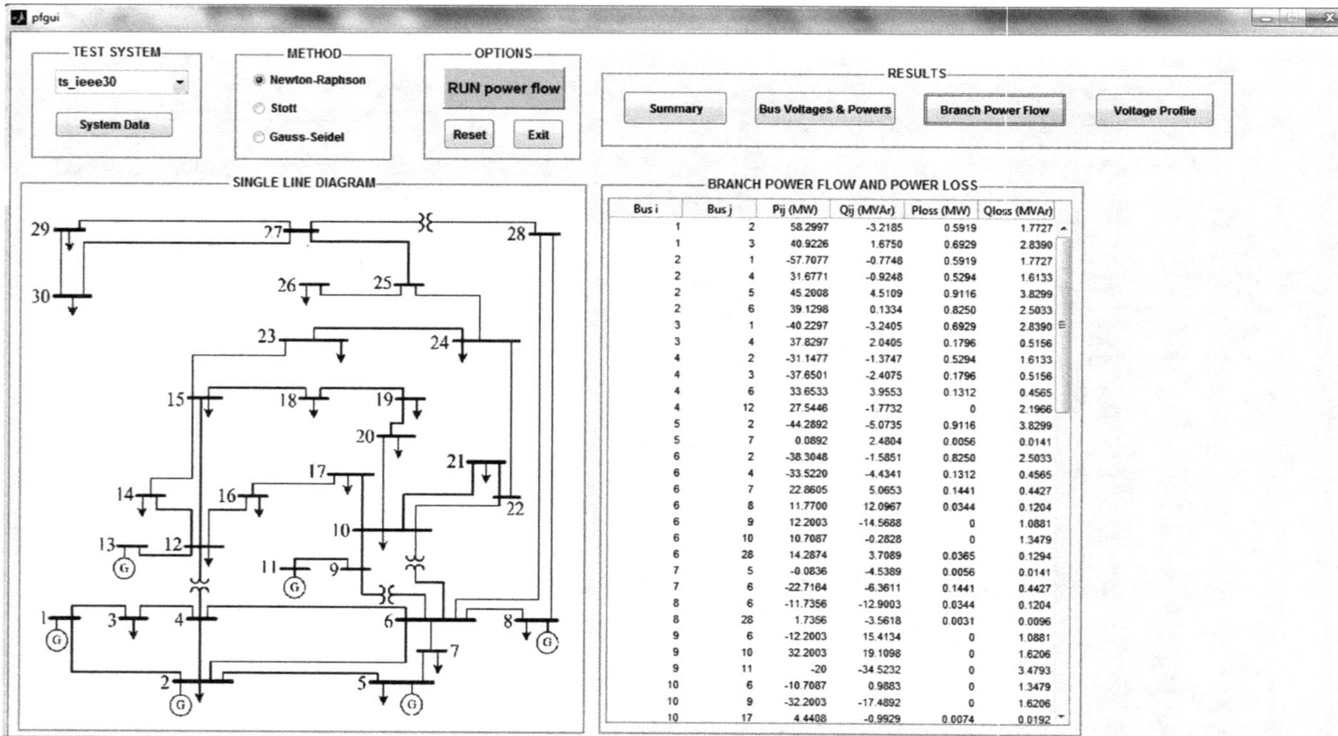

Figure 5.3 Display of the pfgui results: branch power flows and losses

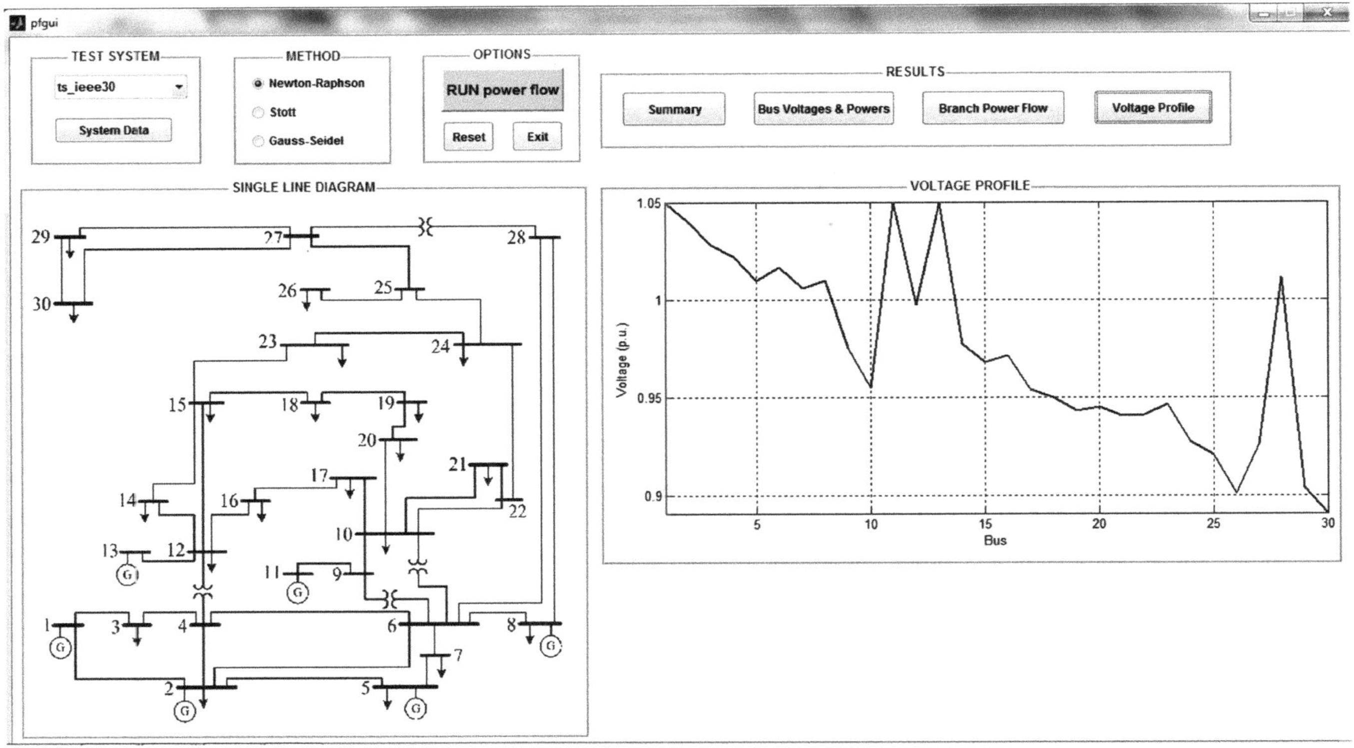

Figure 5.4 Display of the pfgui results: voltage profile

It is well known that distribution networks have certain specifics in relation to transmission networks. These are the radial structure, the high ratio r/x of the network elements, the large number of buses, as well as the asymmetry of network components and the unbalance of the loads. These properties which make most of the classic iterative methods for calculating power flows in transmission networks (Newton–Raphson, and others) are ineffective in solving the power-flow problem in distribution networks. Starting from the distribution networks properties, specialized methods have been developed for the power-flow calculation in distribution networks, including DG units. These methods, often referred to as *backward/ forward sweep*, are branch-oriented methods based on the application of Kirchhoff's laws and iterative passage along the branch of the distribution network.

5.3.1 Backward/forward sweep power-flow algorithm

Initial development of such a power flow for radial distribution networks was addressed in [3]. This method involves two sweeps of calculations. In the forward sweep, the end voltages are initialized for the first iteration, and currents are calculated starting at the buses at the load end of the radial branch and solved up to the source bus by applying the current summation method. The backward sweep starts at the source bus and calculates voltages using the current calculated from the forward sweep until the load end of the radial branches. The voltages from the backward sweep are used for the next iteration in the forward sweep calculations [4].

The power-flow calculation for a radial distribution network (Figure 5.5) using backward/forward sweep method is carried out through the following iterative steps:

Step 1. Read input data: Define network configuration, parameters (r, x, b) of network elements, load data, and DG units data;

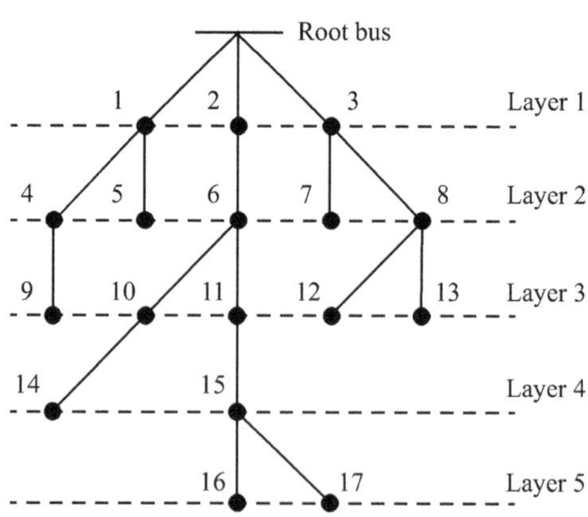

Figure 5.5 A typical radial configuration of distribution network

Step 2. Initialization: Set the voltage at the root bus and assuming a flat profile for the initial voltages at all other network buses;

$$V_i^{(0)} = 1; \qquad \theta_i^{(0)} = 0; \quad i = 1, \ldots, NB \tag{5.41}$$

where V_i and θ_i are the magnitude and angle of voltage at the bus i, NB is the total number of buses.

Step 3. Backward sweep: At iteration k, starting from the branch in the last layer (branch corresponding to last bus N_{bus}) and moving toward the branches connected to the root bus, the current in branch i, $\underline{J}_i$, according to Figure 5.6, is calculated as follows:

$$\underline{J}_i^{(k)} = \underline{I}_{Li}^{(k)} + \underline{I}_{Ci}^{(k)} + \underline{Y}_i^o \cdot \underline{V}_i^{(k-1)} + \sum_{\substack{\ell \in \alpha_{\ell i} \\ \ell \neq i}} \underline{J}_\ell^{(k)}$$
$$i = NB,\ NB-1, \ldots, 0; \quad k = 1, 2, \ldots \tag{5.42}$$

where k is the index of iteration; $\underline{I}_{Li}$ is the complex current of load at bus i; $\underline{I}_{Ci}$ is the complex current of a Var compensator at bus i; $\underline{J}_\ell$ is the complex current in branch ℓ emanating from bus i; $\underline{Y}_i^o$ is the shunt admittance at bus i; $\alpha_{\ell i}$ is the set of buses connected with bus i; $\underline{V}_i$ is the complex voltage at bus i. For a specified power of load, $\underline{S}_{Li} = P_{Li} + jQ_{Li}$, the current of load $\underline{I}_{Li}$ is calculated in accordance with load type, as follows:

$$\text{Constant power type (PQ):} \quad \underline{I}_{Li}^{(k)} = \left(\frac{\underline{S}_{Li}}{\underline{V}_i^{(k-1)}} \right)^* \tag{5.43}$$

$$\text{Constant current type (I):} \quad \underline{I}_{Li}^{(k)} = \left| \frac{\underline{S}_{Li}}{V_{\text{nom}}} \right| \cdot e^{j\left(\theta^{(k)} - \phi\right)} \tag{5.44}$$

where V_{nom} is the nominal voltage; $\delta^{(k)}$ is the voltage angle at bus i in kth iteration; and φ is the phase angle of the complex power $\underline{S}_{Li}$.

$$\text{Constant impedance (admittance) type (Z):} \quad \underline{I}_{Li}^{(k)} = \underline{V}_i^{(k-1)} \cdot \frac{\underline{S}_{Li}^*}{V_{\text{nom}}^2} \tag{5.45}$$

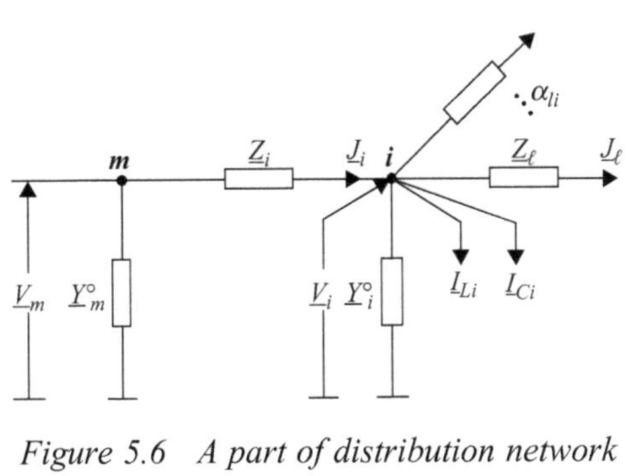

Figure 5.6 A part of distribution network

The current $\underline{I}_{Ci}^{(k)}$ in (5.42) is calculated in an analogous way.

Step 4. Forward sweep: Bus voltages are updated in a forward sweep starting from branches connected to the root bus toward those in the last. For each branch, i, the voltage at bus i is calculated using the updated voltage at bus m (see Figure 5.6) and the branch current calculated in the preceding backward sweep:

$$\underline{V}_i^{(k)} = \underline{V}_m^{(k)} - \underline{Z}_i \underline{J}_i^{(k)} \tag{5.46}$$

where $\underline{Z}_i$ is the series impedance of branch i.

Step 5. Check convergence criteria: Steps 2 and 3 are repeated until convergence is achieved. For example, a convergence criteria can be adopted as follows:

$$\max\left|V_i^{(k)} - V_i^{(k-1)}\right| \le \varepsilon_V = \left(10^{-3} - 10^{-5}\right) \tag{5.47}$$

5.3.1.1 DG units treatment

Depending on the control, the DG may be set to output power at either constant power factor for small DG or constant voltage for large DG. Therefore, the bus at which the DG is connected should be modeled as a PQ bus for the first case, and a PV bus for the second.

Figure 5.7 shows a part of the distribution network with DG connected at node i. In accordance with Figure 5.7, the current in line section i at iteration k is

$$\underline{J}_i^{(k)} = \underline{I}_{Li}^{(k)} + \underline{I}_{Ci}^{(k)} - \underline{I}_{\text{DG}i}^{(k)} + \underline{Y}_i^o \cdot \underline{V}_i^{(k-1)} + \sum_{\substack{\ell \in \alpha_{\ell i} \\ \ell \neq i}} \underline{J}_\ell^{(k)}; \tag{5.48}$$

$$i = NB, NB-1, \ldots, 1; \quad k = 1, 2, \ldots$$

where $\underline{I}_{\text{DG}i}^{(k)}$ is the current injections at node i corresponding to DG. In case the node i is modeled as a PQ node, the DG unit should be treated as negative PQ load.

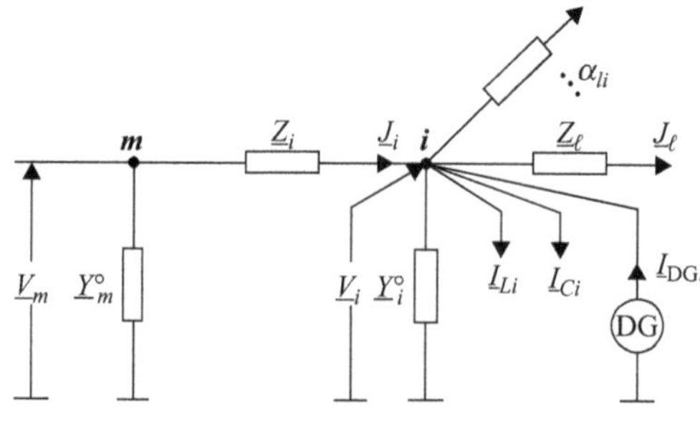

Figure 5.7 A part of distribution network with DG

Accordingly, the current injection of the DG at node i can be calculated as follows:

$$\underline{I}_{\mathrm{DG}i}^{(k)} = \left(\frac{P_{\mathrm{DG}i}^{sp} + jQ_{\mathrm{DG}i}^{sp}}{\underline{V}_i^{(k-1)}} \right)^* \qquad P_{\mathrm{DG}i}^{sp} = const; \quad Q_{\mathrm{DG}i}^{sp} = const; \tag{5.49}$$

where $P_{\mathrm{DG}i}^{sp}$ and $Q_{\mathrm{DG}i}^{sp}$ are specified active and reactive power outputs of the DG connected at node i.

In case the bus i in Figure 5.7 is modeled as a PV bus, handling the PV buses in the power-flow algorithm requires some additional processes, which are as follows:

1. Initially the DG active power outputs, reactive power generation limits, and voltage magnitudes at PV nodes are specified. The reactive power of the DGs is initialized to zero. Perform power-flow calculation considering the DOGs as negative PQ loads.
2. Calculate the voltage at each PV bus:

$$\underline{V}_i^{\mathrm{cal}(k)} = V_i^{\mathrm{cal}(k)} \left[\cos\left(\theta_i^{(k)}\right) + j \sin\left(\theta_i^{(k)}\right) \right] \quad i = 1, \ldots, N\mathrm{PV} \tag{5.50}$$

3. Calculate complex value of the specified voltage for all PV buses:

$$\underline{V}_i^{sp(k)} = V_i^{sp} \left[\cos\left(\theta_i^{(k)}\right) + j \sin\left(\theta_i^{(k)}\right) \right] \quad i = 1, \ldots, N\mathrm{PV} \tag{5.51}$$

4. Calculate the vector of the voltage mismatches for PV buses:

$$\Delta\underline{\mathbf{V}}^{(k)} = \underline{\mathbf{V}}^{sp(k)} - \underline{\mathbf{V}}^{\mathrm{cal}(k)} \tag{5.52}$$

If any of the mismatches is greater than a threshold, i.e., $\left|\Delta V_i^{(k)}\right| > \varepsilon$, then perform the next step.

5. Calculate the required reactive power generation of the DGs for all PV nodes:

$$\mathbf{Q}_{\mathrm{DG}}^{(k)} = \mathbf{Q}^{(k)} + \mathbf{Q}_L \tag{5.53}$$

where $\mathbf{Q}_L$ is the vector of reactive power loads at PV nodes and $\mathbf{Q}^{(k)}$ is the vector of the new reactive power injections at PV nodes. The vector of the new reactive power injections at PV nodes is

$$\mathbf{Q}^{(k)} = \mathbf{Q}^{(k-1)} + \Delta\mathbf{Q}^{(k)} \tag{5.54}$$

$\Delta\mathbf{Q}^{(k)}$ is the vector of the reactive power increments at PV nodes:

$$\Delta\mathbf{Q}^{(k)} = \mathrm{Im}\left\{ \underline{\mathbf{V}}^{sp(k)} \cdot \left(\underline{\mathbf{Z}}_{\mathrm{PV}}^{-1} \cdot \Delta\underline{\mathbf{V}}^{(k)} \right)^* \right\} \tag{5.55}$$

$\underline{\mathbf{Z}}_{\mathrm{PV}}$ is the PV bus sensitivity matrix. The dimension of $\underline{\mathbf{Z}}_{\mathrm{PV}}$ is equal to the number of PV buses. The diagonal entry, $\underline{Z}_{\mathrm{PV}ii}$, in $\underline{\mathbf{Z}}_{\mathrm{PV}}$ is equal to the sum positive sequence impedance of all branches between PV bus i and the root bus (substation bus). If two PV buses, i and j, have completely different paths to the root bus, then the off-diagonal entry $\underline{Z}_{\mathrm{PV}ij}$ is zero. If i and j share a piece of common path to the root bus, then $\underline{Z}_{\text{PV}ij}$ is equal to the positive sequence impedances of all branches on this common path.

6. $Q_{\text{DG}i}^{(k)}$ then is compared with the reactive power generation limits. If $Q_{\text{DG}i}^{(k)}$ is within the limits, i.e., $Q_{\text{DG}i}^{\min} < Q_{\text{DG}i}^{(k)} < Q_{\text{DG}i}^{\max}$, then the DG current injected to PV bus i is calculated as

$$\underline{\mathbf{I}}_{\text{DG}i}^{(k)} = \left(\frac{P_{\text{DG}i}^{sp} + jQ_{\text{DG}i}^{(k)}}{\underline{V}_i^{(k-1)}} \right)^* \tag{5.56}$$

In subsequent iterations, these currents will be combined with other bus current injections according (5.48). Otherwise, if $Q_{\text{DG}i}^{(k)}$ violates any reactive power generation limit, it will be set to that limit, and combined with the reactive load at this bus. Subsequently, the rows and columns in the PV node sensitivity matrix, $\underline{\mathbf{Z}}_{\text{PV}}$, corresponding to this bus are removed and the LU factors of $\underline{\mathbf{Z}}_{\text{PV}}$ are updated.

Now, the backward/forward sweep algorithm is performed again. The power-flow solution will be terminated when the voltage magnitudes of all PV buses converge to their specified values.

5.3.2 Power-flow software—pfdngui

The described b/f algorithm was implemented in the software package *pfdngui* which is available at [6]. It uses the GUI capabilities of MATLAB and enables power-flow computation in distribution network with connected DG units.

Figure 5.8 shows the main window of *pfdngui* while it operates on a Windows-based computer. As can be seen from the figure, the *pfdngui* is organized in following components:

1. Panel TEST SYSTEM. It consists of one pop-up menu and one button. Pop-up menu enables selection of a test system to be considered. It is available in six test systems; three of which are standard IEEE distribution test systems such as IEEE 31-bus, 33-bus, modified 34-bus, and 63-bus test systems. The button System Data opens the input file of the selected test system. The input file contains branch data and bus data of the selected test system. The data are given in matrix form. The user can change the data and save those modifications.
2. After selection test system, the user can start the program by clicking the button RUN power flow.
3. Panel RESULTS. It manages the access to the different segments of the power flow results generated from the program. The summary of results, such as the accuracy, number of iterations, total power of load, total power of generation, total power loss, maximum voltage, minimum voltage etc., are displayed immediately after execution of the program. The buttons Bus Voltages & Powers and Branch Power Flow enables to display the calculated bus voltages and bus powers and branch power flows and power losses in the distribution network. The distribution network's voltage profile is displayed by clicking the button Voltage Profile.

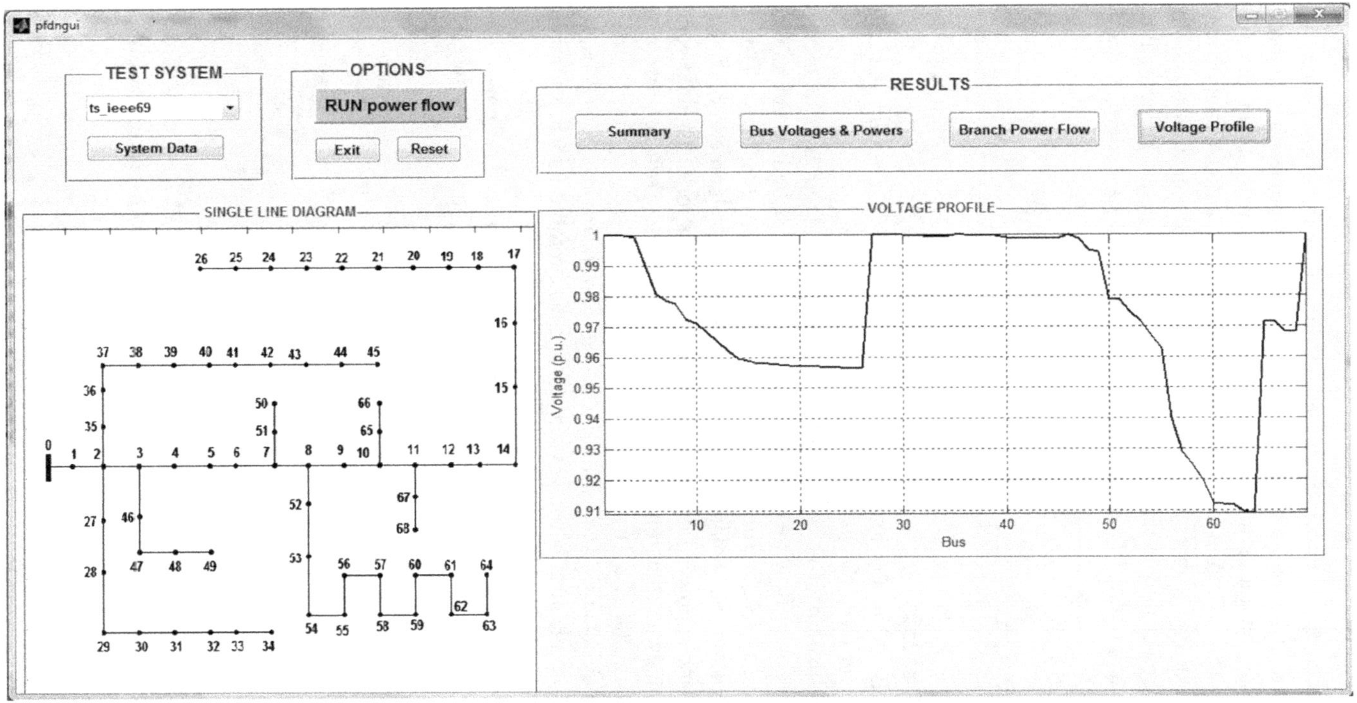

Figure 5.8 Main window of the pfdngui program

5.4 Conclusion

This chapter describes the classical methods for power-flow computation in transmission and distribution networks. The efficiency of the power-flow algorithm has crucial impact on the efficiency and robustness of methods in the functions ensuring optimal operation and planning of transmission and distribution networks. In this context, this chapter presents an introduction in the next chapters related to the optimal power flow in transmission and distribution networks, optimal reactive power dispatch, optimal siting and sizing of distributed generation in distribution networks, etc.

References

[1] Kundur P. *Power system stability and control*. New York: McGraw-Hill; 1994. p. 257

[2] Stott B., Alsac O. 'Fast decoupled load flow'. *IEEE Transactions on Power Apparatus and Systems*. 1974;**PAS-93**(3):859–69

[3] Shirmohammadi D., Hong H.W., Semlyen A., Luo G.X. 'A compensation-based power flow method for weakly meshed distribution and transmission networks'. *IEEE Transactions on Power Systems*. 1988;**3**(2):753–62

[4] Khushalani S., Solanki J.M., Shulz N.N. 'Development of three-phase unbalanced power flow using PV and PQ models for distributed generation and study of the impact of DG models'. *IEEE Transactions on Power Systems*. 2007;**22**(3):1019–25

[5] Radosavljević J. *pfgui—Matlab GUI based software for power flow analysis*. 2017. Available at https://www.researchgate.net/profile/Jordan_Radosavljevic2/contributions

[6] Radosavljević J. *pfdngui—Matlab GUI based software for power flow analysis in distribution networks*. 2017. Available at https://www.researchgate.net/profile/Jordan_Radosavljevic2/contributions

Chapter 6
Optimal power flow in transmission networks

6.1 Introduction

The electric power system is a complex, dynamic system of large dimensions, consisting of numerous generators, transformers, lines, and consumers. The main purpose of power system is to provide consumers with sufficient quantities of quality electricity in a secure, reliable, and economical manner.

Some of the system variables including the generator active power outputs, the generator bus voltages, the transformer tap settings, the reactive power output of shunt Var compensators, etc. can be controlled independently and thus have a direct impact on the power flows, that is, on the steady-state performance of the power system. These variables are defined as control variables. Other system variables such as the load bus voltages, the generator reactive powers, and the power flows through branches are considered as the dependent variables. The power-balance equation can be achieved under different combinations of the control variables, but only some of these combinations enable economical, secure, and reliable operation of the power system. This brings us to the optimal power-flow (OPF) problem. The OPF problem can be defined as follows: finding the optimal combinations of the control variables to obtain a desired condition of the power system.

Historically, the OPF was introduced for the first time by Carpentier [1] as an extension of conventional economic dispatch. Later, Dommel and Tinney [2] worked out the problem and defined it as the OPF.

The OPF is one of most important tools for achieving the economic and secure operation of the power system. The OPF problem solution aims to optimize a chosen objective function, such as fuel cost, power loss, voltage deviation (VD), etc. through optimal adjustment of the power-system control variables, while at the same time satisfying various operating constraints. In its most general formulation, the OPF is a nonlinear, nonconvex, large-scale, static optimization problem with both continuous and discrete control variables.

The OPF problem can be solved by using deterministic (classical) and meta-heuristic optimization methods [3,4].

Classical optimization methods that have been applied to OPF problems such as gradient methods, Newton's methods, the simplex method, sequential linear programing, sequential quadratic programing, and interior point methods are based on an estimation of the global optimum. However, due to difficulties of differentiability, nonlinearity, and nonconvexity, these methods failed to provide the global optimum and only reached the local one. Moreover, these methods exhibit some limitations, depending on the type of problem, e.g., when the objective function is not available in algebraic form. It becomes essential to develop optimization methods able to overcome these drawbacks and handle such difficulties [5].

This chapter has been focused on application of metaheuristic population-based optimization methods to solution of the OPF problem.

In the past two decades, various population-based metaheuristic optimization methods have been suggested for solving the OPF problem. Their main advantage compared to the classical (deterministic) optimization methods is that they are not limited with requirements for differentiability, nonconvexity, and continuity of the objective function or types of control variables. Moreover, these methods can be used for practical power systems taking into account various types of objective function and constraints. The essence of metaheuristic methods is iterative correction of solutions, i.e., generating new populations by applying stochastic search operators on individuals from the current population. The main performances of metaheuristics are fast search of large solution spaces, ability to find global solutions, and avoiding local optimum.

The rest of the chapter is organized as follows. A comprehensive review of recent literature of population-based algorithms applied to OPF problems is given in Section 6.2. The OPF problem is mathematically formulated in Section 6.3. Section 6.4 explains the solution methodology based on the particle swarm optimization (PSO), gravitational search algorithm (GSA), and hybrid PSOGSA algorithms. Section 6.5 presents the results of the optimization and compares the algorithms used to solve the case studies of OPF problems with IEEE 30-bus and IEEE 118-bus system. Section 6.6 describes a MATLAB® Graphical User Interface (GUI)-based software for OPF using different metaheuristic optimization methods. Finally, Section 6.7 outlines the conclusions.

6.2 Literature overview

Genetic algorithm (GA). The GA is one of the most widely used metaheuristic methods to the OPF problems. One of first application of GA on the OPF problem is described in [6]. The authors proposed an improved GA with the dynamical hierarchy of the coding system. This improvement means that the GA has a high potential in solving the OPF for large-scale systems under both normal and contingency conditions. Bakirtzis *et al.* [7] proposed an enhanced GA that incorporated problem-specific operators to solve larger OPF problems. The approach is similar to the one in [6] with the following improvements: Switchable shunt devices and transformer taps are modeled as discrete control variables. In GA, the control variables are self-constrained. Dependent variables (state variables), such as slack bus power, load bus voltages, generator reactive power outputs, and transmission line loadings, are restricted in [6,7] by adding them as the quadratic penalty terms on the objective function. The authors of [8] discussed about the impact of these penalty terms on the accuracy and convergence of GA in solving the OPF problem. The authors in [9] proposed two-stage methodology for solving the OPF problem using GA. The first part employs the GA to obtain a feasible solution subject to desired load convergence, while the other part employs GA to obtain the optimal solution. A GA with the representation of the control variables in their natural form was successfully applied in [10] for solving the OPF that aims for security enhancement of the power system. In [11], the authors used the GA in solving the OPF problem with the voltage angles at generator buses as control variables. A real-coded mixed-integer GA is proposed in [12] for solving the OPF problems with considering transmission security and bus voltage constraints under normal and contingent states. An improved GA with an orthogonal array-based crossover operator has been proposed in [13] for solving multicontingency transient-stability-constrained OPF problems. A GA-based approach for solving the OPF problem considering flexible alternating current (AC) transmission system (FACTS) devices is proposed in [14]. The FACTS devices are used to eliminate line overloads in the power system following single-line outages. To overcome the high CPU time execution and improve the qualities of the solution in practical large-scale OPF problems, the authors [15] proposed an efficient parallel GA for the solution of large-scale OPF with consideration shunt FACTS devices under severe loading conditions. The OPF problem is decomposed in two subproblems: (i) the active power dispatch to minimize the total generation cost, and (ii) the reactive power dispatch to adjust the voltage source and reactive power compensation within their specified constraints limits to reduce VD and the thermal transmission line. These two subproblems are solved separately by efficient parallel GA. The authors in [16] proposed enhanced GA in combination with a new decoupled quadratic load flow routine to solve the multiobjective OPF problem. The multiobjective OPF problem is defined as simultaneous optimization of three conflicting objectives: total generation cost, active power losses, and voltage-stability index. An adapted GA with adjusting population size is used in [17] to solve OPF with different objectives that reflect generation costs minimization and voltage-profile improvement. Non-dominated sorting GA II (NSGA II) approach is proposed in [18,19] for solving the VSC OPF. In [20,21], OPF problem of purely AC power systems is extended to an OPF problem of integrated AC–DC power system by taking into consideration power-transfer control characteristics of high-voltage direct current (HVDC) links. Then, the OPF of integrated AC–DC power systems is solved by GA. Reference [22] presents an alternative approach to OPF problem using line flow-based formulation and decoupled approach. The OPF solution was obtained using GA toolbox available in MATLAB. To overcome the excessive execution time, the authors [23] proposed an efficient evolutionary algorithm (EEA)-based OPF using the concept of incremental power-flow model. The proposed approach was implemented using enhanced GA. Based on simulation results, the authors concluded that the proposed EEA approach is approximately five times faster than any conventional evolutionary algorithm, with insignificant difference in the objective function value. In recent years, the authors pay attention to the OPF solution with an aim of transient-stability enhancement. A successful application of parallel NSGA II to solve the multiobjective OPF considering transient stability as an objective function is presented in [24]. The authors [25] prove excellent performances of a recently developed GA with new multiparent crossover in solving the OPF with different objective functions and constraints. The integration of renewable energy sources into power system is a very challenging problem due to their intermittent nature. Reference [26] proposes a new strategy based on GA and two-point estimate method for solving the OPF problem taking into account uncertainties of output power of wind generators and photovoltaic generators, and load forecasts.

PSO. First application of PSO for solving the OPF problem is described in [27]. The authors [28] applied an improved PSO (IPSO) to solve the OPF problems. This improvement is based on incorporating a biology concept "passive congregation" in the standard PSO. Zhao *et al.* [29] used standard PSO for OPF problem with nonstationary multistage assignment penalty function. Wang *et al.* [30] proposed a modified PSO algorithm and its application on the OPF problem. In the modified PSO algorithm, particles are not only studied from itself and the best one, but also from other individuals. In this way, the opportunity to find the global optimum is increased and the influence of the initial position of the particles is decreased. In order to solve the constrained OPF with both continuous and discrete control variables, the author [31] proposed an efficient mixed-integer PSO with two mutation schemes. Esmin *et al.* [32] expanded the standard PSO to a hybrid PSO by introducing the arithmetic mutation from the GA field. The hybrid PSO is applied in [32] to solve the OPF problem with objective to power-loss minimization. Dutta and Sinha [33] used PSO to solve the OPF problem that is decoupled into two subproblems, namely, the economic dispatch and the reactive power dispatch. The authors [34] proved the feasibility and robustness of PSO in solving a challenging OPF problem, namely, the transient-stability-constrained OPF problem. Yumbla *et al.* [35] used the PSO algorithm with reconstruction operators to solve the OPF problem taking into account both operating security constraints and system capacity requirements. In this approach, the reconstruction operators and an external penalty are adopted to handle the constraints of the OPF problem. The use of reconstruction operators allows increasing the suitable particles in the searching space. Abido [36] proposed a multiobjective version of the conventional PSO technique by redefining of global best and local best individuals in multiobjective optimization domain. The proposed multiobjective PSO is applied to OPF problem. Liu *et al.* [37] introduced a multistart technique to overcome premature convergence of the PSO algorithm in solving the OPF problem. In [38], fuzzy logic is applied in combination with PSO algorithm for solving OPF to enhance the power-system security. Niknam *et al.* [39] proposed an IPSO with a chaos algorithm to compute the inertia weight and a self-adaptive method to compute acceleration coefficients for the multiobjective OPF problem. In this approach, a fuzzy decision-based mechanism is used to select the best compromise solution of Pareto set obtained by the IPSO. The authors [40] incorporated one new index named, iteration best, in the equation for speed updating of standard PSO. By adding this new index, the OPF solution quality is improved. Kumar and Chaturvedi [41] employed the integration of fuzzy system and PSO algorithm to solve the OPF problems. The simulation results show the effectiveness and robustness of proposed PSO–fuzzy approach. Similar applications of the standard PSO to solve the OPF problem are described in [42–44], and a comparison of the performance of PSO along with evolutionary programing (EP) and GA in solving the multiobjective OPF problem is given in [45]. Chang *et al.* [46] proposed an evolutionary PSO approach for the OPF problem considering upspinning reserves, downspinning reserves, and the operational constraints of the generation unit in a power system with both thermal generators and wind turbine generators.

GSA. One of the first applications of the GSA in solving the OPF problem is presented in [47]. The authors considered different objective functions such as the voltage-stability enhancement in both nominal and contingency conditions, the fuel cost minimization including piecewise quadratic cost function, and cost function with valve-point effect. Bhattacharya and Roy [48] used GSA for solution of multiobjective OPF problems. The multiobjective OPF problem is formulated as simultaneous minimization of fuel cost, power loss, and VD. Sonmez *et al.* [49] applied GSA to solve the OPF considering FACTS devices. The author concludes that the GSA has fast computing time and finds better results than other popular optimization techniques. The authors [50] proposed an enhanced GSA for solving security-constrained OPF. A new mutation operator is introduced in GSA for continuously generating of new masses throughout the solution space in order to enhance the exploration capability and avoiding being trapped in local minima. The solution approach has two levels: Economic dispatch is solved to generate an initial solution, and security-constrained OPF is solved to find global optimum. The nondominated sorting multiobjective GSA is proposed and successfully applied in [51] for solving different multiobjective OPF problems. The authors used fast elitism nondominated sorting and crowding distance to locate and manage the Pareto optimal solutions.

Differential evolution (DE). Many of the authors have applied the DE algorithm to solve the OPF problem. Basu [52,53] has used a DE algorithm to minimize the generator fuel cost in OPF control with FACTS devices. In [54], a DE approach has been used for multiobjective OPF with multiple objectives, where the authors considered two subproblems of OPF, namely, the active power dispatch with generation cost and total emission as the competing objectives, and the reactive power dispatch with power loss and VD as the competing objectives. Sayah and Zehar [55] proposed an efficient modified DE algorithm for solving OPF with nonsmooth and nonconvex generator fuel cost curves. Cai *et al.* [56] developed a robust and efficient method for solving transient-stability-constrained OPF problems based on DE. In two identical articles in [5,57], Abou El Ela *et al.* have proposed DE to solve the OPF problem with different objectives including fuel-cost minimization, voltage-profile improvement, and voltage-stability enhancement. Amjady and Sharifzadeh [58] proposed a novel robust DE algorithm owning a new recombination operator to solve OPF problem, considering a detailed generator model including valve loading effects, multiple fuel options, and prohibited operating zones of units. The authors [59] proposed integration of the multiagent systems and

DE in a multiagent-based DE for solving OPF problem with nonsmooth and nonconvex generator fuel-cost curves. Recently, Shaheen *et al.* [60] proposed a multiobjective DE algorithm based on forced initialization to solve the OPF problem. The considered OPF objectives are fuel-cost minimization, power-loss minimization, voltage-profile improvement, and voltage-stability enhancement.

EP. Yuryevich and Wong [61] proposed an efficient and reliable EP algorithm to solve the OPF problem. In [62,63], the authors proposed EP algorithm to solve the OPF problems with nonsmooth and nonconvex fuel-cost curves. Sood [64] used an EP-based approach to solve OPF for the analysis of deregulated model of power system.

Ant colony optimization (ACO). Allaoua and Laoufi [65,66] applied ACO for solving the OPF problem to minimize the total fuel cost of generating units while also retaining an acceptable system performance level in terms of limits on generator active and reactive power outputs, bus voltages, shunt capacitors/reactors, transformer tap settings, and power flow on transmission lines. Gasbaoui and Allaoua [67] also used ACO to solve the OPF problem with multiple objectives, including fuel-cost minimization, voltage-profile improvement, and voltage-stability enhancement. In [68], a touring ACO is applied to solve the OPF problem of power system with FACTS and showed that the FACTS can provide wider operating margin and higher voltage stability. Joshi and Ghanchi [69] applied ACO to minimize total generation cost, to reduce transmission loss, and to improve the bus voltage profile under normal as well as contingent states.

Artificial bee colony (ABC). The ABC algorithm proved to be very effective in solving the OPF problems. Adaryani and Karami [70] successfully employed ABC algorithm as the main optimizer for optimal adjustments of the power-system control variables of the OPF problem considering both continuous and discrete variables under different objective functions. Medina *et al.* [71] proposed a multiobjective variant of ABC based on a decomposition approach, where the multiobjective OPF problem is decomposed into a number of scalar optimization subproblems that are simultaneously optimized. The OPF problem considering the stochastic nature of the wind generation power is solved in [72] by using Gbest-guided ABC optimization algorithm. Based on the simulation results obtained on IEEE 30-bus system, the authors have concluded that the solution method based on Gbest-guided ABC algorithm gives better accuracy of results compared to other well-established methods tried in the past with faster convergence and better solution quality. Abaci *et al.* [73] presented a simple and efficient solution method based on the ABC algorithm for the OPF problem in power systems with static Var compensator devices. Recently, the authors [74] proposed a chaotic ABC algorithm based on chaos theory for solving security and transient-stability-constrained OPF problems.

Backtracking search algorithm (BSA). Kılıç [75] proposed and successfully applied the BSA for solving the OPF problems considering the valve-point effect and prohibited operating zones of generators in power systems. The author has concluded that the BSA is superior to other algorithms in terms of minimization of total fuel cost.

Colliding bodies optimization (CBO). The authors [76,77] successfully applied CBO algorithm to solve the OPF problem considering different test systems and objective functions. Based on simulation results, the authors noted the high efficiency and robustness of the CBO algorithm and its applicability for a variety of cases with complex objective functions, security constraints, prohibited zones, and different test systems.

Black-hole optmization (BH). Hasan and El-Hawary [78] and Bouchekara [79] applied the BH optimization algorithm to solve different types of the OPF problem under various types of constraints. The author [79] concluded that the BH algorithm gives better results than some well-known methods like GA and PSO.

Gray wolf optimizer (GWO). El-Fergany and Hasenian [80] applied the GWO and DE algorithms to solve the OPF problem. These algorithms are used to optimize single objective functions under the system constraints. Then, the DE algorithm is employed to solve multiobjective OPF problems. Mahdad and Srairi [81] used GWO along with pattern search as a solving tool for the OPF problems to minimize the shedding of critical loads at critical situations considering loading margin stability and faults at generation units to reinforce defense planning strategy of practical power system against blackout.

Firefly algorithm (FA). Herbadji *et al.* used FA for the solution of the OPF problem to minimize the total fuel cost of generation and environmental pollution caused by fossil-based thermal generating units [82] and to minimize the total fuel cost of generation with consideration the placement of FACTS devices in the power system [83]. The authors [84–86] proposed FA-based methods to solve OPF in power systems that have a unified power-flow controller (UPFC).

Cuckoo search (CS). CS algorithm is proposed in [87] for solving the OPF in power systems incorporating wind power. Balasubbareddy *et al.* [88] proposed a hybridization of the conventional CS algorithm and crossover operations to solve single- and multiobjective OPF problems with generation fuel cost, emission, and total power losses as objectives. Similar approach is applied for multiobjective OPF problem with FACTS devices [89].

Moth swarm algorithm (MSA). MSA was developed by Mohamed *et al.* [90] and successfully applied to solve several OPF objective functions in three standard IEEE test power systems.

Krill herd algorithm (KHA). Roy and Paul [91] used KHA for the solution of multiobjective OPF with the objective of fuel-cost minimization, VD minimization, and voltage-stability improvement. Mukherjee *et al.* proposed an opposition-based KHA for solving the OPF problem of power system incorporating FACTS devices [92] and the transient-stability-constrained OPF [93]. Recently, Pulluri *et al.* [94] presented a creature-inspired stud KHA consisting of stud selection and crossover operators, which are augmented in the KHA to solve the OPF problems in a power system.

Shuffled frog leaping algorithm (SFLA). Niknam *et al.* [95] proposed an algorithm based on the SFLA to solve the multiobjective OPF problem. Similar approach based on SFLA algorithm combined with a new mutation is applied for multiobjective OPF with FACTS devices [96]. Rahiminejad *et al.* [97] used SFLA to solve the OPF dispatch in an AC–DC power system, including both high-voltage AC and HVDC transmission lines.

Bacterial foraging algorithm (BFA). BFA was applied for solving OPF problems with considering transmission security for practical application. The objective of OPF is to minimize total generation cost, enhance transmission security, reduce transmission loss, and improve the bus voltage profile under normal or contingent states [98]. Panda and Tripathy [99] proposed a modified BFA to solve multiobjective OPF in a power system with wind turbines aiming to minimize the cost of wind-thermal generation system, to reduce the real power loss, and to effectively operate the system in a voltage-secure manner with the incorporation of the shunt-connected Var sources simultaneously.

Biogeography-based optimization (BBO). Bhattacharya and Chattopadhyay [100] implemented the BBO method to solve different types of OPF problems with various types of complexities. By using adaptive real-coded BBO, the authors [101] presented the optimization of various objective functions of an OPF problem in a power system. Based on simulation results on standard IEEE test systems, the authors concluded that the adaptive real-coded BBO approach is effective, robust, and more accurate than other methods of power-flow optimization in literature.

Teaching–learning-based optimization (TLBO). In [102], the TLBO technique is used to solve the OPF problem with different complexities. Medina *et al.* [71] presented a multiobjective variant of the TLBO based on a decomposition approach, where the multiobjective OPF problem is decomposed into a number of scalar OPF subproblems which are simultaneously optimized. An improved TLBO algorithm using Lévy mutation strategy is proposed by Ghasemi *et al.* [103] and applied to OPF problem.

League championship algorithm (LSA). This interesting algorithm is proposed for solving the OPF problem and applied to the Algerian power-system network for different objectives [104].

Sine cosine algorithm (SCA). Mahdad and Srairi [105] used a new developed optimization technique called SCA for OPF problem to optimize the loading margin stability in coordination with total cost, total power loss, total VD, and voltage-stability index.

Imperialist competitive algorithm (ICA). Ranjbar *et al.* [106] used an ICA-based approach for solving the single-objective OPF problem to minimize the system total generation cost, whereas Ghasemi *et al.* [107,108] proposed a multi-objective modified ICA for the multiobjective OPF problem with the fuel cost, emission, VD, and active power losses as the objective functions.

Differential search algorithm (DSA). In addition to CBO, BH, TLBO, and LSA methods, Bouchekara and Abido proposed another one solution method for the OPF problem, namely, DSA algorithm [109]. As in the case of the aforementioned methods, the authors concluded that the DSA has better performances in terms of efficiency, robustness, and quality of solution in comparison with other methods reported in the literature. The question arises, if all methods are best, which method is really the best? Abaci and Yamacli [110] have used the DSA for multiobjective OPF problems. The authors also concluded that DSA provides better solution performances, robustness, and superiority in comparison with other methods reported in the literature.

The Jaya algorithm. Warid *et al.* [111] presented application of an effective parameter-free metaheuristic optimization method, namely, the Jaya algorithm in solving the OPF problem with three goal functions: generation cost minimization, active power-loss minimization, and voltage-stability improvement.

Harmony search (HS). Sinsuphan *et al.* [112] proposed an improved HS method to solve OPF problems with smooth and nonsmooth fuel-cost characteristics. Based on simulation results on the IEEE test systems, the authors concluded that the improved HS can be the most accurate method in solving large-scale power system in comparison with GA and sequential programing.

BAT search algorithm (BAT). Rao and Kumar [113] proposed BAT algorithm to solve OPF problem in the presence of UPFC with an aim to minimize active power losses in a power system. The authors in [114] compared the performances of BAT algorithm with PSO, HS, and CS on standard IEEE test systems and concluded that the BAT algorithm provide better results for the OPF problem over the other algorithms.

Hybrid algorithms. In recent years, some hybrid algorithms have been proposed to solve the OPF problem. Combination of several population-based algorithms in hybrid algorithm allows them to exploit the strength of each algorithm. In [115], the authors investigated the possibility of using recently emerged evolutionary-based approach as a solution for the OPF problem that is based on the hybrid-modified imperialistic competitive algorithm (MICA) and teaching–learning algorithm (TLA) for optimal settings of OPF control variables. Narimani *et al.* presented a new hybrid algorithm based on the PSO and SFLA for solving the OPF in power systems [116]. Niknam *et al.* proposed a hybrid algorithm based on SFLA and simulated annealing (SA) for solving the OPF problem with nonsmooth and nonconvex generator fuel-cost characteristics [117]. In [118], a hybrid algorithm consisting of BBO with and adaptive mutation scheme and the concept of predator–prey optimization technique has been proposed for solving the multi-objective OPF problems. Vaisakh and Srinivas [119] proposed an evolving ant direction DE approach for solving the OPF problem with nonsmooth and nonconvex generator fuel-cost curves with different inequality constraints.

Radosavljević *et al.* [120,121] proposed a hybrid algorithm based on the PSO and GSA for solving OPF in power systems. The proposed hybrid PSOGA algorithm takes advantages of both PSO and GSA algorithms by combining the ability for social thinking in PSO with the local search capability of GSA. Simulation results show that the hybrid PSOGSA algorithm provides effective and robust high-quality solution of the OPF problem.

6.3 Formulation of the OPF problem

The OPF problem solution aims to optimize a chosen objective function though optimal adjustment of the power-system control variables, while simultaneously satisfying various system operations, such as power-flow equations and inequality constraints. Mathematically, the OPF problem can be formulated as follows:

$$\text{optimize} \quad F(\mathbf{x}, \mathbf{y}) \tag{6.1}$$

Subject to

$$g(\mathbf{x}, \mathbf{y}) = 0 \tag{6.2}$$

$$h(\mathbf{x}, \mathbf{y}) \leq 0 \tag{6.3}$$

$$\mathbf{x} \in \mathbf{X} \tag{6.4}$$

where F is the objective function to be optimized (minimized or maximized) and $\mathbf{x}$ is the vector of control variables, consisting of generator active power outputs P_G (except at the slack bus, supposed P_{G1}), generator voltages V_G, transformer tap settings T, and shunt Var compensations Q_C. Hence, $\mathbf{x}$ can be expressed as

$$\mathbf{x} = [P_{G2} \ldots P_{GNG}, V_{G1} \ldots V_{GNG}, T_1 \ldots T_{NT}, Q_{C1} \ldots Q_{CNC}] \tag{6.5}$$

where NG, NT, and NC are the number of generators, regulating transformer, and Var compensators, respectively.

$\mathbf{y}$ is the vector of dependent variables consisting of slack bus power P_{Gsl}, load bus voltages V_L, generator reactive power outputs Q_G, and transmission line loadings S_l. Accordingly, vector $\mathbf{y}$ can be expressed as

$$\mathbf{y} = [P_{Gsl}, V_{L1} \ldots V_{LNL}, Q_{G1} \ldots Q_{GNG}, S_{l1} \ldots S_{lNTL}] \tag{6.6}$$

where NL and NTL are the number of load buses and transmission lines, respectively.

6.3.1 Equality constraints

The equality constraints (6.2) are the typical nonlinear power-flow equations.

$$P_{Gi} - P_{Di} - V_i \sum_{j=1}^{NB} V_j \left(G_{ij} \cos\, \theta_{ij} + B_{ij} \sin\, \theta_{ij} \right) = 0 \tag{6.7}$$

$$Q_{Gi} - Q_{Di} - V_i \sum_{j=1}^{NB} V_j \left(G_{ij} \sin\, \theta_{ij} - B_{ij} \cos\, \theta_{ij} \right) = 0 \tag{6.8}$$

where $i = 1, \ldots, NB$; NB is the number of busses, P_{Gi} is the active power generation, Q_{Gi} is the reactive power generation, P_{Di} is the active load demand, Q_{Di} is the reactive load demand, θ_{ij} is the voltage angle between busses i and j, and G_{ij} and B_{ij} are the real and imaginary terms of bus admittance matrix corresponding to the ith row and jth column, respectively.

6.3.2 Inequality constraints

Inequality constraints (6.3) are the functional operating constraints, such as load bus voltage magnitude limits, generator reactive power output limits, and branch flow limits.

$$V_{Li}^{\min} \leq V_{Li} \leq V_{Li}^{\max}, \quad i = 1, \ldots, NL \tag{6.9}$$

$$Q_{Gi}^{\min} \leq Q_{Gi} \leq Q_{Gi}^{\max} \quad i = 1, \ldots, NG \tag{6.10}$$

$$S_{li} \leq S_{li}^{\max} \quad i = 1, \ldots, NTL \tag{6.11}$$

The constraints related to voltage at load buses (6.9) and power flow through network branches (6.11) are also known as the security constraints.

Constraints (6.4) define the feasibility region of the problem control variables, such as generator active power output limits, generator bus voltage magnitude limits, transformer tap setting limits, and shunt Var compensation limits. Actually, those define the space of possible solutions for the OPF problem.

$$P_{Gi}^{\min} \leq P_{Gi} \leq P_{Gi}^{\max}, \quad i = 1, \ldots, NG \tag{6.12}$$

$$V_{Gi}^{\min} \leq V_{Gi} \leq V_{Gi}^{\max}, \quad i = 1, \ldots, NG \tag{6.13}$$

$$T_{i}^{\min} \leq T_{i} \leq T_{i}^{\max}, \quad i = 1, \ldots, NT \tag{6.14}$$

$$Q_{Ci}^{\min} \leq Q_{Ci} \leq Q_{Ci}^{\max}, \quad i = 1, \ldots, NC \tag{6.15}$$

It is worth mentioning that the control variables ($\mathbf{x}$) are self-constrained. Inequality constraints of the dependent variables ($\mathbf{y}$) contain the active power generation at the slack bus P_{Gsl}, load bus voltages V_L, reactive power generations Q_G, and branch apparent power flows S_l are restricted by adding them as the quadratic penalty terms to the objective function [6]. The new expanded objective function to be minimized becomes

$$F_p(\mathbf{x}, \mathbf{y}) = F(\mathbf{x}, \mathbf{y}) + \lambda_P \left(P_{Gsl} - P_{Gsl}^{\lim}\right)^2 + \lambda_V \sum_{i=1}^{NL} \left(V_{Li} - V_{Li}^{\lim}\right)^2$$

$$+ \lambda_Q \sum_{i=1}^{NG} \left(Q_{Gi} - Q_{Gi}^{\lim}\right)^2 + \lambda_S \sum_{i=1}^{NTL} \left(S_{li} - S_{li}^{\lim}\right)^2 \tag{6.16}$$

where λ_P, λ_V, λ_Q, and λ_S are the penalty factors. $P_{Gsl}^{\lim}$, $V_{Li}^{\lim}$, $Q_{Gi}^{\lim}$, and $S_{li}^{\lim}$ are defined as follows:

$$P_{Gsl}^{\lim} = \begin{cases} P_{Gsl}^{\min}, & P_{Gsl} < P_{Gsl}^{\min} \\ P_{Gsl}^{\max}, & P_{Gsl} > P_{Gsl}^{\max} \\ P_{Gsl}, & P_{Gsl}^{\min} \leq P_{Gsl} \leq P_{Gsl}^{\max} \end{cases} \tag{6.17}$$

$$V_{Li}^{\lim} = \begin{cases} V_{Li}^{\min}, & V_{Li} < V_{Li}^{\min} \\ V_{Li}^{\max}, & V_{Li} > V_{Li}^{\max} \\ V_{Li}, & V_{Li}^{\min} \leq V_{Li} \leq V_{Li}^{\max} \end{cases} \tag{6.18}$$

$$Q_{Gi}^{\lim} = \begin{cases} Q_{Gi}^{\min}, & Q_{Gi} < Q_{Gi}^{\min} \\ Q_{Gi}^{\max}, & Q_{Gi} > Q_{Gi}^{\max} \\ Q_{Gi}, & Q_{Gi}^{\min} \leq Q_{Gi} \leq Q_{Gi}^{\max} \end{cases} \tag{6.19}$$

$$S_{Gsl}^{\lim} = \begin{cases} S_{li}^{\max}, & S_{li} > S_{l\max} \\ S_{li}, & S_{li} \le S_{li}^{\max} \end{cases} \tag{6.20}$$

6.3.3 Objective function

The objective function can take different forms. Classical objective function for the OPF is the total fuel cost in the system. However, many other objectives are possible, such as minimization of system losses, voltage-profile improvement, voltage-stability enhancement, etc. In addition, simultaneous optimization of different objective functions is often stated as objective of the OPF.

6.3.3.1 Minimization of fuel cost

The basic objective function F considered in OPF problems is that the total fuel cost of the system $Fcost$, where the generator cost characteristics f are defined as quadratic cost function of generator power output P_G. The OPF solution aims to minimize the total generation cost in the system:

$$\min_{\mathbf{x}} F(\mathbf{x}, \mathbf{y}) = \min_{\mathbf{x}} Fcost(\mathbf{x}, \mathbf{y}) = \min_{\mathbf{x}} \sum_{i=1}^{NG} f_i(P_{Gi})$$

$$= \min_{\mathbf{x}} \sum_{i=1}^{NG} \left(a_i + b_i P_{Gi} + c_i P_{Gi}^2\right) \tag{6.21}$$

where $\mathbf{x}$ is the vector of control variables, NG is the number of generators; a_i, b_i, and c_i are the cost coefficients of the ith generator, and P_{Gi} is the corresponding active power output.

Large steam turbine generators will have a number of steam admission valves that are opened in sequence to control the power output of the unit. As the unit loading increases, the input to the unit increases and the incremental heat rate decreases between the opening points for any two valves. However, when a valve is first opened, the throttling losses increase rapidly and the incremental heat rate rises suddenly. This is called valve-point effect that leads to nonsmooth, nonconvex input–output characteristics as illustrated in Figure 6.1. Usually, valve-point effect is modeled by adding a recurring rectified sinusoid to the basic quadratic cost function:

$$\min_{\mathbf{x}} F(\mathbf{x}, \mathbf{y}) = \min_{\mathbf{x}} Fcost\,(\mathbf{x}, \mathbf{y}) \min_{\mathbf{x}} \sum_{i=1}^{NG} a_i + b_i P_{Gi} + c_i P_{Gi}^2$$

$$+ \left|d_i \sin\left(e_i\left(P_{Gi}^{\min} - P_{Gi}\right)\right)\right| \tag{6.22}$$

where d_i and e_i are fuel cost coefficients for valve-point effects.

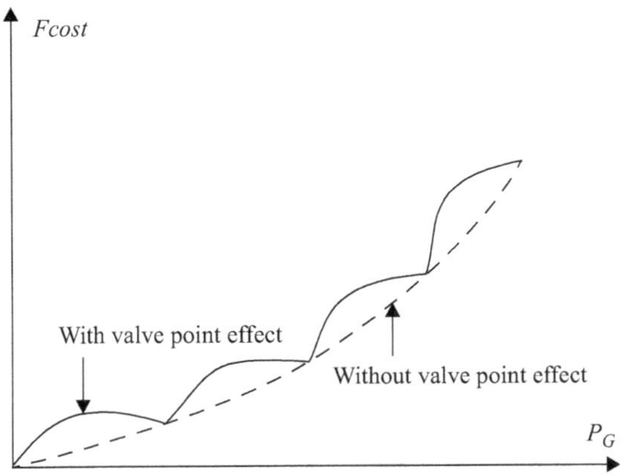

Figure 6.1 Generator cost functions

6.3.3.2 Minimization of active power loss

In this case, the objective is to minimizing the total active power loss (*Ploss*) in the system. The objective function has the following form:

$$\min_{\mathbf{x}} F(\mathbf{x}, \mathbf{y}) = \min_{\mathbf{x}} Ploss(\mathbf{x}, \mathbf{y}) = \min_{\mathbf{x}} \sum_{L=1}^{NTL} P_{loss,L} \tag{6.23}$$

where $P_{loss,L}$ is the active power loss at line L, and NTL is the number of transmission lines. The active power loss through a transmission line can be calculated by

$$P_{loss,ij} = g_{ij}\left[V_i^2 + V_j^2 - 2V_iV_j\cos\left(\theta_i - \theta_j\right)\right] \tag{6.24}$$

where g_{ij} is the conductance of transmission line connecting the ith and jth bus; V_i, V_j, θ_i, and θ_j are the voltage magnitudes and voltage angles at bus i and j, respectively.

6.3.3.3 Voltage-profile improvement

One of the most important and significant safety and service quality indices is bus voltage. In this case, the objective is minimization of the load bus *VDs*:

$$\min_{\mathbf{x}} F(\mathbf{x}, \mathbf{y}) = \min_{\mathbf{x}} VD(\mathbf{x}, \mathbf{y}) = \min_{\mathbf{x}} \sum_{i=1}^{NL} \left|V_i - V_i^{ref}\right| \tag{6.25}$$

where NL is the number of load buses, V_i is the voltage magnitude at bus i, and V_i^{ref} is the reference value of the voltage magnitude of the ith bus, which is usually set to 1 p.u.

6.3.3.4 Voltage-stability enhancement

One important characteristic of the power system is its ability to conserve constantly acceptable bus voltage at each node under normal operating conditions, after load increase, following system configuration changes or when the system is being subjected to a disturbance. The nonoptimized control variables may lead to progressive and uncontrollable drop in voltage resulting in an eventual wide-spread voltage collapse. The static approach for voltage-stability analysis involves determination of an index known as voltage collapse proximity indicator. This index is an approximate measure of closeness of the system operating point to voltage collapse. There are various methods of determining the voltage collapse proximity indicator. One such method is the voltage-stability indicator L-index of the load buses in the system. The L-index gives a scalar number to each load bus. This index uses information on a normal power flow and is in the range of 0 (no load case) to 1 (voltage collapse).

For a power system with *NB*, *NG*, and *NL* buses representing the total number of buses, the total number of generators, and the total number of load buses, respectively, the relationship between voltage and current can be expressed as follows [122,123]:

$$\begin{bmatrix} \underline{\mathbf{I}}_G \\ \underline{\mathbf{I}}_L \end{bmatrix} = \begin{bmatrix} \underline{\mathbf{Y}}_{GG} & \underline{\mathbf{Y}}_{GL} \\ \underline{\mathbf{Y}}_{LG} & \underline{\mathbf{Y}}_{LL} \end{bmatrix} \begin{bmatrix} \underline{\mathbf{V}}_G \\ \underline{\mathbf{V}}_L \end{bmatrix} \tag{6.26}$$

where $\underline{\mathbf{I}}_G$ and $\underline{\mathbf{V}}_G$ are vectors of complex currents and voltages at generator buses; $\underline{\mathbf{I}}_L$ and $\underline{\mathbf{V}}_L$ are vectors of complex currents and voltages at load buses; $\underline{\mathbf{Y}}_{GG}$, $\underline{\mathbf{Y}}_{GL}$, $\underline{\mathbf{Y}}_{LG}$, and $\underline{\mathbf{Y}}_{LL}$ are the submatrices of the system bus admittance matrix $\underline{\mathbf{Y}}_{bus}$.

After rearranging (6.26), the following hybrid system of equation is obtained:

$$\begin{bmatrix} \underline{\mathbf{V}}_L \\ \underline{\mathbf{I}}_G \end{bmatrix} = \begin{bmatrix} \underline{\mathbf{Z}}_{LL} & \underline{\mathbf{F}}_{LG} \\ \underline{\mathbf{K}}_{GL} & \underline{\mathbf{Y}}_{GG} \end{bmatrix} \begin{bmatrix} \underline{\mathbf{I}}_L \\ \underline{\mathbf{V}}_G \end{bmatrix} \tag{6.27}$$

where $\underline{\mathbf{F}}_{LG} = -\underline{\mathbf{Y}}_{LL}^{-1}\underline{\mathbf{Y}}_{LG}$

Therefore, the L-index of the jth bus is given by the following equation:

$$L_j = \left| 1 - \sum_{i=1}^{NG} \underline{F}_{ji} \frac{\underline{V}_i}{\underline{V}_j} \right|, \quad j = 1, 2, \ldots, NL \tag{6.28}$$

If the condition $L < 1$ satisfied for each load bus, the voltage stability is guaranteed. Hence, a global indicator $L_{\max}$ describing the voltage stability of the entire system can be defined by

$$L_{\max} = \max\{L_j\}, \quad j = 1, \ldots, NL \tag{6.29}$$

Therefore, the enhancement of the voltage stability can be achieved by minimizing the value of L-index at each bus of the power system. In terms of the OPF objective, this condition can be expressed as follows:

$$\min_{\mathbf{x}} F(\mathbf{x}, \mathbf{y}) = \min_{\mathbf{x}} L_{\max}(\mathbf{x}, \mathbf{y}) \tag{6.30}$$

6.3.4 *Multiobjective function*

6.3.4.1 Minimization of fuel cost and active power loss

In this case, the minimizations of the system fuel cost and active power loss are approached. These two competing objectives are optimized simultaneously. The objective function balances the two objectives in such a way that one objective should not dominate the other. The new objective function may be expressed as follows:

$$\min_{\mathbf{x}} F(\mathbf{x}, \mathbf{y}) = \min_{\mathbf{x}} \{Fcost(\mathbf{x}, \mathbf{y}) + w_{Ploss} Ploss(\mathbf{x}, \mathbf{y})\} \tag{6.31}$$

where w_{Ploss} is a suitable weighting factor for the real power loss, to be chosen by the user.

6.3.4.2 Minimization of fuel cost and voltage deviation

Though the minimization of total generation fuel cost may give a feasible solution, the voltage profile may not be acceptable. Therefore, in the present case, a two-fold objective function is taken in consideration to minimize the fuel cost and improve voltage profile by minimizing the load bus voltage deviations from a reference value. The objective function can be expressed as

$$\min_{\mathbf{x}} F(\mathbf{x}, \mathbf{y}) = \min_{\mathbf{x}} \{Fcost(\mathbf{x}, \mathbf{y}) + w_V VD(\mathbf{x}, \mathbf{y})\} \tag{6.32}$$

where w_V is a weighting factor for the voltage deviation.

6.3.4.3 Minimization of fuel-cost and voltage-stability index

In order to minimize total the fuel cost and enhance the voltage stability to move the system far from the voltage collapse point, the objective function chosen in this case is a linear combination of total fuel cost and $L_{\max}$. Hence, the objective function can be described as

$$\min_{\mathbf{x}} F(\mathbf{x}, \mathbf{y}) = \min_{\mathbf{x}} \{Fcost(\mathbf{x}, \mathbf{y}) + w_L L_{\max}(\mathbf{x}, \mathbf{y})\} \tag{6.33}$$

where w_L is a suitable weighting factor.

6.3.4.4 Other multiobjective functions

In addition to previously stated multiobjective functions, the OPF problem can be considered for another variant of multiobjectives, as follows:

- Minimization of fuel cost, active power loss, and voltage deviation

$$\min_{\mathbf{x}} F(\mathbf{x}, \mathbf{y}) = \min_{\mathbf{x}} \{Fcost(\mathbf{x}, \mathbf{y}) + w_P Ploss(\mathbf{x}, \mathbf{y}) + w_V VD(\mathbf{x}, \mathbf{y})\} \tag{6.34}$$

- Minimization of fuel cost, active power loss, and voltage-stability index

$$\min_{\mathbf{x}} F(\mathbf{x}, \mathbf{y}) = \min_{\mathbf{x}} \{Fcost(\mathbf{x}, \mathbf{y}) + w_P Ploss(\mathbf{x}, \mathbf{y}) + w_L L_{\max}(\mathbf{x}, \mathbf{y})\} \tag{6.35}$$

- Minimization of active power loss and voltage deviation

$$\min_{\mathbf{x}} F(\mathbf{x}, \mathbf{y}) = \min_{\mathbf{x}} \{w_P Ploss(\mathbf{x}, \mathbf{y}) + w_V VD(\mathbf{x}, \mathbf{y})\} \tag{6.36}$$

- Minimization of active power loss and voltage-stability index

$$\min_{\mathbf{x}} F(\mathbf{x}, \mathbf{y}) = \min_{\mathbf{x}} \{w_P Ploss(\mathbf{x}, \mathbf{y}) + w_L L_{\max}(\mathbf{x}, \mathbf{y})\} \tag{6.37}$$

etc.

6.3.5 Transient-stability-constrained OPF

Transient stability is the ability of the power system to maintain synchronism under a severe transient disturbance such as a fault on transmission facilities, loss of generation, or loss of a large load. After the disturbance, if the resulting angular separation between the machines in the system remains within certain bounds, the system maintains synchronism [124].

Transient-stability-constrained OPF can be considered as a conventional OPF with additional inequality constraints imposed by the rotor angle limits. The power-flow solution should meet the constraints defined by (6.9)–(6.15) related to solution of the conventional OPF problem and the dynamic constraints imposed on the rotor angles during the transient period under undesirable conditions.

In the transient-stability analysis, the synchronous generator is modeled by a constant voltage source behind a transient reactance. The mechanical power input is considered constant during the transient period. The transient-stability problem of power systems is described with a set of differential-algebraic equations [124]. The swing equation set for ith generator is

$$\frac{d\delta_i}{dt} = \omega_i - \omega_0 \tag{6.38}$$

$$2H_i \frac{d\omega_i}{dt} = \omega_0 \left(P_{mi} - P_{gi} - D_i \omega_i\right) \quad i = 1, \ldots, NG \tag{6.39}$$

where δ_i is the rotor angle of ith generator, ω_i is the rotor speed of ith generator, ω_0 is the synchronous speed, D_i is the damping constant of ith generator, H_i is the inertia constant of ith generator, P_{mi} is the mechanical input power of ith generator, and P_{gi} is the electrical output power of ith generator. The electrical output power of ith generator is calculated as

$$P_{gi} = E_i^2 G'_{ii} + \sum_{\substack{j=1 \\ j \neq i}}^{NG} \left[E_i E_j B'_{ij} \sin\left(\delta_i - \delta_j\right) + E_i E_j G'_{ij} \cos\left(\delta_i - \delta_j\right)\right]$$
$$i = 1, \ldots, NG \tag{6.40}$$

where E_i and E_j are the constant voltages behind a transient reactance of the ith and jth generators obtained from the power-flow solution before contingency; δ_i and δ_j are the rotor angles of ith and jth generators; G'_{ij} and B'_{ij} are transfer conductance and susceptance between buses i and j of the reduced $\mathbf{Y}_{bus}$; and G'_{ii} is the self-conductance of bus i of the reduced $\mathbf{Y}_{bus}$. The reduced $\mathbf{Y}_{bus}$ before, during, and after a contingency can be obtained by eliminating all buses except for generator buses.

In order to calculate the rotor angles of the generators, the swing (6.38) and (6.39) should be solved by using a numerical integration method such as Euler or Runge–Kutta methods.

The transient-stability index (*TSI*) is usually defined as the rotor angle deviation with respect to the center of inertia (COI). The position of COI at time t is defined as follows:

$$\delta^t_{COI} = \frac{\sum_{i=1}^{NG} H_i \delta_i}{\sum_{i=1}^{NG} H_i} \tag{6.41}$$

Therefore, the *TSI* of ith generator, TSI_i can be expressed as

$$TSI_i = \left|\delta^t_i - \delta^t_{COI}\right|_{\max} \tag{6.42}$$

where $\left|\delta^t_i - \delta^t_{COI}\right|_{\max}$ represents the maximum rotor angle deviation of ith generator from COI.

Finally, the transient-stability constraints can be expressed by the following equation [74]:

$$TSI^k_i \le \delta_{\max}, \quad \{k \in C\} \quad i = 1, \ldots, NG \tag{6.43}$$

where TSI^k_i is the transient-stability index of ith generator in case contingency k; C is a predefined contingency list; $\delta_{\max}$ is the maximum allowable rotor angle deviation. The *TSI* and $\delta_{\max}$ define the stability requirement. The setting of $\delta_{\max}$ is often based on operational experience. Most utilities would have it set to 100°–120° to allow the system to have sufficient stability margin.

In order to enforce the inequality constraints (6.43) in solving the OPF problem, the objective function (6.16) is added by the penalty function of transient-stability limit as follows:

$$\begin{aligned} F_p(\mathbf{x},\mathbf{y}) = {} & F(\mathbf{x},\mathbf{y}) + \lambda_P\left(P_{Gsl} - P^{\lim}_{Gsl}\right)^2 + \lambda_V \sum_{i=1}^{NL}\left(V_{Li} - V^{\lim}_{Li}\right)^2 \\ & + \lambda_Q \sum_{i=1}^{NG}\left(Q_{Gi} - Q^{\lim}_{Gi}\right)^2 + \lambda_S \sum_{i=1}^{NTL}\left(S_{li} - S^{\lim}_{li}\right)^2 \\ & + \lambda_T \sum_{i=1}^{NG}\left(TSI_i - TSI^{\lim}_{li}\right)^2 \end{aligned} \tag{6.44}$$

where:

$$TSI^{\lim}_i = \begin{cases} \delta_{\max}, & TSI_i > \delta_{\max} \\ TSI_i, & TSI_i \le \delta_{\max} \end{cases} \tag{6.45}$$

6.4 Solution methodology for OPF problem

This section describes applications of PSO, GSA, and hybrid PSOGSA algorithm for solving the OPF problem in power systems.

6.4.1 Overview of PSO

PSO is basically developed through simulation of bird flocking in two-dimension space [125]. Bird flocking optimizes a certain objective function. PSO uses a number of particles (search agents) that fly around in the search space to find best solution. Meanwhile, the particles all look at the best particle (best solution) in their paths. In other words, particles consider their own best solutions as well as the best solution found so far.

The basic elements of the PSO method are briefly stated and defined as follows [27]:

Particle, $\mathbf{x}(t)$: It is a candidate solution represented by an n-dimensional vector, where n is the number of control variables. At time (iteration) t, the ith particle $\mathbf{x}_i(t)$ can be described as $\mathbf{x}_i(t) = \left[x_i^1(t), \ldots, x_i^d(t), \ldots, x_i^n(t)\right]$, where $x_i^d(t)$ is the position of the ith particle with respect to the dth dimension, i.e., the value of the dth control variable in the ith candidate solution.

Population, $\mathbf{POP}(t)$: It is a set of N particles at time (iteration) t, i.e., $\mathbf{POP}(t) = [\mathbf{x}_1(t), \ldots, \mathbf{x}_N(t)]^T$.

Swarm: It is an apparently disorganized population of moving particles that tend to cluster together while each particle seems to be moving in a random direction.

Particle velocity, $\mathbf{v}(t)$: It is the velocity of the moving particles represented by an n-dimensional vector. At time (iteration) t, the ith particle velocity $\mathbf{v}_i(t)$ can be described as $\mathbf{v}_i(t) = \left[v_i^1(t), \ldots, v_i^d(t), \ldots, v_i^n(t)\right]$, where $v_i^d(t)$ is the velocity component of the ith particle with respect to the dth dimension.

Individual best, $\mathbf{pbest}_i(t)$: It is individual best. As a particle moves through the search space, it compares its fitness value at the current position to the best fitness value it has ever attained at any time up to the current time. The best position that is associated with the best fitness encountered so far is called the individual best, $\mathbf{pbest}(t)$. For the ith particle, individual best can be expressed as $\mathbf{pbest}_i(t) = \left[pbest_i^1(t), \ldots, pbest_i^d(t), \ldots, pbest_i^n(t)\right]$.

Global best, $\mathbf{gbest}(t)$: It is the best position among all individual best position achieved so far. Hence, global best can be expressed as $\mathbf{gbest}(t) = \left[gbest^1(t), \ldots, gbest^d(t), \ldots, gbest^n(t)\right]$.

Fitness is a direct metric of the performance of the individual population member (agent). The fitness of each agent of the population is calculated from the value of the function being optimized.

Stopping criteria: These are the conditions under which the search process will terminate. It is often stated that the search will terminate if the number of iterations t reaches the maximum allowable number $t_{\max}$.

In PSO, each agent tries to modify its position using the following information: the current position, the current velocity, the distance between the current position and **pbest**, and the distance between the current position and **gbest**. This modification can be represented by the concept of velocity. Velocity of each agent can be modified by the following equation:

$$\mathbf{v}_i(t+1) = w \cdot \mathbf{v}_i(t) + C_1 \cdot r_1 \cdot (\mathbf{pbest}_i(t) - \mathbf{x}_i(t)) + C_2 \cdot r_2 \cdot (\mathbf{gbest}(t) - \mathbf{x}_i(t)) \tag{6.46}$$

where C_1 and C_2 are positive constants defined as acceleration coefficients, r_1 and r_2 are uniformly distributed random numbers in [0,1], and w is a inertia weight. This w plays the role of balancing the global search and local search.

After updating the velocity, the current position of particle (searching point in the solution space) can be modified by the following equation:

$$\mathbf{x}_i(t+1) = \mathbf{x}_i(t) + \mathbf{v}_i(t+1) \tag{6.47}$$

The first term of (6.46), $w \cdot \mathbf{v}_i(t)$, provides exploration ability for PSO. For initial stages of the search process, large w to enhance the global exploration is recommended while, for last stages, the weight factor is reduced for better local search. The second term of (6.46), $C_1 \cdot r_1 \cdot (\mathbf{pbest}_i(t) - \mathbf{x}_i(t))$, represents the cognitive part of PSO where the particle changes its velocity based on its own thinking and memory. The third term of (6.46), $C_2 \cdot r_2 \cdot (\mathbf{gbest}(t) - \mathbf{x}_i(t))$, represents the social part of PSO where the particle changes its velocity based on the social–psychological adaptation of knowledge.

The PSO starts by randomly placing the particles in a problem space. In each iteration, the velocities of particles are calculated using (6.46). After defining the velocities, the positions of particles can be calculated as (6.47). The process of changing particles' positions will continue until an end criterion is met.

6.4.2 Application of PSO to the OPF problem

The following steps describe the procedure of applying the PSO algorithm to solve the OPF problem:

Step 1: Define the power-system configuration, lines data, transformers data, shunt Var compensators data, loads data, and generation units data. Use the per unit system.

Step 2: Specify the control variables **x** (6.5) and their lower and upper limits (6.12–6.15), specify the dependent variables **y** (6.6) and their lower and upper limits (6.9–6.11), specify the objective function to be optimized $F(\mathbf{x},\mathbf{y})$ (6.21–6.43) and the penalty factors ($\lambda_{P,V,Q,S,T}$) to form the expanded objective function $F_p(\mathbf{x},\mathbf{y})$ (6.44).

Step 3: Initialize PSO parameters, such as the population size (N), maximum iteration number (t_{max}), inertia weight (w), cognitive constant (C_1), and social constant (C_2).

Step 4: Generate an initial random population of N particles.

$$\mathbf{POP}(0) = \begin{bmatrix} \mathbf{x}_1(0) \\ \mathbf{x}_2(0) \\ \vdots \\ \mathbf{x}_i(0) \\ \vdots \\ \mathbf{x}_N(0) \end{bmatrix} = \begin{bmatrix} x_1^1(0), \ldots, x_1^d(0), \ldots, x_1^n(0) \\ x_2^1(0), \ldots, x_2^d(0), \ldots, x_2^n(0) \\ \vdots \\ x_i^1(0), \ldots, x_i^d(0), \ldots, x_i^n(0) \\ \vdots \\ x_N^1(0), \ldots, x_N^d(0), \ldots, x_N^n(0) \end{bmatrix} \tag{6.48}$$

The initial positions of each particle are randomly selected between minimum and maximum values of the control variables. Therefore, the initial value of the dth control variable in the ith particle (candidate solution) is obtained as follows:

$$x_i^d(0) = x_{i,\min}^d + r \cdot \left(x_{i,\max}^d - x_{i,\min}^d\right) \tag{6.49}$$

where r is a random number in [0,1], and $x_{i,\min}^d$ and $x_{i,\max}^d$ are the lower and upper limits of the dth control variable. According to (6.5), the ith particle for the OPF problem can be written as follows:

$$\mathbf{x}_i = \left[P_{G2i}^1, \ldots, P_{GNGi}^{\cdots}, V_{G1i}^{\cdots}, \ldots, V_{GNGi}^d, T_{T1i}^{d+1}, \ldots, T_{NT1i}^{\cdots}, Q_{C1i}^{\cdots}, \ldots, Q_{CNCi}^n\right] \tag{6.50}$$

Initialize the velocity of particles by a zero matrix of dimension $N \times n$, where N is the number of particles (population size) and n is the number of control variables ($n = 2NG - 1 + NT + NC$).

Step 5: Run the power-flow program based on fast decoupled power-flow method [126] for each particle $\mathbf{x}_i(t)$ from the current population $\mathbf{POP}(t)$ and calculate the corresponding values of the objective function.

Step 6: Update $\mathbf{pbest}(t)$ for each particle and determine $\mathbf{gbest}(t)$.

Step 7: Update the velocity of all particles using (6.46).

Step 8: Update the position of each particle using (6.47).

Step 9: Repeat steps 5–8 until the stop criteria are reached, that is, the maximum number of iterations $t_{\max}$.

Step 10: Return best solution obtained in the last iteration; Stop.

6.4.3 Overview of GSA

The GSA is a metaheuristic optimization algorithm where the search agents are considered as a collection of masses that interact with each other based on the Newtonian gravity and the laws of motion. The position of the mass corresponds to the solution of the problem, and its gravitational and inertial masses are determined using a fitness function. In other words, each mass presents a potential solution. The algorithm is navigated by properly adjusting the gravitational and inertial masses [127]. In a system with N agents (masses), the position of the ith agent is defined by

$$\mathbf{x}_i = \left[x_i^1, \ldots, x_i^d, \ldots, x_i^n\right] \quad i = 1, \ldots, N \tag{6.51}$$

where n is the number of control variables and x_i^k defines the position of the ith agent in the kth dimension, i.e., the value of kth control variable in the ith potential solution.

After evaluating the current population fitness, the mass of each agent is calculated as follows:

$$M_i(t) = \frac{m_i(t)}{\sum_{j=1}^{N} m_j(t)} \tag{6.52}$$

where:

$$m_i(t) = \frac{fit_i(t) - worst(t)}{best(t) - worst(t)} \tag{6.53}$$

where $fit_i(t)$ represent the fitness value of the agent i at iteration t; $best(t)$ and $worst(t)$ are the best and worst fitness of all agents, respectively.

According to Newton gravitation theory, the total force that acts on the ith agent at iteration t is specified as follows:

$$\mathbf{F}_i(t) = \sum_{j \in Kbest, j \neq i} r \cdot G(t) \frac{M_j(t) \cdot M_i(t)}{R_{i,j}(t) + \varepsilon} \left(\mathbf{x}_j(t) - \mathbf{x}_i(t)\right) \tag{6.54}$$

where r is a random number within the interval [0,1]. $G(t)$ is gravitational constant at iteration t, $M_i(t)$, and $M_j(t)$ are masses of agents i and j, ε is a small constant, and $R_{ij}(t)$ is the Euclidian distance between the two agents i and j, given by following equation:

$$R_{ij}(t) = \|\mathbf{x}_i(t), \mathbf{x}_j(t)\|_2 \tag{6.55}$$

$Kbest$ is the set of first K agents with the best fitness value and biggest mass, which is a function of time, initialized to K_0 at the beginning and decreased with time. In such a way, at the beginning, all agents apply the force, and as time passes, $Kbest$ is decreased linearly and at the end, there will be just one agent applying force to the others.

According to the law of motion, the acceleration of the ith agent, at iteration t is given by the following equation:

$$\mathbf{a}_i(t) = \frac{\mathbf{F}_i(t)}{M_i(t)} \tag{6.56}$$

The searching strategy on this notion can be defined to find the next velocity and next position of an agent. Next velocity of an agent is defined as a function of its current velocity added to its current acceleration. Hence, the next position and next velocity of an agent can be computed as follows:

$$\mathbf{v}_i(t+1) = r_i \cdot \mathbf{v}_i(t) + \mathbf{a}_i(t) \tag{6.57}$$

$$\mathbf{x}_i(t+1) = \mathbf{x}_i(t) + \mathbf{v}_i(t+1) \tag{6.58}$$

where r_i is a uniform random number in the interval [0,1]; $\mathbf{x}_i(t)$, $\mathbf{v}_i(t)$, and $\mathbf{a}_i(t)$ are the position, velocity, and acceleration of agent i at iteration t, respectively.

The gravitational constant $G(t)$ in (6.54) is a function of the initial value G_0 and time (iteration) t:

$$G(t) = G_0 \cdot e^{-\alpha(t/t_{\max})} \tag{6.59}$$

The parameters of maximum iteration $t_{\max}$, population size N, initial gravitational constant G_0, and constant α control the performance of GSA.

6.4.4 Application of GSA to the OPF problem

The described GSA approach has been applied to solve the OPF problem in following steps:

Step 1: Load the power-system configuration, lines data, transformers data, shunt Var compensators data, loads data, and generation units data. Use the per unit system.

Step 2: Specify the control variables $\mathbf{x}$ (6.5) and their lower and upper limits (6.12–6.15), specify the dependent variables $\mathbf{y}$ (6.6) and their lower and upper limits (6.9–6.11), and specify the objective function to be optimized $F(\mathbf{x},\mathbf{y})$ (6.21–6.43) and the penalty factors ($\lambda_{P,V,Q,S,T}$) to form the expanded objective function $F_p(\mathbf{x},\mathbf{y})$ (6.44).

Step 3: Initialize GSA parameters, such as the population size (N), maximum iteration number ($t_{\max}$), initial constant K_0, initial gravitational constant G_0, and constant α.

Step 4: Generate an initial random population of N agents. The initial positions of each agent are randomly selected between minimum and maximum values of the control variables (6.49). Initialize the velocity of agents by a zero matrix of dimension $N \times n$, where N is the number of agents (population size) and n is the number of control variables.

Step 5: Run the power-flow program based on fast decoupled power-flow method for each agent $\mathbf{x}_i(t)$ from the current population $\mathbf{POP}(t)$ and calculate the corresponding values of the objective function (fitness values).

Step 6: Calculate the gravitational constant $G(t)$ (6.59), the best and worst fitness value, and the mass of each agent $M_i(t)$ (6.52), $i = 1, \ldots N$.

Step 7: Calculate the total force in different directions using (6.54) and the acceleration of each agent using (6.56).

Step 8: Update the velocity of each agent using (6.57).

Step 9: Update the position of each agent using (6.58).

Step 10: Repeat steps 5–9 until the stop criteria are reached, that is, the maximum number of iterations $t_{\max}$.

Step 11: Return best solution obtained in the last iteration; Stop.

6.4.5 Overview of hybrid PSOGSA

The basic idea of PSOGSA is to combine the ability for social thinking (**gbest**) in PSO with the local search capability of GSA [128]. In order to combine these algorithms, the equation for updating the agents' velocity is proposed as follows [120]:

$$\mathbf{v}_i(t+1) = r_1 \cdot \mathbf{v}_i(t) + C_1 \cdot r_2 \cdot \mathbf{a}_i(t) + C_2 \cdot r_3 \cdot (\mathbf{gbest}(t) - \mathbf{x}_i(t)) \quad i = 1, \ldots, N \tag{6.60}$$

where $\mathbf{v}_i(t)$, $\mathbf{a}_i(t)$, and $\mathbf{x}_i(t)$ are the velocity, acceleration, and position of the ith agent at iteration t, respectively; $\mathbf{gbest}(t)$ is the best position (solution) so far at iteration t; C_1 and C_2 are positive constants; r_1, r_2, and r_3 are uniformly distributed random numbers in [0,1]; N is the size of population (the number of agents), where each agent consists of n elements, i.e., n control variables.

In each iteration, the positions of agents are updated as follows:

$$\mathbf{x}_i(t+1) = \mathbf{x}_i(t) + \mathbf{v}_i(t+1) \quad i = 1, \ldots, N \tag{6.61}$$

In PSOGSA, at first, all agents are randomly initialized. Each agent is considered as a candidate solution. After initialization, the gravitational constant and resultant forces among agents are calculated using (6.59) and (6.54), respectively. After that, the accelerations of agents are defined as (6.56). In each iteration, the best solution so far should be updated. After calculating the accelerations and updating the best solution so far, the velocities of all agents can be calculated using (6.60). Finally, the positions of agents are updated by (6.61). The process of updating velocities and positions will be stopped by meeting an end criterion.

6.4.6 Application of PSOGSA to the OPF problem

The proposed hybrid PSOGSA approach has been applied to solve the OPF problem. Different steps to solve the OPF problem using PSOGSA are listed as follows:

Step 1: Read the input data including the power-system configuration, lines data, transformers data, shunt Var compensators data, loads data, and generation units data. Use the per unit system.

Step 2: Specify the control variables **x** (6.5) and their lower and upper limits (6.12–6.15), specify the dependent variables **y** (6.6) and their lower and upper limits (6.9–6.11), and specify the objective function to be optimized $F(\mathbf{x},\mathbf{y})$ (6.21–6.43) and the penalty factors ($\lambda_{P,V,Q,S,T}$) to form the expanded objective function $F_p(\mathbf{x},\mathbf{y})$ (6.44).

Step 3: Initialize PSOGSA parameters, such as the population size (N), maximum iteration number ($t_{\max}$), initial constant K_0, initial gravitational constant G_0, and constants α, C_1, and C_2.

Step 4: Generate an initial random population of N agents. The initial positions of each agent are randomly selected between minimum and maximum values of the control variables (6.49). Initialize the velocity of agents by a zero matrix of dimension $N \times n$, where N is the number of agents (population size) and n is the number of control variables.

Step 5: Run the power-flow program based on fast decoupled power-flow method for each agent $\mathbf{x}_i(t)$ from the current population $\mathbf{POP}(t)$ and calculate the corresponding values of the objective function (fitness values).

Step 6: Calculate the gravitational constant $G(t)$ (6.59), the best and worst fitness value, and the mass of each agent $M_i(t)$ (6.52), $i = 1, \ldots, N$.

Step 7: Calculate the total force in different directions using (6.54), and the acceleration of each agent using (6.56).

Step 8: Update the best solution so far $\mathbf{gbest}(t)$.

Step 9: Update the velocity of each agent using (6.60).

Step 10: Update the position of each agent using (6.61).

Step 11: Repeat steps 5–10 until the stop criteria is reached, that is, the maximum number of iterations $t_{\max}$.

Step 12: Return best solution obtained in the last iteration; Stop.

6.5 Simulation results

The PSO, GSA, and hybrid PSOGSA have been implemented in a MATLAB GUI-based software tool for solving the OPF problem. The computer program, called *opfgui*, has been developed to present the efficiency of different metaheuristic optimization methods in solving the OPF. The program is described in Section 6.6.

This section presents OPF results obtained by PSO, GSA, and hybrid PSOGSA on the IEEE 30-bus and IEEE 118-bus power systems. These systems are established as the standard test systems for evaluating of the OPF methods.

The *opfgui* program was run in the MATLAB 2011b (The MathWorks, Natick, Massachusetts, USA) computing environment on a 2.20-GHz PC with 3.0-GB RAM.

6.5.1 IEEE 30-bus test system

The IEEE 30-bus test system has 41 transmission lines; 6 generators at the buses 1, 2, 5, 8, 11, and 13; 4 transformers with off-nominal tap ratio at lines 6–9, 6–10, 4–12, and 28–27; and 9 shunt Var compensators at the buses 10, 12, 15, 17, 20, 21, 23, 24, and 29. The total load demands are $P_{load} = 2.834$ p.u. and $Q_{load} = 1.262$ p.u. at 100 MVA base. There are 24 control variables as follows: 5 generator active power outputs, 6 generator voltage magnitudes, 4 transformer tap settings, and 9 shunt Var compensator power injections. The single-line diagram of the network containing the lower and upper limits of the control variables is shown in Figure 6.2. The system data are taken from [129]. The system line and bus data are provided in Tables 6.1 and 6.2, whereas active and reactive power limits and cost-coefficient data for generators are listed in Tables 6.3 and 6.4.

The different algorithms' parameters used for the simulation are adopted as follows: For PSO, C_1 and C_2 are set to 2 and w decreases linearly from 0.9 to 0.4;

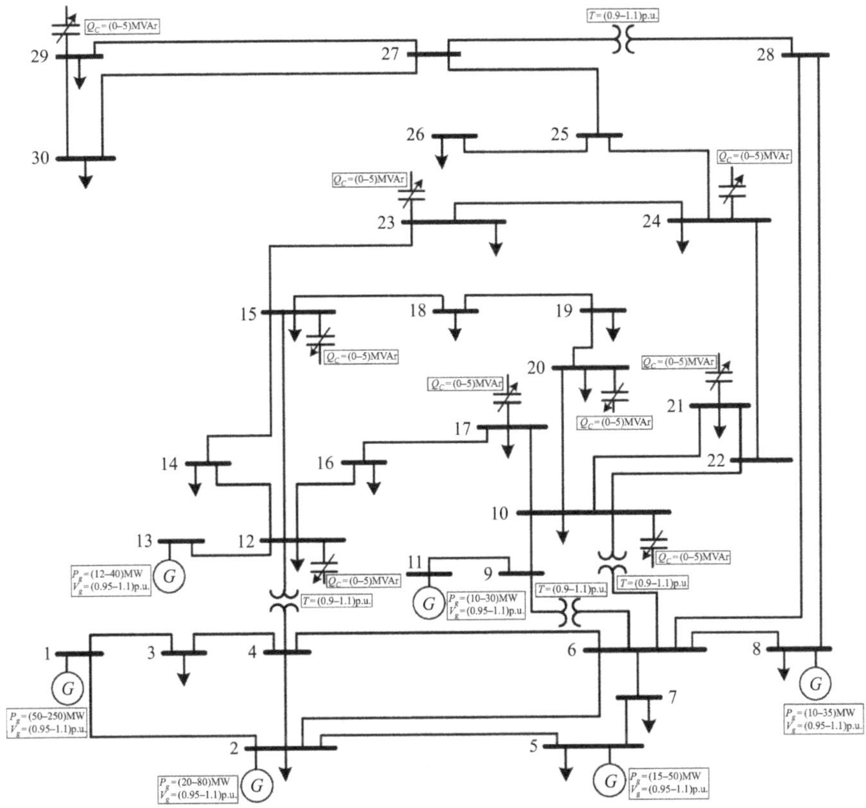

Figure 6.2 Single-line diagram of IEEE 30-bus test system

for GSA, α is set to 10 and G_0 is set to 100; for hybrid PSOGSA algorithm, C_1 and C_2 are set to 2, α is set to 20, and G_0 is set to 1. The population size (N) and maximum iteration number (t_{max}) are set to 50 and 200, respectively, for all case studies. These values were selected after a number of careful experimentation.

Twenty consecutive test runs have been performed for each case examined. The results shown are the best values obtained over these 20 runs.

6.5.1.1 Case 1: minimization of fuel cost

The generator cost characteristics are defined as quadratic cost function of generator power output P_G, given by (6.21). The objective is minimization of fuel cost for all generators in the system.

Most of authors considered the lower and upper voltage limits for the load buses to be 0.95 and 1.05 in p.u. [7,16,17,27,47,70,100,130]. However, in some recent papers [5,56,57,91,102,118], the load bus voltage limits are taken as 0.95 and 1.10 p.u. In order to demonstrate the effectiveness of the PSOGSA algorithm

Table 6.1 Line data for IEEE 30-bus test system

Line no.	From bus	To bus	*R* (p.u.)	*X* (p.u.)	*B* (p.u.)	Tap settings
1	1	2	0.0192	0.0575	0.0264	–
2	1	3	0.0452	0.1852	0.0204	–
3	2	4	0.0570	0.1737	0.0184	–
4	3	4	0.0132	0.0379	0.0042	–
5	2	5	0.0472	0.1983	0.0209	–
6	2	6	0.0581	0.1763	0.0187	–
7	4	6	0.0119	0.0414	0.0045	–
8	5	7	0.0460	0.1160	0.0102	–
9	6	7	0.0267	0.0820	0.0085	–
10	6	8	0.0120	0.0420	0.0045	–
11	6	9	0	0.2080	0	1.078
12	6	10	0	0.5560	0	1.069
13	9	11	0	0.2080	0	–
14	9	10	0	0.1100	0	–
15	4	12	0	0.2560	0	1.032
16	12	13	0	0.1400	0	–
17	12	14	0.1231	0.2559	0	–
18	12	15	0.0662	0.1304	0	–
19	12	16	0.0945	0.1987	0	–
20	14	15	0.2210	0.1997	0	–
21	16	17	0.0824	0.1932	0	–
22	15	18	0.1070	0.2185	0	–
23	18	19	0.0639	0.1292	0	–
24	19	20	0.0340	0.0680	0	—
25	10	20	0.0936	0.2090	0	–
26	10	17	0.0324	0.0845	0	–
27	10	21	0.0348	0.0749	0	–
28	10	22	0.0727	0.1499	0	–
29	21	22	0.0116	0.0236	0	–
30	15	23	0.1000	0.2020	0	–
31	22	24	0.1150	0.1790	0	–
32	23	24	0.1320	0.2700	0	–
33	24	25	0.1885	0.3292	0	–
34	25	26	0.2544	0.3800	0	–
35	25	27	0.1093	0.2087	0	–
36	28	27	0	0.3960	0	1.068
37	27	29	0.2198	0.4153	0	–
38	27	30	0.3202	0.6027	0	–
39	29	30	0.2399	0.4533	0	–
40	8	28	0.0636	0.2000	0.0214	–
41	6	28	0.0169	0.0599	0.0065	–

and fair comparison of the obtained optimal solution with those that reported in the literature recently, two different versions of Case 1 have been considered:

Case 1a: The voltages of all load bus have been constrained within limits of 0.95 and 1.05 p.u.

Table 6.2 Load data for IEEE 30-bus test system (100-MVA base)

Bus no.	**Load**		**Bus no.**	**Load**		**Bus no.**	**Load**	
	***P* (p.u.)**	***Q* (p.u.)**		***P* (p.u.)**	***Q* (p.u.)**		***P* (p.u.)**	***Q* (p.u.)**
1	0	0	11	0	0	21	0.175	0.112
2	0.217	0.127	12	0.112	0.075	22	0	0
3	0.024	0.012	13	0	0	23	0.032	0.016
4	0.076	0.016	14	0.062	0.016	24	0.087	0.067
5	0.942	0.190	15	0.082	0.025	25	0	0
6	0	0	16	0.035	0.018	26	0.035	0.023
7	0.228	0.109	17	0.090	0.058	27	0	0
8	0.300	0.300	18	0.032	0.009	28	0	0
9	0	0	19	0.095	0.034	29	0.024	0.009
10	0.058	0.020	20	0.022	0.007	30	0.106	0.019

Table 6.3 Generator data for IEEE 30-bus test system (100 MVA base)

Bus no.	P_{Gmin} **(p.u.)**	P_{Gmax} **(p.u.)**	Q_{Gmin} **(p.u.)**	Q_{Gmax} **(p.u.)**	**Cost coefficients**				
					a	*b*	*c*	*d*	*e*
1	0.5	2.5	−0.2	2	0	200	37.5	18	3.7
2	0.2	0.8	−0.2	1	0	175	175	16	3.8
5	0.15	0.5	−0.15	0.8	0	100	625	14	4
8	0.1	0.35	−0.15	0.6	0	325	83.4	12	4.5
11	0.1	0.3	−0.1	0.5	0	300	250	13	4.2
13	0.12	0.4	−0.15	0.6	0	300	250	13.5	4.1

Table 6.4 Generator data for IEEE 30-bus test system in Case 4b

Bus no.	P_{Gmin} **(p.u.)**	P_{Gmax} **(p.u.)**	Q_{Gmin} **(p.u.)**	Q_{Gmax} **(p.u.)**	**Cost coefficients**				
					a	*b*	*c*	*d*	*e*
1	0.5	2.0	−0.2	2	150	200	16	50	6.3
2	0.2	0.8	−0.2	1	25	250	100	40	9.8

Case 1b: The voltages of all load bus have been constrained within limits of 0.95 and 1.10 p.u.

The optimal settings of control variables are given in Table 6.5. It appears that the total fuel cost is highly reduced in both cases (Case 1a and Case 1b) compared to the initial case. The convergence of algorithms for the minimum fuel cost is shown in Figures 6.3 and 6.4. It is clear that the proposed hybrid PSOGSA algorithm can converge to its global best in less iteration compared with PSO and GSA algorithms.

Table 6.5 Optimal solution obtained by PSOGSA for different test cases in IEEE 30-bus system

	Initial	**Case 1a**	**Case 1b**	**Case 2a**	**Case 2b**	**Case 3**	**Case 4**	**Case 5**
P_{G1} (MW)	99.22227	177.22184	177.05846	219.79893	199.56361	127.61738	176.17236	177.91887
P_{G2} (MW)	80	48.74698	48.69769	27.76368	20.00000	51.72966	48.42573	49.01737
P_{G5} (MW)	50	21.39255	21.30436	15.77727	20.81501	29.83039	21.60725	21.49628
P_{G8} (MW)	20	21.10061	21.08006	10.00000	27.94175	34.99994	22.60044	22.23327
P_{G11} (MW)	20	11.97192	11.88402	10.00000	12.84380	24.75936	12.61330	10.00000
P_{G13} (MW)	20	12.00006	12.00000	12.00000	12.06233	19.93146	12.00000	12.00000
V_{G1} (p.u.)	1.05	1.08459	1.10000	1.09264	1.08836	1.10000	1.02986	1.06960
V_{G2} (p.u.)	1.04	1.06513	1.08785	1.06727	1.06504	1.09058	1.01837	1.05536
V_{G5} (p.u.)	1.01	1.03370	1.06166	1.03262	1.03381	1.06752	1.01569	1.03487
V_{G8} (p.u.)	1.01	1.03848	1.06940	1.03505	1.04052	1.07835	1.00700	1.04568
V_{G11} (p.u.)	1.05	1.08592	1.10000	1.10000	1.10000	1.10000	1.04449	1.10000
V_{G13} (p.u.)	1.05	1.04114	1.10000	1.06462	1.06198	1.10000	1.00053	1.05126
$T_{11(6-9)}$ (p.u.)	1.078	1.02559	1.04473	1.02475	1.03152	1.04293	1.06366	1.04235
$T_{12(6-10)}$ (p.u.)	1.069	0.95857	0.90000	0.90000	0.90000	0.90000	0.90116	0.95619
$T_{15(4-12)}$ (p.u.)	1.032	0.96478	0.98630	0.97032	0.96529	0.98097	0.95846	0.99467
$T_{36(28-27)}$ (p.u.)	1.068	0.97458	0.96570	0.95795	0.96724	0.96771	0.96830	0.98518
Q_{C10} (MVAr)	0	3.92906	5.00000	–	–	5.00000	4.67903	4.99995
Q_{C12} (MVAr)	0	3.68477	5.00000	–	–	5.00000	0.00592	5.00000
Q_{C15} (MVAr)	0	4.57095	5.00000	–	–	5.00000	4.98146	5.00000
Q_{C17} (MVAr)	0	5.00000	5.00000	–	–	4.99999	0.00000	4.99965
Q_{C20} (MVAr)	0	4.20562	5.00000	–	–	5.00000	4.99879	5.00000
Q_{C21} (MVAr)	0	5.00000	5.00000	–	–	5.00000	5.00000	5.00000
Q_{C23} (MVAr)	0	3.17397	3.82855	–	–	3.84156	4.99997	5.00000
Q_{C24} (MVAr)	0	5.00000	5.00000	–	–	5.00000	5.00000	5.00000
Q_{C29} (MVAr)	0	2.64020	2.74001	–	–	2.87091	2.58595	4.99994
Cost ($/h)	901.94942	**800.49859**	**799.07055**	**824.70959**	**919.65785**	**822.40631**	**804.43123**	**801.22928**
Ploss (MW)	5.82225	9.03390	8.62457	11.93989	9.82651	**5.46816**	10.01913	9.26567
Volt. dev. (p.u.)	1.14966	0.91373	1.85735	0.73034	0.71741	1.99222	**0.09638**	1.01011
L_{max}	0.17233	0.12674	0.11640	0.13882	0.13965	0.11513	0.13655	**0.12393**

Bold indicates the objective function that optimized.

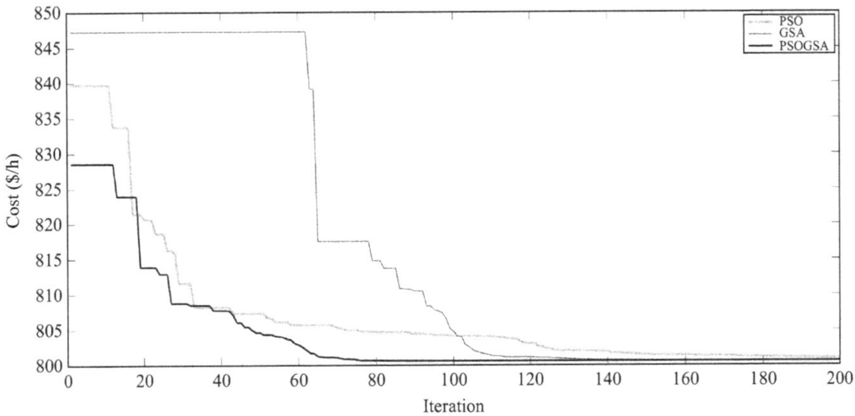

Figure 6.3 Convergence of PSO, GSA, and PSOGSA for Case 1a

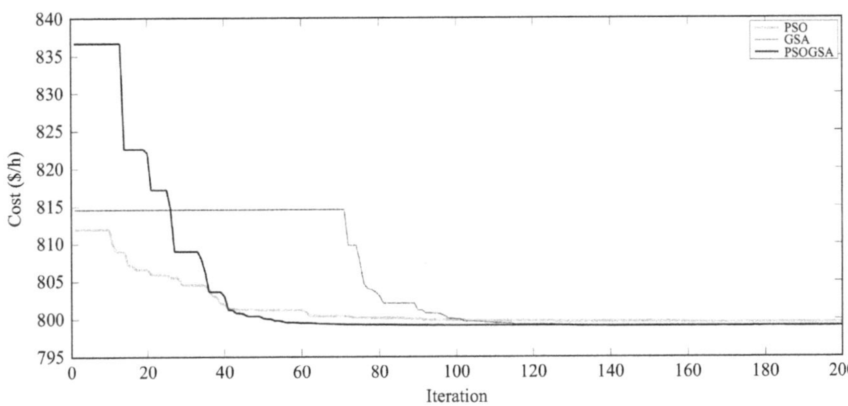

Figure 6.4 Convergence of PSO, GSA, and PSOGSA for Case 1b

6.5.1.2 Case 2: minimization of fuel cost with considering the valve-point effect

For this case, the objective function is defined by (6.22). In order to comparison with the reported results, two different cases for the IEEE 30-bus test system have been considered:

Case 2a: Minimization of fuel cost with considering the valve-point effect for all generators. The fuel cost coefficients of the generators are given in Table 6.3.

Case 2b: Minimization of fuel cost with considering the valve-point effect for generators 1 and 2 only. In this case, the fuel cost coefficients of generators 1 and 2 are listed in Table 6.4, whereas the fuel cost coefficients of remaining generators (at buses 5, 8, 11, and 13) have the same values given in Table 6.3.

In these cases, the shunt Var compensators in busses 10, 12, 15, 17, 20, 21, 23, 24, and 29 are ignored and the voltages at load buses are considered within the range 0.95–1.05 p.u.

The obtained optimal settings of control variables found using the proposed PSOGSA approach are given in Table 6.5. Figures 6.5 and 6.6 illustrate the convergence characteristics of the algorithms for Case 2a and Case 2b, respectively. As can been seen, the proposed hybrid PSOGSA algorithm can converge to its global optima in fewer iterations compared with PSO and GSA.

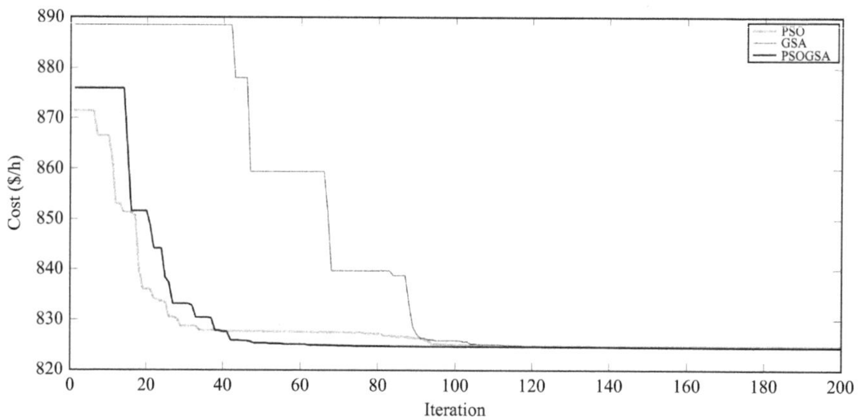

Figure 6.5 Convergence of PSO, GSA, and PSOGSA for Case 2a

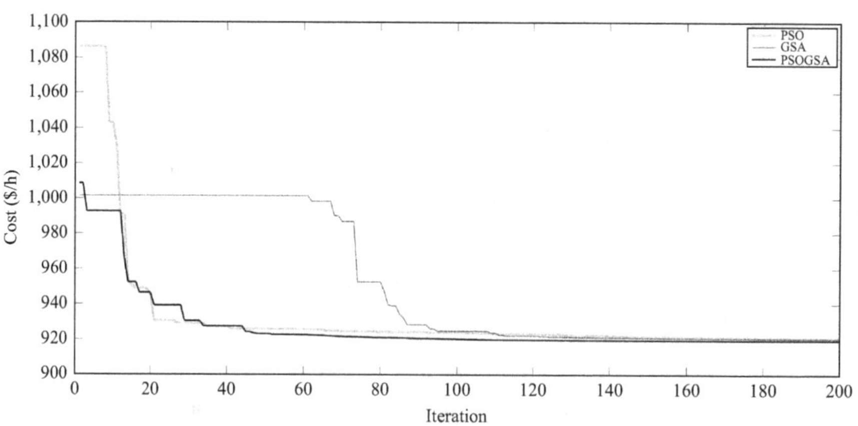

Figure 6.6 Convergence of PSO, GSA, and PSOGSA for Case 2b

6.5.1.3 Case 3: minimization of fuel cost and active power loss

The objective function in the case simultaneous minimization of total fuel cost and active power loss is expressed by (6.31). The value of w_{Ploss} in the present study is selected as 1,950. For fair comparison with the results reported in [16,118], the lower and upper limits of the voltage at load buses were assumed to be 0.95 and 1.10 p.u., respectively. After applying the PSOGSA algorithm, it appears from Table 6.5 that the value of power loss is significantly reduced in this case compared to the Case 1b, but the fuel cost and voltage deviation is increased.

6.5.1.4 Case 4: minimization of fuel cost and voltage deviation

Simultaneous minimization of the total fuel cost and voltage deviation is considered in this case. The objective function is described by (6.32). In this study, the value of weighting factor w_V in (6.32) is selected as 200. The proposed algorithms have been applied to search for the optimal solution of Case 4. These optimal results are given in Table 6.5. The total fuel cost and voltage deviations are 804.43123 \$/h and 0.09638 p.u. for this case compared to 800.49859 \$/h and 0.91373 p.u. for Case 1a. Although the cost has been increased by 0.4913%, the voltage deviation has been reduced by 89.45%. The system voltage profile of this case is compared to that of Case 1a and Case 1b as shown in Figure 6.7. It is evident that the voltage profile is greatly improved compared to that of Case 1a, and especially to the Case 1b, which is quite expected considering upper voltage limits for load buses in this case.

6.5.1.5 Case 5: minimization of fuel cost and voltage-stability index

The objective function for this case is defined by (6.33). The value of w_L in this case is chosen as 6,000. As can be seen from Table 6.5, the minimum fuel cost and index L_{max} found by the PSOGSA algorithm were 801.22928 and 0.12393, respectively, which show a 2.22% reduction in the L-index and about a 0.09% increase in the fuel cost in comparison with Case 1a.

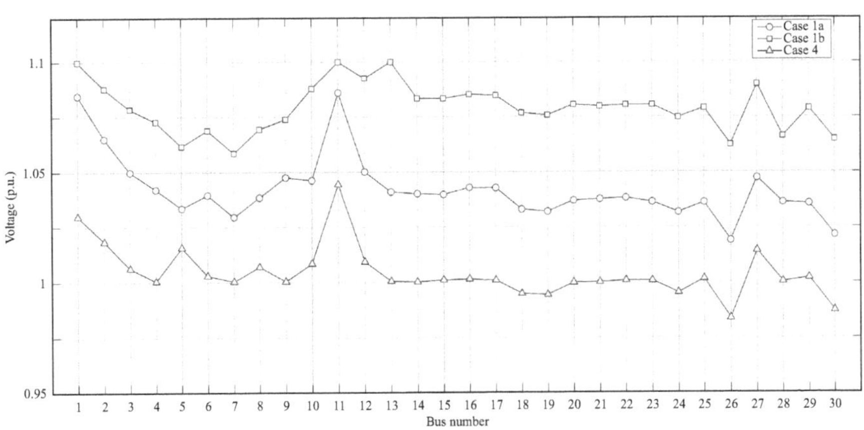

Figure 6.7 System-voltage profile

The value of L_{max} in Case 5 is higher than the values of L_{max} in Cases 1b and 3, because in those cases, the upper voltage limit at the load buses is considered to be 1.10 p.u.

6.5.1.6 Comparative study

Under the same system data, control variable limits, and constraints, the results obtained using the hybrid PSOGSA approach reported herein are compared to some other algorithms reported in the literature, as shown in Table 6.6. Bold font has been used to indicate the proposed algorithm and better performance.

Case 1a: For Case 1a, it can be seen in Table 6.6 that the methods reported in [16,17,27,47,100,130] have resulted in lesser minimum fuel cost than proposed PSOGSA approach. However, after the power-flow computation with control variables reported in these references, those results can be considered as infeasible solutions. Reasons for infeasibility of those results are summarized as follows:

- the best solution mentioned in [16] is an infeasible solution because there were bus voltage violations at all load buses except at buses 7, 18, 19, 20, 24, 25, 26, 29, and 30;
- for the best solution given in [17], the voltage limits are violated at the load buses 3 and 9; the true value of P_{G1} is 178.10537 MW, and the total fuel cost is 800.48421 \$/h for this violated case;
- for the best solution given in [27], there were voltage violations at the load bus 3;
- for the results reported in [47], the voltage limits are violated at all load buses except at buses 4, 6, 7, and 28; in addition, the reactive power limits of generators at buses 2 and 8 are violated also; the true value of P_{G1} is 177.78284 MW and exact fuel cost is 805.43661 \$/h for this violated case;
- for the optimum control variables given in [100], the voltage limits are violated at all load buses; and
- for the best solution given in [130], the voltage limits are violated at the load buses 3, 4, 26, 29, and 30.

Case 1b: The results in Case 1b obtained from the proposed PSOGSA approach were compared to the results reported in [5,57,91,102,118]. In these works, the load bus voltage limits are assumed to be 0.95 and 1.10 p.u., respectively. Some of those results are marked as incorrect solutions:

- for the optimum control variables given in [91], the true value of P_{G1} is 177.05766 MW and the total fuel cost for this case is 799.08296 \$/h;
- for the optimum control variables given in [102], the true value of P_{G1} is 177.33037 MW and the total fuel cost for this case is 799.97946 \$/h; and
- for the optimum control variables given in [118], the true value of P_{G1} is 177.01450 MW, as obtained from the power-flow computations, and the total fuel cost for this case is 799.10122 \$/h.

Case 2a: For Case 2a, the results obtained from the proposed PSOGSA approach were compared to the methods reported in [53,115,117]. The best solution given in

Table 6.6 Comparison of the simulation results for IEEE 30-bus test system

Case	Method	Cost ($/h)	*Ploss* (MW)	*VD* (p.u.)	L_{max}
Case 1a	EGA [7]	802.06	–	–	–
	EGA [16]	799.56	8.697	–	0.111
	AGA [17]	799.8441	8.9166	0.8043	–
	PSO [27]	800.41	–	0.8765	
	GSA [47]	798.6751	8.386	0.87286	0.130759
	ABC [70]	800.66	9.0328	0.9209	0.1381
	BBO [100]	799.1116	8.63	–	–
	GPM [130]	804.853	10.486	–	–
	PSOGSA	**800.49859**	9.03390	0.91373	0.12674
Case 1b	DE [5,57]	799.2891	8.6150	1.5306	0.1226
	KHA [91]	799.0311	–	1.9485	0.1265
	TLBO [102]	799.0715	8.6260	1.8925	0.1159
	ABPPO [118]	799.0566	8.615	1.8206	0.11796
	PSOGSA	**799.07055**	8.62457	1.85735	0.11640
Case 2a	DE [53]	826.54	–	–	–
	RCGA [53]	831.04	–	–	–
	MICA–TLA [115]	824.6023	11.9122	–	–
	SFLA–SA [117]	825.6922	–	–	–
	PSOGSA	**824.70959**	11.93989	0.73034	0.13882
Case 2b	GSA [47]	929.72405	15.14585	0.577974	0.156484
	EP [61]	919.89	–	–	–
	IEP [63]	953.573	–	–	–
	ABC [70]	945.4495	14.0928	0.5284	0.1431
	BBO [100]	919.7647	12.18	–	–
	TLBO [102]	923.4147	10.9599	1.0282	0.1253
	PSOGSA	**919.65785**	9.82651	0.71741	0.13965
Case 3	EGA [16]	822.87	5.613	–	–
	ABPPO [118]	822.7693	5.452	2.05503	0.11546
	PSOGSA	**822.40631**	**5.46816**	1.99222	0.11513
Case 4	DE [5,57]	805.2619	10.4412	0.1357	0.1316
	PSO [27]	806.38	–	0.0891	0.1392
	GSA [47]	804.3148	9.7659	0.093269	0.13578
	KHA [91]	804.6337	–	0.0996	0.1466
	BBO [100]	804.9982	9.95	0.1020	–
	TLBO [102]	803.7871	9.8641	0.0945	0.1369
	MOMICA [108]	804.9611	9.8212	0.0952	–
	ABPPO [118]	804.7339	10.072	0.09232	0.13648
	PSOGSA	**804.43123**	10.01913	**0.09638**	0.13655
Case 5	EGA [16]	802.06	–	–	0.1056
	PSO [27]	801.16	–	0.9607	0.1246
	GSA [47]	806.6013	9.9161	0.9000	0.116247
	ABC [70]	801.6650	9.2954	0.9380	0.1379
	BBO [100]	805.7252	10.21	–	0.1104
	PSOGSA	**801.22928**	9.26567	1.01011	**0.12393**

[115] is an infeasible solution because the upper voltage limit at bus 3 is violated. From Table 6.6, it is obvious that the proposed PSOGSA algorithm has resulted in less fuel cost than those reported in [53,115,117].

Case 2b: The best solution in Case 2b obtained from the proposed PSOGSA approach was compared to the methods reported in [47,61,63,70,100,102]. Some of those reported results are incorrect or infeasible solutions:

- for the optimum control variables given in [47], the reactive power limits of generators at buses 2, 5, and 11 are violated; in addition, the voltages at buses 18, 19, 23, 24, 26, 29, and 30 are less than 0.95 p.u.; the true value of P_{G1} is 202.28001 MW, and the exact fuel cost is 945.21117 \$/h for this violated case;
- for the optimum control variables given in [61], the lower reactive power limit of generator at bus 1 is violated; the true value of P_{G1} is 200.66618 MW, and the total fuel cost for this violated case is 926.07046 \$/h;
- for the optimum control variables given in [100], the true value of P_{G1} is 200.13339 MW, and the total fuel cost for this case is 955.80191 \$/h; and
- for the optimum control variables given in [102], the exact fuel cost is 962.12485 \$/h, as obtained from the power-flow computations.

Case 3: From Table 6.6, it can be seen that the OPF results in Case 3 obtained by the PSOGSA algorithm are either better or comparable to those obtained using methods reported in [16,118].

Case 4: The results obtained from the proposed PSOGSA algorithm were compared to the methods reported in [5,27,47,57,91,100,102,118]. Some of those results are incorrect solutions, as obtained from the power-flow computations:

- for the optimum control variables given in [27], the exact voltage deviation is 0.09622 p.u., and the total fuel cost is 806.40971 \$/h;
- for the optimum control variables given in [47], the exact voltage deviation is 0.10700 p.u.;
- for the optimum control variables given in [102], the exact cost is 804.99389 \$/h, and exact voltage deviation is 0.55276 p.u.; and
- for the optimum control variables given in [118], the exact voltage deviation is 0.11302 p.u.

Therefore, it is clear that the OPF result in Case 4 obtained by the PSOGSA algorithm is better than those reported in the referenced literature.

Case 5: The results obtained from the PSOGSA algorithm have been compared with other algorithms reported in [16,27,47,70,100]. Some of those algorithms have resulted in better performance than the proposed PSOGSA approach. However, it could be shown that those results are indeed infeasible solutions. Reasons for infeasibility of those results are summarized as follows:

- the best solution given in [16] is an infeasible solution because reactive powers of the generators at buses 1, 8, 11, and 13 were given as −30.63, 82.61, −12.64, and −17.59 MVAr, respectively, which violate their corresponding limits as reported Table 6.3; in addition, the bus voltages at all load buses, except bus 30, are higher than their upper limits;

- the optimum control variables given in [27] represent an infeasible solution because the reactive powers of the generators at buses 11 and 13 are −10.75 and −15.45 MVAr, respectively, which violate their lower limits;
- the best result given in [47] is an infeasible solution because reactive powers of the generators at buses 8, 11, and 13 are 81.35, −21.63, and −24.45 MVAr, respectively, which violate their limits; in addition, the voltage limits are violated at all load buses; and
- for the optimum control variables given in [100], the upper limits of the voltages at buses 27 and 29 are violated.

Based on the comparative study of the simulation results presented in this section, it can be noted that the proposed PSOGSA approach outperforms many techniques used to solve different OPF problems. This highlights its ability to find better quality solution.

In addition, the comparison of minimum, maximum, and standard deviation of the results obtained by PSO, GSA, and hybrid PSOGSA over 20 runs is presented in Table 6.7. It is clear that the hybrid PSOGSA approach provides better and more stable solutions compared to the original PSO and GSA.

6.5.2 IEEE 118-bus test system

To evaluate the efficiency of the proposed PSOGSA approach in solving a larger power system, the IEEE 118-bus test system is considered. The single-line diagram of IEEE 118-bus test system is shown in Figure 6.8. The system data, with generator active and reactive power limits and cost coefficients, are taken from [131]. The system data are available in [132] as well. The test system has 54 generators, 186 branches, 9 transformers, 2 reactors, and 12 capacitors. It has a total of 129 control variables as follows: 54 generator active powers and bus voltages, 9 transformer tap settings, and 12 shunt capacitor reactive power injections. The voltage limits of all busses are between 0.94 and 1.06 p.u. The transformer tap settings are considered within the interval 0.90–1.10 p.u. The available reactive powers of shunt capacitors are within the range 0–30 MVAr.

One optimization case was considered: the minimization of the total fuel cost of all generators. The objective is as in Case 1 for the IEEE 30-bus test system. The test results obtained using proposed PSOGSA along with PSO and GSA are presented in Table 6.8. The total fuel cost obtained by the proposed PSOGSA is reduced to be 129,733.58 $/h compared to the initial case fuel cost of 131,220.52 $/h. The control variable settings corresponding to fuel cost-based OPF result in a reduction of 1.13% in fuel cost and significant loss reduction of 44.9%. It can be seen that the results obtained from the PSOGSA are better than those obtained from the PSO and GSA methods. This clearly indicates that the proposed PSOGSA algorithm can be effectively used to solve the OPF problem for large-scale power systems.

Table 6.7 Minimum, maximum, and standard deviation of the results obtained over 20 independent runs with PSO, GSA, and PSOGSA

	PSO			GSA			PSOGSA		
	Min	**Max**	**Std dev**	**Min**	**Max**	**Std dev**	**Min**	**Max**	**Std dev**
Case 1a	800.6199	808.1951	2.1235	800.5356	804.7186	1.1577	800.4981	800.6111	0.0346
Case 1b	799.3586	805.2647	1.9553	799.07123	808.1124	2.0531	799.0705	799.1703	0.0259
Case 2a	824.8412	830.6802	1.9884	824.7196	837.7094	2.8764	824.7096	824.9921	0.0725
Case 2b	920.6937	946.8402	6.1272	920.6366	955.3907	13.2097	919.6579	924.2865	0.9384
Case 3	931.2734	958.5769	8.5905	929.6336	966.9396	8.3591	929.0354	930.0418	0.0983
Case 4	826.4021	867.7560	10.1548	823.9962	876.4744	10.5366	823.7072	829.6988	1.8387
Case 5	1571.0	1619.3	12.2823	1548.9	1590.2	10.9834	1547.2	1552.5	1.6108

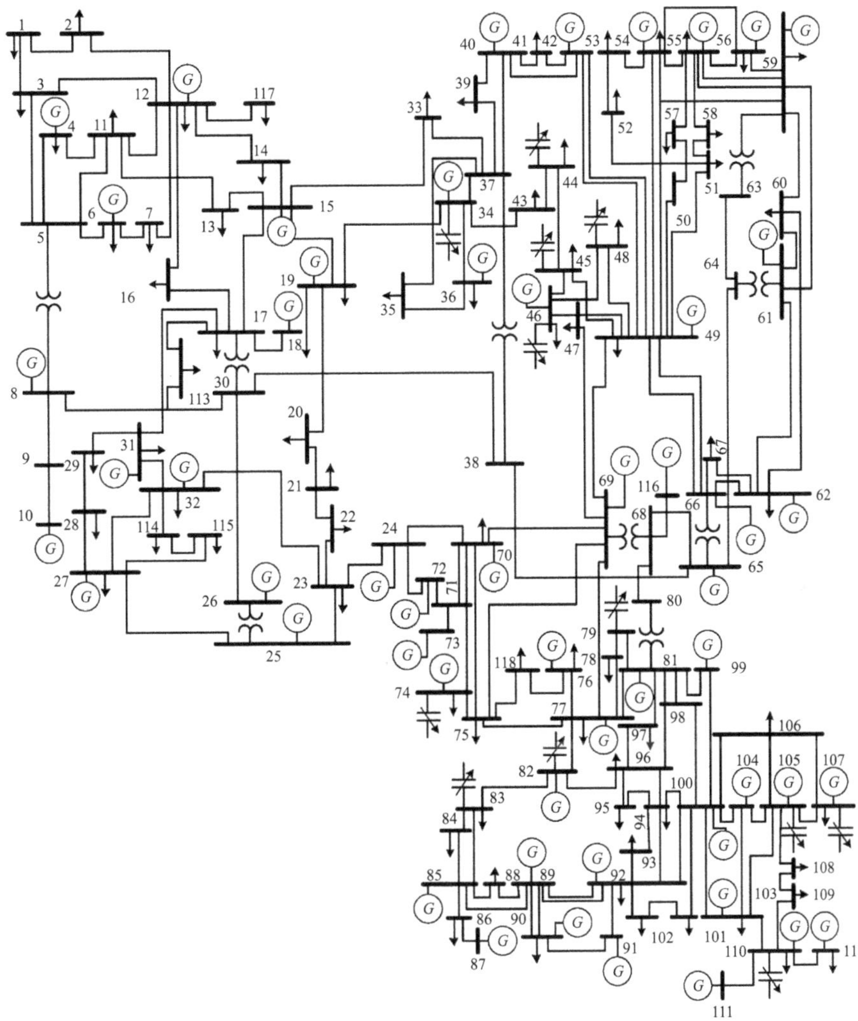

Figure 6.8 Single-line diagram of IEEE 118-bus test system

6.6 Solution software—opfgui

OPF graphical user interface (*opfgui*) is an educational software tool for the OPF solution using different metaheuristic optimization algorithms. The *opfgui* has been developed to present the efficiency of different metaheuristic optimization methods in solving the OPF problem. The *opfgui* can be freely downloaded from [132].

General flowchart for the OPF solution using metaheuristic optimization methods is shown in Figure 6.9.

The aim of this program is to encompass the main steps in solving the OPF problem using metaheuristic methods. These steps include (i) selection of test system, display single-line diagram, and edit system data; (ii) selection of objective

Table 6.8 Optimal settings of control variables for the IEEE 118-bus test system

Control variables	Initial	PSO	GSA	PSOGSA
P_{G1} (MW)	0	0	18.0203	26.3512
P_{G4} (MW)	0	0	0	0
P_{G6} (MW)	0	0	0	0.0383
P_{G8} (MW)	0	0	0	0
P_{G10} (MW)	450	402.6074	396.7158	398.8872
P_{G12} (MW)	85	86.3411	82.9682	85.4599
P_{G15} (MW)	0	0	100.0000	18.6446
P_{G18} (MW)	0	0	2.3084	9.2256
P_{G19} (MW)	0	0	0	20.1511
P_{G24} (MW)	0	0	0	0.0008
P_{G25} (MW)	220	194.2105	194.5240	193.3415
P_{G26} (MW)	314	275.7963	277.3266	279.4154
P_{G27} (MW)	0	0	0	12.9775
P_{G31} (MW)	7	7.3522	7.2690	7.2086
P_{G32} (MW)	0	6.1085	0	15.5474
P_{G34} (MW)	0	0	0	0.4701
P_{G36} (MW)	0	100.0000	0.1184	6.4111
P_{G40} (MW)	0	99.5142	89.4233	55.4906
P_{G42} (MW)	0	27.6644	45.9372	61.4402
P_{G46} (MW)	19	18.9795	19.4743	19.1722
P_{G49} (MW)	204	192.7700	191.7629	194.3087
P_{G54} (MW)	48	49.7514	50.3958	50.0011
P_{G55} (MW)	0	33.1114	36.0990	41.8031
P_{G56} (MW)	0	37.3379	46.8795	45.9483
P_{G59} (MW)	155	145.8597	141.9034	148.2103
P_{G61} (MW)	160	145.5823	146.1287	146.7431
P_{G62} (MW)	0	0	0	0.0000
P_{G65} (MW)	391	348.2607	347.1688	349.6582
P_{G66} (MW)	392	341.7118	343.5346	343.9029
P_{G69} (MW)	513.86	434.03	438.33	440.31
P_{G70} (MW)	0	0	0	0
P_{G72} (MW)	0	0	0	0
P_{G73} (MW)	0	0	0	0.2592
P_{G74} (MW)	0	17.1178	15.9548	17.3527
P_{G76} (MW)	0	0	23.7543	25.1604
P_{G77} (MW)	0	100.0000	0	0
P_{G80} (MW)	477	420.8776	426.4570	427.6091
P_{G85} (MW)	0	0	0.0000	0
P_{G87} (MW)	4	3.0771	3.4142	3.4759
P_{G89} (MW)	607	461.6651	464.2279	470.5221
P_{G90} (MW)	0	0	0	0.0001
P_{G91} (MW)	0	0	0.0052	0.0031
P_{G92} (MW)	0	0	0	0
P_{G99} (MW)	0	0	0	0.0000
P_{G100} (MW)	252	237.6351	235.2962	233.9947
P_{G103} (MW)	40	40.0098	38.5887	38.4579
P_{G104} (MW)	0	23.4303	0	0
P_{G105} (MW)	0	0	27.0793	9.4232

Table 6.8 (Continued)

Control variables	Initial	PSO	GSA	PSOGSA
P_{G107} (MW)	0	0	27.6829	33.1790
P_{G110} (MW)	0	35.3934	0	10.4795
P_{G111} (MW)	36	36.4896	35.3759	35.2538
P_{G112} (MW)	0	0	40.1863	38.8605
P_{G113} (MW)	0	0	0	0.0519
P_{G116} (MW)	0	0	0	0
V_{G1} (p.u.)	0.955	0.9980	1.0413	1.0227
V_{G4} (p.u.)	0.998	1.0340	1.0600	1.0487
V_{G6} (p.u.)	0.990	1.0255	1.0600	1.0392
V_{G8} (p.u.)	1.015	1.0574	1.0600	1.0600
V_{G10} (p.u.)	1.050	1.0600	1.0600	1.0600
V_{G12} (p.u.)	0.990	1.0208	1.0600	1.0371
V_{G15} (p.u.)	0.970	1.0254	1.0459	1.0311
V_{G18} (p.u.)	0.973	1.0306	1.0397	1.0334
V_{G19} (p.u.)	0.962	1.0263	1.0363	1.0302
V_{G24} (p.u.)	0.992	1.0389	1.0600	1.0444
V_{G25} (p.u.)	1.050	1.0535	1.0600	1.0600
V_{G26} (p.u.)	1.015	1.0600	1.0600	1.0600
V_{G27} (p.u.)	0.968	1.0252	1.0600	1.0366
V_{G31} (p.u.)	0.967	1.0195	1.0468	1.0290
V_{G32} (p.u.)	0.963	1.0222	1.0522	1.0343
V_{G34} (p.u.)	0.984	1.0520	1.0395	1.0287
V_{G36} (p.u.)	0.980	1.0547	1.0374	1.0263
V_{G40} (p.u.)	0.970	1.0411	1.0294	1.0110
V_{G42} (p.u.)	0.985	1.0291	1.0213	1.0052
V_{G46} (p.u.)	1.005	1.0335	1.0156	1.0426
V_{G49} (p.u.)	1.025	1.0474	1.0304	1.0577
V_{G54} (p.u.)	0.955	1.0157	0.9941	1.0555
V_{G55} (p.u.)	0.952	1.0135	0.9942	1.0545
V_{G56} (p.u.)	0.954	1.0150	0.9939	1.0538
V_{G59} (p.u.)	0.985	1.0557	1.0072	1.0600
V_{G61} (p.u.)	0.995	1.0599	1.0001	1.0600
V_{G62} (p.u.)	0.998	1.0566	1.0028	1.0552
V_{G65} (p.u.)	1.005	1.0272	1.0600	1.0600
V_{G66} (p.u.)	1.050	1.0600	1.0392	1.0600
V_{G69} (p.u.)	1.035	1.0600	1.0600	1.0600
V_{G70} (p.u.)	0.984	1.0269	1.0389	1.0344
V_{G72} (p.u.)	0.980	1.0291	1.0460	1.0362
V_{G73} (p.u.)	0.991	1.0262	1.0379	1.0335
V_{G74} (p.u.)	0.958	1.0145	1.0205	1.0169
V_{G76} (p.u.)	0.943	1.0040	1.0090	1.0050
V_{G77} (p.u.)	1.006	1.0460	1.0396	1.0344
V_{G80} (p.u.)	1.040	1.0600	1.0600	1.0483
V_{G85} (p.u.)	0.985	1.0545	1.0521	1.0502
V_{G87} (p.u.)	1.015	1.0600	1.0488	1.0479
V_{G89} (p.u.)	1.005	1.0600	1.0600	1.0600
V_{G90} (p.u.)	0.985	1.0343	1.0373	1.0311
V_{G91} (p.u.)	0.980	1.0334	1.0406	1.0264

(Continues)

Table 6.8 (Continued)

Control variables	Initial	PSO	GSA	PSOGSA
V_{G92} (p.u.)	0.990	1.0333	1.0382	1.0204
V_{G99} (p.u.)	1.010	1.0600	1.0600	1.0286
V_{G100} (p.u.)	1.017	1.0600	1.0600	1.0276
V_{G103} (p.u.)	1.010	1.0483	1.0542	1.0154
V_{G104} (p.u.)	0.971	1.0436	1.0474	1.0041
V_{G105} (p.u.)	0.965	1.0400	1.0460	1.0003
V_{G107} (p.u.)	0.952	1.0600	1.0341	0.9955
V_{G110} (p.u.)	0.973	0.9986	1.0487	0.9925
V_{G111} (p.u.)	0.980	0.9942	1.0600	1.0005
V_{G112} (p.u.)	0.975	0.9797	1.0417	0.9851
V_{G113} (p.u.)	0.993	1.0361	1.0533	1.0409
V_{G116} (p.u.)	1.005	1.0253	1.0600	1.0600
T_{5-8} (p.u.)	0.985	1.0046	0.9750	0.9929
T_{26-25} (p.u.)	0.960	1.0592	1.0626	1.0484
T_{30-17} (p.u.)	0.960	0.9948	0.9925	1.0041
T_{38-37} (p.u.)	0.935	0.9666	0.9968	1.0129
T_{63-59} (p.u.)	0.960	0.9559	1.0335	0.9812
T_{64-61} (p.u.)	0.985	0.9725	1.0605	0.9968
T_{65-66} (p.u.)	0.935	0.9524	0.9882	0.9819
T_{68-69} (p.u.)	0.935	0.9000	0.9494	0.9621
T_{81-80} (p.u.)	0.935	0.9564	0.9894	0.9985
Q_{C34} (MVAr)	0	30	30	1.77
Q_{C44} (MVAr)	0	0	0	0
Q_{C45} (MVAr)	0	30	0	10.20
Q_{C46} (MVAr)	0	0	30	14.85
Q_{C48} (MVAr)	0	0.81	0	0
Q_{C74} (MVAr)	0	30	0	30.00
Q_{C79} (MVAr)	0	30	2.83	25.18
Q_{C82} (MVAr)	0	0	0.3	0.01
Q_{C83} (MVAr)	0	30	18.03	29.79
Q_{C105} (MVAr)	0	19.16	30	23.14
Q_{C107} (MVAr)	0	0	30	23.58
Q_{C110} (MVAr)	0	30	0	6.27
Cost ($/h)	131,220.52	130,305.53	129,873.89	129,733.58
Ploss (MW)	132.86	80.69	72.31	73.21

function; (iii) selection of solution method and setting the algorithm parameters; (iv) program execution; and (v) display of the results.

The *opfgui* was developed in MATLAB, because it integrates computation, programing, analyze data, and producing graphical displays and graphical user interfaces in an easy-to-use environment where problems and solutions are expressed in familiar mathematical notation. When designing the program, special care was paid to its graphical user interface, so the *opfgui* is very friendly to the user.

The *opfgui* program offers a choice of 7 standard IEEE test systems, 6 objective functions, and 12 optimization methods. The program generates not only optimal solution, that is, optimum control variables and objective function, but also important results such as, convergence profile, bus voltages and bus powers, brunch

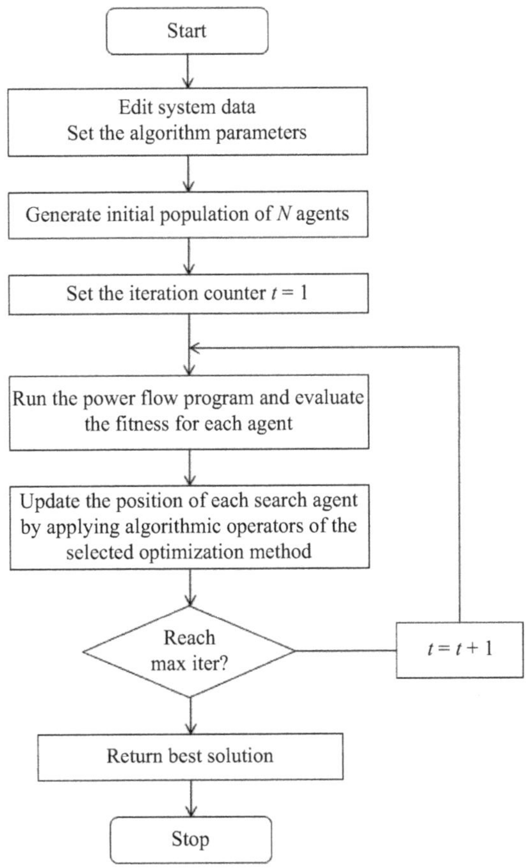

Figure 6.9 Flowchart of metaheuristic optimization method based OPF solution

power flows and losses, violating constraints (if exist), and statistical evaluation of the results. Using *opfgui*, the user can compare the performances of different optimization methods based on statistical evaluation of the results.

Figure 6.10 shows the main window of *opfgui* while it operates on a Windows-based computer. Shown is an example of the execution of the program on IEEE 30-bus test system. Minimization of the fuel cost (*Fcost*) is selected as the objective function, and hybrid PSOGSA algorithm is selected as the solution method.

As can be seen from Figure 6.10, the *opfgui* is organized in following components:

1. Panel TEST SYSTEM. It consists of one pop-up menu and two buttons. Pop-up menu enables the selection of a test system of seven available test systems. It is available in 6-bus, 9-bus, 14-bus, 30-bus, 39-bus, 57-bus, and 118-bus standard IEEE test systems. When clicking on the button "Single-Line Diagram," the selected test system will be displayed. The button "System Data" opens the input file of the selected test system. The input file contains branch data, bus

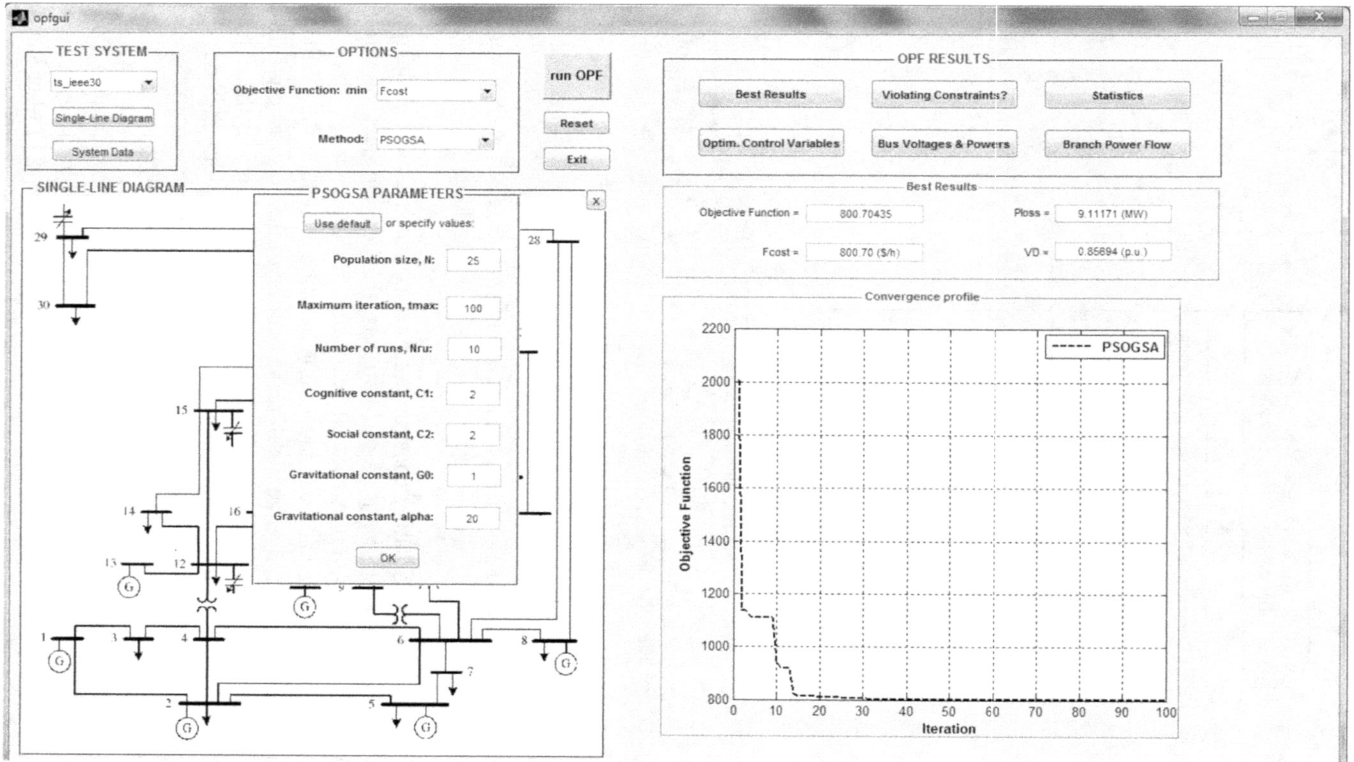

Figure 6.10 Main window of the opfgui

data, generator data, transformer data, and Var compensation data for the selected test system. The data are given in matrix form. The user can change the data and save those modifications.

2. Panel OPTIONS. It consists of two pop-up menus. In the first pop-up menu, the user performs selection of an objective function between six offered OPF objectives, such as minimization of fuel cost (Fcost), minimization of real power loss (Ploss), voltage-profile improvement (VD), simultaneous minimization of fuel cost and real power loss (Fcost_Ploss), simultaneous minimization of fuel cost and voltage deviation (Fcost_VD), and simultaneous minimization of fuel cost, real power loss, and voltage deviation (Fcost_Ploss_VD). The pop-up menu method enables to choose a metaheuristic optimization method as a solution of the OPF problem. The user can select one of methods, such as PSO, GSA, PSOGSA, ABC, WDO, FFA, GWO, CS, MSA, BSA, TLBO, etc. When clicking on the optimization method, the panel to settings of the algorithm parameters will be opened. The user can use the default parameters by clicking on button "Use default", or to enter the desired value for each of algorithm parameters.
3. After selection of appropriate options on the panels TEST SYSTEM and OPTIONS, the user can start the program by clicking the button "run OPF."
4. Panel OPF RESULTS. It manages the access to the different segments of the OPF results obtained from the program. The best results, that is objective function values in the last iteration, and convergence profile of the optimization algorithm will be displayed immediately after execution of the program (Figure 6.10). Besides, the program generates other important results as well. When clicking the button "Optim. Control Variables," the obtained optimal values of control variables such as generator active power outputs, generator voltages, transformer tap settings, and shunt Var compensations will be displayed, as shown in Figure 6.11.

   The button "Bus Voltages & Powers" enables to display bus voltages and bus powers in the system under optimum control variables (Figure 6.12).

   The button "Branch Power Flow" enables to display branch power flows and losses in the system under optimum control variables (Figure 6.13).

   The button "Violating Constraints?" is used to check if operating constraints are violated (Figure 6.14), as well as which operating constraints are violated.

   Because of the stochastic nature of metaheuristic optimization methods, a very important segment of the OPF program is statistical evaluation of the results. When the user clicks on the button "Statistics," a statistical evaluation of the optimal results that are obtained using the selected optimization method in specified number of program execution will be displayed (Figure 6.15). These statistical indicators are number of program execution, the minimum value of the objective function, the maximum value of the objective function, the mean value of the objective function, the standard deviation of the objective function, and the mean time of execution of the program. On the basis of these statistical indicators, a relevant comparison of different metaheuristic optimization methods can be achieved.

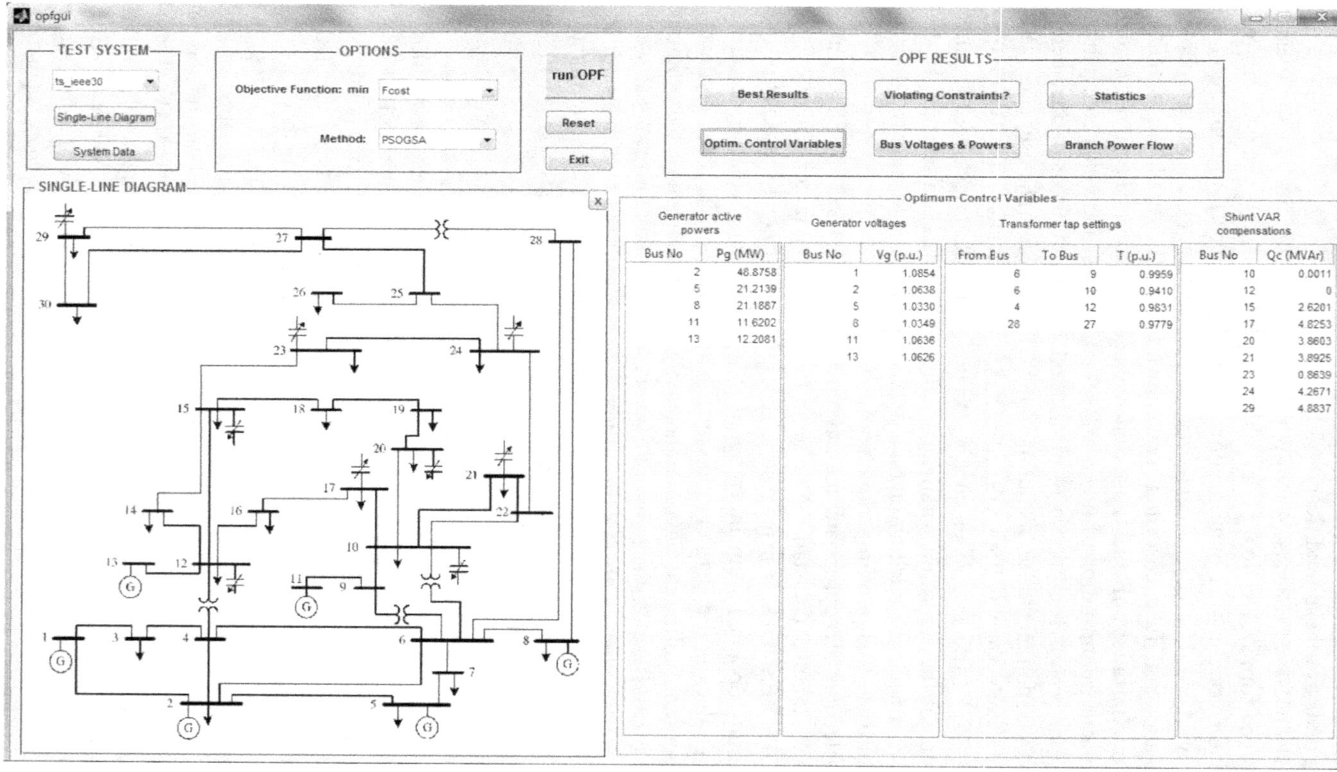

Figure 6.11 Display of the results: optimum control variables

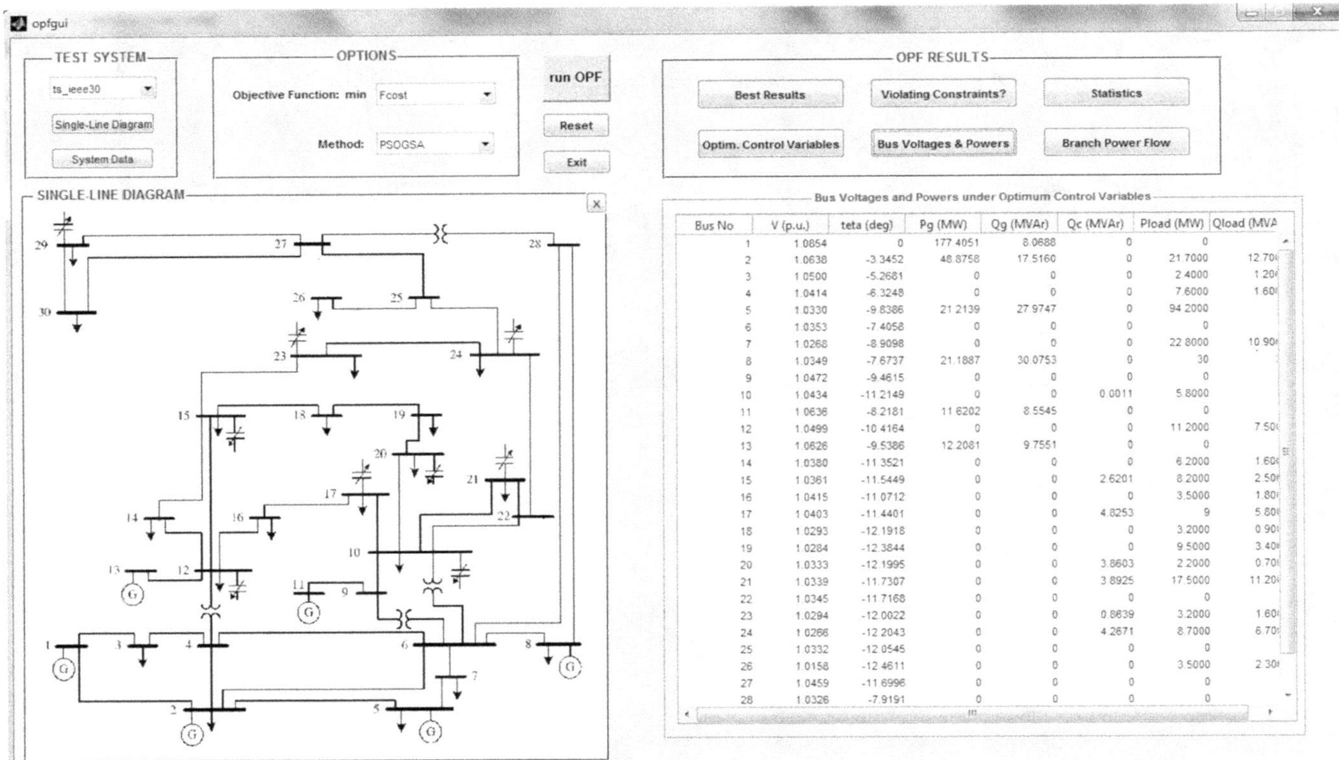

Figure 6.12 Display of the results: bus voltages and powers

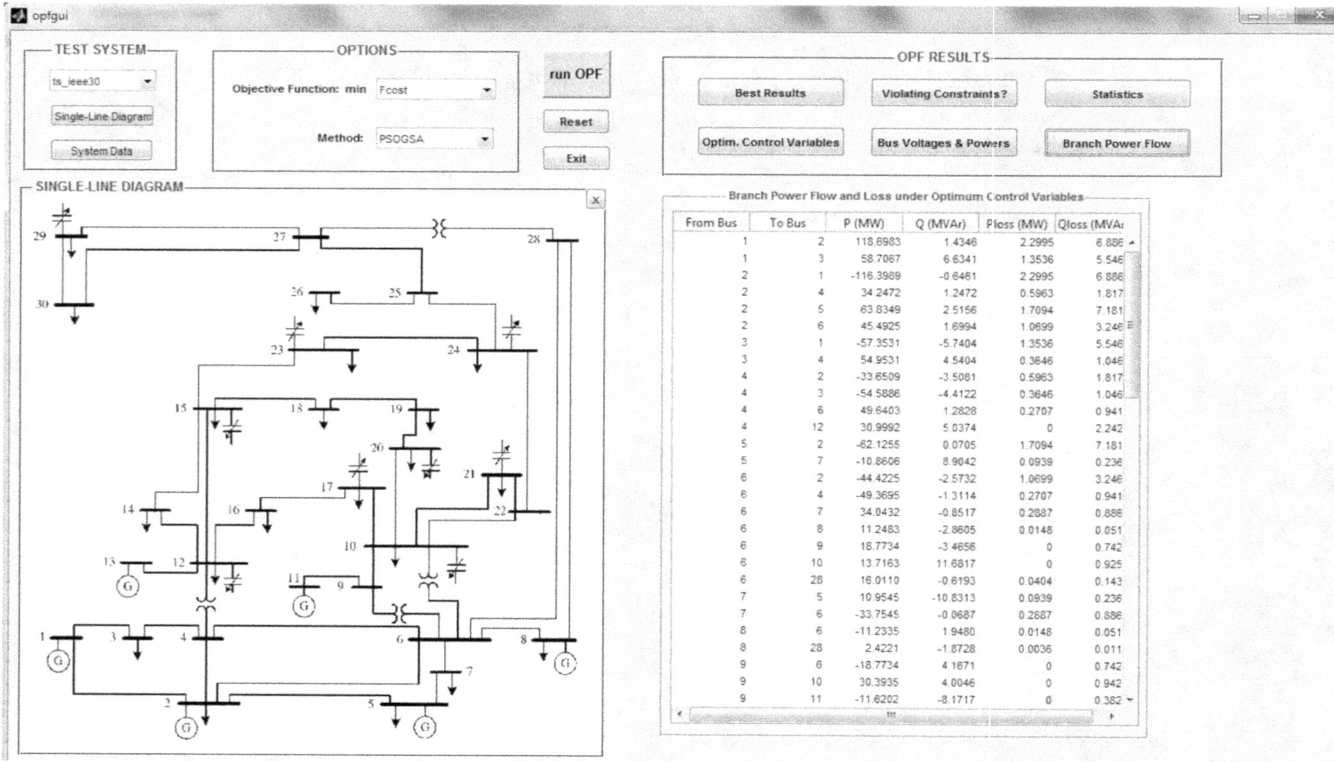

Figure 6.13 Display of the results: branch power flow and loss

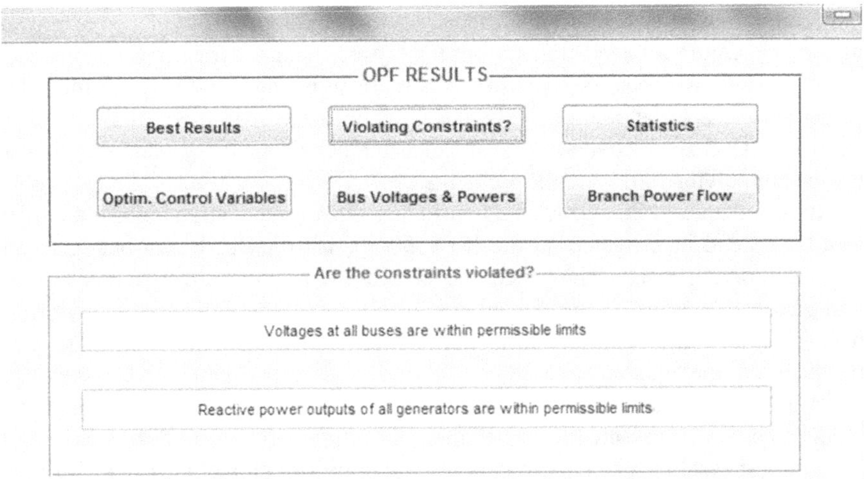

Figure 6.14 Display of the results: checking of the operating constraints

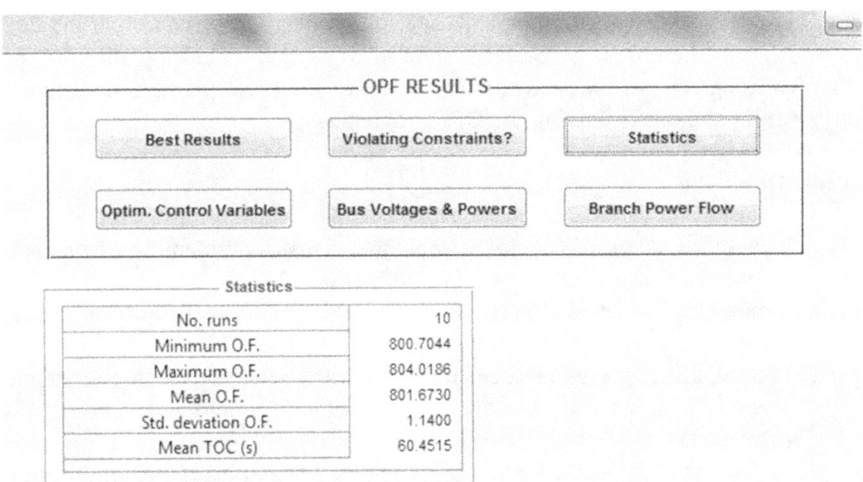

Figure 6.15 Statistical evaluation of the results

6.7 Conclusion

The OPF problem solution aims to optimize a chosen objective function through optimal adjustment of the power-system-control variables while at the same time satisfying various operating constraints. Mathematically, the OPF problem can be considered as a nonlinear, nonconvex, large-scale, static optimization problem with both continuous and discrete control variables including generator active power outputs, generator voltages, transformer tap settings, shunt Var compensations, etc.

In the last two decades, many of metaheuristic population-based optimization methods have been applied to solve the OPF problem. A comprehensive review given in this chapter covers more of 25 different metaheuristic methods implemented for solving various OPF problems.

In this chapter, PSO, GSA, and hybrid PSOGSA optimization algorithms have been tested and investigated on the IEEE 30-bus and the IEEE 118-bus test systems. Simulation results show that the algorithms provide effective and robust high-quality solutions. For practical applications in large-scale power systems, there is need to improving the computational speed. This is main disadvantage of metaheuristic methods in general.

Additionally, the *opfgui* program to solve OPF using different metaheuristic methods has been presented in this chapter. The *opfgui* program offers a choice of different test systems, objective functions, and optimization methods. The program generates not only optimal solution, that is, optimum control variables and objective function, but also important results such as convergence profile, bus voltages and bus powers, brunch power flows and losses, violating constraints (if exist), and statistical evaluation of the results. The software can be useful for researches in area of power-system optimization, and especially it may serve as a supporting tool for students in the course of power-system analysis that includes studies of the OPF. By using *opfgui*, the students can compare the performances of different optimization methods in solving the OPF problem.

References

[1] Carpentier J. 'Contribution to the economic dispatch problem'. *Bulletin de la Societe Francaise Electricians*. 1962;**8**(3):431–47

[2] Dommel H., Tinney W. 'Optimal power flow solutions'. *IEEE Transactions on Power Apparatus and Systems*. 1968;**87**(10):1866–76

[3] Frank S., Steponavice I., Rebennack S. 'Optimal power flow: a bibliographic survey II Non-deterministic and hybrid methods'. *Energy Systems*. 2012; **3**(3):259–89

[4] Niu M., Wan C., Xu Z. 'A review on applications of heuristic optimization algorithms for optimal power flow in modern power systems'. *Journal of Modern Power Systems and Clean Energy*. 2014;**2**(4):289–97

[5] Abou El Ela A.A., Abido M.A., Spea S.R. 'Optimal power flow using differential evolution algorithm'. *Electrical Engineering*. 2009;**91**:69–78

[6] Lai L.L., Ma J.T., Yokoyama R., Zhao M. 'Improved genetic algorithms for optimal power flow under both normal and contingent operation states'. *International Journal of Electrical Power & Energy Systems*. 1997;**19**(5): 287–92

[7] Bakirtzis A., Biskas P., Zoumas C., Petridis V. 'Optimal power flow by enhanced genetic algorithm'. *IEEE Transactions on Power Systems*. 2002; **17**(2):229–36

[8] Sood Y., Padhy N., Gupta H. 'Discussion of optimal power flow by enhanced genetic algorithm'. *IEEE Transactions on Power Systems*. 2003; **18**(3):1219

[9] Osman M.S., Abo-Sinna M.A., Mousa A.A. 'A solution to the optimal power flow using genetic algorithm'. *Applied Mathematics and Computation*. 2004;**155**:391–45

[10] Devaraj D., Yegnanarayana B. 'Genetic-algorithm-based optimal power flow for security enhancement'. *IEE Proceedings—Generation, Transmission and Distribution*. 2005;**152**(6):899–905.

[11] Todorovski M., Rajicic D. 'An initialization procedure in solving optimal power flow by genetic algorithm'. *IEEE Transactions on Power Systems*. 2006;**21**(2):480–7

[12] Gaing Z.L., Chang R.F. 'Security-constrained optimal power flow by mixed-integer genetic algorithm with arithmetic operators'. *Proceedings of IEEE Power Engineering Society General Meeting*; 2006. pp. 8

[13] Chan K.Y., Chan K.W., Ling S.H., Iu H.H.C., Pong G.T.Y. 'Solving multi-contingency transient stability constrained optimal power flow problems with an improved GA'. *Proceedings of IEEE Congress on Evolutionary Computation (CEC 2007)*; Singapore, Sep 2007. pp. 2901–8

[14] Banu R., Devaraj D. 'Genetic algorithm approach for optimal power flow with FACTS devices'. *Proceedings of 4th International IEEE Conference Intelligent Systems*; Varna, Bulgaria, Sep 2008. IEEE; 2008. pp. 11–6

[15] Mahdad B., Bouktir T., Srairi K., Benbouzid M.E. 'Dynamic strategy based fast decomposed GA coordinated with FACTS devices to enhance the optimal power flow'. *Energy Conversion and Management*. 2010;**51**:1370–80

[16] Kumari M.S., Maheswarapu S. 'Enhanced genetic algorithm based computation technique for multi-objective optimal power flow solution'. *Electrical Power and Energy Systems*. 2010;**32**:736–42

[17] Attia A.F., Al-Turki Y.A., Abusorrah A.M. 'Optimal power flow using adapted genetic algorithm with adjusting population size'. *Electric Power Components and Systems*. 2012;**40**:1285–99

[18] Nithya C., Roselyn J.P., Devaraj D., Dash S.S. 'Voltage stability constrained optimal power flow using non-dominated sorting genetic algorithm-II (NSGA II)'. *Proceedings of the 3rd International Conference on Swarm, Evolutionary, and Memetic Computing*; Bhubaneswar, India, Dec 2012. pp. 793–801

[19] Panuganti S., Roselyn J.P., Devaraj D., Dash S. S. 'Voltage stability constrained optimal power flow using NSGA-II'. *Computational Water, Energy, and Environmental Engineering*. 2013;**2**:1–8

[20] Kılıç U., Ayan K. 'Optimal power flow solution of two-terminal HVDC systems using genetic algorithm'. *Electrical Engineering*. 2014;**96**(1):65–77

[21] Kılıç U., Ayan K. 'Transient stability constrained optimal power flow solution of ac–dc systems using genetic algorithm'. *Proceedings of 3rd International Conference on Electric Power and Energy Conversion Systems*; Istanbul, Turkey, Oct 2013. IEEE; 2014. pp. 1–6

[22] Balasubramanian S., Gokhale R.V., Sekar A. 'A new AC optimal power flow formulation and solution using genetic algorithm based on P–Q decomposition'. *Proceedings of North American Power Symposium (NAPS)*; Charlotte, USA, Oct 2015. IEEE; 2015. pp. 1–5

[23] Reddy S.S., Bijwe P.R., Abhyankar A.R. 'Faster evolutionary algorithm based optimal power flow using incremental variables'. *Electrical Power and Energy Systems*. 2014;**54**: 198–210

[24] Ye C.-J., Huang M.-X. 'Multi-objective optimal power flow considering transient stability based on parallel NSGA-II'. *IEEE Transaction on Power Systems*. 2015;**30**(2):857–66

[25] Bouchekara H.R.E.H., Chaib A.E., Abido M.A. 'Optimal power flow using GA with a new multi-parent crossover considering: prohibited zones, valve-point effect, multi-fuels and emission'. *Electrical Engineering*. 2016; doi: 10.1007/s00202-016-0488-9

[26] Reddy S.S. 'Optimal power flow with renewable energy resources including storage'. *Electrical Engineering*. 2017;**99**(2):685–95

[27] Abido M.A. 'Optimal power flow using particle swarm optimization'. *Electrical Power and Energy Systems*. 2002;**24**:563–71

[28] He S., Wen J.Y., Prempain E., Wu Q.H., Fitch J., Mann S. 'An improved particle swarm optimization for optimal power flow'. *Proceedings of 2004 International Conference on Power System Technology*; Singapore, Singapore, Nov 2004. IEEE; 2005. pp. 1633–7

[29] Zhao B., Guo C., Cao Y. 'Improved particle swam optimization algorithm for OPF problems'. *Proceedings of IEEE PES Power Systems Conference and Exposition*; New York, NY, USA, Oct 2004. IEEE; 2005. pp. 233–8

[30] Wang C.R., Yuan H.J., Huang Z.Q., Zhang J.W., Sun C.J. 'A modified particle swarm optimization algorithm and its application in optimal power flow problem'. *Proceedings of the 4th International Conference on Machine Learning and Cybernetics*; Guangzhou, China, Aug 2005. IEEE; 2005. pp. 2885–9

[31] Gaing Z.L. 'Constrained optimal power flow by mixed-integer particle swarm optimization'. *2005 IEEE Power Engineering Society General Meeting*; San Francisco, CA, USA, Jun 2005. IEEE; 2005. pp. 1–8

[32] Esmin A.A.A, Lambert-Tores G., de Souza A.C.Z. 'A hybrid particle swarm optimization applied to loss power minimization'. *IEEE Transactions on Power Systems*. 2005;**20**(2):859–66

[33] Dutta P., Sinha A.K. 'Voltage stability constrained multi-objective optimal power flow using particle swarm optimization'. *1st International Conference on Industrial and Information Systems*; Peradeniya, Sri Lanka, Aug 2006. IEEE; 2007. pp. 161–6

[34] Mo N., Zou Z., Chan K., Pong T. 'Transient stability constrained optimal power flow using particle swarm optimisation'. *IET Generation, Transmission & Distribution*. 2007:**1**(3): 476–83

[35] Yumbla P., Ramirez J., Coello C. 'Optimal power flow subject to security constraints solved with a particle swarm optimizer'. *IEEE Transactions on Power Systems*. 2008;**23**(1):33–40

[36] Abido M.A. 'Multiobjective particle swarm optimization for optimal power flow problem'. *Proceedings of 12th International Middle-East Power System Conference*; Aswan, Egypt, Mar 2008. IEEE; 2008. pp. 392–6

[37] Liu W., Li M., Wang X. 'An improved particle swarm optimization algorithm for optimal power flow'. *Proceedings of IEEE 6th International Power Electronics and Motion Control Conference*; Wuhan, China, May 2009. IEEE; 2009. pp. 2448–50

[38] Swaroopan N.M.J, Somasundaram P. 'Optimal power flow for security enhancement using fuzzified particle swarm optimization'. *Proceedings of International Conference on Sustainable Energy and Intelligent Systems*; Chennai, India, Jul 2011. IEEE; 2012. pp. 474–9

[39] Niknam T., Narimani M.R., Aghaei J., Azizipanah-Abarghooee R. 'Improved particle swarm optimisation for multi-objective optimal power flow considering the cost, loss, emission and voltage stability index'. *IET Generation, Transmission & Distribution*. 2012;**6**(6):515–27

[40] Ghanizadeh T., Babaei E. 'Optimal power flow using iteration particle swarm optimization'. *Proceedings of IEEE 5th India International Conference on Power Electronics*; Delhi, India, Dec 2012. IEEE; 2013. pp. 1–6

[41] Kumar S., Chaturvedi D.K. 'Optimal power flow solution using fuzzy evolutionary and swarm optimization'. *Electrical Power and Energy Systems*. 2013;**47**:416–23

[42] Turkay B.E., Cabadag R.I. 'Optimal power flow solution using particle swarm optimization algorithm'. *EuroCon 2013*; Zagreb, Croatia, Jul 2013. IEEE; 2013. pp. 1418–24

[43] Choudhury D., Patra S. 'Multi objective optimal power flow using particle swarm optimization technique'. *International conference on Signal Processing, Communication, Power and Embedded System*; Odisha, India, Oct 2016. IEEE; 2016. pp. 1261–6

[44] Reddy S.S., Rathnam S. 'Optimal power flow using glowworm swarm optimization'. *Electrical Power and Energy Systems*. 2016;**80**:128–39

[45] Kahouradze S., Mahmoudi A., Mokhlis H.B. 'A comparative study of multi-objective optimal power flow based on particle swarm, evolutionary programming, and genetic algorithm'. *Electrical Engineering*. 2015;**97**(1):1–12

[46] Chang Y.C., Lee T.Y., Chen C.L., Jan R.M. 'Optimal power flow of a wind-thermal generation system'. *Electrical Power and Energy Systems*. 2014;**55**:312–20

[47] Duman S., Guvenc U., Sonmez Y., Yorukeren N. 'Optimal power flow using gravitational search algorithm'. *Energy Conversion and Management*. 2012;**59**:86–95

[48] Bhattacharya A., Roy P.K. 'Solution of multi-objective optimal power flow using gravitational search algorithm'. *IET Generation, Transmission & Distribution*. 2012;**6**(8):751–63

[49] Sonmez Y., Duman S., Guvenc U., Yorukeren N. 'Optimal power flow incorporating FACTS devices using gravitational search algorithm'.

Proceedings of International Symposium on Innovations in Intelligent Systems and Applications; Trabzon, Turkey, Jul 2012. IEEE; 2012. pp. 1–5

[50] Jahan M.S., Amjady N. 'Solution of large-scale security constrained optimal power flow by a new bi-level optimization approach based on enhanced gravitational search algorithm'. *IET Generation, Transmission & Distribution*. 2013;**7**(12):1481–91

[51] Bhowmik A.R., Chakraborty A.K. 'Solution of optimal power flow using nondominated sorting multi objective gravitational search algorithm'. *Electrical Power and Energy Systems*. 2014;**62**:323–34

[52] Basu M. 'Optimal power flow with FACTS devices using differential evolution'. *Electrical Power and Energy Systems*. 2008;**30**:150–6

[53] Basu M. 'Multi-objective optimal power flow with FACTS devices'. *Energy Conversion and Management*. 2011;**52**:903–10

[54] Varadarajan M., Swarup K.S. 'Solving multi-objective optimal power flow using differential evolution'. *IET Generation, Transmission & Distribution*. 2008;**2**(5):720–30

[55] Sayah S., Zehar K. 'Modified differential evolution algorithm for optimal power flow with non-smooth cost functions'. *Energy Conversion and Management*. 2008;**49**:3036–42

[56] Cai H.R., Chung C.Y., Wong K.P. 'Application of differential evolution algorithm for transient stability constrained optimal power flow'. *IEEE Transactions on Power Systems*. 2008;**23**(2):719–28

[57] Abou El Ela A.A., Abido M.A., Spea S.R. 'Optimal power flow using differential evolution algorithm'. *Electric Power System Research*. 2010; **80**:878–85

[58] Amjady N., Sharifzadeh H. 'Security constrained optimal power flow considering detailed generator model by a new robust differential evolution algorithm'. *Electric Power System Research*. 2011;**81**:740–9

[59] Sivasubramani S, Swarup K.S. 'Multiagent based differential evolution approach to optimal power flow'. *Applied Soft Computing*. 2012;**12**:735–40

[60] Shaheen A., El-Sahiemy R.A., Farrag S.M. 'Solving multi-objective optimal power flow problem via forced initialised differential evolution algorithm'. *IET Generation, Transmission & Distribution*. 2016;**10**(7):1634–47

[61] Yuryevich J., Wong K.P. 'Evolutionary programming based optimal power flow algorithm'. *IEEE Transactions on Power Systems*. 1999;**14**(4):1245–50

[62] Gnanadass R., Venkatesh P., Padhy N.P. 'Evolutionary programming based optimal power flow for units with non-smooth fuel cost functions'. *Electric Power Components and Systems*. 2005;**33**:349–61

[63] Ongsakul W., Tantimaporn T. 'Optimal power flow by improved evolutionary programming'. *Electric Power Components and Systems*. 2006; **34**:79–95

[64] Sood Y.R. 'Evolutionary programming based optimal power flow and its validation for deregulated power system analysis'. *Electrical Power and Energy Systems*. 2007;**29**:65–75

[65] Allaoua B., Laoufi A. 'Collective intelligence for optimal power flow solution using ant colony optimization'. *Leonardo Electronic Journal of Practices and Technologies*. 2008;**13**:88–105

[66] Allaoua B., Laoufi A. 'Optimal power flow solution using ant manners for electrical network'. *Advances in Electrical and Computer Engineering*. 2009;**9**:34–40

[67] Gasbaoui B., Allaoua B. 'Ant colony optimization applied on combinatorial problem for optimal power flow solution'. *Leonardo Journal of Sciences*. 2009;**14**:1–17

[68] Sreejith S., Chandrasekaran K., Simon S.P. 'Touring ant colony optimization technique for optimal power flow incorporating thyristor controlled series compensator'. *Proceedings of World Congress on Nature & Biologically Inspired Computing*; Coimbatore, India, Dec 2009. IEEE; 2010. pp. 1127–32

[69] Joshi S.L., Ghanchi V.H. 'Solution of optimal power flow subject to security constraints by an ant colony optimization'. *Proceedings of 3rd International Conference on Computational Intelligence and Information Technology*; Mumbai, India, Oct 2013. IEEE; 2014. pp. 590–7

[70] Adaryani M.R., Karami A. 'Artificial bee colony algorithm for solving multi-objective optimal power flow problem'. *Electrical Power and Energy Systems*. 2013;**53**:219–30

[71] Medina M.A., Das S., Coello C.A.C., Ramírez J.M. 'Decomposition-based modern metaheuristic algorithms for multi-objective optimal power flow—a comparative study'. *Engineering Applications of Artificial Intelligence*. 2014;**32**:10–20

[72] Roy R., Jadhav H.T. 'Optimal power flow solution of power system incorporating stochastic wind power using Gbest guided artificial bee colony algorithm'. *Electrical Power and Energy Systems*. 2015;**64**:562–78

[73] Abaci K., Yamacli V., Akdagli A. 'Optimal power flow with SVC devices by using the artificial bee colony algorithm'. *Turkish Journal of Electrical Engineering & Computer Sciences*. 2016;**24**:341–53

[74] Ayan K., Kilic U., Barakli B. 'Chaotic artificial bee colony algorithm based solution of security and transient stability constrained optimal power flow'. *Electrical Power and Energy Systems*. 2015;**64**:136–47

[75] Kılıç U. 'Backtracking search algorithm-based optimal power flow with valve point effect and prohibited zones'. *Electrical Engineering*. 2015;**97**(2):101–10

[76] Pulluri H., Naresh R., Sharma V., Preeti. 'A new colliding bodies optimization for solving optimal power flow problem in power system'. *Proceedings of 6th International Conference on Power Systems*; New Delhi, India, Mar 2016. IEEE; 2016. pp. 1–6

[77] Bouchekara H.R.E.H., Chaib A.E., Abido M.A., El-Sehiemy R.A. 'Optimal power flow using an improved colliding bodies optimization algorithm'. *Applied Soft Computing*. 2016;**42**:119–31

[78] Hasan Z., El-Hawary M.E. 'Optimal power flow by black hole optimization algorithm'. *Proceedings of 2014 IEEE Electrical Power and Energy Conference*; Calgary, Canada, Nov 2014. IEEE; 2015. pp. 134–41

[79] Bouchekara H.R.E.H. 'Optimal power flow using black-hole-based optimization approach'. *Applied Soft Computing*. 2014;**24**:879–88

[80] El-Fergany A.A., Hasanien H.M. 'Single and multi-objective optimal power flow using grey wolf optimizer and differential evolution algorithms'. *Electric Power Components and Systems*. 2015;**43**(13):1548–59

[81] Mahdad B., Srairi K. 'Blackout risk prevention in a smart grid based flexible optimal strategy using Grey Wolf-pattern search algorithms'. *Energy Conversion and Management*. 2015;**98**:411–29

[82] Herbadji O., Nadhir K., Slimani L., Bouktir T. 'Optimal power flow with emission controlled using firefly algorithm'. *Proceedings of 5th International Conference on Modeling, Simulation and Applied Optimization*; Hammamet, Tunisia, Apr 2013. IEEE; 2013. pp. 1–6

[83] Herbadji O., Bouktir T. 'Optimal power flow using firefly algorithm with consideration of FACTS devices "UPFC"'. *International Journal on Electrical Engineering and Informatics*. 2015;**7**(1):1–17

[84] Balachennaiah P., Suryakalavathi M., Nagendra P. 'Firefly algorithm based solution to minimize the real power loss in a power system'. *Ain Shams Engineering Journal*. 2015; https://doi.org/10.1016/j.asej.2015.10.005.

[85] Manoranjitham E.G., Alagarsamy S. 'Application of firefly algorithm on optimal power flow control incorporating simplified impedance UPFC model'. *Electrical Power and Energy Systems*. 2015;**71**:358–63

[86] Hariharan T., Sundaram K.M. 'Optimal power flow using firefly algorithm with unified power flow controller'. *Circuits and Systems*. 2016;**7**:1934–42.

[87] Mishra C., Singh S.P., Rokadia J. 'Optimal power flow in the presence of wind power using modified cuckoo search'. *IET Generation, Transmission & Distribution*. 2015;**9**(7):615–26

[88] Balasubbareddy M., Sivanagaraju S., Suresh C.V. 'Multi-objective optimization in the presence of practical constraints using non-dominated sorting hybrid cuckoo search algorithm'. *Engineering Science Technology, an International Journal*. 2015;**18**:603–15

[89] Balasubbareddy M., Sivanagaraju S., Suresh C.V., Babud A.V.N., Srilathaa D. 'A non-dominated sorting hybrid cuckoo search algorithm for multi-objective optimization in the presence of FACTS devices'. *Russian Electrical Engineering*. 2017;**88**(1):44–53

[90] Mohamed A.A., Mohamed Y.S., El-Gaafary A.A.M., Hemeida A.M. 'Optimal power flow using moth swarm algorithm'. *Electric Power System Research*. 2017;**142**:190–206

[91] Roy P.K., Paul C. 'Optimal power flow using krill herd algorithm'. *International Transactions on Electrical Energy Systems*. 2015;**25**(8):1397–419

[92] Mukherjee A., Mukherjee V. 'Solution of optimal power flow with FACTS devices using a novel oppositional krill herd algorithm'. *Electrical Power and Energy Systems*. 2016;**78**:700–14

[93] Mukherjee A., Roy P.K., Mukherjee V. 'Transient stability constrained optimal power flow using oppositional krill herd algorithm'. *Electrical Power and Energy Systems*. 2016;**83**:283–97

[94] Pulluri H., Naresh R., Sharma V. 'Application of stud krill herd algorithm for solution of optimal power flow problems'. *International Transactions on Electrical Energy Systems*. 2017;**27**(6):1–18

[95] Niknam T., Narimani M.R., Jabbari M., Malekpour A.R. 'A modified shuffle frog leaping algorithm for multi-objective optimal power flow'. *Energy*. 2011;**36**:6420–32

[96] Azizipanah-Abarghooee R., Narimani M.R., Bahmani-Firouzi B., Niknam T. 'Modified shuffled frog leaping algorithm for multi-objective optimal power flow with FACTS devices'. *Journal of Intelligent & Fuzzy Systems*. 2014;**26**:681–92

[97] Rahiminejad A., Alimardani A., Vahidi B. Hosseinian S.H. 'Shuffled frog leaping algorithm optimization for AC–DC optimal power flow dispatch'. *Turkish Journal of Electrical Engineering & Computer Sciences*. 2014;**22**:874–92

[98] Vahedi H., Hosseini S. H., Noroozian R. 'Bacterial foraging algorithm for security constrained optimal power flow'. *Proceedings of 7th International Conference on the European Energy Market*; Madrid, Spain, Jun 2010. IEEE; 2010. pp. 1–6

[99] Panda A., Tripathy M. 'Security constrained optimal power flow solution of wind-thermal generation system using modified bacteria foraging algorithm'. *Energy*. 2015;**93**:816–27

[100] Bhattacharya A., Chattopadhyay P.K. 'Application of biogeography-based optimisation to solve different optimal power flow problems'. *IET Generation, Transmission & Distribution*. 2011;**5**(1):70–80

[101] Kumar A.R., Premalatha L. 'Optimal power flow for a deregulated power system using adaptive real coded biogeography-based optimization'. *Electrical Power and Energy Systems*. 2015;**73**:393–9

[102] Bouchekara H.R.E.H., Abido M.A., Boucherma M. 'Optimal power flow using teaching-learning-based optimization technique'. *Electric Power Systems Research*. 2014;**114**:49–59

[103] Ghasemi M., Ghavidel S., Gitizadeh M., Akbari E. 'An improved teaching–learning-based optimization algorithm using Lévy mutation strategy for non-smooth optimal power flow'. *Electrical Power and Energy Systems*. 2015;**65**:375–84

[104] Bouchekara H.R.E.H., Abido M.A., Chaib A.E., Mehasni R. 'Optimal power flow using the league championship algorithm: a case study of the Algerian power system'. *Energy Conversion and Management*. 2014; **87**:58–70

[105] Mahdad B., Srairi K. 'A new interactive sine cosine algorithm for loading margin stability improvement under contingency'. *Electrical Engineering*. 2017; https://doi.org/10.1007/s00202-017-0539-x

[106] Ranjbar H., Zarei S.F., Hosseini S.H. 'Imperialistic competitive algorithm based optimal power flow'. *Proceedings of 22nd Iranian Conference on Electrical Engineering*; Tehran, Iran, May 2014. IEEE; 2015. pp. 746–50

[107] Ghasemi M., Ghavidel S., Ghanbarian M.M., Massrur H.R., Gharibzadeh M. 'Application of imperialist competitive algorithm with its modified techniques for multi-objective optimal power flow problem: a comparative study'. *Information Sciences*. 2014;**281**(10):225–47

[108] Ghasemi M., Ghavidel S., Ghanbarian M.M., Massrur H.R., Vahed A.A. 'Multi-objective optimal power flow considering the cost, emission, voltage deviation and power losses using multi-objective modified imperialist competitive algorithm'. *Energy*. 2014;**78**:276–89

[109] Bouchekara H.R.E.H., Abido M.A. 'Optimal power flow using differential search algorithm'. *Electric Power Components and Systems*. 2014;**42**:1683–99

[110] Abaci K., Yamacli V. 'Differential search algorithm for solving multi-objective optimal power flow problem'. *Electrical Power and Energy Systems*. 2016;**79**:1–10

[111] Warid W., Hizam H., Mariun N., Abdul-Wahab N.I. 'Optimal power flow using the Jaya algorithm'. *Energies*. 2016;9(9)**678**:1–18

[112] Sinsuphan N., Leeton U., Kulworawanichpong T. 'Optimal power flow solution using improved harmony search method'. *Applied Soft Computing*. 2013;**13**:2364–74

[113] Rao B.V., Kumar G.V.N. 'Optimal power flow by BAT search algorithm for generation reallocation with unified power flow controller'. *Electrical Power and Energy Systems*. 2015;**68**:81–8

[114] Abatari H.D., Abad M.S.S., Seifi H. 'Application of bat optimization algorithm in optimal power flow'. *Proceedings of 24th Iranian Conference on Electrical Engineering*; Shiraz, Iran, May 2016. IEEE; 2016. pp. 793–8

[115] Ghasemi M., Ghavidel S., Rahmani S., Roosta A., Falah H. 'A novel hybrid algorithm of imperialist competitive algorithm and teaching learning algorithm for optimal power flow problem with non-smooth cost function'. *Engineering Applications of Artificial Intelligence*. 2014;**29**:54–69

[116] Narimani M.R., Azizipanah A.R., Zoghdar M.S.B., Gholami K. 'A novel approach to multi-objective optimal power flow by a new hybrid optimization algorithm considering generator constraints and multi-fuel type'. *Energy*. 2013;**49**:119–36

[117] Niknam T., Narimani M.R., Abarghooee A.R. 'A new hybrid algorithm for optimal power flow considering prohibited zones and valve point effect'. *Energy Conversion and Management*. 2012;**58**:197–206

[118] Christy A.A., Raj P.A.D.V. 'Adaptive biogeography based predator-prey optimization technique for optimal power flow'. *Electrical Power and Energy Systems*. 2014;**62**:344–52

[119] Vaisakh K., Srinivas L.R. 'Evolving ant direction differential evolution for OPF with non-smooth cost functions'. *Engineering Applications of Artificial Intelligence*. 2011;**24**:426–36

[120] Radosavljević J., Klimenta D., Jevtić M., Arsić N. 'Optimal power flow using a hybrid optimization algorithm of particle swarm optimization and gravitational search algorithm'. *Electric Power Components and Systems*. 2015;**43**(17):1958–70

[121] Radosavljević J., Arsić N., Jevtić M. 'Optimal power flow using hybrid PSOGSA algorithm'. *Proceedings of 55th International Scientific Conference of Riga Technical University on Power and Electrical Engineering;* Riga, Latvia, Oct 2014. IEEE; 2014. pp. 136–40

[122] Kessel P., Glavitsch H. 'Estimating the voltage stability of a power system'. *IEEE Transactions on Power Delivery*. 1986;**1**(3):346–52

[123] Thukaram B.D., Parthasarathy K. 'Optimal reactive power dispatch algorithm for voltage stability improvement'. *International Journal of Electrical Power & Energy Systems*. 1996;**18**:461–8

[124] Kundur P. *Power system stability and control*. New York: McGraw-Hill; 1994. p. 257

[125] Kennedy J., Eberhart R.C. 'Particle swarm optimization'. *Proceedings of IEEE International Conference on Neural Networks*; Perth, Australia, 1995. IEEE; 1995. pp. 1942–8

[126] Stott B., Alsac O. 'Fast decoupled load flow'. *IEEE Transactions on Power Apparatus and Systems*. 1974;PAS-**93**(3):859–69

[127] Rashedi E., Nezamabadi-pour H., Saryazdi S. 'GSA: a gravitational search algorithm'. *Information Sciences*. 2009;**179**:2232–48

[128] Mirjalili S., Hashim S.Z.M. 'A new hybrid PSOGSA algorithm for function optimization'. *Proceedings of International Conference on Computer and Information Application*; Tianjin, China, Dec 2010. IEEE; 2012. pp. 374–7

[129] Alsac O., Stott B. 'Optimal load flow with steady-state security'. *IEEE Transactions on Power Apparatus and Systems*. 1974;**93**(3):745–51

[130] Lee K.Y., Park Y.M., Ortiz J.L. 'A united approach to optimal real and reactive power dispatch'. *IEEE Transactions on Power Apparatus and Systems*. 1985;**104**(5):1147–53

[131] Zimmerman R.D., Murillo-Sanchez C.E., *MATPOWER MATLAB toolbox*. 2014. Available at: http://www.pserc.cornell.edu//matpower/

[132] Radosavljević J. *opfgui—Matlab GUI based software for optimal power flow using metaheuristic optimization methods*. 2017. Available at https://www.researchgate.net/profile/Jordan_Radosavljevic2/contributions

Chapter 7

Optimal reactive power dispatch in transmission networks

7.1 Introduction

One of the most important conditions for economic and secure operation of electric power system is the optimal reactive power dispatch (ORPD). The ORPD is achieving by appropriate coordination of the equipment which manage the reactive power flows. The ORPD problem solution aims to minimize a chosen objective function, such as power loss or voltage deviation (VD), through optimal adjustment of the power system control variables, under specified active power outputs of all generators (except at the slack bus P_{Gslack}), while at the same time satisfying various operating constraints [1,2]. Usually, the active power outputs of generators are obtained as the solution of the economic dispatch problem. The economic dispatch problem solution aims to minimize the cost of generating electric power. Mathematically, the ORPD problem can be formulated as a nonlinear optimization problem with constraints.

In recent years, various population-based metaheuristic optimization methods such as differential evolution (DE) [3–5], genetic algorithm (GA) [6,7], biogeography-based optimization (BBO) [8], particle swarm optimization (PSO) [9–12], gravitational search algorithm (GSA) [13–16], krill herd algorithm (KHA) [17,18], harmony search algorithm (HSA) [19], teaching–learning-based optimization [20], differential search algorithm [21], ant colony optimization algorithm [22], artificial bee colony algorithm (ABC) [23], gray wolf optimizer [24], and enhanced marked algorithm [25] have been suggested for solving the ORPD problem. Their main advantage compared to the classical (deterministic) optimization methods is that they are not limited with requirements for differentiability, nonconvexity, and continuity of the objective function or types of control variables. Moreover, these methods can be used for practical power systems taking into account various types of objective function and constraints. The essence of metaheuristic methods is iterative correction of solutions, i.e., generating new populations by applying stochastic search operators on individuals from the current population. The main performances of metaheuristics are fast search of large solution spaces, ability to find global solutions, and avoiding local optimum.

Hybridization two or more optimization algorithms can produce a synergy of theirs main advantages. Some of those hybrid algorithms, such as hybrid PSO and

GSA algorithm (PSOGSA) [1], hybrid GSA and sequential programming algorithm (GSA–SQP) [2], hybrid-modified imperialist competitive algorithm (ICA) and invasive weed optimization (MICA–IWO) [26], hybrid-shuffled frog leaping algorithm and Nelder–Mead simplex search [27], hybrid firefly algorithm and Nelder–Mead simplex method [28], hybrid PSO and ICA [29], hybrid GA and interior point method [30], and hybrid chaotic ABC and DE algorithm [31], have been applied to solve the ORPD problem.

This chapter presents the applications of hybrid PSOGSA algorithm and hybrid GSA–SQP algorithm to the ORPD problem. Performances of the algorithms for the ORPD problem are studied and evaluated on standard IEEE 30-bus and IEEE-118 test systems. The simulation results are compared with those of other metaheuristic optimization algorithms reported in the literature recently. In addition, a critical analysis of the ORPD results is also presented.

7.2 Problem formulation

Mathematically, the ORPD problem can be expressed as follows [1,2]:

$$\min F(\mathbf{x}, \mathbf{y}) \tag{7.1}$$

with constraints:

$$g(\mathbf{x}, \mathbf{y}) = 0 \tag{7.2}$$

$$h(\mathbf{x}, \mathbf{y}) \leq 0 \tag{7.3}$$

$$\mathbf{x} \in \mathbf{X} \tag{7.4}$$

where F is objective function to be minimized; $\mathbf{x}$ is the vector of control variables, consisting of generator voltages V_G, transformer tap settings T, and shunt Var compensations Q_C. Hence, $\mathbf{x}$ can be expressed as

$$\mathbf{x} = [V_{G1} \ldots V_{GNG}, T_1 \ldots T_{NT}, Q_{C1} \ldots Q_{CNC}] \tag{7.5}$$

where NG, NT, and NC are the number of generators, regulating transformer, and Var compensators, respectively.

$\mathbf{y}$ is the vector of dependent variables consisting of slack bus power P_{Gsl}, load bus voltages V_L, generator reactive power outputs Q_G, and transmission line loadings S_l. Accordingly, vector $\mathbf{y}$ can be expressed as

$$\mathbf{y} = [P_{Gsl}, V_{L1} \ldots V_{LNL}, Q_{G1} \ldots Q_{GNG}, S_{l1} \ldots S_{lNTL}] \tag{7.6}$$

where NL, NG, and NTL are number of load buses, number of generators, and number of transmission lines, respectively.

The equality constraints (7.2) are the typical nonlinear power flow equations described as follows:

$$P_{Gi} - P_{Di} - V_i \sum_{j=1}^{NB} V_j (G_{ij} \cos \theta_{ij} + B_{ij} \sin \theta_{ij}) = 0 \tag{7.7}$$

$$Q_{Gi} - Q_{Di} - V_i \sum_{j=1}^{NB} V_j \left(G_{ij} \sin\theta_{ij} - B_{ij} \cos\theta_{ij} \right) = 0 \tag{7.8}$$

where $i = 1, \ldots, NB$; NB is the number of busses; P_{Gi} is the active power generation, Q_{Gi} is the reactive power generation, P_{Di} is the active load demand, Q_{Di} is the reactive load demand, θ_{ij} is the voltage angle difference between bus i and bus j, G_{ij} and B_{ij} are the real and imaginary terms of bus admittance matrix corresponding to ith row and jth column, respectively.

Inequality constraints (7.3) are the functional operating constraints, such as (i) slack bus active power output limits, (ii) load bus voltage magnitude limits, (iii) generator reactive power output limits, and (iv) branch flow limits (MVA or A). These constraints can be mathematically expressed as

$$P_{Gsl}^{\min} \le P_{Gsl} \le P_{Gsl}^{\max}, \tag{7.9}$$

$$V_{Li}^{\min} \le V_{Li} \le V_{Li}^{\max} \quad i = 1, \ldots, NL \tag{7.10}$$

$$Q_{Gi}^{\min} \le Q_{Gi} \le Q_{Gi}^{\max} \quad i = 1, \ldots, NG \tag{7.11}$$

$$S_{li} \le S_{li}^{\max} \quad i = 1, \ldots, NTL \tag{7.12}$$

Constraints (7.4) define the feasibility region of the problem control variables such as (i) generator bus voltage magnitude limits, (ii) transformer tap setting limits, and (iii) shunt Var compensation limits.

$$V_{Gi}^{\min} \le V_{Gi} \le V_{Gi}^{\max}, \quad i = 1, \ldots, NG \tag{7.13}$$

$$T_i^{\min} \le T_i \le T_i^{\max}, \quad i = 1, \ldots, NT \tag{7.14}$$

$$Q_{Ci}^{\min} \le Q_{Ci} \le Q_{Ci}^{\max}, \quad i = 1, \ldots, NC \tag{7.15}$$

It is worth mentioning that the control variables are self-constrained. The hard inequality constraints of dependent variables, such as V_L and Q_G, can be incorporated in the objective function (7.1) as quadratic penalty terms. The new expanded objective function to be minimized becomes

$$F_p = F + \lambda_V \sum_{i=1}^{NL} \left(V_{Li} - V_{Li}^{\lim} \right)^2 + \lambda_Q \sum_{i=1}^{NG} \left(Q_{Gi} - Q_{Gi}^{\lim} \right)^2 \tag{7.16}$$

where λ_V and λ_Q are defined as penalty factors. $x^{\lim}$ is the limit value of the dependent variable x and given as

if $x > x^{\max}$, then $x^{\lim} = x^{\max}$

elseif $x < x^{\min}$, then $x^{\lim} = x^{\min}$

else

$x^{\lim} = x$

end

7.3 ORPD using hybrid PSOGSA

7.3.1 Overview of PSOGSA

The hybrid PSOGSA is developed by Mirjalili and Hashim [32]. The authors are hybridized PSO with GSA using low-level coevolutionary heterogeneous hybrid. The hybrid PSOGSA is of low level because it combines the functionality of both algorithms. It is coevolutionary because it does not use both algorithm one after another, that is, they run in parallel. It is heterogeneous because there are two different algorithms that involved to produce final results. The effectiveness of the hybrid PSOGSA for solving various optimization problems is demonstrated and confirmed in [1,32–37].

The position of the ith search agent in a system with N agents is defined by

$$\mathbf{x}_i = \left[x_i^1, \ldots, x_i^d, \ldots, x_i^n\right] \quad \text{for } i = 1, 2, \ldots, N \tag{7.17}$$

where N is the number of agents (the size of population), n is the search space dimension of the problem, i.e., the number of control variables, and x_i^d defines the position of the ith agent in the dth dimension, that is, value of dth control variable in ith candidate solution.

7.3.1.1 Particle swarm optimization

The PSO algorithm is based on simulation of bird flocking in two-dimensional space. It uses a number of particles (candidate solutions) which fly around in the search space to find best solution. Each particle tries to modify its position using the following information: the current position, the current velocity, the distance between the current position and its own best solution (**pbest**), and the distance between the current position and the best solution found so far (**gbest**).

In each iteration, the velocities of particles are calculated using the following equation:

$$\mathbf{v}_i(t+1) = w \cdot \mathbf{v}_i(t) + C_1 \cdot r_1 \cdot (\mathbf{pbest}_i(t) - \mathbf{x}_i(t)) + C_2 \cdot r_2 \cdot (\mathbf{gbest}(t) - \mathbf{x}_i(t)) \tag{7.18}$$

After updating the velocities, the positions of particles can be calculated as

$$\mathbf{x}_i(t+1) = \mathbf{x}_i(t) + \mathbf{v}_i(t+1) \tag{7.19}$$

where $\mathbf{v}_i(t)$ is velocity of particle i at iteration t, w is a weighting function, C_1 and C_2 are positive constants, r_1 and r_2 are uniformly distributed random numbers in [0,1], $\mathbf{x}_i(t)$ is the current position of particle i at iteration t, $\mathbf{pbest}_i$ is the individual best of particle i at iteration t, and **gbest** is the best solution found so far. The first part of (7.18) provides exploration ability for PSO. The second and third parts represent private thinking and collaboration of particles, respectively.

7.3.1.2 Gravitational search algorithm

In GSA, the search agents are a collection of masses which interact with each other based on the Newtonian gravity and the laws of motion. The position of the mass corresponds to the solution of the problem, and its gravitational and inertial masses are determined using a fitness function. In other words, each mass presents a solution. The algorithm is navigated by properly adjusting the gravitational and inertial masses. After evaluating the current population fitness, the mass of each agent is calculated as follows:

$$M_i(t) = \frac{m_i(t)}{\sum_{j=1}^{N} m_j(t)} \tag{7.20}$$

where

$$m_i(t) = \frac{fit_i(t) - worst(t)}{best(t) - worst(t)} \tag{7.21}$$

where $fit_i(t)$ represents the fitness value of the agent i at iteration t. $best(t)$ and $worst(t)$ are the best and worst fitness of all agents, respectively. According to Newton gravitation theory, the total force that acts on the ith agent at iteration t is specified as follows:

$$\mathbf{F}_i(t) = \sum_{j \in Kbest, j \neq i} r \cdot G(t) \frac{M_j(t) \cdot M_i(t)}{R_{i,j}(t) + \varepsilon} \left(\mathbf{x}_j(t) - \mathbf{x}_i(t) \right) \tag{7.22}$$

where r is a random number within the interval [0,1]. $G(t)$ is gravitational constant at iteration t, $M_i(t)$ and $M_j(t)$ are masses of agents i and j, ε is a small constant, and $R_{ij}(t)$ is the Euclidian distance between the two agents i and j, $Kbest$ is the set of first K agents with the best fitness value and biggest mass. According to the law of motion, the acceleration of the ith agent, at iteration t, is given by the following equation:

$$\mathbf{a}_i(t) = \frac{\mathbf{F}_i(t)}{M_i(t)} \tag{7.23}$$

The velocity and position of an agent are updated as follows:

$$\mathbf{v}_i(t+1) = r_i \cdot \mathbf{v}_i(t) + \mathbf{a}_i(t) \tag{7.24}$$

$$\mathbf{x}_i(t+1) = \mathbf{x}_i(t) + \mathbf{v}_i(t+1) \tag{7.25}$$

where r_i is a uniform random variable in the interval [0,1].

The gravitational constant $G(t)$ in (7.22) is a function of the initial value G_0 and time t:

$$G(t) = G_0 \cdot e^{-\alpha(t/t_{\max})} \tag{7.26}$$

The parameters of maximum iteration $t_{\max}$, population size N, initial gravitational constant G_0, and constant α control the performance of GSA.

7.3.1.3 Hybridization of PSO and GSA

The basic idea of PSOGSA is to combine the ability for social thinking (**gbest**) in PSO with the local search capability of GSA [32]. In order to combine these algorithms, the equation for updating the agents' velocity is proposed as follows [35]:

$$\mathbf{v}_i(t+1) = r_1 \cdot \mathbf{v}_i(t) + C_1 \cdot r_2 \cdot \mathbf{a}_i(t) + C_2 \cdot r_3 \cdot (\mathbf{gbest}(t) - \mathbf{x}_i(t)) \quad i = 1, \ldots, N \tag{7.27}$$

where $\mathbf{v}_i(t)$, $\mathbf{a}_i(t)$, and $\mathbf{x}_i(t)$ are the velocity, acceleration, and position of the ith agent at iteration t, respectively; $\mathbf{gbest}(t)$ is the best position (solution) so far at iteration t; C_1 and C_2 are positive constants; r_1, r_2, and r_3 are uniformly distributed random numbers in [0,1]; N is the size of population (the number of agents), where each agent consists of n elements, i.e., n control variables.

In each iteration, the positions of agents are updated as follows:

$$\mathbf{x}_i(t+1) = \mathbf{x}_i(t) + \mathbf{v}_i(t+1) \quad i = 1, \ldots, N \tag{7.28}$$

In PSOGSA, at first, all agents are randomly initialized. Each agent is considered as a candidate solution. After initialization, the gravitational constant and resultant forces among agents are calculated using (7.26) and (7.22), respectively. After that, the accelerations of agents are defined as (7.23). In each iteration, the best solution so far should be updated. After calculating the accelerations and updating the best solution so far, the velocities of all agents are calculated using (7.27). Finally, the positions of agents are updated by (7.28). The process of updating velocities and positions will be stopped by meeting an end criterion.

7.3.2 Application of PSOGSA to the ORPD problem

The following steps describe the procedure of applying the hybrid PSOGSA algorithm to solve the ORPD problem:

Step 1: Read the input data including the power system configuration, lines data, transformers data, shunt Var compensators data, loads data, and generation units data; specify the active power outputs of generators (except slack generator). Use the per unit system.

Step 2: Specify the control variables **x** (7.5) and their lower and upper limits (7.13)–(7.15); specify the dependent variables **y** (7.6), and their lower and upper limits (7.9)–(7.12); specify the objective function to be optimized $F(\mathbf{x},\mathbf{y})$, and the penalty factors ($\lambda_{V,Q}$) to form the expanded objective function $F_p(\mathbf{x},\mathbf{y})$ (7.16).

Step 3: Initialize PSOGSA parameters, such as the population size (N), maximum iteration number ($t_{\max}$), initial gravitational constant G_0, and constants α, C_1, and C_2.

Step 4: Generate an initial random population of N agents:

$$\mathbf{POP}(0) = \begin{bmatrix} \mathbf{x}_1(0) \\ \mathbf{x}_2(0) \\ \vdots \\ \mathbf{x}_i(0) \\ \vdots \\ \mathbf{x}_N(0) \end{bmatrix} = \begin{bmatrix} x_1^1(0), \ldots, x_1^d(0), \ldots, x_1^n(0) \\ x_2^1(0), \ldots, x_2^d(0), \ldots, x_2^n(0) \\ \vdots \\ x_i^1(0), \ldots, x_i^d(0), \ldots, x_i^n(0) \\ \vdots \\ x_N^1(0), \ldots, x_N^d(0), \ldots, x_N^n(0) \end{bmatrix} \tag{7.29}$$

The initial positions of each agent are randomly selected between minimum and maximum values of the control variables. Therefore, the initial value of the dth control variable in the ith agent (candidate solution) is obtained as follows:

$$x_i^d(0) = x_{i,\min}^d + r \cdot \left(x_{i,\max}^d - x_{i,\min}^d\right) \tag{7.30}$$

where r is a random number in [0,1], $x_{i,\min}^d$ and $x_{i,\max}^d$ are the lower and upper limits of the dth control variable. According to (7.5), the ith agent for the ORPD problem can be written as follows:

$$\mathbf{x}_i = \left[V_{G1i}^1, \ldots, V_{GNGi}^{\cdots}, T_{T1i}^{\cdots}, \ldots, T_{NT1i}^d, Q_{C1i}^{d+1}, \ldots, Q_{CNCi}^n\right] \tag{7.31}$$

Initialize the velocity of agents by a zero matrix of dimension $N \times n$; where N is the number of particles (population size) and n is the number of control variables ($n = NG + NT + NC$).

Step 5: Run the power flow program based on fast decoupled power flow method for each agent $\mathbf{x}_i(t)$ from the current population $\mathbf{POP}(t)$ and calculate the corresponding values of the objective function (fitness values).

Step 6: Calculate the gravitational constant $G(t)$ (7.26), the best and worst fitness value, and the mass of each agent $M_i(t)$ (7.20), $i = 1, \ldots, N$.

Step 7: Calculate the total force in different directions using (7.22), and the acceleration of each agent using (7.23).

Step 8: Update the best solution so far **gbest**(t).

Step 9: Update the velocity of each agent using (7.27).

Step 10: Update the position of each agent using (7.28).

Step 11: Repeat steps 5–10 until the stop criterion is reached, that is, maximum number of iterations $t_{\max}$.

Step 12: Return best solution obtained in the last iteration; Stop.

7.3.3 *Simulation results of PSOGSA*

The hybrid PSOGSA algorithm is applied to IEEE 30-bus and IEEE 118-bus standard test systems for the solution of ORPD problem. The algorithm has been implemented in the *ORPD* package of MATLAB® m-files described in Section 7.5.

Table 7.1 Algorithms parameters

Parameters	IEEE-30 bus test system			IEEE-118 bus test system		
	PSO	GSA	PSOGSA	PSO	GSA	PSOGSA
C_1	2	–	2	2	–	2
C_2	2	–	2	2	–	2
α	–	10	20	–	10	20
G_0	–	100	1	–	100	1
N	50	50	50	100	100	100
t_{max}	200	200	200	500	500	500

Table 7.2 Limits of generation reactive power for IEEE 30-bus test system

Bus	1	2	5	8	11	13
Q_G^{min} (MVAr)	−20	−20	−15	−15	−10	−15
Q_G^{max} (MVAr)	200	100	80	60	50	60

The different algorithms parameters used for the simulation are given in Table 7.1. "No Free Lunch" theorem states that no single method is best in solving all optimization problems. Similarly, there are no existing universal values of the algorithm parameters for different optimization problems. The algorithms parameters given in Table 7.1 are adopted based on numerous experiments performed for both test systems. The parameter values proposed in [32] were used as initial values in these experimentations. The results shown are the best values obtained over 10 consecutive test runs.

7.3.3.1 IEEE 30-bus test system

The IEEE 30-bus test system has 41 transmission lines, 6 generators at the buses 1, 2, 5, 8, 11, and 13, 4 transformers with off-nominal tap ratio at lines 6–9, 6–10, 4–12, and 28–27, and 9 shunt capacitors at the busses 10, 12, 15, 17, 20, 21, 23, 24, and 29. The total load demands are $P_{load} = 2.834$ p.u. and $Q_{load} = 1.262$ p.u. at 100 MVA base. There are 19 control variables for the ORPD problem as follows: six generator voltage magnitudes, four transformer tap settings, and nine shunt capacitor reactive power injections. The voltages of the generator buses as well as load buses have been constrained within limits of 0.95 and 1.10 p.u. The lower and upper limits of the transformer tap settings are 0.9 and 1.1 p.u., respectively. The range of the shunt capacitor reactive power injections is between 0 and 5 MVAr. The reactive power generation limits are given in Table 7.2. All the system data have been adopted from [13,38,39]. The active power outputs of generators are specified as follows: $P_{G2} = 80$ MW, $P_{G5} = 50$ MW, $P_{G8} = 20$ MW, $P_{G11} = 20$ MW, and $P_{G13} = 20$ MW. Bus 1 is selected as the slack bus.

Two cases are considered, namely, minimization of active power loss and voltage profile improvement.

Case 1: Minimization of active power loss. In this case, the objective function (7.1) has the following form:

$$\min_{\mathbf{x}} F(\mathbf{x}, \mathbf{y}) = \min_{\mathbf{x}} P_{loss}(\mathbf{x}, \mathbf{y}) = \min_{\mathbf{x}} \sum_{L=1}^{NTL} P_{loss,L} \tag{7.32}$$

where $P_{loss,L}$ is the active power loss at line L, and NTL is the number of transmission lines. The active power loss through a transmission line ($L \equiv ij$) can be calculated by

$$P_{loss,ij} = g_{ij}\left[V_i^2 + V_j^2 - 2V_iV_j\cos(\theta_i - \theta_j)\right] \tag{7.33}$$

where g_{ij} is the conductance of transmission line connecting the ith and jth bus; V_i, V_j, θ_i, and θ_j are the voltage magnitudes and voltage angles at bus i and j, respectively.

The proposed hybrid PSOGSA is applied to minimize the active power loss which defined in (7.32) as objective together with the penalty terms defined in (7.16).

The optimal results are presented in Table 7.3. It appears that the active power loss corresponding to hybrid PSOGSA is highly reduced (22.18%) in comparison with the initial case. The results of this table show that hybrid PSOGSA leads to lower active power loss than the standard PSO and GSA. Figure 7.1 shows the comparative convergence profiles of PSO, GSA, and proposed hybrid PSOGSA for Case 1. It may be observed from this figure that the proposed hybrid PSOGSA algorithm has better convergence characteristic than PSOGSAs. Furthermore, the statistical evaluation of the results presented in Table 7.4 clearly indicates that the proposed hybrid PSOGSA approach provides better and stable solutions compared to the original PSO and GSA.

Case 2: Voltage profile improvement. In this case, the objective is minimization of the load bus *VD*s:

$$\min_{\mathbf{x}} F(\mathbf{x}, \mathbf{y}) = \min_{\mathbf{x}} VD(\mathbf{x}, \mathbf{y}) = \min_{\mathbf{x}} \sum_{i=1}^{NL} \left|V_i - V_i^{ref}\right| \tag{7.34}$$

where NL is the number of load buses, V_i is the voltage magnitude at bus i, and V_i^{ref} is the reference value of the voltage magnitude of the ith bus, which is usually set to 1 p.u.

The optimal results for Case 2 are shown in Table 7.3. The PSOGSA results are compared to those obtained by original PSO and GSA algorithms. The proposed PSOGSA is reduced the VD by 92.14% with respect to the initial case solution. The PSOGSA is resulted the VD which is lesser by 10% compared to PSO and 29.98% compared to GSA.

Table 7.3 Optimal results for IEEE 30-bus test system

Control variables	Initial	Case 1: min P_{loss}			Case 2: min VD		
		PSO	GSA	PSOGSA	PSO	GSA	PSOGSA
V_{G1} (p.u.)	1.05	1.1000	1.1000	1.1000	1.0264	1.0374	1.0153
V_{G2} (p.u.)	1.04	1.0944	1.0944	1.0944	1.0162	1.0400	1.0122
V_{G5} (p.u.)	1.01	1.0750	1.0748	1.0749	1.0185	1.0220	1.0185
V_{G8} (p.u.)	1.01	1.0767	1.0768	1.0767	0.9987	1.0047	1.0107
V_{G11} (p.u.)	1.05	1.1000	1.1000	1.1000	1.0427	0.9885	0.9889
V_{G13} (p.u.)	1.05	1.1000	1.1000	1.1000	0.9965	0.9924	1.0083
$T_{11(6-9)}$ (p.u.)	1.078	1.0473	1.0399	1.0452	1.0598	0.9772	1.0024
$T_{12(6-10)}$ (p.u.)	1.069	0.9000	0.9000	0.9000	0.9144	0.9000	0.9000
$T_{15(4-12)}$ (p.u.)	1.032	0.9831	0.9827	0.9794	0.9580	0.9274	0.9791
$T_{36(28-27)}$ (p.u.)	1.068	0.9664	0.9699	0.9651	0.9758	0.9612	0.9737
Q_{C10} (MVAr)	0	5.0000	3.5717	5.0000	4.9995	1.9778	4.3048
Q_{C12} (MVAr)	0	5.0000	3.0984	5.0000	0.0000	0.4240	2.3931
Q_{C15} (MVAr)	0	5.0000	3.2925	5.0000	5.0000	2.2268	5.0000
Q_{C17} (MVAr)	0	5.0000	4.0166	5.0000	4.9958	2.8945	0.0000
Q_{C20} (MVAr)	0	3.4023	3.0309	3.9792	5.0000	4.0503	5.0000
Q_{C21} (MVAr)	0	5.0000	4.0339	5.0000	5.0000	3.2996	5.0000
Q_{C23} (MVAr)	0	5.0000	2.9946	2.4583	4.9988	2.5926	5.0000
Q_{C24} (MVAr)	0	5.0000	4.3499	5.0000	5.0000	2.6791	5.0000
Q_{C29} (MVAr)	0	2.1038	2.6902	2.1865	4.9994	2.8961	4.1670
P_{loss} (MW)	5.8223	4.5388	4.5515	**4.5309**	5.5192	5.7296	5.7344
VD (p.u.)	1.1497	2.0648	1.9319	2.0504	0.1005	0.1291	**0.0904**

The bold values indicate the best results.

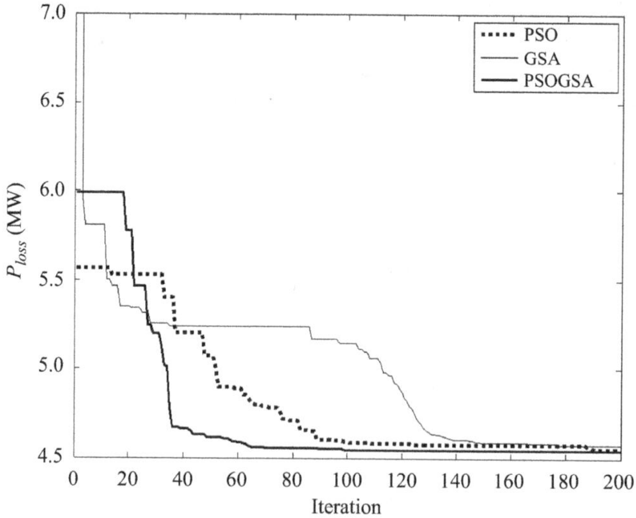

Figure 7.1 Convergence characteristics of PSO, GSA, and hybrid PSOGSA for Case 1 in the IEEE 30-bus test system

Table 7.4 Statistical evaluation of the results for IEEE 30-bus test system

	PSO			GSA			PSOGSA		
	Min	**Max**	**Std. dev.**	**Min**	**Max**	**Std. dev.**	**Min**	**Max**	**Std. dev.**
Case 1	4.5388	5.1327	0.2040	4.5515	4.6408	0.0240	4.5309	4.5379	0.0027
Case 2	0.1005	0.1672	0.0221	0.1291	0.1884	0.0164	0.0904	0.1398	0.0167

Table 7.5 Simulation results of dependent variables obtained by PSOGSA for IEEE 30-bus test system

Dependent variables	**Case 1**	**Case 2**
Q_{G1} (MVAr)	−12.5968	−19.6096
Q_{G2} (MVAr)	12.3901	0.2903
Q_{G5} (MVAr)	22.7519	47.9316
Q_{G8} (MVAr)	30.8136	59.7477
Q_{G11} (MVAr)	11.0156	−4.8238
Q_{G13} (MVAr)	0.2320	−0.1290
V_{L3} (p.u.)	1.0843	1.0033
V_{L4} (p.u.)	1.0799	0.9998
V_{L6} (p.u.)	1.0769	1.0007
V_{L7} (p.u.)	1.0687	1.0000
V_{L9} (p.u.)	1.0798	0.9999
V_{L10} (p.u.)	1.0963	1.0079
V_{L12} (p.u.)	1.1000	1.0089
V_{L14} (p.u.)	1.0912	0.9998
V_{L15} (p.u.)	1.0914	1.0005
V_{L16} (p.u.)	1.0934	1.0009
V_{L17} (p.u.)	1.0936	1.0002
V_{L18} (p.u.)	1.0848	0.9942
V_{L19} (p.u.)	1.0839	0.9937
V_{L20} (p.u.)	1.0885	0.9990
V_{L21} (p.u.)	1.0888	0.9993
V_{L22} (p.u.)	1.0894	1.0000
V_{L23} (p.u.)	1.0878	0.9999
V_{L24} (p.u.)	1.0838	0.9944
V_{L25} (p.u.)	1.0872	1.0005
V_{L26} (p.u.)	1.0707	0.9825
V_{L27} (p.u.)	1.0974	1.0132
V_{L28} (p.u.)	1.0740	1.0000
V_{L29} (p.u.)	1.0861	1.0057
V_{L30} (p.u.)	1.0725	0.9888

The dependent variables of ORPD solutions obtained by proposed hybrid PSOGSA for Cases 1 and 2 are listed in Table 7.5. It can be seen from Table 7.5 that all dependent variables, i.e., the reactive power outputs of generators and the voltages of load buses, remained within their permissible limits. Therefore, the

Table 7.6 Comparison of the simulation results for Case 1, IEEE 30-bus test system

	PSOGSA	DE [3]	QODE [4]	BBO [8]	ALC-PSO [12]	GSA [13]	OGSA [14]	CKHA [17]
V_{G1} (p.u.)	1.1000	1.1000	1.0500	1.1000	1.0500	1.0717	1.0500	1.0500
V_{G2} (p.u.)	1.0944	1.0931	1.0338	1.0944	1.0384	1.0222	1.0410	1.0473
V_{G5} (p.u.)	1.0749	1.0736	1.0058	1.0749	1.0108	1.0401	1.0154	1.0293
V_{G8} (p.u.)	1.0767	1.0756	1.0230	1.0768	1.0210	1.0507	1.0267	1.0350
V_{G11} (p.u.)	1.1000	1.1000	1.0913	1.0999	1.0500	0.9771	1.0082	1.0500
V_{G13} (p.u.)	1.1000	1.1000	1.0400	1.0999	1.0500	0.9677	1.0500	1.0500
$T_{11(6\text{–}9)}$ (p.u.)	1.0452	1.0465	0.9994	1.0435	0.9521	1.0984	1.0585	0.9916
$T_{12(6\text{–}10)}$ (p.u.)	0.9000	0.9097	1.0012	0.9012	1.0299	0.9825	0.9089	0.9538
$T_{15(4\text{–}12)}$ (p.u.)	0.9794	0.9867	0.9983	0.9824	0.9721	1.0959	1.0141	0.9603
$T_{36(28\text{–}27)}$ (p.u.)	0.9651	0.9689	1.0141	0.9692	0.9657	1.0593	1.0182	0.9670
Q_{C10} (MVAr)	5.0000	5.0000	0.3000	4.9998	0.9000	1.6538	3.3000	0.9200
Q_{C12} (MVAr)	5.0000	5.0000	1.9900	4.9870	1.2600	4.3723	2.4900	0
Q_{C15} (MVAr)	5.0000	5.0000	4.7500	4.9906	2.0900	0.1200	1.7700	1.5300
Q_{C17} (MVAr)	5.0000	5.0000	3.3400	4.9970	5.0000	2.0876	5.0000	4.9700
Q_{C20} (MVAr)	3.9792	4.4060	1.8200	4.9901	0.3100	0.3577	3.3400	3.0200
Q_{C21} (MVAr)	5.0000	5.0000	2.5000	4.9946	2.9300	0.2603	4.0300	5.0000
Q_{C23} (MVAr)	2.4583	2.8004	3.4200	3.8753	2.2600	0	2.6900	1.3400
Q_{C24} (MVAr)	5.0000	5.0000	3.4500	4.9867	5.0000	1.3840	5.0000	5.0000
Q_{C29} (MVAr)	2.1865	2.5979	0	2.9098	1.0700	0.0003	1.9400	1.2100
$\boldsymbol{P_{loss}}$ **(MW)**	4.5309	4.5550	2.6867	4.5511	4.4793	4.5143	4.4984	3.2400
$\boldsymbol{VD}$ **(p.u.)**	2.0504	1.9589	0.4609	NR*	0.8425	0.8752	0.8085	1.3364

*NR means not reported.

proposed hybrid PSOGSA approach gives the best solutions of the ORPD problem without violating any operating constraint limits.

7.3.3.2 Critical analysis of the results for IEEE 30-bus test system

Under the same system data, variable limits, and constraints, the ORPD results obtained using the hybrid PSOGSA approach are compared to some other recently reported optimization algorithms such as DE [3], quasioppositional DE (QODE) [4], BBO [8], PSO with an aging leader and challengers (ALC-PSO) [12], GSA [13], opposition-based GSA (OGSA) [14], and chaotic KHA (CKHA) [17]. As can be seen from Tables 7.6 and 7.7, some of these algorithms like QODE [4], ALC-PSO [12], GSA [13], OGSA [14], and CKHA [17] are resulted in lesser minimum of the objective functions than the proposed hybrid PSOGSA approach.

To check these results, the power flow computations with control variables given in Tables 7.6 and 7.7 were performed. It should be noted that the active power outputs of generators are specified as follows: $P_{G2} = 80$ MW, $P_{G5} = 50$ MW, $P_{G8} = 20$ MW, $P_{G11} = 20$ MW, and $P_{G13} = 20$ MW. The power flow results are summarized in Table 7.8. All results were checked and verified by MATPOWER power flow program [40]. The Newton–Raphson method was used for the power flow calculations.

As can be seen from Table 7.8, some of the results reported in the above mentioned references can be considered as infeasible and/or incorrect solutions.

Table 7.7 Comparison of the simulation results for Case 2, IEEE 30-bus test system

	PSOGSA	DE [3]	QODE [4]	ALC-PSO [12]	GSA [13]	OGSA [14]	CKHA [17]
V_{G1} (p.u.)	1.0153	1.0100	1.0500	0.9985	0.98385	0.9746	0.9985
V_{G2} (p.u.)	1.0122	0.9918	1.0340	1.0106	1.04481	1.0273	1.0059
V_{G5} (p.u.)	1.0185	1.0179	1.0059	0.9963	1.02035	0.9965	0.9951
V_{G8} (p.u.)	1.0107	1.0183	1.0235	1.0013	0.99913	0.9982	1.0022
V_{G11} (p.u.)	0.9889	1.0114	1.0917	1.0105	1.07700	0.9826	1.0293
V_{G13} (p.u.)	1.0083	1.0282	1.0396	1.0005	1.04393	1.0403	0.9993
$T_{11(6–9)}$ (p.u.)	1.0024	1.0265	1.0157	1.0103	0.90000	0.9909	1.0303
$T_{12(6–10)}$ (p.u.)	0.9000	0.9038	1.0274	1.0818	1.10000	1.0629	1.0328
$T_{15(4–12)}$ (p.u.)	0.9791	1.0114	1.0087	1.0198	1.05059	1.0762	1.0215
$T_{36(28–27)}$ (p.u.)	0.9737	0.9635	0.9817	1.0151	0.96199	1.0117	1.0171
Q_{C10} (MVAr)	4.3048	4.9420	0.9500	0.9300	0.00000	2.4600	4.1200
Q_{C12} (MVAr)	2.3931	1.0885	0.6800	0.7300	0.47351	1.7500	1.4800
Q_{C15} (MVAr)	5.0000	4.9985	3.0100	0.8800	5.00000	2.8300	0.0000
Q_{C17} (MVAr)	0.0000	0.2393	0.7900	3.9900	0.00000	4.0300	3.4300
Q_{C20} (MVAr)	5.0000	4.9958	2.4700	0.0000	5.00000	0.0000	0.0100
Q_{C21} (MVAr)	5.0000	4.9075	1.7100	4.3200	0.00000	2.7000	4.2100
Q_{C23} (MVAr)	5.0000	4.9863	3.0100	0.0000	4.99983	3.8500	0.0000
Q_{C24} (MVAr)	5.0000	4.9663	5.0000	2.6900	5.00000	2.5700	1.8500
Q_{C29} (MVAr)	4.1670	2.2325	1.7800	0.0000	5.00000	0.0000	0.0200
P_{loss} (MW)	5.7344	6.4755	9.2745	6.2800	6.91176	6.9044	8.2400
VD (p.u.)	0.0904	0.0911	0.0607	0.0437	0.06763	0.0640	0.0416

Table 7.8 The power flow results for control variables given in Tables 7.6 and 7.7

Case	Method	P_{loss} (MW)	VD (p.u.)	Violating constraints
Case 1: min P_{loss}	PSOGSA	4.5309	2.0504	No
	DE [3]	4.5358	1.9972	No
	QODE [4]	5.2953	0.3984	No
	BBO [8]	4.5367	2.0900	No
	ALC-PSO [12]	5.1861	0.3760	No
	GSA [13]	7.1787	1.5626	Q_{G2}, Q_{G8}, V_{L10}, V_{L12}, V_{L14}–V_{L27}, V_{L29}, V_{L30}
	OGSA [14]	5.1676	0.3299	No
	CKHA [17]	5.1163	0.6663	No
Case 2: min VD	PSOGSA	5.7344	0.0904	No
	DE [3]	6.4145	0.1138	Q_{G2}, Q_{G8}
	QODE [4]	5.2427	0.2527	No
	ALC-PSO [12]	6.1120	1.1549	Q_{G1}, V_{L17}–V_{L27}, V_{L29}, V_{L30}
	GSA [13]	8.9090	0.1912	Q_{G1}, Q_{G2}
	OGSA [14]	8.3660	0.9570	Q_{G1}, Q_{G2}, V_{L18}–V_{L22}, V_{L24}–V_{L26}, V_{L29}, V_{L30}
	CKHA [17]	6.0285	1.0610	Q_{G1}, V_{L18}–V_{L27}, V_{L29}, V_{L30}

Case 1: min P_{loss}

- For the optimum control variables reported in [4] (QODE), [12] (ALC-PSO), [14] (OGSA), and [17] (CKHA), the exact values of power loss (P_{loss}) as obtained from the power flow computations are 5.2953, 5.1861, 5.1676, and 5.1163 MW, respectively. Obviously, the calculated values of P_{loss} are substantially different than those reported in the references.
- For the optimum results obtained using GSA reported in [13], the voltage limits are violated at buses 10, 12, 14–27, 29, and 30. In addition, the reactive power limits of generators at buses 2 and 8 are violated also. The true value of P_{loss} is 7.1787 MW for this violated case.

Case 2: min *VD*

- The optimum control variables obtained using DE and given in [3] represent an infeasible solution because the reactive powers of the generators at buses 2 and 8 come −61.03 and 71.83 MVAr, respectively, which violate their corresponding limits as given in Table 7.2 and reported in [38,39].
- For the optimum control variables given in [4] (QODE), the true value of *VD* is 0.2527 p.u., as obtained from the power flow computations.
- The best solution given in [12] (ALC-PSO) is an infeasible solution because reactive power of the generator at bus 1 is given as −50.6 MVAr which violate their lower limit of −20 MW. In addition, the voltages at buses 17–27, 29, and 30 violate their lower limits. The exact value of *VD* is 1.1549 p.u. for this violated case.
- The best solution given in [13] (GSA) is an infeasible solution because reactive powers of the generators at buses 1 and 2 were given as −146.29 and 164.66 MVAr, respectively, which violate their corresponding limits as given in Table 7.2. The exact value of *VD* is 0.1912 for this violated case, as obtained from the power flow computations.
- For the optimum control variables given in [14] (OGSA), the reactive powers of the generators at buses 1 and 2 were given as −131.16 and 139.39 MVAr, respectively, which violate their corresponding limits given in Table 7.2. Additionally, the voltage limits are violated at buses 18–22, 24–26, 29, and 30. The exact value of *VD* is 0.957 for this violated case.
- The best result given in [17] (CKHA) is also an infeasible solution because reactive power of the generator at bus 1 comes −42.58 MVAr, which violate their lower limit of −20 MVAr as specified in Table 7.2. The voltage limits are violated at buses 18–27, 29, and 30. The exact value of *VD* is 1.061 p.u. for this violated case.

There are relatively large differences between the power flow results and the corresponding results reported in the literature [3,4,12–14,17]. Some of the possible explanations for these differences are as follows:

1. The authors solved the ORPD problem with some other values of active power outputs of generators, instead with the values that are specified in this work.

2. The authors solved the ORPD problem as a classical OPF problem. However, the active power outputs of generators are not considered as the control variables in these papers. Moreover, the active power outputs of generators are not specified in these references, and therefore, it is not possible to reliably verify how the reported ORPD results are obtained?

These differences in the results for [3,8] are insignificant and can be explained by the different mathematical models adapted for various components such as transformers during the power flow computations, rounding values of control variables, etc.

7.3.3.3 IEEE 118-bus test system

To evaluate the efficiency of the proposed PSOGSA approach in solving larger power system, the IEEE 118-bus test system is considered. The system branch data, bus data, and the initial values of control variables are given in [40,41,45]. The test system consists of 54 generators, 186 branches, 9 transformers, 2 reactors, and 12 capacitors. The total load demands are $P_{load} = 4{,}242$ MW and $Q_{load} = 1{,}438$ MVAr. The initial power loss and VD are as follows: $P_{loss} = 132.8629$ MW and $VD = 1.4393$ p.u. For the ORPD problem, the system has a total of 77 control variables as follows: 54 generators' bus voltages, 9 transformers' tap settings, and 14 shunt reactive power injections. The voltage limits of all busses are between 0.94 and 1.06 p.u. The transformer tap settings are considered within the interval 0.90–1.10 p.u. The minimum and maximum limits of reactive power sources are given in Table 7.9.

The optimal solutions of the ORPD problem obtained using the proposed PSOGSA are presented in Table 7.10. It is observed that PSOGSA is able to reduce the real power loss by 7.82% with respect to the initial case solution. Minimization of VD as objective in ORPD (Case 2) results in reduction of 49.23% in VD over the initial case.

The system voltage profiles for load buses are shown in Figure 7.2. The reactive power outputs of generators which correspond to the ORPD results listed in Table 7.10 are given in Table 7.11. It can be seen from Figure 7.2 and Table 7.11 that all dependent variables, i.e., the voltages of load buses and the reactive power outputs of generators, are within their allowable limits.

Table 7.9 Limits of reactive power sources in MVAr, IEEE 118-bus test system

Bus	5	34	37	44	45	46	48
Q_C^{min}	−40	0	−25	0	0	0	0
Q_C^{max}	0	14	0	10	10	10	15
Bus	74	79	82	83	105	107	110
Q_C^{min}	0	0	0	0	0	0	0
Q_C^{max}	12	20	20	10	20	6	6

Table 7.10 Results for IEEE 118-bus test system obtained by PSOGSA

Control variables	Initial	Case 1: min P_{loss}	Case 2: min VD
V_{G1} (p.u.)	0.9550	1.0299	0.9809
V_{G4} (p.u.)	0.9980	1.0598	1.0127
V_{G6} (p.u.)	0.9900	1.0529	0.9976
V_{G8} (p.u.)	1.0150	0.9888	1.0091
V_{G10} (p.u.)	1.0500	0.9408	0.9731
V_{G12} (p.u.)	0.9900	1.0508	1.0057
V_{G15} (p.u.)	0.9700	1.0235	1.0011
V_{G18} (p.u.)	0.9730	1.0211	1.0002
V_{G19} (p.u.)	0.9620	1.0187	0.9975
V_{G24} (p.u.)	0.9920	1.0231	1.0240
V_{G25} (p.u.)	1.0500	1.0281	1.0445
V_{G26} (p.u.)	1.0150	1.0599	1.0600
V_{G27} (p.u.)	0.9680	1.0228	0.9765
V_{G31} (p.u.)	0.9670	1.0143	1.0089
V_{G32} (p.u.)	0.9630	1.0194	0.9944
V_{G34} (p.u.)	0.9840	1.0207	1.0072
V_{G36} (p.u.)	0.9800	1.0183	1.0050
V_{G40} (p.u.)	0.9700	0.9935	0.9881
V_{G42} (p.u.)	0.9850	0.9886	1.0581
V_{G46} (p.u.)	1.0050	1.0357	0.9987
V_{G49} (p.u.)	1.0250	1.0538	1.0150
V_{G54} (p.u.)	0.9550	1.0436	1.0467
V_{G55} (p.u.)	0.9520	1.0404	1.0412
V_{G56} (p.u.)	0.9540	1.0410	1.0409
V_{G59} (p.u.)	0.9850	1.0600	1.0600
V_{G61} (p.u.)	0.9950	1.0600	0.9753
V_{G62} (p.u.)	0.9980	1.0566	0.9792
V_{G65} (p.u.)	1.0050	1.0239	0.9938
V_{G66} (p.u.)	1.0500	1.0600	1.0463
V_{G69} (p.u.)	1.0350	1.0600	1.0336
V_{G70} (p.u.)	0.9840	1.0350	1.0000
V_{G72} (p.u.)	0.9800	1.0302	0.9546
V_{G73} (p.u.)	0.9910	1.0587	1.0101
V_{G74} (p.u.)	0.9580	1.0066	0.9749
V_{G76} (p.u.)	0.9430	0.9957	0.9633
V_{G77} (p.u.)	1.0060	1.0382	1.0189
V_{G80} (p.u.)	1.0400	1.0542	1.0498
V_{G85} (p.u.)	0.9850	1.0446	1.0080
V_{G87} (p.u.)	1.0150	1.0515	1.0112
V_{G89} (p.u.)	1.0050	1.0600	1.0234
V_{G90} (p.u.)	0.9850	1.0323	1.0550
V_{G91} (p.u.)	0.9800	1.0273	1.0284
V_{G92} (p.u.)	0.9900	1.0431	1.0066
V_{G99} (p.u.)	1.0100	1.0072	0.9417
V_{G100} (p.u.)	1.0170	1.0522	0.9906
V_{G103} (p.u.)	1.0100	1.0480	0.9778
V_{G104} (p.u.)	0.9710	1.0353	0.9741
V_{G105} (p.u.)	0.9650	1.0339	0.9865

Table 7.10 (Continued)

Control variables	Initial	Case 1: min P_{loss}	Case 2: min VD
V_{G107} (p.u.)	0.9520	1.0422	1.0600
V_{G110} (p.u.)	0.9730	1.0196	0.9892
V_{G111} (p.u.)	0.9800	1.0270	1.0595
V_{G112} (p.u.)	0.9750	1.0015	0.9405
V_{G113} (p.u.)	0.9930	1.0337	0.9937
V_{G116} (p.u.)	1.0050	1.0067	1.0087
T_8 (p.u.)	0.9850	0.9182	0.9994
T_{32} (p.u.)	0.9600	1.1000	1.0567
T_{36} (p.u.)	0.9600	0.9790	0.9024
T_{51} (p.u.)	0.9350	0.9759	0.9604
T_{93} (p.u.)	0.9600	0.9000	0.9116
T_{95} (p.u.)	0.9850	0.9287	1.0539
T_{102} (p.u.)	0.9350	1.0057	0.9000
T_{107} (p.u.)	0.9350	0.9715	0.9000
T_{127} (p.u.)	0.9350	0.9459	0.9376
Q_{C5} (MVAr)	−40.0000	−33.5074	−27.3957
Q_{C34} (MVAr)	14.0000	7.6243	7.0607
Q_{C37} (MVAr)	−25.0000	−19.7317	−10.6680
Q_{C44} (MVAr)	10.0000	6.5258	3.9224
Q_{C45} (MVAr)	10.0000	4.5300	5.4634
Q_{C46} (MVAr)	10.0000	3.1784	9.2847
Q_{C48} (MVAr)	15.0000	11.8388	0
Q_{C74} (MVAr)	12.0000	3.8061	2.4011
Q_{C79} (MVAr)	20.0000	13.9863	15.0490
Q_{C82} (MVAr)	20.0000	17.7504	10.4150
Q_{C83} (MVAr)	10.0000	1.9938	0.0106
Q_{C105} (MVAr)	20.0000	6.8200	1.9242
Q_{C107} (MVAr)	6.0000	6.0000	4.0001
Q_{C110} (MVAr)	6.0000	4.4194	1.0668
P_{loss} (MW)	132.8629	122.4709	145.4049
VD (p.u.)	1.4393	1.7792	0.7308

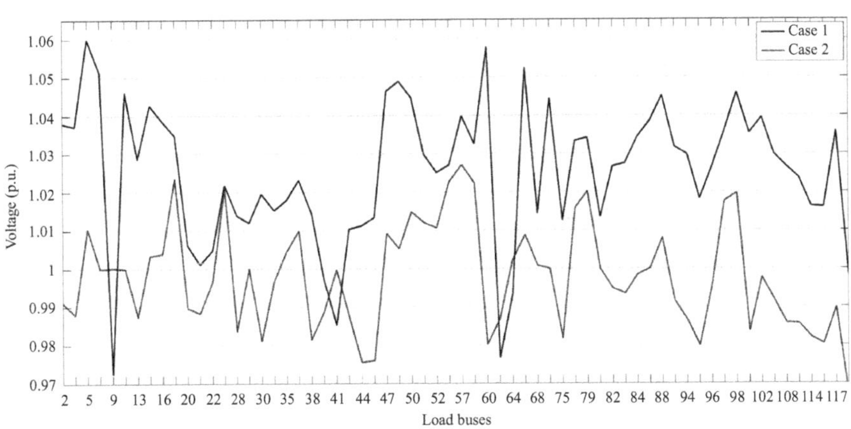

Figure 7.2 Voltage profiles of IEEE 118-bus test system for load buses

Table 7.11 Reactive power outputs of generators for IEEE 118-bus test system. The bold values signify the violated constraints

	Limits [41] (MVAr)		Case 1: min P_{loss}							Case 2: min VD			
	Max	Min	PSOGSA	QODE [4]	CLPSO [11]	ALC-PSO [12]	GSA [13]	OGSA [14]	MICA–IWO [26]	PSOGSA	QODE [4]	ALC-PSO [12]	OGSA [14]
Q_{G1}	15	−5	14.99	−4.11	**82.68**	**32.14**	**47.55**	**37.90**	**35.68**	14.99	−3.05	**44.44**	**114.32**
Q_{G4}	300	−300	13.45	−33.24	**419.03**	43.80	−85.73	66.05	58.27	64.32	−20.20	274.80	−88.92
Q_{G6}	50	−13	24.56	14.53	**−120.53**	**−19.96**	**59.33**	**−28.11**	35.56	−0.55	16.63	−10.24	6.19
Q_{G8}	300	−300	48.94	71.00	**−347.88**	−39.13	42.56	−47.49	−77.53	72.98	62.82	**−379.8**	10.26
Q_{G10}	200	−147	−146.07	−51.22	−63.11	−62.02	−69.22	−89.74	−111.98	−141.32	−51.67	−45.34	−87.92
Q_{G12}	120	−35	84.32	89.18	61.50	86.74	−1.25	96.68	47.34	90.23	89.63	**140.26**	101.45
Q_{G15}	30	−10	29.59	−1.82	**−280.62**	16.02	**129.00**	**−12.98**	26.89	14.79	0.92	**−74.90**	−5.78
Q_{G18}	50	−16	26.35	21.73	**215.11**	4.92	41.39	38.90	35.58	8.38	21.61	**247.12**	**−103.4**
Q_{G19}	24	−8	23.43	1.53	**366.29**	23.39	2.29	**−16.68**	**26.52**	18.76	0.71	**50.54**	**248.02**
Q_{G24}	300	−300	−12.75	−17.71	21.18	−34.12	2.81	9.20	1.82	39.73	−18.79	37.89	58.34
Q_{G25}	140	−47	139.60	48.25	**245.51**	65.13	**−222.9**	74.09	**238.79**	139.53	47.39	**−102.7**	**−129.4**
Q_{G26}	1,000	−1,000	−163.25	9.59	−296.1	−134.9	212.08	−31.83	−296.64	−62.13	9.61	−13.53	15.26
Q_{G27}	300	−300	49.59	−6.21	121.74	33.12	90.51	32.84	35.89	−48.91	−11.01	135.26	101.31
Q_{G31}	300	−300	23.20	27.63	−27.13	50.62	−13.70	15.94	28.19	67.15	24.72	46.12	44.74
Q_{G32}	42	−14	29.38	4.32	**−20.46**	27.58	**−47.15**	**−17.10**	25.58	39.86	15.29	**−55.67**	22.68
Q_{G34}	24	−8	16.10	**−12.45**	**−294.41**	**−46.68**	−2.40	**34.89**	−3.09	22.60	**−34.23**	**325.33**	**149.03**
Q_{G36}	24	−8	22.64	−4.58	**438.85**	**52.18**	**143.76**	**−33.02**	23.23	23.55	13.61	**−241.0**	**−85.96**
Q_{G40}	300	−300	37.98	25.65	−50.03	48.50	26.30	56.10	33.56	−34.02	27.23	44.67	94.89
Q_{G42}	300	−300	2.42	40.99	173.86	61.55	−7.15	41.84	34.81	158.67	40.64	67.73	15.54
Q_{G46}	100	−100	6.56	9.20	20.33	−13.97	−54.28	36.98	10.36	4.92	7.86	93.65	**112.69**
Q_{G49}	210	−85	88.32	122.02	63.61	27.19	177.40	72.92	91.10	−84.92	123.05	**−105.4**	−29.38
Q_{G54}	300	−300	71.78	1.41	**−607.09**	88.44	182.92	12.48	34.46	148.72	−25.28	162.92	160.88
Q_{G55}	23	−8	16.21	−0.45	**−309.27**	**−26.15**	**−318.3**	**32.45**	22.21	19.96	18.68	**187.04**	**25.39**
Q_{G56}	15	−8	10.01	5.56	**1,139.10**	**38.67**	**313.73**	**49.46**	**20.21**	−6.33	**17.58**	**−145.6**	**−81.29**
Q_{G59}	180	−60	59.68	76.73	−38.74	**267.08**	41.59	**−120.7**	102.92	179.99	72.65	155.15	**299.27**

Q_{G61}	300	−100	−51.15	−40.88	**435.87**	**−136.3**	**334.37**	300.00	−1.73	−47.15	−39.63	−38.20	**−348.2**
Q_{G62}	20	−20	16.44	1.26	0.67	**−20.97**	**−178.7**	**75.10**	13.03	−18.85	0.57	**37.23**	**399.51**
Q_{G65}	200	−67	−62.40	80.32	**−687.87**	80.00	**−586.1**	**−384.4**	**264.91**	0.28	84.79	**−552.0**	**−593.5**
Q_{G66}	200	−67	101.03	−2.26	−6.62	−59.92	34.61	130.23	**−369.57**	−66.99	−4.91	**205.61**	30.24
Q_{G69}	300	−300	22.03	−88.31	**−572.36**	−95.12	205.80	−57.88	−206.80	−231.38	−87.88	60.23	**−319.6**
Q_{G70}	32	−10	20.76	8.17	**−68.75**	20.13	**267.00**	**55.18**	16.58	28.47	15.46	**−147.3**	18.15
Q_{G72}	100	−100	−6.24	−11.60	23.04	26.42	−36.19	−49.19	−4.71	−59.09	−11.80	−14.95	−3.02
Q_{G73}	100	−100	33.94	7.77	−25.08	−30.25	−97.58	32.13	−2.27	23.08	5.89	76.94	−8.13
Q_{G74}	9	−6	6.17	**11.83**	**258.01**	−5.82	**−7.49**	**32.28**	**35.62**	8.96	**12.80**	**74.74**	**110.18**
Q_{G76}	23	−8	22.83	0.97	**77.36**	**106.44**	**42.44**	**45.99**	**49.44**	14.26	2.52	**55.79**	**61.16**
Q_{G77}	70	−20	65.66	28.06	**404.87**	64.01	**−122.4**	**92.39**	**86.03**	43.48	21.97	**284.39**	**129.56**
Q_{G80}	280	−165	39.32	146.14	**−335.63**	−72.13	43.43	106.72	21.23	181.44	108.95	**−201.9**	73.21
Q_{G85}	23	−8	22.76	9.98	−4.05	**48.13**	**−8.75**	**−33.32**	**53.41**	22.69	6.73	**77.72**	**52.81**
Q_{G87}	1,000	−100	3.60	9.93	36.13	−0.97	−0.64	−3.98	1.30	2.65	9.83	−7.48	−0.11
Q_{G89}	300	−210	5.63	43.24	−56.71	59.16	47.93	58.13	−78.44	−86.31	−34.83	**−228.5**	−104.5
Q_{G90}	300	−300	43.77	41.92	40.95	41.65	41.99	43.56	50.37	168.88	59.45	**326.11**	25.17
Q_{G91}	100	−100	−18.71	13.22	90.43	−9.76	**−116.3**	−3.46	−2.07	−14.93	−19.93	**−106.3**	−23.52
Q_{G92}	9	−3	8.47	**−43.26**	**−3.92**	**−87.55**	**158.49**	**35.07**	**65.66**	7.92	**31.34**	0.80	**180.39**
Q_{G99}	100	−100	−73.01	42.58	**230.46**	37.02	−41.16	37.95	−1.67	−99.92	−17.71	79.41	−72.27
Q_{G100}	155	−50	97.18	−34.56	**−228.24**	28.80	**241.96**	29.69	19.56	36.53	90.11	89.26	60.36
Q_{G103}	40	−15	39.92	5.06	**−115.57**	**65.81**	**−48.53**	**−16.64**	30.66	−14.85	**44.97**	**134.43**	**−40.76**
Q_{G104}	23	−8	19.56	3.22	**−55.03**	−1.76	**−135.1**	19.38	**23.09**	−7.95	4.63	**−68.56**	−3.49
Q_{G105}	23	−8	22.99	15.56	**407.49**	20.34	**120.43**	**28.93**	**51.75**	19.45	**30.48**	4.06	**78.84**
Q_{G107}	200	−200	28.15	20.62	−56.61	39.04	2.67	48.17	13.30	101.51	4.36	44.58	26.41
Q_{G110}	23	−8	21.72	**−10.65**	**252.49**	**53.27**	**193.12**	**81.39**	**37.14**	21.95	1.60	**−83.90**	**30.40**
Q_{G111}	1,000	−100	−1.09	19.16	−90.93	−0.63	−58.18	−23.57	1.68	87.13	−2.10	63.19	28.53
Q_{G112}	1,000	−100	9.14	66.95	−62.37	7.73	−69.68	12.83	12.97	−34.17	41.66	59.08	12.60
Q_{G113}	200	−100	2.23	4.20	**−106.55**	−34.36	**−124.9**	65.73	−2.85	−99.99	1.85	−61.04	**−109.8**
Q_{G116}	1,000	−1,000	−181.63	51.20	643.57	−11.62	5.02	−212.1	48.09	205.28	49.43	−13.93	304.46

7.3.3.4 Critical analysis of the results for IEEE 118-bus test system

The ORPD results for the IEEE 118-bus test system obtained using the hybrid PSOGSA approach are compared to some other algorithms reported in the literature as shown in Table 7.12. Some of the results reported in the references can be considered as infeasible and/or incorrect solutions as obtained from the power flow computations with optimal settings of control variables reported in these references. Reasons for infeasibility of those results are given below.

Case 1: min P_{loss}

- The optimum control variables given in [4] (QODE) represent an infeasible solution because the reactive powers of the generators at buses 34, 74, 92, and 110 violate their corresponding limits reported in [41], as can be seen from Table 7.11. The true value of P_{loss} is 135.454 MW for this violated case.
- The best solution given in [11] (CLPSO) is an infeasible solution because reactive powers of the generators at buses 1, 4, 6, 8, 15, 18, 19, etc. violate their corresponding limits as can be seen from Table 7.11. The exact value of P_{loss} is 236.17 MW for this violated case.
- For the optimum control variables given in [12] (ALCPSO), the reactive powers of generators at buses 1, 6, 34, 36, 55, 56, 59, 61, etc. violate their limits as can be seen from Table 7.11. The exact value of P_{loss} is 123.89 MW for this violated case.
- The best result given in [13] is an infeasible solution because reactive powers of the generators at many buses such as 1, 6, 15, 25, 32, 36, etc. violate their limits. The exact value of P_{loss} is 152.886 MW for this violated case.
- For the optimum control variables given in [14] (OGSA), the reactive power limits of generators at buses 1, 6, 15, 19, 32, 34, 36, etc. have been violated. The true value of P_{loss} is 130.3507 MW for this violated case.

Table 7.12 Comparison of the simulation results for IEEE 118-bus test system

Case	Method	P_{loss} (MW)	*VD* (p.u.)	Violating constraints
Case 1: min P_{loss}	QODE [4]	80.9257	2.0904	Yes
	CLPSO [11]	130.96	1.8525	Yes
	ALC-PSO [12]	121.53	1.4651	Yes
	GSA [13]	127.7603	NR*	Yes
	OGSA [14]	126.99	NR*	Yes
	MICA–IWO [26]	114.0457	NR*	Yes
	PSOGSA	122.4709	1.7792	No
Case 2: min *VD*	QODE [4]	83.9356	1.6008	Yes
	ALC-PSO [12]	163.13	0.3262	Yes
	OGSA [14]	157.72	0.3666	Yes
	PSOGSA	145.4049	0.7308	No

*NR means not reported.

- As can be seen from Table 7.11, the optimum control variables given in [26] (MICA–IWO) represent an infeasible solution because the reactive power limits of generators at buses 1, 19, 25, 65, 66, 74, etc. have been violated.

Case 2: min VD

- For the optimum control variables given in [4] (QODE), the reactive power limits of generators at buses 34, 56, 74, 92, 103, and 105 have been violated. The true value of *VD* is 1.3567 p.u. for this violated case.
- The optimum control variables given in [12] (ALC-PSO) represent an infeasible solution because the reactive powers of the generators at buses 1, 8, 12, 15, 18, 19, 25, etc. violate their corresponding limits, as can be seen from Table 7.11.
- The best solution given in [14] (OGSA) is an infeasible solution because the reactive power limits of generators at buses 1, 18, 19, 25, 34, 36, 46, etc. have been violated.

7.4 ORPD using hybrid GSA–SQP algorithm

This section describes a hybrid algorithm of GSA with a local search technique based on sequential quadratic programming (SQP) for solving the ORPD problem. In this approach, GSA is used as a global optimizer and SQP as a local optimizer to fine tune the solution.

The SQP method seems to be the best nonlinear programming (NLP) method for constrained optimization problems. SQP is a direct method in which the constraints are handled in an explicit manner. It outperforms every other NLP method in terms of efficiency, accuracy, and percentage of successful solutions over a large number of test problems. The method closely resembles Newton's method for constrained optimization, just as is done for unconstrained optimization. Since the objective function to be minimized is nonconvex, SQP ensures a local minimum for an initial solution [42,43].

GSA as a probabilistic-based multipoint search technique has a drawback of, sometimes, converging to the values which may not be optimum. On the other hand, GSA searches a large solution space with excellent convergence providing approaching to the global optimum. NLP methods, such as SQP, being single-point search methods, have a drawback of being trapped in local optimum point, if the initial choice is nearer to the local optimum. The NLP method gives a global optimum solution, if the proper initial choice is made.

To use of the advantage of GSA and SQP methods, and at the same time to overcome the drawbacks of these methods, a hybridization of GSA with SQP is proposed in [2,44]. The SQP routine is incorporated in GSA as a local search mechanism to improve the convergence of GSA in subsequent iterations. First, the GSA algorithm is executed normally, and the best fitness per generation is selected in each iteration. From that best fitness, the corresponding agent is set as the initial values of variables to the SQP method. Then, the SQP routine is executed depending on the adopted probability of local search α_{LS}, providing improvement of the best fitness obtained from GSA in the current iteration. In this way, the hybrid GSA–SQP algorithm gives the global optimum solution.

7.4.1 Application of hybrid GSA–SQP to the ORPD problem

Proposed hybrid GSA–SQP approach has been applied to solve the ORPD problem. The flowchart of the hybrid GSA–SQP algorithm used to solve the ORPD problem is shown in Figure 7.3.

The hybrid GSA–SQP algorithm has been implemented in MATLAB 2011b computing environment and run on a 2.20 GHz PC with 3.0 GB RAM. The fmincon solver with SQP method available in MATLAB optimization toolbox has been used as the local search mechanism.

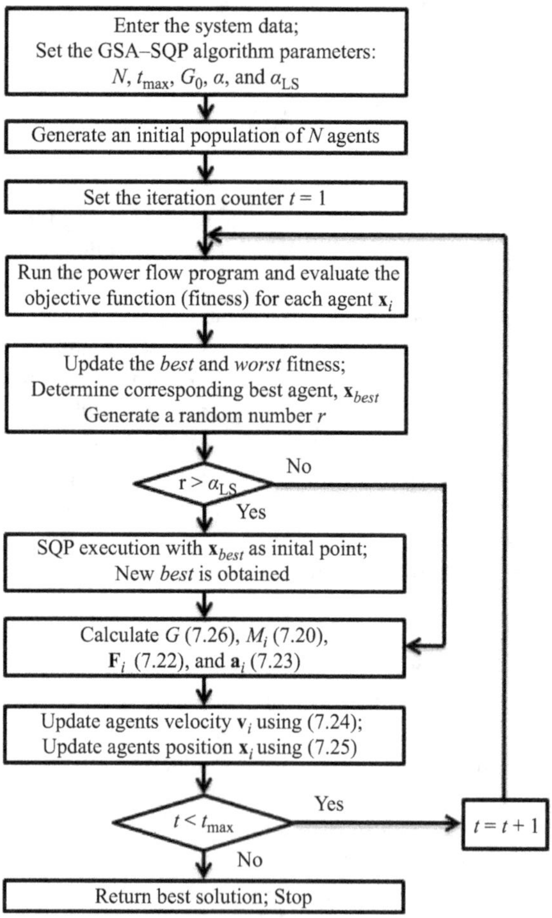

Figure 7.3 Flowchart of hybrid GSA–SQP-based ORPD solution

The algorithm parameters used for the simulation are adopted as follows: α is set to 10 and G_0 is set to 100. The population size (N) and maximum iteration number ($t_{\max}$) are set to 20 and 100, respectively, for all case studies. The probability of local search α_{LS} is set as follows:

$\alpha_{\text{LS}} = 1$ for $t < t_{\max}/2$, and $\alpha_{\text{LS}} = 0.95$ for $t > t_{\max}/2$, where t is the number of iterations.

7.4.2 Simulation results of hybrid GSA–SQP

The proposed hybrid GSA–SQP algorithm has been tested on the standard IEEE 30-bus test system. The system data and the minimum and maximum limits for the control variables are given in [13,38]. The IEEE 30-bus test system has 19 control variables as follows: 6 generator voltage magnitudes, 4 transformer tap settings, and 9 shunt capacitor reactive power injections. The voltages of all load buses have been constrained within limits of 0.95 and 1.10 p.u. The total load demands of this test system are $P_{load} = 2.834$ p.u. and $Q_{load} = 1.262$ p.u. at 100 MVA base. The active power outputs of the generators are specified as follows: $P_{G2} = 80$ MW, $P_{G5} = 50$ MW, $P_{G8} = 20$ MW, $P_{G11} = 20$ MW and $P_{G13} = 20$ MW. Bus 1 is selected as the slack bus.

Two cases have been considered: minimization of active power loss and minimization of *VD*.

Case 1: Minimization of active power loss. The objective function was given by Equation (7.32). The optimal settings of control variables are given in Table 7.13. Application of hybrid GSA–SQP algorithm results in reduction of 21.98% in losses, but *VD* is increased by 74.65% compared to the initial case. Figure 7.4 shows the comparative convergence profiles of GSA and proposed hybrid GSA–SQP for Case 1. It may be observed from this figure that the hybrid GSA–SQP explores the global optima thanks to exploration and exploitation of GSA in the first part of the iterative process and local search ability of SQP in subsequent iterations.

Case 2: Voltage profile improvement, that is, minimization of the load bus *VD*s expressed by (7.34). The optimal results for Case 2 are given in Table 7.13. Using GSA–SQP approach, the *VD* is highly reduced by 92.47% compared to the initial case.

7.5 Educational program package *ORPD*

The *ORPD* is a package of MATLAB m-files for solving ORPD using metaheuristic optimization algorithms. The *ORPD* program package can be downloaded from [45]. The simulation results presented in Sections 7.3 and 7.4 are obtained by the *ORPD* program package on a 2.20 GHz PC with 3.0 GB RAM. This program package can be used by students and researchers to compare the performances of different metaheuristic optimization methods in solving the optimal reactive

Table 7.13 Optimal settings of control variables by GSA–SQP

Control variables	Limits		Initial	Case 1	Case 2
	Lower	Upper			
V_{G1} (p.u.)	0.95	1.1	1.05	1.10000	1.00676
V_{G2} (p.u.)	0.95	1.1	1.04	1.09432	1.00167
V_{G5} (p.u.)	0.95	1.1	1.01	1.07479	1.01752
V_{G8} (p.u.)	0.95	1.1	1.01	1.07671	1.01135
V_{G11} (p.u.)	0.95	1.1	1.05	1.10000	1.00944
V_{G13} (p.u.)	0.95	1.1	1.05	1.10000	1.02033
$T_{6\text{–}9}$ (p.u.)	0.9	1.1	1.078	1.04021	1.02489
$T_{6\text{–}10}$ (p.u.)	0.9	1.1	1.069	0.90000	0.90000
$T_{4\text{–}12}$ (p.u.)	0.9	1.1	1.032	0.97871	1.00152
$T_{28\text{–}27}$ (p.u.)	0.9	1.1	1.068	0.96611	0.96949
Q_{C10} (MVAr)	0	5	0	5.00000	5.00000
Q_{C12} (MVAr)	0	5	0	5.00000	2.51832
Q_{C15} (MVAr)	0	5	0	5.00000	5.00000
Q_{C17} (MVAr)	0	5	0	5.00000	0.00000
Q_{C20} (MVAr)	0	5	0	5.00000	5.00000
Q_{C21} (MVAr)	0	5	0	5.00000	5.00000
Q_{C23} (MVAr)	0	5	0	3.70176	5.00000
Q_{C24} (MVAr)	0	5	0	5.00000	5.00000
Q_{C29} (MVAr)	0	5	0	2.68988	2.61327
P_{loss} (MW)	–	–	5.82225	**4.54271**	5.96236
VD (p.u.)	–	–	1.14966	2.00789	**0.08657**

The bold values indicate the best results.

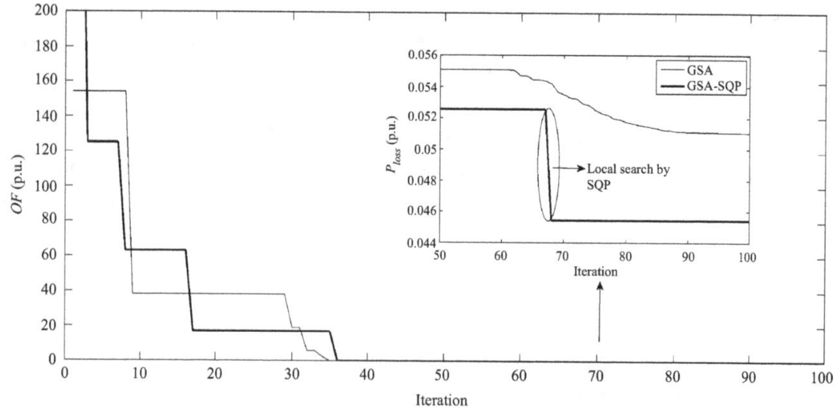

Figure 7.4 Convergence profiles of GSA and hybrid GSA–SQP

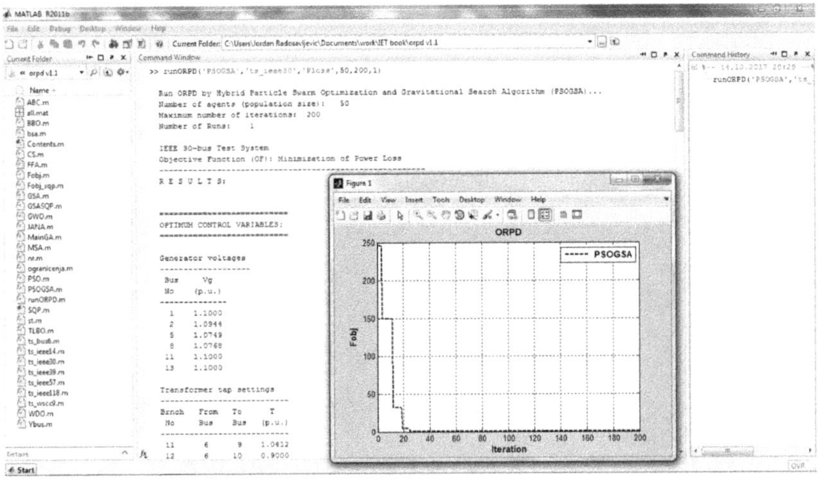

Figure 7.5 Execution of the ORPD program

dispatch problem with different objective functions and test systems. An example of executing this program is shown in Figure 7.5, and other results generated from this program in this example were given below.

```
>>runORPD ('PSOGSA','ts_ieee30','Ploss',50,200,1)

  Run ORPD by Hybrid Particle Swarm Optimization and Gravitational Search
  Algorithm (PSOGSA)...
  Number of agents (population size):   50
  Maximum number of iterations:  200
  Number of Runs:    1
  IEEE 30-bus Test System
  Objective Function (OF): Minimization of Power Loss
  --------------------------------------------------------
  R E S U L T S:

  ============================
  OPTIMUM CONTROL VARIABLES:
  ============================

  Generator voltages
  --------------------
   Bus      Vg
   No     (p.u.)
  --------------
    1    1.1000
    2    1.0944
    5    1.0749
    8    1.0768
   11    1.1000
   13    1.1000
```

```
Transformer tap settings
---------------------------
Branch   From   To      T
 No      Bus    Bus   (p.u.)
---------------------------
 11       6      9    1.0412
 12       6     10    0.9000
 15       4     12    0.9795
 36      28     27    0.9665

Shunt VAR compensations
----------------------
 Bus     Qc
 No    (MVAr)
--------------
 10    5.0000
 12    5.0000
 15    5.0000
 17    5.0000
 20    5.0000
 21    5.0000
 23    3.8264
 24    5.0000
 29    2.6815

=================================================
BEST RESULTS:
=================================================
Objective Function, OF: 0.04543 (units)
Power Loss, Ploss: 4.54271 (MW)
Voltage Deviation, VD: 2.00457 (p.u.)
------------------------------------------------------

======================================================
BALANCE OF ACTIVE POWER:
======================================================
Load: 283.400 (MW)
Generation: 287.943 (MW)
Loss (R*I^2):   4.543 (MW)
IMBALANCE:   0.000 (MW)
------------------------------------------------------

======================================================
BALANCE OF REACTIVE POWER:
======================================================
Load: 126.200 (MVAr)
Generation:  69.001 (MVAr)
Shunt VAR compensations :  41.508 (MVAr)
Shunt admittances and Branch charging (inj):  36.090 (MVAr)
Loss (X*I^2):  20.399 (MVAr)
IMBALANCE:  -0.000 (MVAr)
------------------------------------------------------
========================================================
VIOLATING CONSTRAINTS?
========================================================
Voltages at all buses are within permissible limits
Reactive power outputs of all generators are within permissible limits
------------------------------------------------------
```

BUS VOLTAGES AND POWERS UNDER OPTIMAL CONTROL VARIABLES

Bus No	V (p.u.)	theta (deg)	Pg (MW)	Qg (MVAr)	Qc (MVAr)	Pload (MW)	Qload (MVAr)
1	1.1000	0.000	97.943	-12.258	0.000	0.000	0.000
2	1.0944	-1.647	80.000	12.882	0.000	21.700	12.700
3	1.0839	-3.567	0.000	0.000	0.000	2.400	1.200
4	1.0794	-4.258	0.000	0.000	0.000	7.600	1.600
5	1.0749	-5.999	50.000	23.011	0.000	94.200	19.000
6	1.0764	-5.011	0.000	0.000	0.000	0.000	0.000
7	1.0684	-5.888	0.000	0.000	0.000	22.800	10.900
8	1.0768	-5.289	20.000	32.538	0.000	30.000	30.000
9	1.0795	-6.287	0.000	0.000	0.000	0.000	0.000
10	1.0938	-7.992	0.000	0.000	5.000	5.800	2.000
11	1.1000	-4.279	20.000	11.185	0.000	0.000	0.000
12	1.0982	-7.153	0.000	0.000	5.000	11.200	7.500
13	1.1000	-5.825	20.000	1.643	0.000	0.000	0.000
14	1.0891	-8.054	0.000	0.000	0.000	6.200	1.600
15	1.0891	-8.322	0.000	0.000	5.000	8.200	2.500
16	1.0910	-7.800	0.000	0.000	0.000	3.500	1.800
17	1.0906	-8.181	0.000	0.000	5.000	9.000	5.800
18	1.0826	-8.915	0.000	0.000	0.000	3.200	0.900
19	1.0817	-9.094	0.000	0.000	0.000	9.500	3.400
20	1.0864	-8.930	0.000	0.000	5.000	2.200	0.700
21	1.0858	-8.497	0.000	0.000	5.000	17.500	11.200
22	1.0864	-8.489	0.000	0.000	0.000	0.000	0.000
23	1.0861	-8.869	0.000	0.000	3.826	3.200	1.600
24	1.0806	-9.041	0.000	0.000	5.000	8.700	6.700
25	1.0846	-9.001	0.000	0.000	0.000	0.000	0.000
26	1.0680	-9.369	0.000	0.000	0.000	3.500	2.300
27	1.0951	-8.739	0.000	0.000	0.000	0.000	0.000
28	1.0735	-5.464	0.000	0.000	0.000	0.000	0.000
29	1.0841	-10.011	0.000	0.000	2.681	2.400	0.900
30	1.0703	-10.682	0.000	0.000	0.000	10.600	1.900

BRANCH POWER FLOW AND LOSS UNDER OPTIMAL CONTROL VARIABLES

From Bus	To Bus	P (MW)	Q (MVAr)	Ploss (MW)	Qloss (MVAr)
1	2	57.645	-10.772	0.536	1.606
1	3	40.298	-1.486	0.607	2.487
2	1	-57.109	6.022	0.536	1.606
2	4	30.977	-2.217	0.457	1.392
2	5	45.408	-0.839	0.814	3.419
2	6	39.024	-2.784	0.739	2.242
3	1	-39.691	-0.892	0.607	2.487
3	4	37.291	-0.308	0.156	0.449
4	2	-30.521	-0.738	0.457	1.392
4	3	-37.134	-0.226	0.156	0.449
4	6	36.179	-2.873	0.134	0.467
4	12	23.876	2.237	0.000	1.348
5	2	-44.594	-0.660	0.814	3.419
5	7	0.394	4.671	0.014	0.035
6	2	-38.285	0.620	0.739	2.242

6	4	-36.044	2.294	0.134	0.467
6	7	22.539	2.333	0.120	0.367
6	8	12.124	-4.941	0.017	0.060
6	9	11.948	-22.604	0.000	0.271
6	10	12.238	22.283	0.000	0.698
6	28	15.481	0.014	0.035	0.124
7	5	-0.381	-6.979	0.014	0.035
7	6	-22.419	-3.921	0.120	0.367
8	6	-12.107	3.958	0.017	0.060
8	28	2.107	-1.421	0.003	0.010
9	6	-11.948	23.876	0.000	0.271
9	10	31.948	-13.593	0.000	1.138
9	11	-20.000	-10.283	0.000	0.903
10	6	-12.238	-19.771	0.000	0.698
10	9	-31.948	14.731	0.000	1.138
10	17	5.459	2.109	0.009	0.024
10	20	9.228	-0.161	0.067	0.149
10	21	15.987	4.376	0.080	0.172
10	22	7.711	1.716	0.038	0.078
11	9	20.000	11.185	0.000	0.903
12	4	-23.876	-1.025	0.000	1.348
12	13	-20.000	-1.177	0.000	0.466
12	14	7.518	0.352	0.058	0.120
12	15	18.058	-1.272	0.180	0.354
12	16	7.100	0.622	0.040	0.084
13	12	20.000	1.643	0.000	0.466
14	12	-7.460	-0.232	0.058	0.120
14	15	1.260	-1.368	0.006	0.006
15	12	-17.878	1.626	0.180	0.354
15	14	-1.253	1.374	0.006	0.006
15	18	5.791	0.416	0.030	0.062
15	23	5.141	-0.917	0.023	0.046
16	12	-7.060	-0.539	0.040	0.084
16	17	3.560	-1.261	0.010	0.023
17	10	-5.450	-2.085	0.009	0.024
17	16	-3.550	1.285	0.010	0.023
18	15	-5.761	-0.354	0.030	0.062
18	19	2.561	-0.546	0.004	0.008
19	18	-2.557	0.553	0.004	0.008
19	20	-6.943	-3.953	0.019	0.037
20	10	-9.161	0.310	0.067	0.149
20	19	6.961	3.990	0.019	0.037
21	10	-15.907	-4.204	0.080	0.172
21	22	-1.593	-1.996	0.001	0.001
22	10	-7.673	-1.638	0.038	0.078
22	21	1.594	1.997	0.001	0.001
22	24	6.080	-0.359	0.036	0.056
23	15	-5.118	0.963	0.023	0.046
23	24	1.918	1.263	0.006	0.012
24	22	-6.044	0.416	0.036	0.056
24	23	-1.912	-1.251	0.006	0.012
24	25	-0.744	-0.865	0.002	0.004
25	24	0.746	0.868	0.002	0.004
25	26	3.539	2.358	0.039	0.058
25	27	-4.285	-3.227	0.027	0.051
26	25	-3.500	-2.300	0.039	0.058
27	25	4.312	3.278	0.027	0.051
27	28	-17.549	-3.843	0.000	1.125
27	29	6.183	-0.301	0.070	0.133

```
27    30      7.054     0.867     0.135     0.254
28     6    -15.446    -1.392     0.035     0.124
28     8     -2.104    -3.517     0.003     0.010
28    27     17.549     4.909     0.000     1.125
29    27     -6.113     0.434     0.070     0.133
29    30      3.713     1.347     0.032     0.060
30    27     -6.919    -0.613     0.135     0.254
30    29     -3.681    -1.287     0.032     0.060
------------------------------------------------------
```

7.6 Conclusion

In this chapter, two hybrid metaheuristic population-based optimization algorithms, namely, hybrid PSOGSA and hybrid GSA–SQP algorithm, have been proposed and successfully applied to solve ORPD problem.

The proposed PSOGSA approach has been tested an investigated on the IEEE 30-bus and the IEEE 118-bus test systems. Simulation results show that the hybrid PSOGSA algorithm has the ability to reduce the real power loss and *VD* significantly without violating any constraints of the dependent variables. The results also demonstrate better effectiveness and convergence characteristics of hybrid PSOGSA compared to standard PSOGSAs.

Hybrid GSA–SQP approach has been tested an investigated on the IEEE 30-bus test system. The results indicate that the hybrid GSA–SQP algorithm provides effective and robust high-quality solution. The hybrid GSA–SQP explores the global optima thanks to exploration and exploitation of GSA in the first part of the iterative process and local search ability of SQP in subsequent iterations.

In addition, a critical review of the ORPD results obtained by other optimization methods reported in the recent state-of-the-art literature was given in this chapter. It has been shown that some of the reported results can be considered as infeasible solutions because it violates many of the dependent variable constraints such as Q_G limits.

Based on all simulation results and comparative study, it can be concluded that the proposed metaheuristic optimization algorithms enables high-quality solutions of the ORPD problem.

References

[1] Radosavljević J., Jevtić M., Milovanović M. 'A solution to the ORPD problem and critical analysis of the results'. *Electrical Engineering.* 2016; doi: 10.1007/s00202-016-0503-1

[2] Radosavljević J., Jevtić M. 'Solution of optimal reactive power dispatch by a hybrid GSA–SQP algorithm'. *Elektronika ir Elektrotechnika.* 2016; **22**(3):3–6

[3] Abou El-Ela A.A., Abido M.A., Spea S.R. 'Diferential evolution algortithm for optimal reactive power dispatch'. *Electric Power System Research.* 2011;**81**(2):458–64

[4] Basu M. 'Quasi-oppositional differential evolution for optimal reactive power dispatch'. *Electrical Power and Energy Systems*. 2016;**78**:29–40

[5] Sakr W.S., El-Sehiemy R.A., Azmy A.M. 'Adaptive differential evolution algorithm for efficient reactive power management'. *Applied Soft Computing*. 2017;**53**:336–51

[6] Wu Q.H., Cao Y.J., Wen J.Y. 'Optimal reactive power dispatch using an adaptive genetic algorithm'. *Electrical Power and Energy Systems*. 1998; **20**(8):563–9

[7] Subbaraj P., Rajnarayanan P.N. 'Optimal reactive power dispatch using self-adaptive real coded genetic algorithm'. *Electric Power Systems Research*. 2009;**79**:374–81

[8] Bhattacharya A., Chattopadhyay P.K. 'Solution of optimal reactive power flow using biogeography-based optimization'. *International Journal of Electrical, Computer, Energetic, Electronic and Communication Engineering*. 2010;**4**(3):621–9

[9] Yoshida H., Kawata K., Fukuyama Y., Takayama S., Nakanishi Y. 'A particle swarm optimization for reactive power and voltage control considering voltage security assessment'. *IEEE Transactions on Power Systems*. 2000; **15**(4):1232–9

[10] Zhao B., Guo C.X., Cao Y.J. 'A multiagent-based particle swarm optimization approach for optimal reactive power dispatch'. *IEEE Transactions on Power Systems*. 2005;**20**(2):1070–8

[11] Mahadevan K., Kannan P.S. 'Comprehensive learning particle swarm optimization for reactive power dispatch'. *Applied Soft Computing*. 2010;**10**:641–52

[12] Singh R.P., Mukherjee V., Ghosal S.P. 'Optimal reactive power dispatch by particle swarm optimization with an aging leader and challengers'. *Applied Soft Computing*. 2015;**29**:298–309

[13] Duman S., Sonmez Y., Guvenc U., Yorukeren N. 'Optimal reactive power dispatch using a gravitational search algorithm'. *IET Generation, Transmission & Distribution*. 2012;**6**(6):563–76

[14] Shaw B., Mukherjee V., Ghoshal S.P. 'Solution of reactive power dispatch of power systems by an opposition-based gravitational search algorithm'. *Electrical Power and Energy Systems*. 2014;**55**:29–40

[15] Roy P.K., Mandal B., Bhattacharya K. 'Gravitational search algorithm based optimal reactive power dispatch for voltage stability enhancement'. *Electric Power Components and Systems*. 2012;**40**:965–76

[16] Niknam T., Narimani M.R., Azizipanah-Abarghooee R., Bahmani-Firouzi B. 'Multiobjective optimal reactive power dispatch and voltage control: a new opposition-based self-adaptive modified gravitational search algorithm'. *IEEE Systems Journal*. 2013;**7**(4):742–53

[17] Mukherjee A., Mukherjee V. 'Solution of optimal reactive power dispatch by chaotic krill herd algorithm'. *IET Generation, Transmission & Distribution*. 2015;**9**(15):2351–62

[18] Dutta S., Mukhopadhyay P., Roy P.K., Nandi D. 'Unified power flow controller based reactive power dispatch using oppositional krill herd algorithm'. *Electrical Power and Energy Systems*. 2016;**80**:10–25
[19] Khazali A.H., Kalantar M. 'Optimal reactive power dispatch based on harmony search algorithm'. *Electrical Power and Energy Systems*. 2011;**33**:684–92
[20] Mandal B., Roy P.K. 'Optimal reactive power dispatch using quasi-oppositional teaching learning based optimization'. *Electrical Power and Energy Systems*. 2013;**53**:123–34
[21] Abaci K., Yamacli V. 'Optimal reactive-power dispatch using differential search algorithm'. *Electrical Engineering*. 2016; doi: 10.1007/s00202-016-0410-5
[22] Abou El-Ela A.A., Kinawy A.M., El-Sehiemy R.A., Mouwafi M.T. 'Optimal reactive power dispatch using ant colony optimization algorithm'. *Electrical Engineering*. 2011;**93**:103–16
[23] Ayan K., Kilic U. 'Artificial bee colony algorithm solution for optimal reactive power flow'. *Applied Soft Computing*. 2012;**12**:1477–82
[24] Sulaiman M.H., Mustaffa Z., Mohamed M.R., Aliman O. 'Using the gray wolf optimizer for solving optimal reactive power dispatch problem'. *Applied Soft Computing*. 2015;**32**:286–92
[25] Rajan A., Malakar T. 'Exchange market algorithm based optimum reactive power dispatch'. *Applied Soft Computing*. 2016;**43**:320–36
[26] Ghasemi M., Ghavidal S., Ghanbarian M.M., Habibi A. 'A new hybrid algorithm for optimal reactive power dispatch problem with discrete and continuous control variables'. *Applied Soft Computing*. 2014;**22**:126–40
[27] Khorsandi A., Alimardani A., Vahidi B., Hosseinian S.H. 'Hybrid shuffled frog leaping algorithm and Nelder–Mead simplex search for optimal reactive power dispatch'. *IET Generation, Transmission & Distribution*. 2011;**5**(2):249–56
[28] Rajan A., Malakar T. 'Optimal reactive power dispatch using hybrid Nelder–Mead simplex based firefly algorithm'. *Electrical Power and Energy Systems*. 2015;**66**:9–24
[29] Mehdinejad M., Mohammadi-Ivatloo B., Dadashzadeh-Bonab R., Zare K. 'Solution of optimal reactive power dispatch of power systems using hybrid particle swarm optimization and imperialist competitive algorithms'. *Electrical Power and Energy Systems*. 2016;**83**:104–16
[30] Yan W., Liu F., Chung C.Y., Wong K.P. 'A hybrid genetic algorithm-interior point method for optimal reactive power flow'. *IEEE Transactions on Power Systems*. 2006;**21**(3):1163–9
[31] Li Y., Li X., Li Z. 'Reactive power optimization using hybrid CABC-DE algorithm'. *Electric Power Components and Systems*. 2017;**45**(9):980–9
[32] Mirjalili S., Hashim S.Z.M. 'A new hybrid PSOGSA algorithm for function optimization'. *Proceedings of International Conference on Computer and Information Application*; Tianjin, China, Dec 2010. IEEE; 2012. pp. 374–7

[33] Mirjalli S., Hashim S.Z.M., Sardroudi H.M. 'Training feedforward neural networks using hybrid particle swarm and gravitational search algorithm'. *Applied Mathematics and Computation*. 2012;**218**:11125–37

[34] Khadanga R.K., Satapathy J.K. 'Time delay approach for PSS and SSSC based coordinated controller design using hybrid PSO–GSA algorithm'. *Electrical Power and Energy Systems*. 2015;**71**:262–73

[35] Radosavljević J., Klimenta D., Jevtić M., Arsić N. 'Optimal power flow using a hybrid optimization algorithm of particle swarm optimization and gravitational search algorithm'. *Electric Power Components and Systems*. 2015;**43**(17):1958–70

[36] Radosavljević J., Arsić N., Jevtić M. 'Optimal power flow using hybrid PSOGSA algorithm'. *Proceedings of 55th International Scientific Conference of Riga Technical University on Power and Electrical Engineering*; Riga, Latvia, Oct 2014. IEEE; 2014. pp. 136–40

[37] Radosavljević J. 'A solution to the combined economic and emission dispatch using hybrid PSOGSA algorithm'. *Applied Artificial Intelligence*. 2016;**30**(5):445–74

[38] Lee K.Y., Park Y.M., Ortiz J.L. 'A united approach to optimal real and reactive power dispatch'. *IEEE Transactions on Power Apparatus and Systems*. 1985;**104**(5):1147–53

[39] Lee K.Y., Park Y.M., Ortiz J.L. 'Optimal real and reactive power dispatch'. *Electric Power System Research*. 1984;**7**:201–12

[40] Zimmerman R.D., Murillo-Sánchez C.E., Thomas R.J. 'MATPOWER: steady-state operations, planning and analysis tools for power systems research and education'. *IEEE Transactions on Power Systems*. 2011; **26**(1):12–9. Available at http://www.pserc.cornell.edu/matpower/

[41] Christie R., *The IEEE* 118-bus test system [online]. Available at https://www.ee.washington.edu/research/pstca/pf118/pg_tca118bus.htm. 1993

[42] Sivasubramani S., Swarup K.S. 'Hybrid DE-SQP algorithm for non-convex short term hydrothermal scheduling problem'. *Energy Conversion and Management*. 2011;**52**:757–61

[43] Bedekar P.P., Bhide S.R. 'Optimum coordination of directional overcurrent relays using the hybrid GA-NLP approach'. *IEEE Transactions on Power Delivery*. 2011;**26**(1):109–19

[44] Radosavljević J., Jevtić M. 'Hybrid GSA–SQP algorithm for optimal coordination of directional overcurrent relays'. *IET Generation, Transmission & Distribution*. 2016;**10**(8):1928–37

[45] Radosavljević J. *ORPD—program package of Matlab files for optimal reactive power dispatch using metaheuristic optimization methods*. 2017. Available at https://www.researchgate.net/profile/Jordan_Radosavljevic2/contributions

Chapter 8

Combined economic and emission dispatch

8.1 Introduction

The economic dispatch (ED) problem has a significant importance in the power system's operation, planning, economic scheduling, and security. The ED problem solution aims to minimize the cost of generation of electric power though optimal adjustment of the committed generating unit outputs, while at the same time satisfying all unit and system constraints. It is a large-scale and nonlinear constrained optimization problem. With the increased public awareness of the environmental pollution, the traditional ED, which ignores the pollutant emissions of fossil fuels used by the thermal plants, no longer satisfies the environmental need. When the environmental concerns are combined with the ED, then the problem becomes a combined economic and emission dispatch (CEED) problem. This problem considers two objectives: minimization of the fuel cost and emission from the thermal power plants with both equality and inequality constraints. So, the CEED problem is a multiobjective mathematical problem in which conflicting objectives are optimized simultaneously.

The environmental aspect adds complexity to the solution of the ED problem due to the nonlinear characteristics of the mathematical models used to represent emissions. In addition, the CEED problem can be complicated even further if nonsmooth and nonconvex fuel cost functions are used to model generators, such as valve-point-loading effects. All these considerations make the CEED problem a highly nonlinear and a multimodal optimization problem.

Generally, three approaches to handle the CEED problem have been reported in the literature. In the first approach, the emission is treated as a constraint with a permissible limit. However, this formulation has a severe difficulty in getting the trade-off relations between cost and emission. The second approach treats the emission as another objective in addition to the cost objective. In this case, the CEED problem is converted into a single-objective optimization problem either by linear combination of both objectives or by considering one objective at a time for optimization. In the third approach, simultaneously conflicting objectives are evaluated together in the solution of the CEED problem. Both the fuel cost and the emission are minimized together [1].

In practice, the ED problem has been solved by using deterministic (classical) and metaheuristic population-based optimization methods. In the past few decades, many classical optimization methods, such as gradient method, Newton's method, linear programing, nonlinear programing, dynamic programing, goal programing technique, and Lagrangian relaxation algorithm, have been applied to various ED problems. However, most of them have difficulties in solving ED problems due to nonlinearity and nonconvexity fuel cost and emission characteristics. The conventional optimization methods are highly sensitive to the starting point and frequently converge to local optimum solution. Moreover, these methods are not able to find a solution with a significant computational time for medium or large-scale CEED problem. It becomes essential to develop optimization methods that are able to overcome these drawbacks and handle such difficulties.

Recently, many metaheuristic population-based methods have been used to solve complex constrained optimization problems. Generally, achieving an optimal or near optimal solution for a specific problem will require multiple trials as well as appropriate tuning of associated parameters. A wide variety of population-based techniques in their basic, modified, multiobjective, or oppositional versions have been applied in solving the nonlinear ED and/or CEED problems with different objective functions. Some of these methods are genetic algorithm [2,3], niched Pareto genetic algorithm (NPGA) [4], nondominated sorting genetic algorithm (NSGA II) [5–8], fuzzy logic-controlled genetic algorithm [9], particle swarm optimization (PSO) [10–19], multiobjective particle swarm optimization (MOPSO) [20], modified adaptive θ-particle swarm optimization (MA θ-PSO) [21], gravitational search algorithm (GSA) [22–27], artificial bee colony (ABC) algorithm [28], spiral optimization algorithm (SOA) [29], evolutionary programing [3], multiobjective evolutionary programing (MOEP) [30], differential evolution (DE) [31,32], multiobjective differential evolution (MODE) [33–35], modified bacterial foraging algorithm (MBFA) [36], biogeography-based optimization (BBO) [37], ant colony optimization [38], cuckoo search [39], firefly algorithm [40], harmony search (HS) [41], etc.

Today, some hybrid algorithms have been proposed to solve the CEED problem. A combination of several population-based algorithms in hybrid algorithm allows them to exploit the strength of each algorithm. In [42], the authors proposed a hybrid HS with arithmetic crossover operation (ACHS), for solving five different types of ED problems. Bhattacharya and Chattopadhyay proposed hybrid DE and BBO algorithm, DE/BBO, for solving economic emission load dispatch problem [43]. To solve the highly constrained environmental/ED problem involving conflicting objectives, the authors in [44] proposed a hybrid multiobjective optimization algorithm based on PSO and DE (MODE/PSO). In [45], a hybrid differential evolution algorithm with harmony search has been proposed to solve the multi-objective environmental/ED problems that feature nonsmooth cost curves. Parouha and Das [46] proposed a hybrid optimizer of DE and PSO (DEPSO), whereas Coelho and Mariani [47] combined the DE with the generator of chaos sequences and SQP to form a hybrid DEC-SQP algorithm, for solving ED problem considering valve-point effect. Recently, Neto *et al.* [48] used a hybrid algorithm of continuous greedy randomized adaptive search procedure and self-adaptive differential evolution for solving nonsmooth ED. A comprehensive review of hybrid optimization methods used for CEED problem can be found in [49].

This study deals with the application of hybrid PSO and GSA (hybrid PSOGSA) algorithm to solve the CEED problem [1]. The PSOGSA algorithm profits from the abilities of both PSO and GSA algorithms. The performance of the proposed algorithm is tested on three standard test systems, with different constraints and various cost curve natures. Numerical results obtained by the proposed approach were compared with other optimization results reported in the literature recently.

The rest of the chapter is organized as follows. In Section 8.2, the biobjective CEED problem has been formulated and converted into single-objective optimization problem by weighted sum method (the second approach was used to handle the CEED problem). In Section 8.3, the proposed hybrid PSOGSA algorithm is presented. Section 8.4 presents the results of simulation. A MATLAB® Graphical User Interface (GUI)-based software developed for solving the CEED problem using different metaheuristic optimization methods is presented in Section 8.5. Finally, the conclusions are given in Section 8.6.

8.2 Problem formulation

The solution of the CEED problem is achieved by minimizing the objective function combined with the weighted sum method under the system constraints. Mathematically, the objective function can be expressed as follows:

$$\min_{\mathbf{x}} F(\mathbf{x}, y) = \min_{\mathbf{x}} \left\{ w \sum_{n \in N_G} Fcost(\mathbf{x}, y) + (1 - w)\gamma \sum_{n \in N_G} Emission(\mathbf{x}, y) \right\} \tag{8.1}$$

In (8.1), the fuel cost rate is shown with $Fcost(\mathbf{x}, y)$ and emission rate with $Emission(\mathbf{x}, y)$. Scaling factor, weight factor, and the set of all the thermal generation units are denoted as γ, w $(0 \leq w \leq 1)$, and N_G, respectively; $w = 1$ corresponds to the minimization of total fuel cost only; likewise, $w = 0$ corresponds to the minimization of total emission only.

$\mathbf{x}$ is the vector of control variables, consisting of generator active power outputs P_G (except at the slack bus P_{G1}). Hence, $\mathbf{x}$ can be expressed as

$$\mathbf{x} = [P_{G1} \ldots P_{GNG-1}] \tag{8.2}$$

y is the of dependent variable consisting of slack bus power P_{Gsl}

$$y = P_{Gsl} \tag{8.3}$$

where NG is the number of generators.

8.2.1 Fuel cost function

Basic fuel cost function of each generator in the system is represented as a quadratic function of real-power generation:

$$Fcost(P_{Gi}) = a_i + b_i P_{Gi} + c_i P_{Gi}^2 \tag{8.4}$$

where a_i, b_i, and c_i are the cost coefficients of the ith generator.

Large steam turbine generators will have a number of steam admission valves that are opened in sequence to control the power output of the unit. As the unit loading increases, the input to the unit increases, and the incremental heat rate decreases between the opening points for any two valves. However, when a valve is first opened, the throttling losses increase rapidly and the incremental heat rate rises suddenly. This is called *valve point effect* that leads to nonsmooth, nonconvex input–output characteristics. Usually, valve point effect is modeled by adding a recurring rectified sinusoid to the basic quadratic cost function:

$$Fcost(P_{Gi}) = a_i + b_i P_{Gi} + c_i P_{Gi}^2 + \left| d_i \sin\left(e_i\left(P_{Gi}^{\min} - P_{Gi}\right)\right)\right| \tag{8.5}$$

where d_i and e_i are fuel cost coefficients for valve point effects.

From practical point of view, in a power system, there are some generating units supplied with multiple fuels like coal, oil, or natural gas. Modeling multifuel effect makes the ELD problem nonsmooth and more complicated. The cost function of a unit with multiple fuels is a superposition of this piecewise quadratic function as follows [46]:

$$\begin{aligned} &Fcost_i(P_{Gi}) = a_{ij} + b_{ij} P_{Gi} + c_{ij} P_{Gi}^2 + \left| d_{ij} \sin\left(e_{ij}\left(P_{Gi}^{\min} - P_{Gi}\right)\right)\right| \\ &\text{if} \quad P_{Gij}^{\min} \le P_{Gi} \le P_{Gij}^{\max}, \quad j = 1, \ldots, NF, \end{aligned} \tag{8.6}$$

where i is the index of generator; j is the index of fuel type; a_{ij}, b_{ij}, c_{ij}, e_{ij}, and d_{ij} are cost coefficients of the generator i for fuel type j; $P_{Gij}^{\min}$ and $P_{Gij}^{\max}$ are the minimum and maximum power output of generator i with fuel option j, respectively; and NF is the number of fuel types of each generator.

8.2.2 Emission function

Fossil-fueled thermal units cause atmospheric waste emission composed of gases and particles such as carbon dioxide (CO_2), sulfur dioxide (SO_2), and nitrogen oxide (NO_x). Different mathematical models were proposed to represent the emission function of thermal-generating units [43]. In this study, the emission function of each thermal unit is defined as the sum of a quadratic function and an exponential function:

$$Emission(P_{Gi}) = \alpha_i + \beta_i P_{Gi} + \eta_i P_{Gi}^2 + \xi_i \exp(\lambda_i P_{Gi}) \tag{8.7}$$

where α_i, β_i, η_i, ξ_i, and λ_i are coefficients of the ith generator emission characteristics.

8.2.3 Constraints

During the minimization process, some equality and inequality constraints must be satisfied. In this process, an equality constraint is called a power balance and an inequality constraint is called a generation capacity constraint.

8.2.3.1 Power balance constraint

The total power generation must cover the total load demand P_{load} and the active power loss in transmission lines P_{loss}. Accordingly, the power balance constraint can be represented as follows:

$$\sum_{i \in NG} P_{Gi} - P_{load} - P_{loss} = 0 \tag{8.8}$$

The transmission losses of the system are represented by loss coefficients (B_{nj}), normally referred to as **B**-loss matrices. The **B**-loss matrices approximate the system losses as a quadratic function of the generator real powers:

$$P_{loss} = \sum_{i \in NG} \sum_{j \in NG} P_{Gi} B_{ij} P_{Gj} + \sum_{i \in NG} B_{0i} P_{Gi} + B_{00} \tag{8.9}$$

where B_{ij}, B_{0i}, and B_{00} are the coefficients of the **B**-loss matrices.

8.2.3.2 Generation capacity constraint

For stable operation, real-power output of each generator is restricted by minimum P_{Gi}^{min} and maximum P_{Gi}^{max} power output limits as follows:

$$P_{Gi}^{min} \le P_{Gi} \le P_{Gi}^{max} \qquad i = 1, \ldots, NG \tag{8.10}$$

Inclusion of *ramp rate limits*. The actual operating range of all the online units is restricted by their corresponding ramp rate limits. The ramp-up and ramp-down constraints can be written as follows [14]:

$$P_{Gi} - P_{Gi}^{0} \le UR_i \tag{8.11}$$

$$P_{Gi}^{0} - P_{Gi} \le DR_i \tag{8.12}$$

where P_{Gi}^{0} is the previous power output of the ith generator; UR_i and DR_i are the up-ramp and down-ramp limits of generator i, respectively.

Due to ramp rate limits, (8.10) should be modified as follows:

$$\max\{P_{Gi}^{min}, P_{Gi}^{0} - DR_i\} \le P_{Gi} \le \min\{P_{Gi}^{max}, P_{Gi}^{0} + UR_i\} \tag{8.13}$$

Usually, the ramp rate limits are considered in the case of dynamic ED/CEED problem. The dynamic ED/CEED aims to optimize the operation of thermal-power-generating units on a scheduled time horizon consisted of T successive time intervals. For each time interval t from the time horizon T, classical (static) ED/CEED is solved.

Considering *prohibited operating zones limits*. In some cases, the entire operating range of a generator is not always available due to physical operation limitations. A typical conventional thermal-generating unit may have prohibited operating zones (POZs) due to vibration in a shaft bearing caused by a steam valve or due to faults in the machines themselves or the associated auxiliary equipment, such as boilers and feed pumps. Such events may lead to instability in certain ranges of generator power output. Therefore, for units with POZs, there are additional constraints on the unit operating range as follows [14]:

$$P_{Gi} \in \begin{cases} P_{Gi}^{\min} \le P_{Gi} \le P_{Gi,1}^{l} \\ P_{Gi,k-1}^{u} \le P_{Gi} \le P_{Gi,k}^{l}, & k = 2,3,\ldots,pz_i; \quad i = 1,2,\ldots,NPZ \\ P_{Gi,pz_i}^{u} \le P_{Gi} \le P_{Gi}^{\max} \end{cases} \tag{8.14}$$

where $P_{Gi,k}^{l}$ and $P_{Gi,k}^{u}$ are the lower and upper bounds of POZ of generator i, respectively; pz_i is the number of prohibited zones of generator i, and NPZ is the number of generators which have POZs.

8.2.4 Slack generator calculation

To enforce active power balance constraint given in (8.8), a dependent generator (slack generator) should be selected. The value of generation power of the slack generator, P_{Gsl}^{old}, is calculated by using (8.15) where the initial value of power loss is set to zero ($P_{loss}^{old} = P_{loss}^{first} = 0$)

$$P_{Gsl}^{old} = P_{load} - \sum_{\substack{i=1 \\ i \ne sl}}^{NG} P_{Gi} \tag{8.15}$$

where sl denotes index of the slack generator.

After obtaining P_{Gsl}^{old}, new power loss, P_{loss}^{new}, is determined from (8.9). According to this, P_{Gsl}^{new} is calculated using the following equation:

$$P_{Gsl}^{new} = P_{load} + P_{loss}^{new} - \sum_{\substack{i=1 \\ i \ne sl}}^{NG} P_{Gi} \tag{8.16}$$

The result of this equation is controlled in (8.17) and if the error value (ε) is below error tolerance value, TOL_ε (e.g., $TOL_\varepsilon = 10^{-6}$), the equation satisfies the power balance constraint.

$$\varepsilon = \left| P_{loss}^{new} - P_{loss}^{old} \right|, \qquad \varepsilon \le TOL_\varepsilon, \tag{8.17}$$

The obtained P_{Gsl} is checked to learn whether it satisfies the constraint defined in (8.10). Consequently, the variable $P_{Gsl}^{\lim}$ is defined as

$$P_{Gsl}^{\lim} = \begin{cases} P_{Gsl}^{\max} & \text{if} \quad P_{Gsl} > P_{Gsl}^{\max} \\ P_{Gsl}^{\min} & \text{if} \quad P_{Gsl} < P_{Gsl}^{\min} \\ P_{Gsl} & \text{if} \quad P_{Gsl}^{\min} \leq P_{Gsl} \leq P_{Gsl}^{\max} \end{cases} \tag{8.18}$$

Inequality constraint of the dependent variable, that is, P_{Gsl}, is added to the objective function as a quadratic penalty terms. The new expanded objective function to be minimized becomes

$$F_p = F + \lambda_P \left(P_{Gsl} - P_{Gsl}^{\lim}\right)^2 \tag{8.19}$$

where λ_p is the penalty factor.

8.3 Solution method

8.3.1 Overview of PSOGSA

The hybrid PSOGSA is developed by Mirjalili and Hashim [50]. The authors are hybridized PSO with GSA using low-level coevolutionary heterogeneous hybrid. The hybrid PSOGSA is low level because it combines the functionality of both algorithms. It is coevolutionary because it does not use both algorithm one after another, that is they run in parallel. It is heterogeneous because there are two different algorithms that involved to produce final results.

The position of the ith search agent in a system with N agents is defined by

$$\mathbf{x}_i = \left[x_i^1, \ldots, x_i^d, \ldots, x_i^n\right] \quad \text{for } i = 1, 2, \ldots, N \tag{8.20}$$

where N is the number of agents (the size of population), n is the search space dimension of the problem, i.e., the number of control variables, and x_i^d defines the position of the ith agent in the dth dimension, that is value of dth control variable in ith candidate solution.

8.3.1.1 Particle swarm optimization

The PSO algorithm is based on simulation of bird flocking in two-dimensional space. It uses a number of particles (candidate solutions) which fly around in the search space to find best solution. Each particle tries to modify its position using the following information: the current position, the current velocity, the distance between the current position and its own best solution (**pbest**), and the distance between the current position and the best solution found so far (**gbest**).

In each iteration, the velocities of particles are calculated using the following equation:

$$\mathbf{v}_i(t+1) = w \cdot \mathbf{v}_i(t) + C_1 \cdot r_1 \cdot (\mathbf{pbest}_i(t) - \mathbf{x}_i(t)) + C_2 \cdot r_2 \cdot (\mathbf{gbest}(t) - \mathbf{x}_i(t)) \tag{8.21}$$

After updating the velocities, the positions of particles can be calculated as

$$\mathbf{x}_i(t+1) = \mathbf{x}_i(t) + \mathbf{v}_i(t+1) \tag{8.22}$$

where $\mathbf{v}_i(t)$ is velocity of particle i at iteration t, w is a weighting function, C_1 and C_2 are positive constants, r_1 and r_2 are uniformly distributed random numbers in [0,1], $\mathbf{x}_i(t)$ is the current position of particle i at iteration t, $\mathbf{pbest}_i$ is the individual best of particle i at iteration t, and **gbest** is the best solution found so far. The first part of (8.21) provides exploration ability for PSO. The second and third parts represent private thinking and collaboration of particles, respectively.

8.3.1.2 Gravitational search algorithm

In GSA, the search agents are a collection of masses which interact with each other based on the Newtonian gravity and the laws of motion. The position of the mass corresponds to the solution of the problem, and its gravitational and inertial masses are determined using a fitness function. In other words, each mass presents a solution. The algorithm is navigated by properly adjusting the gravitational and inertial masses. After evaluating the current population fitness, the mass of each agent is calculated as follows:

$$M_i(t) = \frac{m_i(t)}{\sum_{j=1}^{N} m_j(t)} \tag{8.23}$$

where

$$m_i(t) = \frac{fit_i(t) - worst(t)}{best(t) - worst(t)} \tag{8.24}$$

where $fit_i(t)$ represents the fitness value of the agent i at iteration t. $best(t)$ and $worst(t)$ are the best and worst fitness of all agents, respectively. According to Newton gravitation theory, the total force that acts on the ith agent at iteration t is specified as follows:

$$\mathbf{F}_i(t) = \sum_{j \in Kbest, j \neq i} r \cdot G(t) \frac{M_j(t) \cdot M_i(t)}{R_{i,j}(t) + \varepsilon} \left(\mathbf{x}_j(t) - \mathbf{x}_i(t)\right) \tag{8.25}$$

where r is a random number within the interval [0,1]. $G(t)$ is gravitational constant at iteration t, $M_i(t)$ and $M_j(t)$ are masses of agents i and j, ε is a small constant, $R_{ij}(t)$ is the Euclidian distance between the two agents i and j, and $Kbest$ is the set of first K agents with the best fitness value and biggest mass. According to the law of motion, the acceleration of the ith agent at iteration t is given by the following equation:

$$\mathbf{a}_i(t) = \frac{\mathbf{F}_i(t)}{M_i(t)} \tag{8.26}$$

The velocity and position of an agent are updated as follows:

$$\mathbf{v}_i(t+1) = r_i \cdot \mathbf{v}_i(t) + \mathbf{a}_i(t) \tag{8.27}$$

$$\mathbf{x}_i(t+1) = \mathbf{x}_i(t) + \mathbf{v}_i(t+1) \tag{8.28}$$

where r_i is a uniform random variable in the interval [0,1].

The gravitational constant $G(t)$ in (8.25) is a function of the initial value G_0 and time t:

$$G(t) = G_0 \cdot e^{-\alpha(t/t_{\max})} \tag{8.29}$$

The parameters of maximum iteration $t_{\max}$, population size N, initial gravitational constant G_0, and constant α control the performance of GSA.

8.3.1.3 Hybridization of PSO and GSA

The basic idea of PSOGSA is to combine the ability for social thinking (**gbest**) in PSO with the local search capability of GSA. In order to combine these algorithms, the equation for updating the agents' velocity is proposed as follows [1]:

$$\mathbf{v}_i(t+1) = r_1 \cdot \mathbf{v}_i(t) + C_1 \cdot r_2 \cdot \mathbf{a}_i(t) + C_2 \cdot r_3 \cdot (\mathbf{gbest}(t) - \mathbf{x}_i(t)) \quad i = 1, \ldots, N \tag{8.30}$$

where $\mathbf{v}_i(t)$, $\mathbf{a}_i(t)$, and $\mathbf{x}_i(t)$ are the velocity, acceleration, and position of the ith agent at iteration t, respectively; $\mathbf{gbest}(t)$ is the best position (solution) so far at iteration t; C_1 and C_2 are positive constants; r_1, r_2, and r_3 are uniformly distributed random numbers in [0,1]; N is the size of population (the number of agents), where each agent consists of n elements, i.e., n control variables.

In each iteration, the positions of agents are updated as follows:

$$\mathbf{x}_i(t+1) = \mathbf{x}_i(t) + \mathbf{v}_i(t+1) \quad i = 1, \ldots, N \tag{8.31}$$

In PSOGSA, at first, all agents are randomly initialized. Each agent is considered as a candidate solution. After initialization, the gravitational constant and resultant forces among agents are calculated using (8.29) and (8.25), respectively. After that, the accelerations of agents are defined as (8.26). In each iteration, the best solution so far should be updated. After calculating the accelerations and updating the best solution so far (**gbest**), the velocities of all agents are calculated using (8.30). Finally, the positions of agents are updated by (8.31). The process of updating velocities and positions will be stopped by meeting an end criterion.

8.3.2 PSOGSA implementation to the CEED problem

Proposed hybrid PSOGSA approach has been applied to solve the CEED problem. The control variables of the CEED problem constitute the individual position of several agents that represent a complete solution set. In a system with N agents, the position of the ith agent is defined by

$$\mathbf{x}_i = \left[P_i^1, \ldots, P_i^d, \ldots, P_i^n\right], \quad i = 1, 2, \ldots, N; \quad n = N_G - 1 \tag{8.32}$$

The elements of agent $\mathbf{x}_i$ are real-power outputs of all generation units, except the slack generator. Different steps to solve the CEED problem using PSOGSA are listed as follows.

Step 1: Read the input data including the generation output power limits, fuel cost coefficients, and emission coefficients.

Step 2: Specify the control variables and their lower and upper limits; specify the dependent variable, i.e., slack generator, and its lower and upper limits of active power output; specify the objective function to be optimized F (8.1), and the penalty factors (λ_P) to form the expanded objective function F_p (8.19).

Step 3: Initialize PSOGSA parameters, such as the population size (N), maximum iteration number ($t_{\max}$), initial gravitational constant G_0, and constants α, C_1, and C_2.

Step 4: Generate an initial random population of N agents:

$$\mathbf{POP}(0) = \begin{bmatrix} \mathbf{x}_1(0) \\ \mathbf{x}_2(0) \\ \vdots \\ \mathbf{x}_i(0) \\ \vdots \\ \mathbf{x}_N(0) \end{bmatrix} = \begin{bmatrix} x_1^1(0),\ldots,x_1^d(0),\ldots,x_1^n(0) \\ x_2^1(0),\ldots,x_2^d(0),\ldots,x_2^n(0) \\ \vdots \\ x_i^1(0),\ldots,x_i^d(0),\ldots,x_i^n(0) \\ \vdots \\ x_N^1(0),\ldots,x_N^d(0),\ldots,x_N^n(0) \end{bmatrix} \tag{8.33}$$

The initial positions of each agent are randomly selected between minimum and maximum values of the control variables. Therefore, the initial value of the dth control variable in the ith agent (candidate solution) is obtained as follows:

$$x_i^d(0) = x_{i,\min}^d + r \cdot \left(x_{i,\max}^d - x_{i,\min}^d\right) \tag{8.34}$$

where r is a random number in [0,1], $x_{i,\min}^d$ and $x_{i,\max}^d$ are the lower and upper limits of the dth control variable. Initialize the velocity of agents by a zero matrix of dimension $N \times n$, where N is the number of agents (population size) and n is the number of control variables ($n = NG-1$).

Step 5: Calculate the active power output of slack generator for each agent in current population.

Step 6: Calculate the fitness value for each agent using (8.1) and (8.19).

Step 7: Calculate the gravitational constant $G(t)$ (8.29), the best and worst fitness value, and the mass of each agent $M_i(t)$ (8.23), $i = 1,\ldots,N$.

Step 8: Calculate the total force in different directions using (8.25), and the acceleration of each agent using (8.26).

Step 9: Update the best solution so far **gbest**(t).

Step 10: Update the velocity of each agent using (8.30).

Step 11: Update the position of each agent using (8.31).

Step 12: Repeat steps 5–11 until the stop criterion is reached, that is a predefined number of iterations, $t_{\max}$.

Step 13: Return best solution. Stop.

8.4 Simulation results

The proposed hybrid PSOGSA has been implemented in a MATLAB GUI-based software tool for solving the CEED problem. The computer program, called *ceedgui*, has been developed to present the efficiency of different metaheuristic optimization methods in solving the CEED. The program is described in Section 8.5.

This section presents CEED results obtained by hybrid PSOGSA on three test systems with 3, 6, and 40 generating units. These test systems are widely used as benchmarks in the power system field for solving the CEED problem.

The *ceedgui* program was run in the MATLAB 2011b (The MathWorks, Natick, Massachusetts, USA) computing environment on a 2.20-GHz PC with 3.0-GB RAM.

Twenty consecutive test runs have been performed for each case examined. The results shown are the best values obtained over these 20 runs. The different algorithms' parameters used for the simulation are summarized in Table 8.1. These values of algorithms parameters have been adopted after a number of careful experimentation.

8.4.1 Test system 1

A system with three units considering NO_x and SO_x emission is used to examine the applicability of the proposed algorithm with a total demand of 850 MW. The operating limits, fuel cost coefficients, and emission coefficients for this system are taken from [29] and illustrated in Table 8.2.

The error tolerance value in (8.17) is taken as $TOL_\varepsilon = 10^{-6}$ MW. The expression for transmission line loses is given by

$$P_{loss} = 0.00003P_{G1}^2 + 0.00009P_{G2}^2 + 0.00012P_{G3}^2 \text{ (MW)} \tag{8.35}$$

For this system, the fuel cost and the NO_x and SO_x emission are individually optimized and these individual results are reported in Tables 8.3–8.5, respectively. It can be seen that the fuel cost and emission are conflicting objectives. Emission has maximum value when cost is minimized. Table 8.6 shows the best compromise solution between fuel cost, NO_x emission, and SO_x emission for the test system 1 obtained from proposed approach. In this case, the scaling factor for NO_x emission is taken as $\gamma_{NO_x} = 147,582.78814$ (\$/t) and the scaling factor for SO_x emission is taken as $\gamma_{SO_x} = 970.031569$ (\$/t). These values are taken from [43]. The results obtained by proposed hybrid PSOGSA algorithm were compared to those obtained using other algorithms reported in the literature as shown in Tables 8.3–8.6.

Tables 8.3–8.6 have shown that the results obtained using hybrid PSOGSA are either better or comparable to those obtained using other techniques. It is quite obvious that the minimum fuel cost, minimum NO_x emission, and minimum SO_x emission obtained from the proposed approach more than SOA algorithm [29]. However, after the real-power losses computation by (8.35) with control variables reported in [29], the obtained results are different from the corresponding values

Table 8.1 Simulation parameters used for different algorithms for different test systems

Methods	Test system 1						Test system 2						Test system 3					
	N	T	G_0	a	C_1	C_2	N	T	G_0	a	C_1	C_2	N	T	G_0	a	C_1	C_2
PSO	30	50	–	–	2	2	30	50	–	–	0.5	1.5	100	500	–	–	2	2
GSA	30	50	100	10	–	–	30	50	1	20	–	–	100	500	100	1	–	–
Hybrid PSOGSA	30	50	100	10	2	2	30	50	1	20	0.5	1.5	100	500	100	10	2	2

Table 8.2 Generation limits, fuel cost, and emission coefficients of test system 1

Unit	P_{Gi}^{min}	P_{Gi}^{max}	a_i	b_i	c_i	α_{iNO_x}	β_{iNO_x}	η_{iNO_x}	α_{iSO_x}	β_{iSO_x}	η_{iSO_x}
1	150	600	561	7.9200	0.001562	0.043732540	−9.4868099e−5	1.4721848e−7	0.5783298	0.00816466	1.6103e−6
2	100	400	310	7.8500	0.001940	0.055821713	−9.7252878e−5	3.0207577e−7	0.3515338	0.00891174	2.1999e−6
3	50	200	78	7.9700	0.004820	0.027731524	−3.5373734e−4	1.9338531e−6	0.0884504	0.00903782	5.4658e−6

Table 8.3 Minimum fuel cost for test system 1 (P_{load} = 850 MW)

Method	NSGA-II [5]	SOA [29]	MOEP [30]	BBO [43]	DE/BBO [43]	PSOGSA
P_{G1} (MW)	435.885	445.404	437.87	435.1966	435.1978	435.19729
P_{G2} (MW)	299.989	289.445	299.45	299.9723	299.9696	299.96973
P_{G3} (MW)	129.951	130.427	128.49	130.6600	130.6604	130.66195
Total generation (MW)	865.825	865.27	865.81	865.8289	865.8289	865.82897
P_{loss} (MW)	15.826	15.277	15.81	15.8289	15.8289	15.82897
Fuel cost ($/h)	8,344.598	8,342.658	8,344.69	8,344.5927	8,344.58319	8,344.59272
NO_x emission (t/h)	0.09860	0.09816	0.0987	0.098686	0.098686	0.098686
SO_x emission (t/h)	9.02129	9.00988	9.01951	9.02195	9.02194	9.021955
ΔP (MW)	−7.4950e−4	−0.25694	0.00661	−1.3096e−4	−0.00113	−1.3605e−6

Table 8.4 Minimum NO_x emission for test system 1 (P_{load} = 850 MW)

Method	NSGA-II [5]	SOA [29]	MOEP [30]	BBO [43]	DE/BBO [43]	PSOGSA
P_{G1} (MW)	505.810	511.223	507.26	508.5813	508.5813	508.58037
P_{G2} (MW)	252.951	246.439	253.13	250.4433	250.4433	250.44249
P_{G3} (MW)	106.023	105.751	104.41	105.7212	105.7212	105.72297
Total gen. (MW)	864.784	863.413	864.80	864.7459	864.7459	864.74583
P_{loss} (MW)	14.784	13.414	14.8	14.7459	14.7459	14.74583
Fuel cost ($/h)	8,363.627	8,355.227	8,364.7	8,365.11464	8,365.11464	8,365.11410
NO_x emission (t/h)	0.09593	0.09586	0.09593	0.095923	0.095923	0.095924
SO_x emission (t/h)	8.97472	8.95982	8.97428	8.973667	8.973667	8.973668
ΔP (MW)	0.00120	−1.23536	0.00571	−5.1003e−5	−5.1003e−5	−1.0205e−6

reported in [29]. So, power generation is not matching load plus losses, and this approach did not meet the power balance constraint. Consequently, the results obtained from the SOA algorithm [29] must be declared as infeasible solutions. In this regard, in order to assess the relevance of the results presented in Tables 8.3–8.6, a power balance indicator was introduced as follows:

$$P = P_{G,1} + P_{G,2} + P_{G,3} - P_{load} - \left(0.00003P_{G,1}^2 + 0.00009P_{G,2}^2 + 0.00012P_{G,3}^2\right) \tag{8.36}$$

The comparison min, max, and standard deviation of the results obtained by PSO, GSA, and hybrid PSOGSA, over 20 runs, are presented in Table 8.7. It is obvious that the proposed hybrid PSOGSA approach provides better solutions compared to the original PSO and GSA. Figures 8.1–8.3 illustrate convergence

Table 8.5 Minimum SO_x emission for test system 1 (P_{load} = 850 MW)

Method	NSGA-II [5]	SOA [29]	BBO [43]	DE/BBO [43]	PSOGSA
P_{G1} (MW)	541.308	599.251	552.1111	552.11106	552.11139
P_{G2} (MW)	223.249	177.771	219.4441	219.44402	219.44465
P_{G3} (MW)	99.919	83.629	92.96053	92.95969	92.95979
Total gen. (MW)	864.476	860.651	864.5158	864.51579	864.51583
P_{loss} (MW)	14.476	10.651	14.5158	14.51579	14.51583
Fuel cost ($/h)	8,387.518	8,413.031	8,396.4665	8,396.4573	8,396.46666
NO_x emission (t/h)	0.09638	0.09950	0.096817	0.096817	0.096817
SO_x emission (t/h)	8.96655	8.9371	8.965937	8.965927	8.965937
ΔP (MW)	0.0019	−3.80554	−8.3391e−5	−0.00102	1.7869e−6

Table 8.6 Best compromise solution of fuel cost and NO_x, SO_x emission for test system 1 (P_{load} = 850 MW)

Method	NSGA-II [5]	SOA [29]	BBO [43]	DE/BBO [43]	PSOGSA
P_{G1} (MW)	496.328	513.82835	507.11954	507.11941	507.12164
P_{G2} (MW)	260.426	238.96754	251.64262	251.64185	251.64046
P_{G3} (MW)	108.144	110.40421	106.00042	106.00028	106.00047
Total gen. (MW)	864.898	863.200	864.76258	864.76255	864.76257
P_{loss} (MW)	14.898	13.32001	14.76258	14.76255	14.76257
Fuel cost ($/h)	8,358.896	8,356.27194	8,364.31126	8,364.3019	8,364.31239
NO_x emission (t/h)	0.09599	0.09594	0.0959248	0.095924	0.095925
Equivalent cost of NO_x emission ($/h)	14,166.47183	14,158.52277	14,156.84943	14,156.84686	14,156.854959
SO_x emission (t/h)	8.97870	8.95838	8.974201	8.974190	8.974200
Equivalent cost of SO_x emission ($/h)	8,709.62245	8,689.90946	8,705.25828	8,705.24820	8,705.257558
Total cost ($/h)	31,234.99029	31,204.70417	31,226.41898	31,226.39702	31,226.42491
ΔP (MW)	3.8739e−4	−1.322672	−1.8258e−5	−0.001016	4.4107e−6

Table 8.7 Min, max, and std. dev. of the results obtained over 20 runs for test system 1

Methods	Fuel cost minimization			NO_x emission minimization			SO_x emission minimization		
	Min	Max	Std. dev.	Min	Max	Std. dev.	Min	Max	Std. dev.
PSO	8,344.59272	8,344.59282	2.04e−5	0.095924	0.095924	1.40e−9	8.965937	8.965937	1.94e−8
GSA	8,344.59272	8,344.81452	0.05178	0.095924	0.095924	1.69e−8	8.965937	8.966495	1.24e−4
PSOGSA	8,344.59272	8,344.59272	3.47e−8	0.095924	0.095924	2.59e−12	8.965937	8.965937	1.49e−11

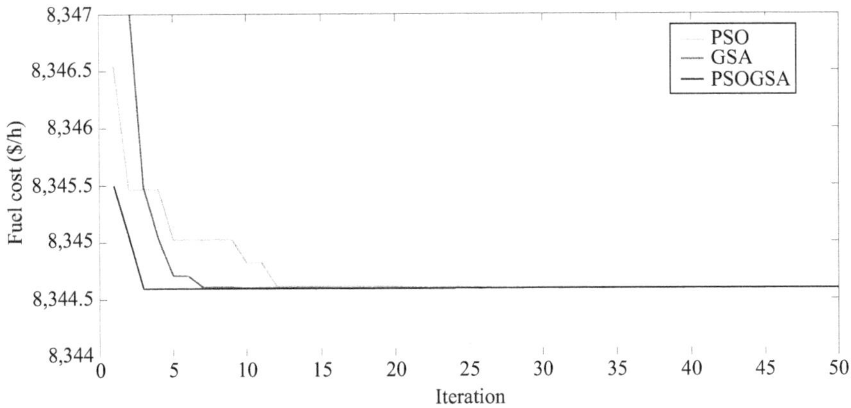

Figure 8.1 Convergence of algorithms for test system 1 in case fuel cost minimization

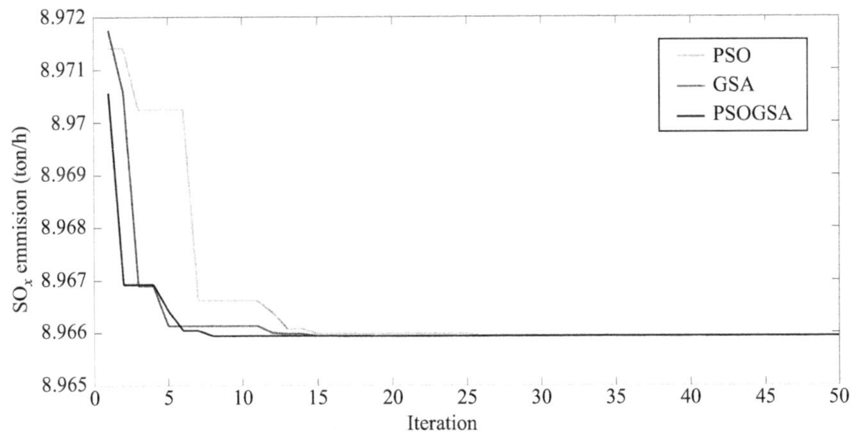

Figure 8.2 Convergence of algorithms for test system 1 in case SO_x emission minimization

characteristics of the algorithms for the fuel cost, SO_x emission and combined fuel cost, and NO_x and SO_x emission minimization, respectively. As can been seen, the proposed hybrid PSOGSA algorithm can converge to its global optimal solution in lesser iteration compared with original PSO and GSA algorithms.

8.4.2 Test system 2

The standard IEEE 30-bus 6-generator system for $P_{load} = 283.4$ MW is considered as test system 2. The fuel cost coefficients and the NO_x emission coefficients, including the limits of generation for the generators, are listed in Table 8.8.

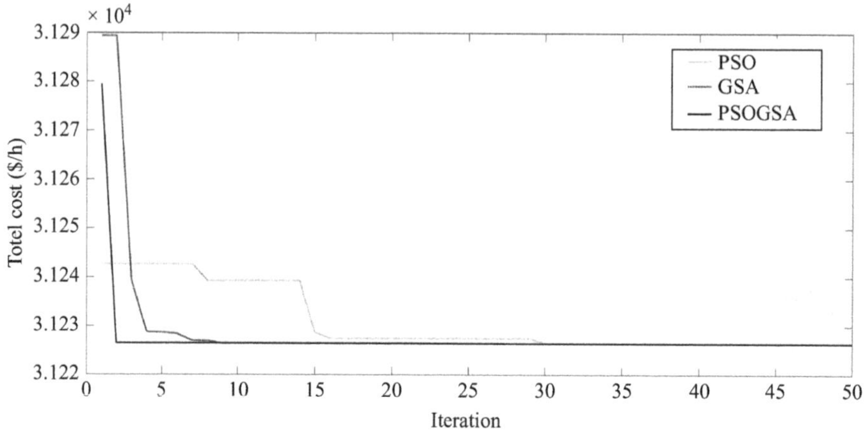

Figure 8.3 Convergence of algorithms for test system 1 in case combine fuel cost, NO_x, and SO_x emission minimization

Table 8.8 Generation limits, fuel cost, and emission coefficients of test system 2

Unit	P_{Gi}^{min}	P_{Gi}^{max}	a_i	b_i	c_i	α_i	β_i	η_i	ξ_i	λ_i
1	5	150	10	200	100	4.091e−2	−5.554e−2	6.940e−2	2.0e−4	2.857
2	5	150	10	150	120	2.543e−2	−6.047e−2	5.638e−2	5.0e−4	3.333
3	5	150	20	180	40	4.258e−2	−5.094e−2	4.586e−2	1.0e−6	8.0
4	5	150	10	100	60	5.326e−2	−3.550e−2	3.380e−2	2.0e−3	2.0
5	5	150	20	180	40	4.258e−2	−5.094e−2	4.586e−2	1.0e−6	8.0
6	5	150	10	150	100	6.131e−2	−5.555e−2	5.151e−2	1.0e−5	6.667

In this study, the scaling factor in (8.1) is taken as $\gamma_{NO_x} = 1,000$ ($/t) and the error tolerance value in (8.17) is $TOL_\varepsilon = 10^{-6}$ MW. The **B**-loss matrix values are shown as follows:

$$\mathbf{B} = \begin{bmatrix} 0.1382 & -0.0299 & 0.0044 & -0.0022 & -0.0010 & -0.0008 \\ -0.0299 & 0.0487 & -0.0025 & 0.0004 & 0.0016 & 0.0041 \\ 0.0044 & -0.0025 & 0.0182 & -0.0070 & -0.0066 & -0.0066 \\ -0.0022 & 0.0004 & -0.0070 & 0.0137 & 0.0050 & 0.0033 \\ -0.0010 & 0.0016 & -0.0066 & 0.0050 & 0.0109 & 0.0005 \\ -0.0008 & 0.0041 & -0.0066 & 0.0033 & 0.0005 & 0.0244 \end{bmatrix} \tag{8.37}$$

$$\mathbf{B}_0 = [-0.0107 \quad 0.0060 \quad -0.0017 \quad 0.0009 \quad 0.0002 \quad 0.0030] \tag{8.38}$$

$$\mathbf{B}_{00} = [0.00098573] \tag{8.39}$$

For the purpose of comparison with the reported results, the test system 2 is considered for two cases.

Table 8.9 The best solution for fuel cost and NO_x emission for test system 2

	w	Generation (MW)						Fuel cost ($/h)	NO_x emission (t/h)	P_{loss} (MW)
		P_{G1}	P_{G2}	P_{G3}	P_{G4}	P_{G5}	P_{G6}			
Case A	1	12.09740	28.63094	58.35308	99.28439	52.39978	35.19063	605.99837	0.220728	2.55623
	0	41.09414	46.36880	54.44345	39.03819	54.44012	51.54843	646.20838	0.194179	3.53313
	0.5	22.55343	35.45531	57.00411	74.54079	54.82271	41.55636	612.25222	0.203571	2.53271
Case B	1	10.97109	29.97677	52.42842	101.61927	52.43249	35.97196	600.11141	0.222145	–
	0	40.60721	45.90711	53.79299	38.29620	53.79338	51.00310	638.27284	0.194203	–
	0.5	23.22992	36.03249	53.88335	74.57626	53.88213	41.79586	606.79829	0.203289	–

Table 8.10 Comparison of best solution of test system 2 for Case A

Methods	Fuel cost minimization (w=1)		NO_x emission minimization (w=0)		CEED minimization (w=0.5)	
	Fuel cost ($/h)	NO_x emission (t/h)	Fuel cost ($/h)	NO_x emission (t/h)	Fuel cost ($/h)	NO_x emission (t/h)
GA [4]	607.7800	0.2199	645.2200	0.1942	–	–
NSGA [4]	607.9800	0.2191	638.9800	0.1947	617.8000	0.2002
NPGA [4]	608.0600	0.2207	644.2300	0.1943	617.7900	0.2004
NSGA II [6]	607.8010	0.2189	644.1330	0.1942	–	–
NSGA II [7]	613.6759	0.2223	648.7090	0.1942	–	–
PSO [15]	607.8400	0.2192	642.9000	0.1942	–	–
MOPSO [20]	607.7900	0.2193	644.7400	0.1942	615.000	0.2021
MA θ-PSO [21]	605.9984	0.2206	649.2070	0.1942	–	–
ABC [28]	605.4258	0.2210	646.0455	0.1942	612.195	0.2035
DE [31]	608.0658	0.2193	645.0850	0.1942	–	–
DE [32]	606.0000	0.2217	645.5900	0.1942	–	–
MODE [33]	606.4160	0.2221	643.5190	0.1942	614.170	0.2043
MBFA [36]	607.6700	0.2198	644.4300	0.1942	616.496	0.2002
MODE/PSO [44]	606.0073	0.2209	646.0243	0.1942	–	–
PSOGSA	605.99837	0.220728	646.20838	0.194179	612.25222	0.203571

Case A: considering P_{loss}, and
Case B: neglecting P_{loss}

Table 8.9 shows the optimum solution values of PSOGSA for the weight factor: $w = 1$ (fuel cost minimization), $w = 0$ (NO_x emission minimization), and $w = 0.5$ (combined fuel cost and NO_x emission minimization—CEED minimization).

Under the same system data, control variable limits, and constraints, the results for Cases A and B obtained using the hybrid PSOGSA approach are compared to some other algorithms reported in the literature as shown in Tables 8.10 and 8.11, respectively. From these tables, it can be seen that the proposed approach

Table 8.11 Comparison of best solution of test system 2 for Case B

Methods	Fuel cost minimization (w=1)		NO_x emission minimization (w=0)		CEED minimization (w=0.5)	
	Fuel cost (\$/h)	NO_x emission (t/h)	Fuel cost (\$/h)	NO_x emission (t/h)	Fuel cost (\$/h)	NO_x emission (t/h)
GA [4]	600.11	0.2221	638.26	0.1942	–	–
NSGA [4]	600.34	0.2241	633.83	0.1946	606.03	0.2041
NPGA [4]	600.31	0.2238	636.04	0.1943	608.90	0.2015
NSGA II [6]	600.155	0.22188	638.269	0.19420	–	–
NSGA II [7]	600.7422	0.2204	636.7316	0.1942	–	–
PSO [15]	600.13	0.2199	636.62	0.1943	–	–
MOPSO [20]	600.12	0.2216	637.42	0.1942	608.65	0.2017
MA θ-PSO [21]	600.1114	0.2221	638.2734	0.1942	–	–
SOA [29]	600.986	0.20889	640.749	0.18729	624.604	0.18708
DE [31]	600.1114	0.2221	638.2907	0.1942	–	–
DE [32]	600.11	0.2231	638.860	0.1952	–	–
MBFA [36]	600.17	0.2200	636.73	0.1942	610.906	0.2000
MODE/PSO [44]	600.115	0.22201	638.270	0.194203	–	–
PSOGSA	600.11141	0.222145	638.27284	0.194203	606.79829	0.203289

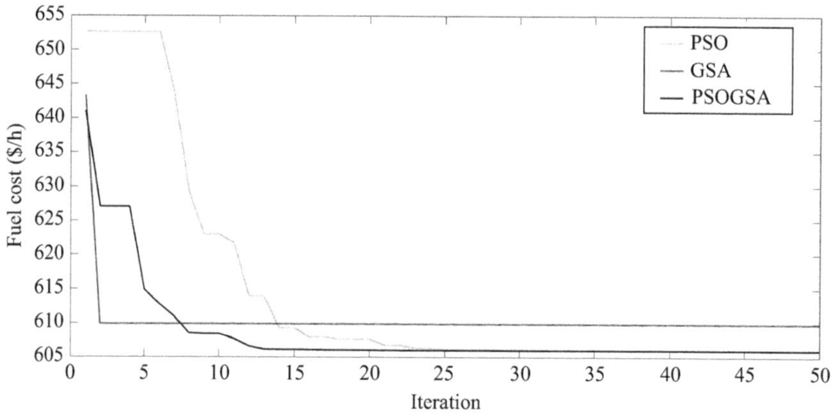

Figure 8.4 Convergence of algorithms for the fuel cost minimization in Case A for test system 2

outperforms many techniques used to solve CEED problems because the results obtained using hybrid PSOGSA are either better or comparable to those obtained using other techniques. This highlights its ability to find better quality solution.

Figure 8.4 illustrates the convergence characteristics of PSO, GSA, and proposed hybrid PSOGSA for the fuel cost minimization ($w = 1$) in Case A for the test system 2.

Table 8.12 Generation limits, fuel cost, and emission coefficients of test system 3

Unit	P_{Gi}^{min} (MW)	P_{Gi}^{max} (MW)	a_i ($/h)	b_i ($/MW h)	c_i ($/(MW)^{2}h)	d_i ($/h)	e_i (rad/MW)	α_i (t/h)	β_i (t/MW h)	η_i (t/(MW)^{2}h)	ξ_i (t/h)	λ_i (1/MW)
1	36	114	94.705	6.73	0.0069	100	0.084	60	−2.22	0.048	1.31	0.0569
2	36	114	94.705	6.73	0.0069	100	0.084	60	−2.22	0.048	1.31	0.0569
3	60	120	309.54	7.07	0.02028	100	0.084	100	−2.36	0.0762	1.31	0.0569
4	80	190	369.03	8.18	0.00942	150	0.063	120	−3.14	0.054	0.9142	0.0454
5	47	97	148.89	5.35	0.0114	120	0.077	50	−1.89	0.085	0.9936	0.0406
6	68	140	222.33	8.05	0.01142	100	0.084	80	−3.08	0.0854	1.31	0.0569
7	110	300	278.71	8.03	0.00357	200	0.042	100	−3.06	0.0242	0.655	0.02846
8	135	300	391.98	6.99	0.00492	200	0.042	130	−2.32	0.031	0.655	0.02846
9	135	300	455.76	6.6	0.00573	200	0.042	150	−2.11	0.0335	0.655	0.02846
10	130	300	722.82	12.9	0.00605	200	0.042	280	−4.34	0.425	0.655	0.02846
11	94	375	635.2	12.9	0.00515	200	0.042	220	−4.34	0.0322	0.655	0.02846
12	94	375	654.69	12.8	0.00569	200	0.042	225	−4.28	0.0338	0.655	0.02846
13	125	500	913.4	12.5	0.00421	300	0.035	300	−4.18	0.0296	0.5035	0.02075
14	125	500	1,760.4	8.84	0.00752	300	0.035	520	−3.34	0.0512	0.5035	0.02075
15	125	500	1,728.3	9.15	0.00708	300	0.035	510	−3.55	0.0496	0.5035	0.02075
16	125	500	1,728.3	9.15	0.00708	300	0.035	510	−3.55	0.0496	0.5035	0.02075
17	220	500	647.85	7.97	0.00313	300	0.035	220	−2.68	0.0151	0.5035	0.02075
18	220	500	649.69	7.95	0.00313	300	0.035	222	−2.66	0.0151	0.5035	0.02075
19	242	550	647.83	7.97	0.00313	300	0.035	220	−2.68	0.0151	0.5035	0.02075
20	242	550	647.81	7.97	0.00313	300	0.035	220	−2.68	0.0151	0.5035	0.02075
21	254	550	785.96	6.63	0.00298	300	0.035	285	−2.22	0.0145	0.5035	0.02075
22	254	550	785.96	6.63	0.00298	300	0.035	285	−2.22	0.0145	0.5035	0.02075
23	254	550	794.53	6.66	0.00284	300	0.035	295	−2.26	0.0138	0.5035	0.02075
24	254	550	794.53	6.66	0.00284	300	0.035	295	−2.26	0.0138	0.5035	0.02075
25	254	550	801.32	7.1	0.00277	300	0.035	310	−2.42	0.0132	0.5035	0.02075
26	254	550	801.32	7.1	0.00277	300	0.035	310	−2.42	0.0132	0.5035	0.02075
27	10	150	1,055.1	3.33	0.52124	120	0.077	360	−1.11	1.842	0.9936	0.0406
28	10	150	1,055.1	3.33	0.52124	120	0.077	360	−1.11	1.842	0.9936	0.0406
29	10	150	1,055.1	3.33	0.52124	120	0.077	360	−1.11	1.842	0.9936	0.0406
30	47	97	148.89	5.35	0.0114	120	0.077	50	−1.89	0.085	0.9936	0.0406
31	60	190	222.92	6.43	0.0016	150	0.063	80	−2.08	0.0121	0.9142	0.0454
32	60	190	222.92	6.43	0.0016	150	0.063	80	−2.08	0.0121	0.9142	0.0454
33	60	190	222.92	6.43	0.0016	150	0.063	80	−2.08	0.0121	0.9142	0.0454
34	90	200	107.87	8.95	0.0001	200	0.042	65	−3.48	0.0012	0.655	0.02846
35	90	200	116.58	8.62	0.0001	200	0.042	70	−3.24	0.0012	0.655	0.02846
36	90	200	116.58	8.62	0.0001	200	0.042	70	−3.24	0.0012	0.655	0.02846
37	25	110	307.45	5.88	0.0161	80	0.098	100	−1.98	0.095	1.42	0.0677
38	25	110	307.45	5.88	0.0161	80	0.098	100	−1.98	0.095	1.42	0.0677
39	25	110	307.45	5.88	0.0161	80	0.098	100	−1.98	0.095	1.42	0.0677
40	242	550	647.83	7.97	0.00313	300	0.035	220	−2.68	0.0151	0.5035	0.02075

8.4.3 Test system 3

This test system consists of 40 generating units with valve-point-loading effects and NO_x emission. The total load demand is set to 10,500 MW and no transmission losses are considered. The input data for this test system are taken from [28] and presented in Table 8.12.

Table 8.13 shows the optimum results obtained from the PSOGSA for test system 3. For the purpose of comparison with the reported results, four different cases have been considered as follows:

- Fuel cost minimization ($w = 1$);
- NO_x emission minimization ($w = 0$);
- CEED minimization, with the weight factor and the scaling factor in (8.1) taken as $w = 0.92$ and $\gamma_{NO_x} = 1$ (\$/t);
- CEED minimization, with the weight factor and the scaling factor in (8.1) taken as $w = 0.5$ and $\gamma_{NO_x} = 0.35198$ (\$/t).

In Table 8.14, the optimization results are compared to other optimization algorithms in the literature. Except in the case of the fuel cost minimization, the minimum values of fuel cost and NO_x emission obtained from the proposed PSOGSA are lesser in comparison to reported best results in the literature. The best fuel cost in the case of the fuel cost minimization obtained by PSOGSA is very close to that found with ABC [28], MBFA [36], and modified adaptive θ-particle swarm optimization (MA θ-PSO) [21] (0.04% difference). The global optimal solution for the 40-generator system in the case of fuel cost minimization is near to 121,412.53 $/h, as discussed in [51]. Therefore, solutions with a total fuel cost lower than this value are probably infeasible or incorrect.

The convergence characteristics of the PSO, GSA, and hybrid PSOGSA for the fuel cost minimization of 40-generator system are shown in Figure 8.5. It is clear that the proposed hybrid PSOGSA algorithm can converge to its global best in less iteration compared with PSO and GSA algorithms.

8.5 Educational software—*ceedgui*

The *ceedgui* is an educational MATLAB-based software tool for the ED/CEED solution using different metaheuristic optimization algorithms. The program is available at [52]. Similarly as the *opfgui* presented in Chapter 6, the *ceedgui* has been developed to present the efficiency of different metaheuristic optimization methods in solving the CEED problems. General flowchart for the CEED solution using metaheuristic optimization methods is shown in Figure 8.6.

The aim of this program is to encompass the main steps in solving the CEED problem using metaheuristic methods. These steps include (i) selection of test system, and edit system data; (ii) selection of objective function; (iii) selection of solution method, setting the algorithm parameters; (iv) program execution; and (v) displaying the results.

Figure 8.7 shows the main window of *ceedgui* while it operates on a Windows-based computer. Shown is an example of the execution of the program on the IEEE 30-bus 6-generator test system. Simultaneous minimization of the fuel cost and the pollutant emission is selected as the objective function (CEED), and wind-driven optimization (WDO) algorithm is selected as the solution method.

Table 8.13 Best dispatch results of the proposed PSOGSA algorithm for test system 3

Unit	Fuel cost minimization ($w = 1$)	NO_x emission minimization ($w = 0$)	CEED minimization ($\gamma_{NO_x} = 1$, $w = 0.92$)	CEED minimization ($\gamma_{NO_x} = 0.35198$, $w = 0.5$)
P_{G1}	113.1715	114.00018	113.99288	114.00008
P_{G2}	112.8790	114.00000	113.99463	114.00000
P_{G3}	97.4091	120.00000	120.00000	120.00000
P_{G4}	179.7441	169.39980	179.73392	178.22399
P_{G5}	96.9970	97.00000	97.00000	97.00000
P_{G6}	140.0000	124.37399	139.99861	129.45096
P_{G7}	259.9108	299.61829	300.00000	300.00000
P_{G8}	284.6112	297.76836	300.00000	299.88151
P_{G9}	284.6168	297.24768	299.43437	299.13243
P_{G10}	130.0000	130.00000	130.00051	130.00000
P_{G11}	94.0000	298.73754	318.39126	307.47843
P_{G12}	94.0000	297.98429	318.40027	307.07323
P_{G13}	214.7602	433.64298	394.29678	433.24767
P_{G14}	394.2856	421.86130	394.27868	408.80268
P_{G15}	394.2773	422.33780	394.27977	411.45448
P_{G16}	394.2807	423.22892	394.27753	410.53116
P_{G17}	489.2805	439.30460	489.28134	452.30768
P_{G18}	489.2813	439.36318	489.26948	452.16100
P_{G19}	511.2746	439.32862	424.05038	437.23697
P_{G20}	511.2790	439.32231	426.23089	437.48707
P_{G21}	523.2826	439.34045	433.56936	437.61601
P_{G22}	523.2804	439.34068	434.43738	437.56132
P_{G23}	523.3031	439.72475	435.06994	438.12224
P_{G24}	523.2796	439.66590	434.30689	438.04826
P_{G25}	523.2825	440.03222	438.95774	437.84051
P_{G26}	523.2828	440.05940	433.81587	437.88104
P_{G27}	10.0000	29.13564	10.87493	19.57977
P_{G28}	10.0000	29.04292	11.47228	19.57858
P_{G29}	10.0016	29.03909	10.64229	19.55932
P_{G30}	88.3025	97.00000	96.99793	97.00000
P_{G31}	190.0000	172.37274	189.99659	175.83358
P_{G32}	189.9976	172.37803	189.99144	175.79553
P_{G33}	190.0000	172.35819	189.99684	175.79774
P_{G34}	199.9994	200.00000	200.00000	200.00000
P_{G35}	199.9987	200.00000	200.00000	200.00000
P_{G36}	164.9176	200.00000	200.00000	200.00000
P_{G37}	110.0000	100.85525	109.99744	104.26519
P_{G38}	89.7335	101.00509	110.00000	104.27832
P_{G39}	110.0000	100.79025	109.99126	104.26062
P_{G40}	511.2794	439.33953	422.97053	437.51262
Fuel cost ($/h)	121,461.42914	129,986.50060	125,710.73492	128,710.90485
NO_x emission (t/h)	358,155.359339	176,678.019377	195,418.786862	178,591.659226

Table 8.14 Comparison of the results for test system 3

Methods	Fuel cost minimization ($w = 1$)		NO_x emission minimization ($w = 0$)		CEED minimization ($\gamma_{NO_x} = 1$, $w = 0.92$)		CEED minimization ($\gamma_{NO_x} = 0.35198$, $w = 0.5$)	
	Fuel cost (\$/h)	**NO_x emiss. (t/h)**	**Fuel cost (\$/h)**	**NO_x emiss. (t/h)**	**Fuel cost (\$/h)**	**NO_x emiss. (t/h)**	**Fuel cost (\$/h)**	**NO_x emiss. (t/h)**
NPSO-LRS [12]	121,664.431	–	–	–	–	–	–	–
MA θ-PSO [21]	121,412.570	359,902.189	129,995.271	176,682.265	–	–	–	–
GSA [22]	–	–	–	–	1.2578e5	2.1093e5	–	–
ABC [28]	121,414.80	356,421.70	129,995.47	176,682.25	–	–	–	–
SOA [29]	125,248.114	334,880.054	152,262.549	120,590.839	–	–	140,044.658	154,291.141
MODE [34]	1.2184e5	3.7479e5	1.2996e5	1.7668e5	1.2579e5	2.1119e5	–	–
NSGA II [34]	–	–	–	–	1.2583e5	2.1095e5	–	–
MODE [35]	121,836.98	374,790.56	129,956.09	176,683.27	–	–	–	–
NSGA II [35]	124,963.50	262,489.27	129,965.89	176,691.96	–	–	–	–
MBFA [36]	121,415.653	356,424.497	1299,950.00	176,682.27	–	–	–	–
DE-HS [45]	121,414.937	356,432.902	129,994.249	176,679.684	–	–	128,713.887	178,634.097
DEC-SQP [47]	121,741.979	–	–	–	–	–	–	–
PSOGSA	121,461.429	358,155.359	129,986.501	176,678.019	125,710.735	195,418.787	128,710.905	178,591.659

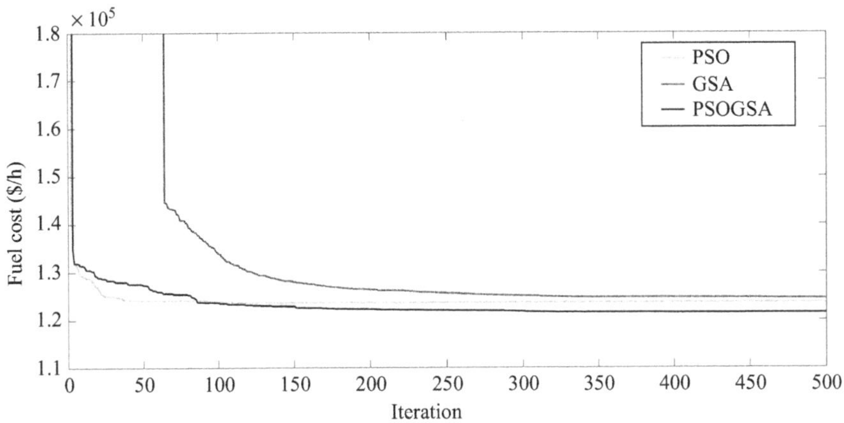

Figure 8.5 Convergence of algorithms for test system 3 in case fuel cost minimization

As can be seen from the figure, the *ceedgui* is organized in the following components:

1. Panel TEST SYSTEM. It consists of one pop-up menu and one buttons. Pop-up menu enables to selection of a test system to be considered. It is available 3-generator, 6-generator, and 40-generator units test systems. The button System Data opens the input file of the selected test system. The input file contains generator data related to the fuel cost and emission characteristics coefficients, total load demand, and transmission loss coefficients (B-loss coefficients) for the selected test system. The data are given in matrix form. The user can change the data and save those modifications.
2. Panel OPTIONS. It consists of two pop-up menus. In the first pop-up menu, the user performs selection of an objective function between four offered objectives, such as minimization of fuel cost (Fcost), minimization of fuel cost considering valve point effect (Fcostvpe), minimization of pollutant emission (Emission), and simultaneous minimization of fuel cost and pollutant emission (CEED). Pop-up-menu method enables choice of a metaheuristic optimization method for the solution of the ED/CEED problem. The user can select one of the methods, such as PSO, GSA, PSOGSA, ABC, WDO, teaching learning based optimization (TLBO), etc. When clicking on the optimization method, the panel to settings of the algorithm parameters will be opened. The user can use the default parameters by clicking on button Use default or to enter the desired value for each of algorithm parameters.
3. After the selection of appropriate options on the panels TEST SYSTEM and OPTIONS, the user can start the program by clicking the button run ED.
4. Panel OPF RESULTS. It manages the access to the different segments of the OPF results obtained from the program. The best results, that is, objective function values in the last iteration, and convergence profile of the

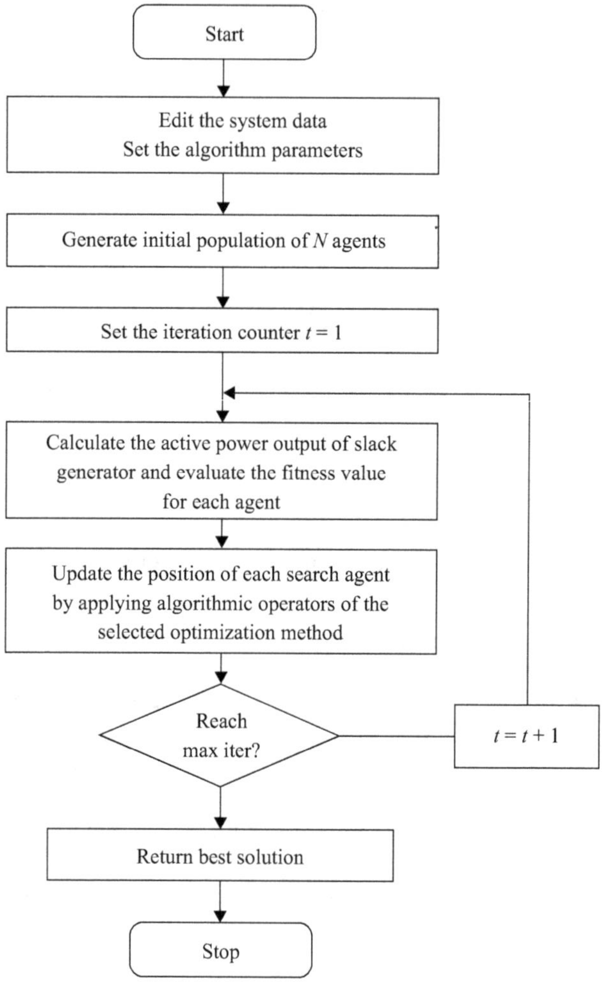

Figure 8.6 Flowchart of metaheuristic optimization method-based CEED solution

optimization algorithm will be displayed immediately after execution of the program (Figure 8.7). Besides, the program generates other important results, also. When clicking the button Optimum Control Variables, the obtained optimal values of control variables, i.e., generator active powers, will be displayed, as shown in Figure 8.8.

5. Because of the stochastic nature of metaheuristic optimization methods, a very important segment of the program is statistical evaluation of the results. When the user clicks on the button Statistics, a statistical evaluation of the optimal results which are obtained using the selected optimization method in specified number of program execution will be displayed (Figure 8.9). These statistical

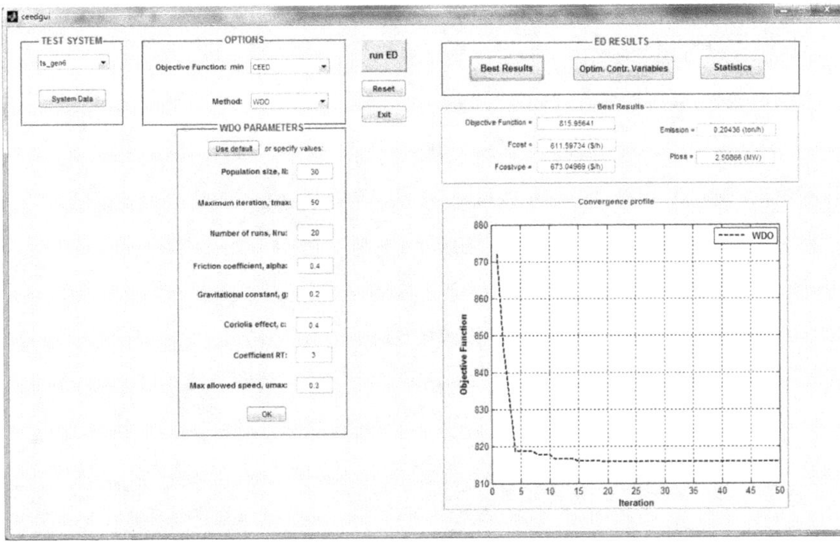

Figure 8.7 Main window of the ceedgui

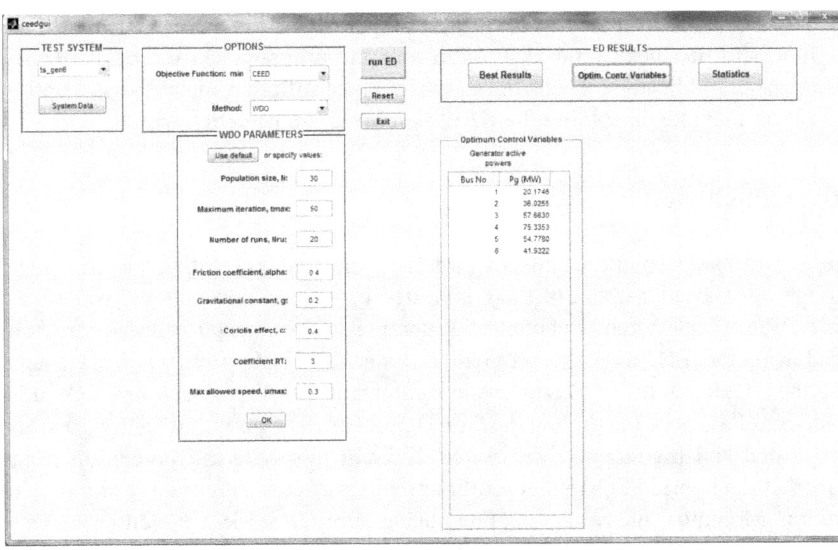

Figure 8.8 Display of the results: optimum control variables

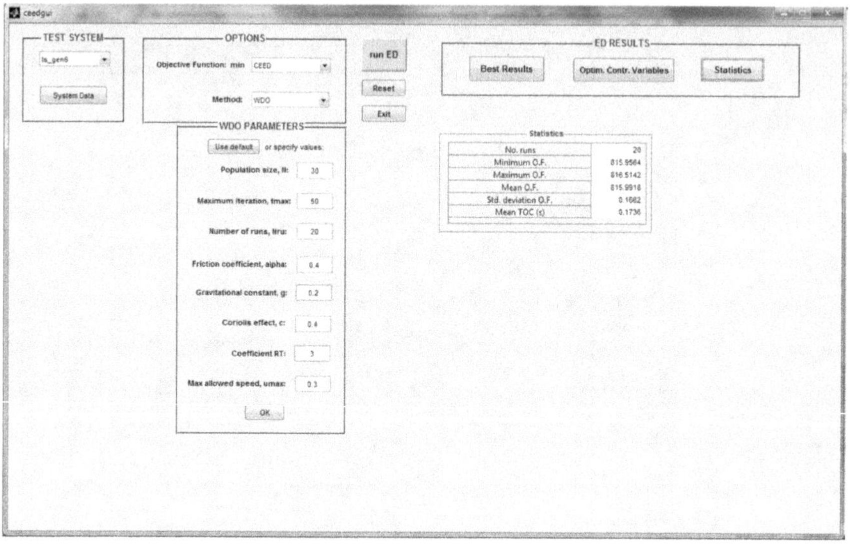

Figure 8.9 Statistical evaluation of the results

indicators are number of program execution, the minimum value of the objective function, the maximum value of the objective function, the mean value of the objective function, the standard deviation of the objective function, and the mean time of execution of the program. On the basis of these statistical indicators, a relevant comparison of different metaheuristic optimization methods in solving the CEED problem can be achieved.

8.6 Conclusion

The ED problem solution aims to minimize the cost generating electric power through optimal adjustment of the committed generating unit outputs, while satisfying all unit and system constraints. When the environmental concerns are combined with the ED, then the problem becomes a CEED problem. This chapter describes applications of metaheuristic optimization methods such as PSO, GSA, and hybrid PSOGSA for solving the CEED problem. The proposed approaches have been tested and investigated with three different test systems. Simulation results show that the hybrid PSOGSA algorithm provides effective and robust high-quality solution. Moreover, the results obtained using hybrid PSOGSA are either better or comparable to those obtained using other techniques reported in the literature. The proposed hybrid PSOGSA algorithm is suitable for solving the complex CEED problem because it benefits from the advantages of both PSO and GSA algorithms.

Additionally, the *ceedgui* software to solve CEED using different metaheuristic methods has been presented in this chapter. This educational software tool can be very useful for students or researchers to compare the performances of different optimization methods in solving the ED/CEED problems.

References

[1] Radosavljević J. 'A solution to the combined economic and emission dispatch using hybrid PSOGSA algorithm'. *Applied Artificial Intelligence.* 2016;**30**(5):445–74

[2] Gjorgiev B., Cepin M. 'A multi-objective optimization based solution for the combined economic-environmental power dispatch problem'. *Engineering Applications of Artificial Intelligence.* 2012;**26**(1):417–29

[3] Raglend I.J., Veeravalli S., Sailaja K., Sudheera B., Kothari D.P. 'Comparison of AI techniques to solve combined economic emission dispatch problem with line flow constraints'. *Electrical Power and Energy Systems.* 2010;**32**:592–8

[4] Abido M.A. 'Multiobjective evolutionary algorithm for electric power dispatch problem'. *IEEE Transactions on Evolutionary Computation.* 2006; **10**(3):315–29

[5] Ah King R.T.F., Rughooputh H.C.S. 'Elitist multiobjective evolutionary algorithm for environmental/economic dispatch'. *Proceedings of Congress on Evolutionary Computation*; Canberra, Australia, Dec 2003. IEEE; 2004. pp. 1108–14

[6] Ah King R.T.F., Rughooputh H.C.S., Deb K. 'Evolutionary multi-objective environmental/economic dispatch: stochastic vs. deterministic approaches'. *Evolutionary Multi-Criterion Optimization Lecture Notes in Computer Science.* 2005;**34**(10):677–91

[7] Alawode K.O., Jubril A.M., Komolafe O.A. 'Multiobjective optimal power flow using hybrid evolutionary algorithm'. *Electrical Power and Energy Systems.* 2010;**4**(7):506–11

[8] Abul'Wafa A.R. 'Optimization of economic/emission load dispatch for hybrid generating systems using controlled Elitist NSGA-II'. *Electric Power Systems Research.* 2013;**105**:142–51

[9] Song Y.H., Wang G.S., Wang P.Y., Johns A.T. 'Environmental/economic dispatch using fuzzy logic controlled genetic algorithms'. *IEE Proceedings—Generation, Transmission, and Distribution.* 1997;**144**(4):377–82

[10] Gaing Z.L. 'Particle swarm optimization to solving the economic dispatch considering the generator constraints'. *IEEE Transactions on Power Systems.* 2003;**18**(3):1187–95

[11] Park J.B., Lee K.S., Shin J.R., Lee K.Y. 'A particle swarm optimization for economic dispatch with nonsmooth cost functions'. *IEEE Transactions on Power Systems.* 2005;**20**(1):34–42

[12] Selvakumar A.I., Thanushkodi K. 'A new particle swarm optimization solution to nonconvex economic dispatch problems'. *IEEE Transactions on Power Systems.* 2007;**22**(1):42–51

[13] Chaturvedi K.T., Pandit M., Srivastava L. 'Self-organizing hierarchical particle swarm optimization for nonconvex economic dispatch'. *IEEE Transactions on Power Systems.* 2008;**23**(3):1079–87

[14] Park J.B., Jeong Y.W., Shin J.R., Lee K.Y. 'An improved particle swarm optimization for nonconvex economic dispatch problems'. *IEEE Transactions on Power Systems*. 2010;**25**(1):156–66
[15] Hazra J., Sinha A.K. 'A multi-objective optimal power flow using particle swarm optimization'. *European Transactions on Electrical Power*. 2011;**21**(1):1028–45
[16] Ghorbani N., Vakili S., Babaei E., Sakhavati A. 'Particle swarm optimization with smart inertia factor for solving non-convex economic load dispatch problems'. *International Transactions on Electrical Energy Systems*. 2014;**24**:1120–33
[17] Gholami A., Ansari J., Jamei M., Kazemi A. 'Environmental/economic dispatch incorporating renewable energy sources and plug-in vehicles'. *IET Generation, Transmission & Distribution*. 2014;**8**(12):2183–98
[18] Hosseinnezhad V., Rafiee M., Ahmadian M., Ameli M.T. 'Species-based quantum particle swarm optimization for economic load dispatch'. *Electrical Power and Energy Systems*. 2014;**63**:311–22
[19] Plathottam S.J., Salehfar H. 'Unbiased economic dispatch in control areas with conventional and renewable generation sources'. *Electric Power Systems Research*. 2015;**119**:313–21
[20] Abido M.A. 'Multiobjective particle swarm optimization for environmental/economic dispatch problem'. *Electric Power Systems Research*. 2009;**79**:1105–13
[21] Niknam T., Doagou-Mojarrad H. 'Multiobjective economic/emission dispatch by multiobjective θ-particle swarm optimisation'. *IET Generation, Transmission & Distribution*, 2012;**6**(5):363–77
[22] Güvenç U., Sönmez Y., Duman S., Yörükerend N. 'Combined economic and emission dispatch solution using gravitational search algorithm'. *Scientia Iranica D*. 2012;**19**(6):1754–62
[23] Duman S., Guvenc U., Yorukeren N. 'Gravitational search algorithm for economic dispatch with valve-point effects'. *International Review of Electrical Engineering*. 2010;**5**(6):2890–5
[24] Shaw B., Mukherjee V., Ghoshal S.P. 'A novel opposition-based gravitational search algorithm for combined economic and emission dispatch problems of power systems'. *Electrical Power and Energy Systems*. 2012;**35**:21–33
[25] Mondal S., Bhattacharya A., nee Dey S.H. 'Multi-objective economic emission load dispatch solution using gravitational search algorithm and considering wind power penetration'. *Electrical Power and Energy Systems*. 2013;**44**:282–92
[26] Radosavljević J. 'Gravitational search algorithm for solving combined economic and emission dispatch'. *Proceedings of XIV International Symposium Infoteh-Jahorina*; Jahorina, Bosnia and Herzegovina, Mar 2015. pp. 148–53
[27] Beigvand S.D., Abdi H., Scala M.L. 'Combined heat and power economic dispatch problem using gravitational search algorithm'. *Electric Power Systems Research*. 2016;**133**:160–72

[28] Aydin D., Ozyon S., Yasar C., Liao T. 'Artificial bee colony algorithm with dynamic population size to combined economic and emission dispatch problem'. *International Journal of Electrical Power and Energy Systems*. 2014;**54**:144–53

[29] Benasla L., Belmadani A., Rahli M. 'Spiral optimization algorithm for solving combined economic and emission dispatch'. *Electrical Power and Energy Systems*. 2014;**62**:163–74

[30] Jeyakumar D.N., Venkatesh P., Kwang Y.L. 'Application of multi objective evolutionary programming to combined economic emission dispatch problem'. *Proceedings of International Joint Conference on Neural Networks*; Orlando, Florida, USA, Aug 2007. IEEE; 2007. pp. 12–7

[31] Abou El Ela A.A., Abido M.A., Spea S.R. 'Differential evolution algorithm for emission constrained economic power dispatch problem'. *Electric Power Systems Research*. 2010;**80**(10):1286–92

[32] Perez-Guerrero R.E., Cedeno-Maldonado J.R. 'Differential evolution based economic environmental power dispatch'. *Proceedings of the 37th Annual North American Power Symposium*. Ames, USA, Oct 2005. IEEE; 2005. pp. 191–7

[33] Wu L.H., Wang Y.N., Yuan X.F., Zhou S.W. 'Environmental/economic power dispatch problem using multi-objective differential evolution algorithm'. *Electric Power Systems Research*. 2010;**80**(9):1171–81

[34] Basu M. 'Economic environmental dispatch using multi-objective differential evolution'. *Applied Soft Computing*. 2011;**11**:2845–53

[35] Sharma R., Samantaray P., Mohanty D.P., Rout P.K. 'Environmental economic load dispatch using multiobjective differential evolution algorithm'. *Proceeding of International Conference on Energy, Automation, and Signal*; Bhubaneswar, India, Dec 2011. IEEE; 2012. pp. 1–7

[36] Hota P.K., Barisal A.K., Chakrabarti R. 'Economic emission load dispatch through fuzzy based bacterial foraging algorithm'. *Electrical Power and Energy Systems*. 2010;**32**(7):794–803

[37] Bhattacharya A., Chattopadhyay P.K. 'Application of biogeography-based optimization for solving multi-objective economic emission load dispatch problems'. *Electric Power Component and Systems*. 2010;**38**:340–65

[38] Karakonstantis I., Vlachos A. 'Ant colony optimization for continuous domains applied to emission and economic dispatch problems'. *Journal of Information and Optimization Sciences*. 2015;**36**:23–42

[39] Chandrasekaran K., Simon S.P., Padhy N.P. 'Cuckoo search algorithm for emission reliable economic multi-objective dispatch problem'. *IETE Journal of Research*. 2014;**60**:128–38

[40] Chandrasekaran K., Simon S.P. 'Firefly algorithm for reliable/emission/economic dispatch multi objective problem'. *International Review of Electrical Engineering*. 2012;**7**:3414–25

[41] Sivasubramani S., Swarup K. 'Environmental/economic dispatch using multiobjective harmony search algorithm'. *Electric Power System Research*. 2011;**81**:1778–85

[42] Niu Q., Zhang H., Wang X., Li K., Irwin G.W. 'A hybrid harmony search with arithmetic crossover operation for economic dispatch'. *Electric Power Systems Research*. 2014;**62**:237–57

[43] Bhattacharya A., Chattopadhyay P.K. 'Solving economic emission load dispatch problems using hybrid differential evolution'. *Applied Soft Computing*. 2011;**11**:2526–37

[44] Gong D.W., Zhang Y., Qi C.L. 'Environmental/economic power dispatch using a hybrid multi-objective optimization algorithm'. *Electrical Power and Energy Systems*. 2010;**32**:607–14

[45] Sayah S., Hamouda A., Bekrar A. 'Efficient hybrid optimization approach for emission constrained economic dispatch with nonsmooth cost curves'. *Electrical Power and Energy Systems*. 2014;**56**:127–39

[46] Parouha R.P., Das K.N. 'A novel hybrid optimizer for solving economic dispatch problem'. *Electrical Power and Energy Systems*. 2016;**78**:108–26

[47] Coelho L.S., Mariani V.C. 'Combining of chaotic differential evolution and quadratic programming for economic dispatch optimization with valve-point effect'. *IEEE Transactions on Power Systems*. 2006;**21**(2):989–96

[48] Neto J.X.V., Meza G.R., Ruppel T.H., Mariani V.C., Coelho L.S. 'Solving non-smooth economic dispatch by a new combination of continuous GRASP algorithm and differential evolution'. *Electrical Power and Energy Systems*. 2017;**84**:13–24

[49] Mahdi F.P., Vasant P., Kallimani V., 'A holistic review on optimization strategies for combined economic emission dispatch problem'. *Renewable and Sustainable Energy Review*. 2017; http://dx.doi.org/10.1016/j.rser.2017.06.111

[50] Mirjalili S., Hashim S.Z.M. 'A new hybrid PSOGSA algorithm for function optimization'. *Proceedings of International Conference on Computer and Information Application;* Tianjin, China, Dec 2010. IEEE; 2012. pp. 374–7

[51] Elsayed W.T., Hegazy Y.G., Bendary F.M., El-Bages M.S. 'A review on accuracy issues related to solving the non-convex economic dispatch problem'. *Electric Power System Research*. 2016;**141**:325–32

[52] Radosavljević J. *Ceedgui—Matlab GUI based software for combined economic and emission dispatch using metaheuristic optimization methods*. 2017. Available at https://www.researchgate.net/profile/Jordan_Radosavljevic2/contributions

Chapter 9

Optimal power flow in distribution networks

9.1 Introduction

Recently, due to the growth of distributed generation (DG) penetration, the nature of distribution network is altered from a passive network to an active one and by a distribution company market which is operated by distribution system operator (DSO). DSO is responsible for operation of the distribution network in an effective manner, maintaining the required reliability of electricity supply and the quality of supply as well as coordinating the operation of the coordinated high voltage (HV) grid in cooperation with the transmission system operator.

Electricity distribution companies tend to integrate different types of distributed DG units in the distribution network. Integration of DG units with renewable energy such as photovoltaic (PV) systems, small wind turbines (WTs) and mini hydropower plants (MHP) is influenced by geographical and meteorological conditions. The DG units with nonrenewable energy, including fuel cells (FCs), micro turbines (MTs), and diesel electric generators (DEGs), can be connected to any point of the distribution network, but the generating cost should be taken into account. Depending on the DG unit technologies, the primary energy of DG units may be injected into the distribution network via either a synchronous or asynchronous electric machine which is directly connected to the grid, a combination of an electric machine and a power electronic interface, or only via a power electronic interface. If electric machine is directly connected to the grid, its operation determines the model of DG unit (constant active and reactive power – PQ bus or constant active power and voltage magnitude – PV bus) for power flow studies. In other cases, the characteristics of the interface control circuit determine the DG unit model. As a general rule, in case when the control circuit of the converter is designed to control active power and voltage independently, the DG unit model shall be as a PV bus and when it is designed to control active and reactive power independently, the DG unit model shall be as a PQ bus [1,2].

If properly planned and controlled, the DG units may offer improved the voltage profile and power losses (Plosses) reduction of the distribution network, better economics, and a reduced dependence on the local utility [3]. Safe, reliable, and economic operation of the active distribution network can be achieved only by efficiently coordination of the DG units operation, voltage regulation (voltage regulators, under-load tap changer (ULTC) transformers), and reactive power compensation

(Var compensators); that is by solving an optimal power flow (OPF) problem. Accordingly, the OPF calculation is one of the important tools for DSO to perform this duty. Main objective of the OPF for a distribution network is to minimize the fuel cost (Fcost) of DG units, voltage profile improvement, Var/Volt coordination, and minimization of Plosses though optimal settings of the control variables while at the same time satisfying various distribution system operating constraints.

The traditional OPF analysis is addressed as a *deterministic OPF problem* based on premise that the input variables and network parameters are accurately known. In other words, it is supposed that the network parameters, power of load, and output power of DG units are deterministic known. Therefore, the deterministic OPF approaches are highly dependent on the accuracy of the input data. However, many random disturbances or uncertain factors, such as the variation of load, the change in network configuration, and especially the stochastic nature of the output power of the DG with renewable energy such as WT and PV units, really exist in distribution network operation. In these cases, it is necessary to incorporate these uncertainties in OPF modeling to overcome inaccuracy in the OPF results. Therefore, the OPF problem is transformed in *probabilistic OPF problem* [4].

In its most general mathematical formulation, the OPF is a nonlinear, nonconvex, large-scale, static optimization problem with both continuous and discrete control variables. Several different approaches, such as methods of linear and quadratic programing (QP), and recently very popular metaheuristic optimization techniques, have been applied to solve the OPF problem in distribution networks. The author in [5] proposed a mixed integer linear programing model for distribution network optimization and analyzes various operating conditions. In [6], an optimal distribution power flow strategy based on decomposition of the overall system problem into two components: economic dispatch for energy at the system level and loss minimization at the distribution level is proposed. The optimization problem was solved by a combined QP and sectioning algorithm. Gan *et al.* [7] proposed conic programing for solving the OPF problem in radial distribution networks with objective to minimize the generation cost of DG units and system Ploss. The authors in [8] used sequential QP method to solve a multiobjective OPF problem for distribution networks with objectives such as minimal Ploss, minimal voltage deviation (VD), and maximum DG active power output.

In recent year, many population-based optimization techniques have been used to solve complex constrained optimization problems. Some of the population-based methods, such as genetic algorithm (GA), particle swarm optimization (PSO), and gravitational search algorithm (GSA), have been successfully applied to find the optimal solution for OPF problem in distribution networks.

The author of this chapter, together with his associates, proposed a GA-based approach for the solution of the OPF in distribution networks with DG units, including FCs, MTs, diesel generators, PV systems, and WTs [9]. Also, a GA was applied in [10] to solve the OPF problem in distribution system considering spatial electrothermal coupling effect.

Niknam [11] used standard PSO to determine active and reactive power generated by DG units, main substation, capacitors, and also tap of tap-changer transformers in distribution networks with objective to minimize the energy cost, considering the physical and technical constraints. In the continuation of the research, Niknam *et al.* [12,13] proposed a multiobjective fuzzy adaptive chaotic PSO for the multiobjective OPF in distribution networks with aim to minimize the total electrical energy losses, the total electrical energy cost, and the total pollutant emission produced by sources. Recent publications, covering the OPF problem in distribution networks, also proposed PSO as an efficient solution method [14,15].

In [16], the author proposed the GSA to solve the OPF problem in distribution networks considering the different DG units and load models. The GSA approach was successfully applied and tested on two distribution systems, and simulation results are compared with GA method. Wang *et al.* [17] proposed an improved GSA for solving the optimal reactive power flow problem in distribution networks including the DGs with randomness power outputs.

In this chapter, a generalized approach for the OPF analysis in distribution networks is presented. This implies the inclusion of different types of DG units, taking into account the uncertainties of the input variables, the multiobjective optimization, and application of efficient metaheuristic methods for solving the OPF problem.

The rest of the chapter is organized as follows. In Section 9.2, the deterministic OPF as a nonlinear multiobjective optimization problem with equality and inequality constraints is mathematically formulated. This formulation includes the DG units with renewable and nonrenewable energy sources. Optimization models of the DG units are given in Section 9.3. In addition to the deterministic, a probabilistic OPF approach based on the $2m + 1$ point estimate method is presented in Section 9.4. The Weibull and normal distributions are employed to model the input random variables, namely, the wind speed, solar irradiance, and load power. The Gram–Charlier expansion series are used to obtain probability density function (PDFs) of the OPF results. Section 9.5 briefly describes two metaheuristic optimization methods proposed to solve the OPF problems, namely, the GA and GSA methods. Simulation results are presented in Section 9.6. Section 9.7 describes a MATLAB® graphical user interface (GUI)-based software for OPF in distribution networks by using different metaheuristic optimization methods. Finally, the main conclusions are summarized in Section 9.8.

9.2 Deterministic optimal power flow

The goal of the OPF is to minimize a selected objective function via optimal settings of the control variables, subjected to various equality and inequality constraints. Generally, the OPF problem can be formulated as follows:

$$\min F(\mathbf{x}, \mathbf{y}) \tag{9.1}$$

Subject to

$$g(\mathbf{x}, \mathbf{y}) = 0 \tag{9.2}$$

$$h(\mathbf{x}, \mathbf{y}) \leq 0 \tag{9.3}$$

$$\mathbf{x} \in \mathbf{X} \tag{9.4}$$

where F is objective function to be minimized, $\mathbf{x}$ and $\mathbf{y}$ are vectors of control and dependent variables, respectively.

For distribution network with DG units the vector of control variables ($\mathbf{x}$) consisting of

- The active power output of the DG units with nonrenewable energy sources P_{DG};
- Root bus voltage V_0;
- The terminal voltage at PV buses V_{PV};
- Transformer (voltage regulation) tap setting T;
- The output of shunt Var compensators Q_C.

Therefore, the vector of control variables can be expressed as

$$\mathbf{x} = [P_{DG1} \ldots P_{DGNNR}, V_0, V_{PV1} \ldots V_{PVNPV}, T_1 \ldots T_{NT}, Q_{C1} \ldots Q_{CNC}] \tag{9.5}$$

where NNR, NPV, NT, and NC are the number of the nonrenewable DG units, number of PV buses (i.e., DG units which modeled as PV buses), number of regulating transformers, and number of Var compensators, respectively.

The vector of dependent variables ($\mathbf{y}$) consisting of

- The active power bought/sold from/to the utility grid P_{gr}. This is active power that is delivered through the root bus of the distribution network;
- Load bus voltage, including DG units which are modeled as PQ buses, V_L;
- The reactive power outputs of the DG units which are modeled as PV buses Q_{DG};
- Branch flow S_l.

Therefore, $\mathbf{y}$ can be expressed as

$$\mathbf{y} = [P_{gr}, \; V_{L1} \ldots V_{LNL}, Q_{DG1} \ldots Q_{DGNPV}, S_{l1} \ldots S_{lN}] \tag{9.6}$$

where NL, NPV, and NB are the number of load buses, number of PV buses, and total number of buses (namely branches) in the distribution network, respectively.

9.2.1 Objective function

The objective function can take different forms. Several cases have been considered herein.

Case 1: Generating cost minimization

$$F = f_{gr}(P_{gr}) + \sum_{i=1}^{NNR} f_i(P_{DGi}) \tag{9.7}$$

where $f_{gr}(P_{gr})$ is the cost characteristics for the active power bought/sold from/to the utility grid, and $f_i(P_{DGi})$ is the generating cost characteristic of the ith DG unit.

Usually, the generating cost characteristics are defined as quadratic cost function of active power, as follows:

$$f_i(P_i) = a_i + b_iP_i + c_iP_i^2 \tag{9.8}$$

where P_i stands for P_{gr} or P_{DGi}. a_i, b_i, and c_i are the appropriate cost coefficients.
Case 2: Plosses minimization:

$$F = \sum_{i=1}^{NB} P_{lossi} \tag{9.9}$$

where P_{lossi} is the Ploss in branch i.
Case 3: Multiobjective OPF as simultaneous minimization of the generating cost and Plosses:

$$F = f_{gr}(P_{gr}) + \sum_{i=1}^{NNR} f_i(P_{DGi}) + w_P \sum_{i=1}^{NB} P_{lossi} \tag{9.10}$$

where w_P is the weighting factor for Plosses.
Case 4: Multiobjective OPF as simultaneous minimization of the generating cost and VD at load buses:

$$F = f_{gr}(P_{gr}) + \sum_{i=1}^{NNR} f_i(P_{DGi}) + w_V \sum_{i=1}^{NL} |1 - V_i| \tag{9.11}$$

where V_i and w_V are the voltage magnitude at load bus i and weighting factor for VD, respectively.
Case 5: Multiobjective OPF as simultaneous minimization of the generating cost, Plosses, and VD at load buses:

$$F = f_{gr}(P_{gr}) + \sum_{i=1}^{NNR} f_i(P_{DGi}) + w_P \sum_{i=1}^{NB} P_{lossi} + w_V \sum_{i=1}^{NL} |1 - V_i| \tag{9.12}$$

9.2.2 Constraints

The equality constraints (9.2) represent typical power balance and power flow equations. The power balance equation in distribution network with DG units with renewable and nonrenewable energy sources can be expressed as follows:

$$\sum_{i=1}^{NNR} P_{DGi} + P_{gr} = \sum_{i=1}^{NL} P_{Li} + \sum_{i=1}^{NB} P_{lossi} - \sum_{i=1}^{NR} P_{DGi} \tag{9.13}$$

where NR are the number of DG units with renewable energy sources.

The backward/forward sweep power flow equations [18,19] are given in Chapter 5 of this book.

The inequality constraints (9.3) are the functional operating constraints that contain: load bus voltage magnitude limits, DG unit reactive power capabilities, and branch flow limits:

$$V_i^{\min} \leq V_i \leq V_i^{\max} \quad i = 1, \ldots, NL \tag{9.14}$$

$$Q_{DGi}^{\min} \leq Q_{DGi} \leq Q_{DGi}^{\max} \quad i = 1, \ldots, NPV \tag{9.15}$$

$$S_i \leq S_i^{\max} \quad i = 1, \ldots, NB \tag{9.16}$$

Constraints (9.4) define the feasibility region of the problem control variables such as: DG unit active power output limits, root bus voltage limits, PV bus voltage limits, transformer tap setting limits, and shunt Var compensation limits:

$$P_{DGi}^{\min} \leq P_{DGi} \leq P_{DGi}^{\max} \quad i = 1, \ldots, NNR \tag{9.17}$$

$$V_0^{\min} \leq V_0 \leq V_0^{\max} \tag{9.18}$$

$$V_{DGi}^{\min} \leq V_{DGi} \leq V_{DGi}^{\max} \quad i = 1, \ldots, NPV \tag{9.19}$$

$$T_i^{\min} \leq T_i \leq T_i^{\max} \quad i = 1, \ldots, NT \tag{9.20}$$

$$Q_{Ci}^{\min} \leq Q_{Ci} \leq Q_{Ci}^{\max} \quad i = 1, \ldots, NC \tag{9.21}$$

It should be noted that in optimization process the control variables are self-constrained. The inequality constraints of dependent variables, such as V_L, Q_{DG}, and S_l, can be incorporated in the objective function F as quadratic penalty terms. The new expanded objective function to be minimized becomes

$$F_p = F + \lambda_V \sum_{i=1}^{NL} \left(V_i - V_i^{\lim}\right)^2 + \lambda_{QDG} \sum_{i=1}^{NPV} \left(Q_{DGi} - Q_{DGi}^{\lim}\right)^2 + \lambda_S \sum_{i=1}^{NB} \left(S_i - S_i^{\lim}\right)^2 \tag{9.22}$$

where λ_V, λ_{QDG}, and λ_S are defined as penalty factors. $x^{\lim}$ is the limit value of the dependent variable x, and given as follows:

if $x > x^{\max}$, then $x^{\lim} = x^{\max}$
elseif $x < x^{\min}$, then $x^{\lim} = x^{\min}$
else
$x^{\lim} = x$; end

9.3 DG units modeling for OPF

This section describes the models of different type of DG units used for the OPF computation in distribution networks. These models are related to the generating cost characteristics of DGs.

9.3.1 Diesel generator

Usually, the diesel fuel consumption data in (L/h) at 25%, 50%, 70%, and 100% of the DEG power rating (kW) are given by the manufacturer. Based on these data, the fuel consumption characteristic of the DEG can be estimated as a quadratic function of its active power output:

$$Fuel_{DEG} = a + bP_{DEG} + cP_{DEG}^2 \tag{9.23}$$

where $Fuel_{DEG}$ is the DEG fuel consumption in (L/h); P_{DEG} is the DEG power output in (kW); a, b, and c are coefficients of the fuel consumption characteristic. Accordingly, the total DEG Fcost f_{DEG} can be calculated as

$$f_{DEG} = c_{DEG} \cdot Fuel_{DEG} = c_{DEG}\left(a + bP_{DEG} + cP_{DEG}^2\right) \tag{9.24}$$

where c_{DEG} is the diesel fuel price to supply the DEG in ($/L). For example, let us to consider the HDJW 410 T6, 369 kW DEG. The diesel fuel consumption data of the DEG are given in Table 9.1 [20]. Based on these data and the fuel price was adopted $c_{DEG} = 0.6$ ($/L), the Fcosts characteristic shown in Figure 9.1 is obtained.

Table 9.1 Fuel consumption data for the HDJW 410 T6, 369 kW DEG

Load level (%)	100	75	50	25
Fuel consumption (L/h)	100.9	74.8	51.9	29.5

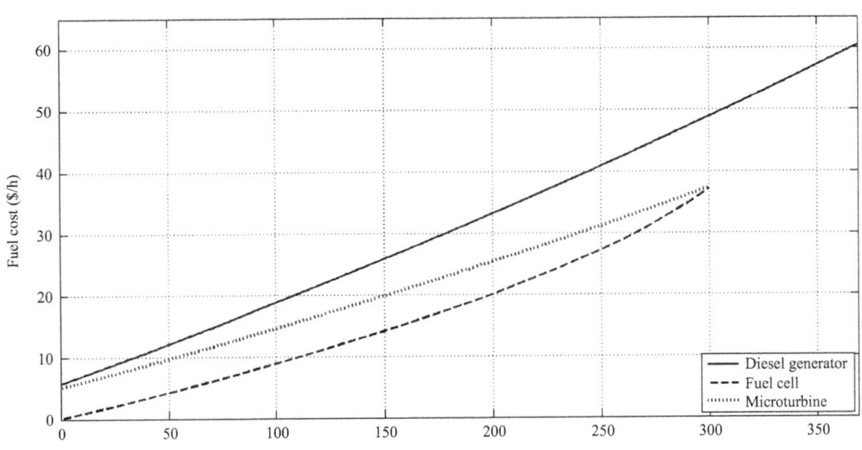

Figure 9.1 Fuel cost characteristics of the DG units

9.3.2 Fuel cell

Fcost for the FC is dependent of the active power output and the FC efficiency [21]:

$$f_{FC} = c_{FC} P_{FC} / \eta_{FC} \tag{9.25}$$

where f_{FC} is the Fcost for the FC in ($/h); P_{FC} is the FC output (electrical) power (kW); η_{FC} is the FC efficiency; c_{FC} is fuel (natural gas) price to supply the FC in ($/kW h). The FC efficiency can be expressed as the ratio of actual operating voltage V and 1.482 (V):

$$\eta_{fc} = \frac{V}{1.482} \tag{9.26}$$

As example, the proton exchange membrane (PEM) FC has been considered. The actual operating voltage of the FC is a function of power level and can be calculated using the following formula [22]:

$$V = \frac{1}{2}\left[V_0 + \sqrt{V_0^2 - 4\left(V_0 V_{Pnom} - V_{Pnom}^2\right) \cdot \xi} \right] \tag{9.27}$$

where V_0 is theoretical or thermodynamic FC potential (0.9 V); V_{Pnom} is selected cell potential at nominal power (0.45–0.75 V); $\xi = P_{fc}/P_{nom}$ is the power level; P_{nom} is nominal power of the FC. Let us assume the PEM FC efficiency characteristic for $V_0 = 0.9$ (V) and $V_{Pnom} = 0.6$ (V). Figure 9.1 shows the Fcost curve of the FC estimated for $P_{nom} = 300$ (kW) and $c_{fc} = 0.05$ ($/kW h).

9.3.3 Microturbine

The MT Fcosts, f_{MT}, can be calculated as follows:

$$f_{MT} = c_{MT} P_{MT} / \eta_{MT} \tag{9.28}$$

where P_{MT} is the MT electrical power output (kW); η_{MT} is the MT electric efficiency; c_{MT} is the fuel price to supply the MT in ($/kW h). The efficiency of the MT increases with the increase of the output power [23]. The electric efficiency characteristic of the MT can be estimated as a quadratic function of its power output:

$$\eta_{MT} = a + bP_{MT} + cP_{MT}^2 \tag{9.29}$$

From a typical electrical efficiency curve of a 300-kW CGT301-302 MT given in [23], the MT Fcosts curve is obtained as shown in Figure 9.1. It is assumed that the fuel price to supply the MT is $c_{MT} = 0.05$ ($/kW h).

9.3.4 Wind turbine

To calculate power output of a WT, two main factors must be known: the wind speed on certain location and the power curve of the WT. According to [24], the power curve of a WT can be modeled by means of a function split into four different parts:

$$P_{WT} = \begin{cases} 0, & v \le v_{ci} \\ \dfrac{v^2 - v_{ci}^2}{v_{nom}^2 - v_{ci}^2} \cdot P_{nom}, & v_{ci} < v \le v_{nom} \\ P_{nom}, & v_{nom} < v \le v_{co} \\ 0, & v > v_{co} \end{cases} \tag{9.30}$$

where P_{nom}, v_{nom}, v_{ci}, and v_{co} are nominal power, nominal wind speed, cut-in wind speed, and cut-out wind speed of the WT, respectively; P_{WT} and v are denoted power output of the WT and wind speed. These data, as well as the experimentally determined power curve (9.30), are given by the manufacturers of WTs. Fcosts at the wind farm are equal to zero, so they usually operate with the maximum available power.

For the OPF simulation tests presented in Section 9.6, it is adopted Vestas V44/600 kW WT model [25]. The parameters used to model the power curve obtained from the owner's manual are as follows: $P_{nom} = 600$ (kW), $v_{ci} = 4$ (m/s), $v_{nom} = 16$ (m/s), and $v_{co} = 20$ (m/s). Figure 9.2 shows the wind speed data used to calculate the power generated by the WT generator in the deterministic OPF algorithm.

9.3.5 Photovoltaic

The power output of the PV module is dependent on the solar irradiance and ambient temperature of the site as well as the characteristics of the module

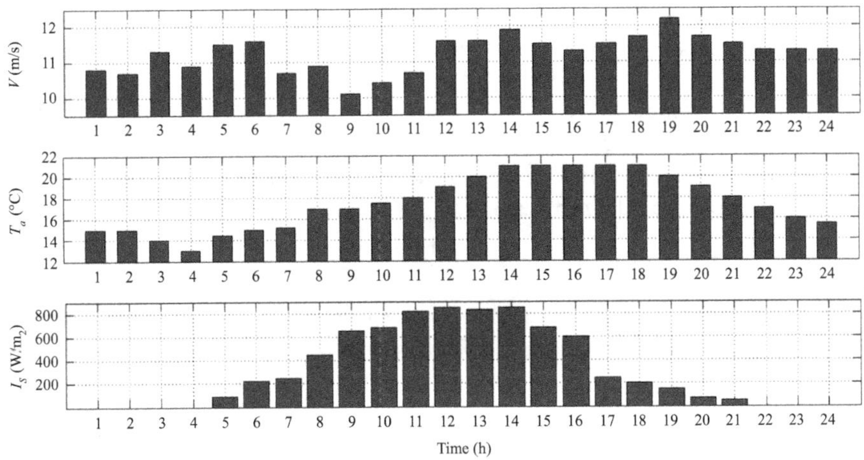

Figure 9.2 The wind speed, ambient temperature, and solar irradiance data

itself [24]. The following equation can be used to calculate the power output of the PV module P_{PV} [26]:

$$P_{PV} = P_{STC} \frac{Is}{1{,}000} [1 + \gamma(T_c - 25)] \tag{9.31}$$

where P_{STC} is the PV module maximum power at standard test condition (STC) (W); I_S is the solar irradiance on the PV module surface (W/m^2); γ is the PV module temperature coefficient for power (°C^{-1}); T_c is the PV cell (module) temperature (°C).

The PV module temperature can be calculated as a function of solar irradiance and ambient temperature based on the module's nominal operating cell temperature (NOCT). The NOCT model equation is [27]

$$T_c = T_a + \frac{Is}{800} \cdot (T_{NOCT} - 20) \tag{9.32}$$

where T_a is the ambient temperature (°C); T_{NOCT} is the NOCT (°C), of the module.

Sunmodule SW 250 mono modules are used in the OPF simulation tests performed in Section 9.6. Their performance characteristics are $P_{STC} = 250$ (W); $\gamma = -0.0045$ (°C^{-1}); $T_{NOCT} = 46$ (°C). Figure 9.2 shows the ambient temperature and solar irradiance data used to calculate the output power generated by the PV modules.

9.3.6 Mini hydropower plants

The power output of MHP is calculated according to the following expression:

$$P_{MHE} = \rho g Q H \eta_T \eta_G \tag{9.33}$$

where ρ is the water density (1,000 kg/m^3); g is the gravitation constant (9.81 m/s^2); Q is the flow rate (m^3/s); H is the head (the difference in height between the inlet and outlet surfaces) (m); η_T is the water turbine efficiency; and η_G is the generator efficiency.

The turbine efficiency depends on the flow rate, $\eta_T = f(Q)$. It can be determined experimentally or by calculation based on data provided by turbine manufacturers. Fcosts for MHP are equal to zero, so they always operate with the maximum available power (according to 9.33).

9.3.7 Electric grid

The market energy costs f_{gr} in (\$/h) can be represented by quadratic function as

$$f_{gr} = a + bP_{gr} + cP_{gr}^2 \tag{9.34}$$

where P_{gr} is electrical power in (kW) from the electric grid and a, b, c are cost coefficients. In the OPF test cases considered in Section 9.6, the coefficients values was adopted as $a = 0$, $b = 0.1$ (\$/kW h), and $c = 5 \cdot 10^{-5}$ (\$/kW^2 h).

9.4 Solution methods

9.4.1 Genetic algorithm

The GA is one of the techniques, that is, one optimization procedure based on the natural evolution process imitation [28]. They belong to the methods of a directed random search of solution domain with the aim of finding a global optimum. The classical optimization methods start from a single possible initial solution, and they reach the optimum by applying the heuristic rules iteratively. The GA starts from the population, which is a group of individuals. Each individual represents a potential solution of the optimization problem. The individuals are presented in the same way, usually through a column or a string of data. The quality of each solution or individual is determined based on the fitness function values. Through a series of GA operations, a new population is obtained, and its individuals are engendered by the individuals from the previous population according to the natural evolution principles: the choice of parents, crossover, and mutation. Basic operations of the GA are:

1. Representation of individuals: All data (variables) that make an individual are written in a string. A string is composed of substrings. Each substring represents a binary encoded variable on which the process of optimization is carried out. The number of substrings, therefore the size of a string, depends on the number of variables that are optimized.
2. Initialization: Individuals with random strings are generated that set up the initial population.
3. Fitness function calculation: It is used to rate the quality of an individual, and it represents an equivalent of the function that should be optimized, that is, objective function.
4. Selection: During the selection process the individuals that will participate in the reproduction (parents) are selected. The point of the selection is to store and transfer good individuals to the next generation.
5. Crossover: The way in which coded column parts (substrings) are crossed over actually makes a GA. Crossover is an exchange process of column parts between two individuals, that is, "parents." One or two new individuals engender by the crossover, that is, a "child." The possibility of inheriting the first parent's characteristics by a child is introduced during this process.
6. Mutation: Mutation is a way to give a new piece of information to an individual. Mutation represents an accidental bit variation of an individual, generally with a constant probability for each bit within a population. The mutation probability can further vary depending on the size of the population, application, and preferences of the explorer. A fixed value which is often kept during the whole GA is used for each generation.
7. Ending conditions: The process of finding the optimal solution by the use of the GA is an iterative process which ends when a maximum number of generations is achieved or when another criterion is fulfilled, such as a minimum

offset from the best fitness value and medium fitness value of all individuals in a current population. If end conditions are fulfilled, the best individual thus obtained is the semioptimal solution in question. Otherwise, return to 3.

9.4.1.1 GA implementation to the deterministic OPF problem

General flowchart which briefly describes the application of GA in solving the OPF in distribution networks is presented in Figure 9.3. The MATLAB program realization of the GA within the *toolbox/gads* module is applied herein. The power flow program for evaluation of the objective function was based on the backward/forward power flow method.

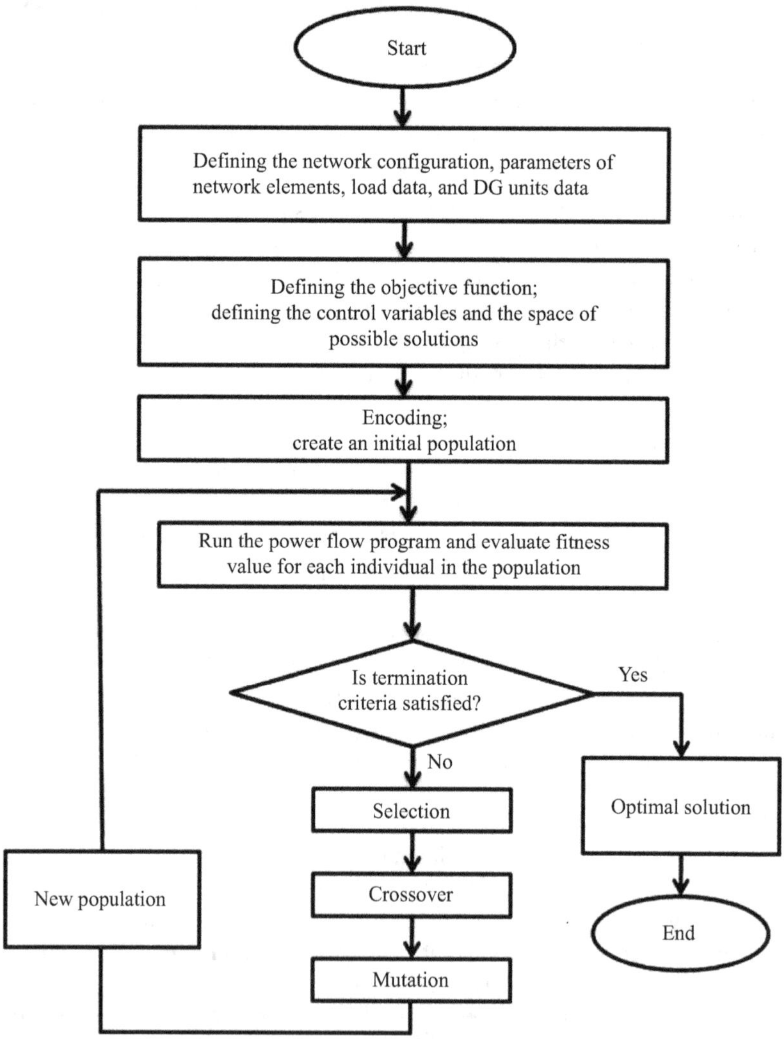

Figure 9.3 Flowchart of GA for OPF in distribution networks

9.4.2 Gravitational search algorithm

The GSA is a metaheuristic optimization algorithm where the search agents are considered as a collection of masses which interact with each other based on the Newtonian gravity and the laws of motion [29]. The position of the mass corresponds to the solution of the problem, and its gravitational and inertial masses are determined using a fitness function. In other words, each mass presents a potential solution. The algorithm is navigated by properly adjusting the gravitational and inertial masses. In a system with N agents (masses), the position of the ith agent is defined by

$$\mathbf{x}_i = \left[x_i^1, \ldots, x_i^d, \ldots, x_i^n\right] \quad i = 1, \ldots, N \tag{9.35}$$

where n is the number of control variables, and x_i^k defines the position of the ith agent in the kth dimension, i.e., the value of kth control variable in the ith potential solution.

After evaluating the current population fitness, the mass of each agent is calculated as follows:

$$M_i(t) = \frac{m_i(t)}{\sum_{j=1}^{N} m_j(t)} \tag{9.36}$$

where

$$m_i(t) = \frac{fit_i(t) - worst(t)}{best(t) - worst(t)} \tag{9.37}$$

where $fit_i(t)$ represents the fitness value of the agent i at iteration t; $best(t)$ and $worst(t)$ are the best and worst fitness of all agents, respectively.

According to Newton gravitation theory, the total force that acts on the ith agent at iteration t is specified as follows:

$$\mathbf{F}_i(t) = \sum_{j \in Kbest, j \neq i} r \cdot G(t) \frac{M_j(t) \cdot M_i(t)}{R_{i,j}(t) + \varepsilon} \left(\mathbf{x}_j(t) - \mathbf{x}_i(t)\right) \tag{9.38}$$

where r is a random number within the interval [0,1]. $G(t)$ is gravitational constant at iteration t, $M_i(t)$ and $M_j(t)$ are masses of agents i and j, ε is a small constant, and $R_{ij}(t)$ is the Euclidian distance between the two agents i and j, given by following equation:

$$R_{ij}(t) = \|\mathbf{x}_i(t), \mathbf{x}_j(t)\|_2 \tag{9.39}$$

Kbest is the set of first K agents with the best fitness value and biggest mass, which is a function of time, initialized to K_0 at the beginning and decreased with time. In such a way, at the beginning, all agents apply the force, and as time passes, *Kbest* is decreased linearly, and at the end, there will be just one agent applying force to the others.

According to the law of motion, the acceleration of the ith agent, at iteration t is given by the following equation:

$$\mathbf{a}_i(t) = \frac{\mathbf{F}_i(t)}{M_i(t)} \tag{9.40}$$

The searching strategy on this notion can be defined to find the next velocity and next position of an agent. Next velocity of an agent is defined as a function of its current velocity added to its current acceleration. Hence, the next position and next velocity of an agent can be computed as follows:

$$\mathbf{v}_i(t+1) = r_i \cdot \mathbf{v}_i(t) + \mathbf{a}_i(t) \tag{9.41}$$

$$\mathbf{x}_i(t+1) = \mathbf{x}_i(t) + \mathbf{v}_i(t+1) \tag{9.42}$$

where r_i is a uniform random number in the interval [0,1]; $\mathbf{x}_i(t)$, $\mathbf{v}_i(t)$, and $\mathbf{a}_i(t)$ are the position, velocity, and acceleration of agent i at iteration t, respectively.

The gravitational constant $G(t)$ in (9.38) is a function of the initial value G_0 and time (iteration) t:

$$G(t) = G_0 \cdot e^{-\alpha(t/t_{\max})} \tag{9.43}$$

The parameters of maximum iteration $t_{\max}$, population size N, initial gravitational constant G_0, and constant α control the performance of GSA.

9.4.2.1 GSA implementation to the deterministic OPF problem

The described GSA approach has been applied to solve the OPF problem in the following steps:

Step 1: Load the distribution network configuration, lines data, transformers data, shunt Var compensators data, loads data, and DG units data. Use the per unit system.

Step 2: Specify the vector of control variables **x** (9.5) and their lower and upper limits (9.17–9.21); Specify the dependent variables **y** (9.6), and their lower and upper limits (9.14–9.16); Specify the objective function to be optimized F(**x**,**y**) (9.7–9.12), and the penalty factors ($\lambda_{V,QDG,S}$) to form the expanded objective function F_p(**x**,**y**) (9.22).

Step 3: Initialize GSA parameters, such as the population size (N), maximum iteration number ($t_{\max}$), initial constant K_0, initial gravitational constant G_0, and constant α.

Step 4: Generate an initial random population of N agents. The initial positions of each agent are randomly selected between minimum and maximum values of the control variables. Initialize the velocity of agents by a zero matrix of dimension $N \times n$, where N is the number of agents (population size) and n is the number of control variables.

Step 5: Run the power flow program based on backward/forward power flow method for each agent $\mathbf{x}_i(t)$ from the current population **POP**(t) and calculate the corresponding values of the objective function (fitness values).

Step 6: Calculate the gravitational constant $G(t)$ (9.43), the best and worst fitness value, and the mass of each agent $M_i(t)$ (9.36), $i = 1, \ldots N$.
Step 7: Calculate the total force in different directions using (9.38), and the acceleration of each agent using (9.40).
Step 8: Update the velocity of each agent using (9.41).
Step 9: Update the position of each agent using (9.42).
Step 10: Repeat steps 5–9 until the stop criteria is reached, that is maximum number of iterations $t_{\max}$.
Step 11: Return best solution obtained in the last iteration; stop.

9.5 Probabilistic optimal power flow

The results of deterministic OPF for a distribution network is highly dependent on the accuracy of the input variables such as the power of load and the power output of DG units. When some of the input variables are uncertain, the OPF problem becomes probabilistic. Because of the probabilistic nature of wind speed and solar irradiance, the power output of the WT and the PV units are random variables as well. Moreover, it is very hard to expect that the forecast of the load demand is exactly correct due to unexpected disturbance, forecast error, or load variation. Such randomly occurred factors would be the main source of uncertainties. Every probabilistic formulation requires statistical characterization of the input random variables and a method for evaluating statistical features of the output variables [30].

Probabilistic approach in solving the OPF problem has two key steps. The first step is the statistical characterization of the input random variables. The uncertainty of independent input variables is the source of the uncertainty of dependent variables in the system. In other words, due to the random character of input variables such as WT's power outputs and power of loads in the distribution network, the obtained OPF results such as bus voltages, power by branches, or Fcosts of DG units should also be considered as random variables. Evaluating the statistical characteristics of the OPF results is the second key step in the probabilistic OPF approach.

9.5.1 Statistical characterization of the input random variables

9.5.1.1 Wind speed modeling

The wind speed probability density function at a certain location is generally described by a Weibull distribution [31]:

$$f_v(v) = \frac{k}{C} \cdot \left(\frac{v}{C}\right)^{k-1} \cdot e^{-(v/C)^k} \tag{9.44}$$

The cumulative density function (CDF) for the Weibull distribution is

$$F_v(v) = 1 - \cdot e^{-(v/C)^k} \tag{9.45}$$

The CDF with its inverse has been utilized to calculate the wind speed:

$$v = C \cdot (-\ln(r))^{1/k} \tag{9.46}$$

where r are the random numbers uniformly distributed on [0,1]. Constants C and k are the scale and shape parameters of the Weibull distribution. Different methods can be used to calculate the Weibull parameters. Usually, parameters k and C are calculated, approximately, using mean wind speed v_m and standard deviation σ as follows [24]:

$$k = \left(\frac{\sigma}{v_m}\right)^{-1.086} \tag{9.47}$$

$$C = \frac{v_m}{\Gamma(1 + 1/k)} \tag{9.48}$$

where $\Gamma(x)$ is the gamma function defined as

$$\Gamma(x) = \int_0^{\infty} t^{x-1} e^{-t} dt \qquad \text{for} \qquad x > 0 \tag{9.49}$$

9.5.1.2 Solar irradiance modeling

Basically, the solar irradiance has stochastic nature. Therefore, probability density function should be adopted. The solar irradiance uncertainty may be characterized by normal probability density function [32–34]:

$$f_{I_S}(I_S) = \frac{1}{\sigma\sqrt{2\pi}} \cdot e^{-(I_S-\mu)^2/2\sigma^2} \tag{9.50}$$

CDF for the normal distribution is

$$F_{I_S}(I_S) = \frac{1}{2}\left[1 + \operatorname{erf}\left(\frac{I_S - \mu}{\sqrt{2}\sigma}\right)\right] \tag{9.51}$$

CDF with its inverse have been utilized to calculate the solar irradiance:

$$I_S = \mu + \sqrt{2} \cdot \sigma \cdot \operatorname{erf}^{-1}(2r - 1) \tag{9.52}$$

where r is a random variable having uniform distribution in [0,1]; μ is hourly mean value of the solar irradiance and σ is standard deviation of the I_S; erf and erf^{-1} are the error function and the inverse error function, respectively, defined as follows:

$$\operatorname{erf}(x) = \frac{2}{\sqrt{\pi}} \int_0^x e^{-t^2} dt; \quad \operatorname{erf}^{-1}(x) = 1 - \operatorname{erf}(x) = \frac{2}{\sqrt{\pi}} \int_x^{\infty} e^{-t^2} dt \tag{9.53}$$

Standard deviation of the hourly mean value of the solar irradiance can be predetermined for the morning period, afternoon period, and hours around the solar noon [32]. For purpose of the OPF simulation tests in Section 9.6, it was assumed that the hourly means of solar irradiance μ are equal to their forecasted values from

Figure 9.3, and values of σ are predetermined with $(\sigma/\mu) \cdot 100 = 8\%$ at morning (5–8 h) and evening hours (17–21 h), and 3% around the solar noon (9–16 h).

9.5.1.3 Probabilistic load model

The load is assumed to be a random variable (L) following the same probability density function within each hour of a given daily load diagram. The PDF of the active/reactive power load will be assumed to be under the normal distribution [35,36]:

$$f_L(L) = \frac{1}{\sigma\sqrt{2\pi}} \cdot e^{-(L-\mu)^2/2\sigma^2} \tag{9.54}$$

where μ is the mean value given as fixed load level on the daily load diagram; σ is standard deviation of L.

Normal CDF (9.55) with its inverse (9.56) has been used to generate the load value:

$$F_L(L) = \frac{1}{2}\left[1 + \text{erf}\left(\frac{L-\mu}{\sqrt{2}\sigma}\right)\right] \tag{9.55}$$

$$L = \mu + \sqrt{2} \cdot \sigma \cdot \text{erf}^{-1}(2r - 1) \tag{9.56}$$

where r is a random variable having uniform distribution in [0,1]; erf and erf^{-1} are the error function and the inverse error function, respectively.

In the simulation tests, it is assumed that the mean values of the active/reactive power load at each load bus are given in Figure 9.3 with a standard deviation of 5%.

9.5.2 *Statistical evaluation of the output variables*

Generally, the probabilistic OPF can be expressed as

$$\mathbf{Y} = F(\mathbf{X}) \tag{9.57}$$

where $\mathbf{X}$ is the vector of input random variables and $\mathbf{Y}$ is the vector of output random variables.

In the technical literature, there are several methods for evaluating the statistical features of the output variables. These methods may be classified into the three main categories [37]: Monte Carlo simulation, analytical methods, and approximate methods.

Monte Carlo simulation [36] randomly generates values for uncertain input variables, and these values are taken into account to solve a deterministic problem. The simulation is repeated many times (several hundred to several thousand) in order to determine with sufficient accuracy the statistical characteristics of the results (i.e., $\mathbf{Y}$). The main drawback of the Monte Carlo method is the great number of simulations required to attain convergence.

Analytical methods [38,39] are based on certain mathematical simplifications of the problem in order to obtain analytically the statistical characteristics of random output variable $\mathbf{Y}$ based on the statistical characteristics of random input variables $\mathbf{X}$.

Approximate methods [30,37,40] provide an approximate description of the statistical properties of output random variables. One of these methods has been explained and applied in this chapter. It is the point estimate method. This method, like Monte Carlo, uses a deterministic approach to solving a probabilistic problem, but with a drastically lower number of simulations.

The point estimate method can guarantee a great reduction of the computational efforts compared to the Monte Carlo simulation procedure [37]. This method concentrates the statistical information provided by the first few central moments of the m input random variables on $K = 2$, 3, or 5 points for each variable, named concentrations. By using these points and function F, which relates input and output variables, statistical moments of the output variables can be obtained. To obtain these moments, function F has to be calculated $2m$, $2m + 1$, or $4m + 1$ times depending on the adopted scheme. In this work, $2m + 1$ scheme was used.

The $2m + 1$ scheme is more accurate than the $2m$ scheme and performances of the $2m + 1$ and $4m + 1$ schemes are practically the same. The $2m+1$ scheme requires $2m + 1$ evaluations of function F and uses only a two-point concentration for each input random variable [30].

9.5.3 Procedure for solving probabilistic OPF

Procedure for computing the moments of the output variables for the probabilistic OPF problem can be summarized by the following steps:

1. Determine number m of input random variables.
2. Set the vector of the jth moment of the output variable equal to zero: $E(\mathbf{Y}^j) = 0$
3. Set $t = 1$, $(t = 1,2,\ldots,m)$
4. Determine two standard locations:

$$\xi_{t,i} = \frac{\lambda_{t,3}}{2} + (-1)^{3-i} \cdot \sqrt{\lambda_{t,4} - \frac{3}{4}\lambda_{t,3}^2} \quad i = 1,2 \tag{9.58}$$

   where $\lambda_{t,3}$ is the skewness and $\lambda_{t,4}$ is the kurtosis of the input random variable x_t.

5. Determine two locations $x_{t,i}$:

$$x_{t,i} = \mu_{x_t} + \xi_{t,i} \cdot \sigma_{x_t} \quad i = 1,2 \tag{9.59}$$

   where μ_{x_t} and σ_{x_t} are the mean and standard deviation of x_t, respectively.

6. Run *the deterministic OPF algorithm* for both locations $x_{t,i}$ using the two input variable vectors:

$$\mathbf{X}_i = \left[\mu_{x1}, \mu_{x2}, \ldots, x_{t,i}, \ldots, \mu_{xm}\right] \quad i = 1,2 \tag{9.60}$$

   where μ_{xk} ($k = 1,2,\ldots,m$; and $k \neq t$) is the mean value of the remaining random input variables.

7. Determine the weight factors:

$$w_{t,i} = \frac{(-1)^{3-i}}{\xi_{t,i}(\xi_{t,1} - \xi_{t,2})} \quad i = 1,2 \tag{9.61}$$

8. Update $E(\mathbf{Y}^j)$:

$$E(\mathbf{Y}^j) = E(\mathbf{Y}^j) + \sum_{i=1}^{2} w_{t,i}[F(\mathbf{X}_i)]^j \tag{9.62}$$

9. Repeat steps 4–8 for $t = t + 1$ until the list of random input variables is exhausted.
10. Run *the deterministic OPF algorithm* using as input variable vector:

$$\mathbf{X}_\mu = [\mu_{x1}, \mu_{x2}, \ldots, \mu_{x,t}, \ldots, \mu_{xm}] \tag{9.63}$$

11. Determine the weight factor of OPF solution of step 10:

$$w_0 = 1 - \sum_{t=1}^{m} \frac{1}{\lambda_{t,4} - \lambda_{t,3}^2} \tag{9.64}$$

12. Update $E(\mathbf{Y}^j)$:

$$\begin{aligned} E(\mathbf{Y}^j) &= E(\mathbf{Y}^j) + w_0 [F(\mathbf{X}_\mu)]^j \\ E(\mathbf{Y}^j) &= \sum_{t=1}^{m}\sum_{i=1}^{2} w_{t,i}\left[F\left(\mu_{x_1}, \mu_{x_1}, \ldots, x_{t,i}, \ldots, \mu_{x_m}\right)\right]^j + w_0 [F(\mathbf{X}_\mu)]^j \end{aligned} \tag{9.65}$$

Knowing statistical moments of the output random variable, the mean, and standard deviation can be computed:

$$\mu_Y = E(\mathbf{Y}) \tag{9.66}$$

$$\sigma_Y = \sqrt{E(\mathbf{Y}^2) - \mu_Y^2} \tag{9.67}$$

Based on statistical moments, it is possible to approximate the PDFs of the output random variables of interest using the Gram–Charlier series approach. For example, let us consider Y as any random variable with a continuous distribution, whose mean value is μ_Y and standard deviation σ_Y. For the standardized variable $x = (Y - \mu_Y)/\sigma_Y$, its cumulative function and density function are denoted as $F(x)$ and $f(x)$, respectively. According to Gram–Charlier expansion, the cumulative and the density functions can be written as [38]

$$F(x) = \Phi(x) + \frac{c_1}{1!}\Phi'(x) + \frac{c_2}{2!}\Phi''(x) + \frac{c_3}{3!}\Phi'''(x) + \cdots \tag{9.68}$$

$$f(x) = \phi(x) + \frac{c_1}{1!}\phi'(x) + \frac{c_2}{2!}\phi''(x) + \frac{c_3}{3!}\phi'''(x) + \cdots \tag{9.69}$$

where $\Phi(x)$ and $\phi(x)$ represent the CDF and PDF of normal distribution with $\mu = 0$ and $\sigma = 1$. The constant coefficients c_k are calculated as follows:

$$c_1 = c_2 = 0 \tag{9.70}$$

$$c_3 = -\frac{E(Y^3) - 3\mu_Y E(Y^2) + 2\mu_Y^3}{\sigma_Y^3} \tag{9.71}$$

$$c_4 = \frac{E(Y^4) - 3[E(Y^2)]^2 - 4\mu_Y E(Y^3) + 12\mu_Y^2 E(Y^2) - 6\mu_Y^4 + 3\sigma_Y^4}{\sigma_Y^4} - 3 \tag{9.72}$$

Note: $E(Y^1), E(Y^2), E(Y^3), E(Y^4), \ldots$ are the *j*th statistical moments of the output random variable *Y*, $E(Y^j), j = 1,2,3,4,\ldots$.

9.6 Simulation results

The proposed OPF approach was tested on a modified IEEE 34-bus test feeder shown in Figure 9.4. The original IEEE 34-bus test feeder [41] is modified using the following assumptions [42]:

- Only the three-phase sections of the original test feeder are included. The single-phase lateral sections were removed and their loads being connected at their respected sending end nodes. The three-phase sections are assumed to be transposed and therefore, considered as only positive sequence impedances. The shunt admittances of the sections are neglected.
- The unbalanced phase loads in each three-phase sections are summed up and taken as total three-phase balanced loads. The loads on laterals have already been

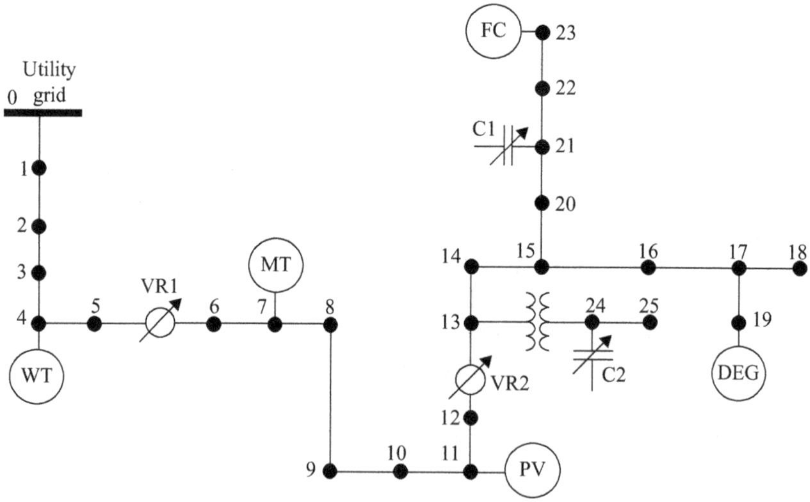

Figure 9.4 Single-line diagram of the modified IEEE 34-bus test system

"absorbed" into the corresponding three-phase "root" sections. The distributed loads in all sections are represented as spot loads. The one-half of the total distributed load is placed at the beginning of the section, while the remaining half of the total distributed load is placed at the end of the section. The branch impedances and the nominal active and reactive power of load are given in Table 9.2. It has been assumed that the active and reactive power of load at each bus varies based on the daily load diagram shown in Figure 9.5 relative to nominal values.

- Five different DG units were introduced in the modified test feeder. It is adopted that the DG units connected at buses 4 (WT), 7 (MT), 11 (PV), and 19 (DEG) operate in PQ mode with 0.9 power factor (produce reactive power). The DG unit at bus 23 (FC) is capable to control active power and voltage independently and therefore, operate in PV mode. The DG units data are listed in Table 9.3. The Fcost coefficients (a,b,c) are determined according to the Fcost and output power characteristics of the DG units given in Section 9.4. The power outputs of WT and PV are calculated in accordance with the forecasted values of wind speed, ambient temperature, and solar irradiance given in Figure 9.3.
- It is assumed, there are two shunt Var compensators connected at nodes 21 and 24.

Table 9.2 Line data and load data of the modified IEEE 34-bus test feeder

Send. bus	Rec. bus	Section parameters		Load at receiving bus		
		R (Ω)	X (Ω)	P_L (kW)	Q_L (kVAr)	Load model
0	1	0.5473	0.4072	27.5	14.5	PQ
1	2	0.3670	0.2730	27.5	14.5	PQ
2	3	6.8373	5.0864	16	8	I
3	4	7.9553	5.9181	0	0	PQ
4	5	6.3069	4.6919	0	0	PQ
5	6	0.0032	0.0016	0	0	PQ
6	7	0.0992	0.0494	171.5	88	PQ
7	8	3.2681	1.6265	44.5	22	I
8	9	0.2689	0.1338	5.5	2.5	PQ
9	10	6.5426	3.2562	48.5	21.5	Z
10	11	0.1664	0.0828	4	2	PQ
11	12	11.7889	5.8672	0	0	PQ
12	13	0.0032	0.0016	7.5	3.5	Z
13	14	1.5684	0.7806	25.5	13	PQ
14	15	1.8661	0.9288	89	45	Z
15	16	0.6466	0.3218	174	106	PQ
16	17	0.8578	0.4269	61	31.5	PQ
17	18	0.2753	0.1370	47	31	I
17	19	0.0896	0.0446	28	14	PQ
15	20	0.0896	0.0446	4.5	2.5	PQ
20	21	0.4321	0.2151	432	329	Z
21	22	1.1651	0.5799	34	17	PQ
22	23	0.1696	0.0844	71.5	53.5	PQ
13	24	11.7800	25.2864	0	0	PQ
24	25	80.2600	59.7059	450	225	I

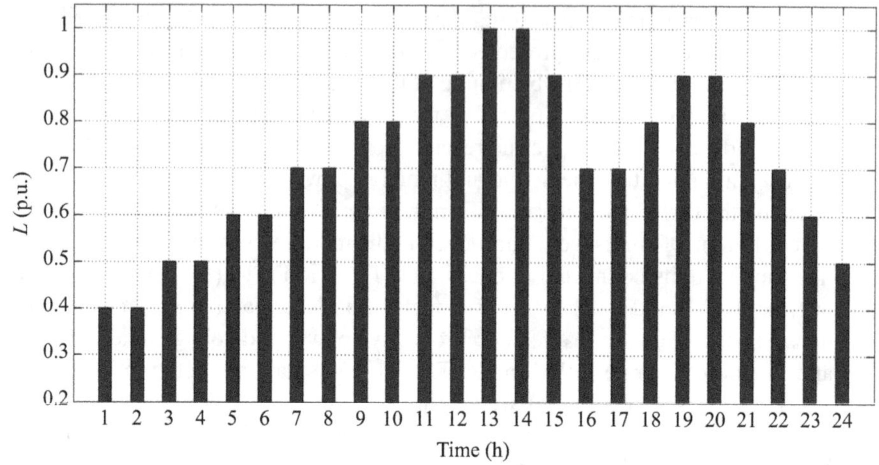

Figure 9.5 Daily load diagram

Table 9.3 DG unit's data

Location	Type	Mode	P_{DGmin} (kW)	P_{DGnom} (kW)	Q_{DG} (kVAr)	Cost coefficients		
						a (\$/h)	b (\$/kW h)	c (\$/kW^2 h)
4	WT	PQ	0	600	$\cos\varphi = 0.9$	0	0	0
7	MT	PQ	90	300	$\cos\varphi = 0.9$	5.3	$88 \cdot 10^{-3}$	$61 \cdot 10^{-6}$
11	PV	PQ	0	250	$\cos\varphi = 0.9$	0	0	0
19	DEG	PQ	0	369	$\cos\varphi = 0.9$	5.8	$124 \cdot 10^{-3}$	$65 \cdot 10^{-6}$
23	FC	PV	0	300	−200 to 225	0.57	$62 \cdot 10^{-3}$	$190 \cdot 10^{-6}$
0	Electric grid	Slack	–	–	–	0	$100 \cdot 10^{-3}$	$50 \cdot 10^{-6}$

There are a total of nine control variables for OPF problem as follows: three DG unit active power outputs, one root bus voltage magnitude, one PV bus voltage magnitude, two voltage regulation tap settings, and two shunt Var compensator reactive power outputs. The lower and upper limits for the control variables are given in Table 9.4. The lower and upper limits of the load bus voltages, including the DG units which are modeled as PQ buses, are 0.95 and 1.05 p.u., respectively.

9.6.1 Deterministic OPF analysis

In the deterministic OPF, the output powers of WT and PV units as well as the load level are equal to their specified (forecasted) values. Three optimization cases were considered.

Table 9.4 The limits of control variables

Control variables	Min	Max
P_{MT} (kW)	0	300
P_{FC} (kW)	0	300
P_{DEG} (kW)	0	369
V_0 (p.u.)	0.97	1.05
V_{23} (p.u.)	0.98	1.05
t_{VR1} (p.u.)	0.90	1.10
t_{VR2} (p.u.)	0.90	1.10
Q_{C1} (kVAr)	0	300
Q_{C2} (kVAr)	0	300

Table 9.5 Optimal settings of control variables for max load level (time = 14 h)

	GA			GSA		
	Case 1	**Case 2**	**Case 3**	**Case 1**	**Case 2**	**Case 3**
P_{MT} (kW)	300.0	300.0	292.3	300	300	300
P_{FC} (kW)	261.2	291.8	220.7	253.9	300	298.2
P_{DEG} (kW)	297.1	369.0	367.3	305.2	369	369
V_0 (p.u.)	0.9995	1.0500	1.0113	1.0017	1.0500	1.0027
V_{23} (p.u.)	1.0498	1.0490	0.9891	1.0328	1.0301	1.0297
t_{VR1} (p.u.)	0.9444	0.9888	0.9650	0.9793	0.9889	0.9870
t_{VR2} (p.u.)	0.9667	0.9865	0.9528	0.9469	0.9927	0.9462
Q_{C1} (kVAr)	148.4	115.1	296.6	270.2	191.5	113.6
Q_{C2} (kVAr)	215.4	300.0	267.7	300	227	300
Fcost (\$/h)	**178.36**	179.75	179.15	**177.18**	179.19	178.47
Ploss (kW)	76.36	**64.35**	79.12	77.95	**62.80**	65.47
VD (p.u.)	0.7252	0.9802	0.6185	0.5183	0.9416	0.4719

Case 1: Fcost minimization for DG units;
Case 2: simultaneous minimization of the Fcost and Plosses;
Case 3: simultaneous minimization of the Fcost, Plosses, and VD.

Table 9.5 shows the optimal settings of the control variables and appropriate values of the Fcost, Plosses, and VD for the maximum load level (T = 14 h) obtained by using GA and GSA. The bold values indicate the optimum value of the Fcost, Plosses, and VD in accordance with the optimization cases. Figure 9.6 shows voltage profiles of the above cases. As it could be seen, values of voltage at all buses are within the allowable limits.

The optimal settings of the control variables for other load levels are determined too and presented in Figures 9.7–9.10. Daily values of the Fcost, energy loss, and max VD in accordance with these optimal settings of control variables are shown in Figure 9.11. Minimization of the Fcost in Case 1 causes the maximum

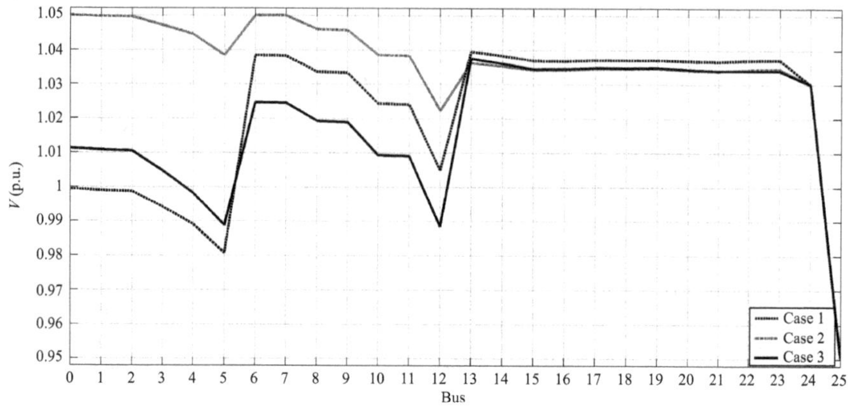

Figure 9.6 Voltage profile for the maximum load level (T = 14 h)

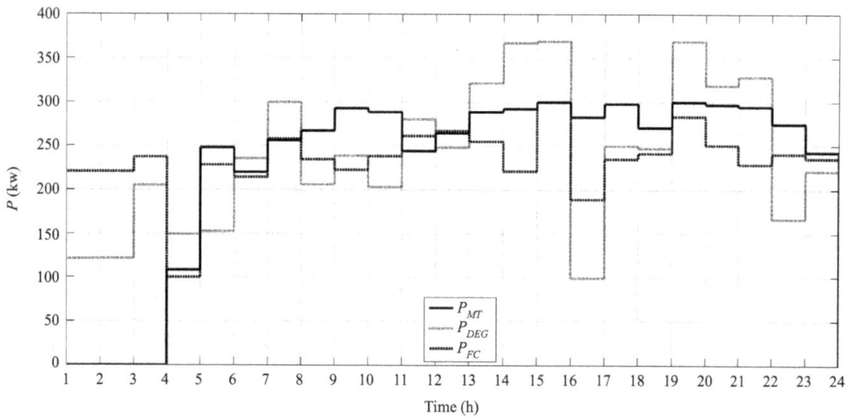

Figure 9.7 Optimal power outputs of DG units

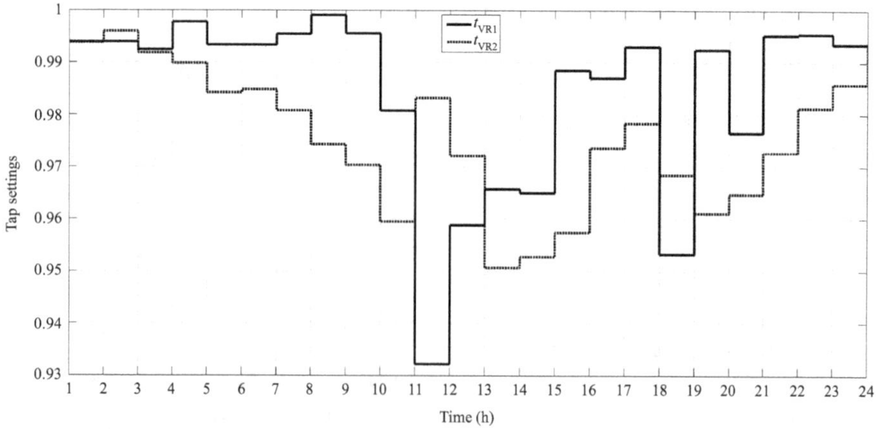

Figure 9.8 Optimal tap settings of voltage regulation transformers

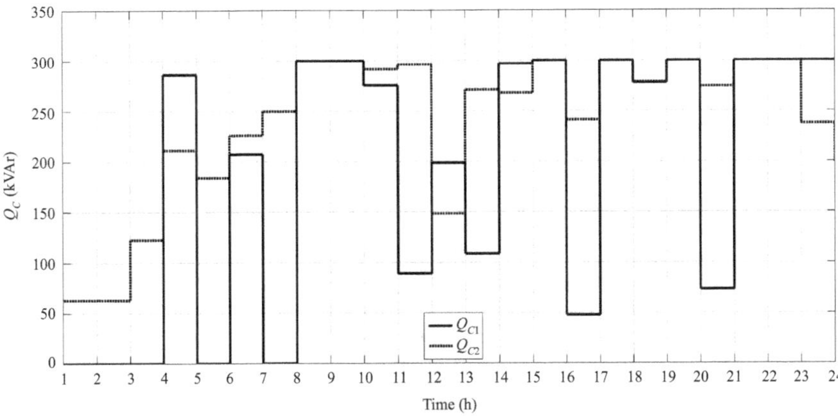

Figure 9.9 Optimal reactive powers of shunt Var compensators

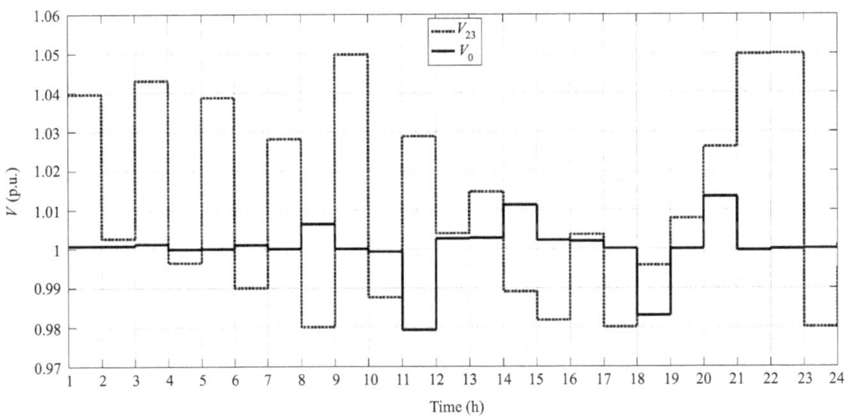

Figure 9.10 Optimal settings of voltage magnitude at the root bus and PV bus

energy loss in relation to other cases. In Case 2, the minimum energy loss is obtained, but the Fcost and VD is increased. The results obtained in Case 3 can be adopted as a compromise solution of the OPF problem.

9.6.2 Probabilistic OPF analysis

In this section, the WT and PV units power generation as well as the load level are adopted as uncorrelated input random variables. Using proper PDFs modeling for the hourly wind speed and solar irradiance given in Sections 9.5.1.1 and 9.5.1.2, the output power of WT and PV units are evaluated. All loads are fully correlated and follow the same PDF given in Section 9.5.1.3.

The $2m + 1$ point estimate method is implemented along with the GA to find the optimal solution of the probabilistic OPF. Table 9.6 reports the mean and

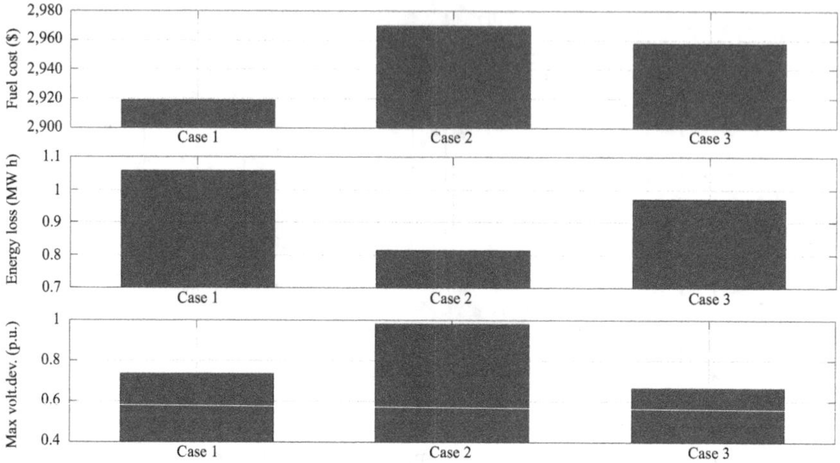

Figure 9.11 *Daily values of energy costs, energy losses, and maximum voltage deviation*

Table 9.6 *Mean and standard deviation results of the probabilistic OPF*

		Time = 14 h		Time = 8 h		Time = 2 h	
		Case 1	**Case 4**	**Case 1**	**Case 4**	**Case 1**	**Case 4**
P_{MT} (kW)	μ	297.1	289.3	271.0	244.7	58.7	38.7
	σ	4.2	8.8	11.6	60.5	96.1	48.8
P_{FC} (kW)	μ	250.4	240.0	209.3	184.6	199.7	183.9
	σ	19.9	17.8	20.5	47.8	16.0	20.7
P_{DG} (kW)	μ	318.9	317.7	159.5	145.6	49.2	101.5
	σ	32.7	48.3	51.0	71.3	26.9	52.1
V_{800} (p.u.)	μ	1.0083	1.0035	1.0050	0.9960	0.9938	0.9997
	σ	0.0260	0.0226	0.0180	0.0118	0.0132	0.0032
V_{848} (p.u.)	μ	1.0186	1.0073	1.0068	0.9980	0.9970	1.0046
	σ	0.0166	0.0143	0.0163	0.0142	0.0036	0.0090
t_{VR1} (p.u.)	μ	0.9645	0.9621	0.9945	0.9675	1.0092	0.9975
	σ	0.0273	0.0238	0.0277	0.0143	0.0111	0.0050
t_{VR2} (p.u.)	μ	0.9554	0.9548	0.9698	0.9779	0.9823	0.9952
	σ	0.0152	0.0119	0.0190	0.0232	0.0116	0.0030
Q_{C1} (kVAr)	μ	197.8	231.9	216.1	196.2	247.5	178.8
	σ	81.4	20.3	33.2	30.8	53.1	70.3
Q_{C2} (kVAr)	μ	236.8	254.3	209.8	209.9	197.3	181.6
	σ	66.2	33.1	76.4	39.3	62.3	23.2
Fcost ($/h)	μ	178.9161	179.1978	111.9691	115.0566	55.4705	59.2724
	σ	16.5964	17.1368	9.7492	10.6029	5.5932	5.5934
Ploss (kW)	μ	77.7	79.2	38.1	45.4	13.6	17.5
	σ	9.4	10.3	4.1	12.7	2.6	4.2
VD (p.u.)	μ	0.7302	0.6936	0.3603	0.3280	0.5550	0.0576
	σ	0.1402	0.1380	0.0576	0.1148	0.0523	0.0075

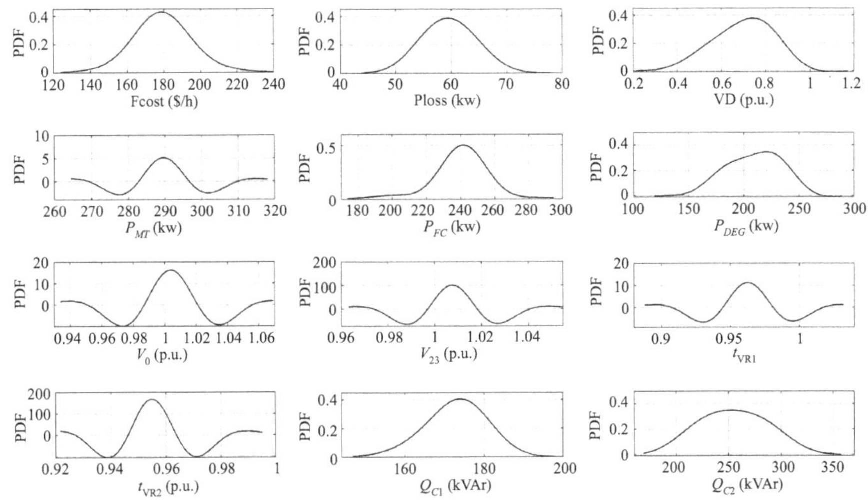

Figure 9.12 PDF functions of the probabilistic OPF results for Case 3 in time = 14 h

Table 9.7 Mean and standard deviation results for different combination of the input random variables

		Case 4, time = 14 h				
		Input random variables				
		P_{WT}, P_{PV}, L	P_{WT}, P_{PV}	P_{wt}	P_{pv}	L
Fcost ($/h)	μ	179.1978	178.8243	178.8353	178.8601	179.2445
	σ	17.1368	4.8933	4.6168	1.6218	16.4223
Ploss (kW)	μ	79.2	76.4	72.7	75.0	74.0
	σ	10.3	5.0	2.8	5.2	10.4
VD (p.u.)	μ	0.6936	0.7144	0.7351	0.6956	0.6955
	σ	0.1380	0.0672	0.0458	0.0406	0.1208

standard deviation of the OPF results. Optimization Case 1 and Case 3 were considered for time = 14, 8, and 2 h. Figure 9.12 shows the PDFs of the results for optimization Case 3 and time = 14 h. To obtain these PDFs, the Gram–Charlier expansion was used.

In order to determine the impact of individual input random variables on the statistical characteristics of the results, the probabilistic OPF was performed for five different combinations of the input random variables. Results are given in Table 9.7. From these results, it is clear that input random variable L has a higher impact on the standard deviation of the output variables than input random variables P_{WT} and P_{PV}. The reason for this is that the total power load in the distribution network is much higher than the power outputs of WT and PV units.

9.7 Solution software—*opfdngui*

Computer program *opfdngui* is an educational software tool for solving the OPF problems in distribution networks using metaheuristic optimization methods. The *opfdngui* has been developed to present the efficiency of different metaheuristic optimization methods in solving the OPF problem for distribution networks with DG units. The program can be downloaded from [43].

Figure 9.13 shows a general flowchart for solving the OPF problem using metaheuristic optimization methods.

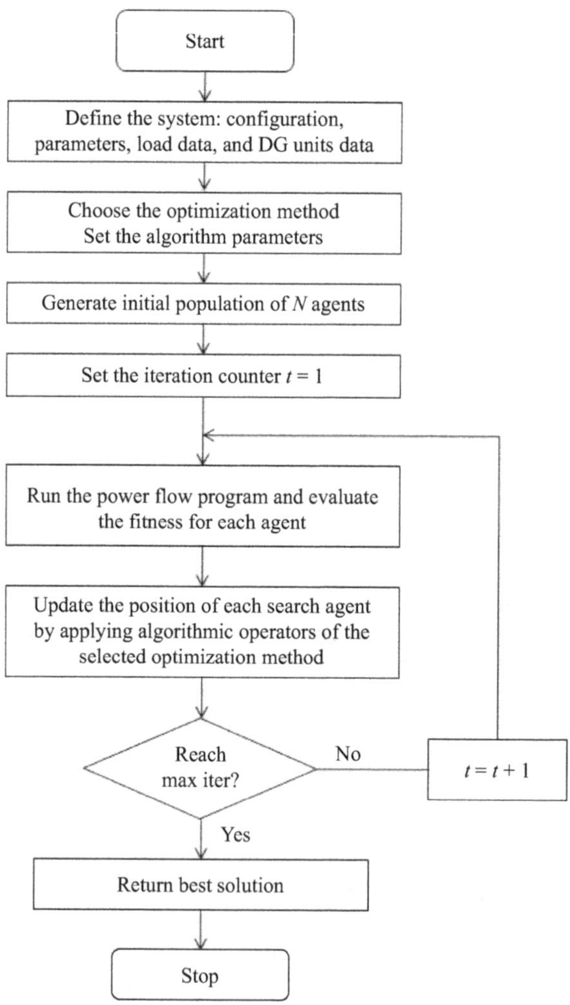

Figure 9.13 OPF using metaheuristic optimization

The aim of this program is to encompass the main steps in solving the OPF problem using metaheuristic methods. These steps include the following: (i) selection of test system, display single-line diagram, and edit system data; (ii) selection of objective function; (iii) selection of solution method, setting the algorithm parameters; (iv) program execution; and (v) display of the results.

The *opfdngui* was developed in MATLAB (The MathWorks, Natick, Massachusetts, USA), because it integrates computation, programing, analyze data, and producing graphical displays and graphical user interfaces in an easy-to-use environment where problems and solutions are expressed in familiar mathematical notation. When designing the program, special care was paid to its graphical user interface, so the *opfdngui* is very friendly to the user.

The *opfdngui* program offers a choice of five test systems, six objective functions, and twelve optimization methods. The program generates not only optimal solution, that is, optimum control variables and objective function, but also important results such as, convergence profile, bus voltages and bus powers, brunch power flows and losses, violating constraints (if exist), and statistical evaluation of the results. Using *opfdngui*, the user can compare the performances of different optimization methods based on statistical evaluation of the results.

Figure 9.14 shows the main window of *opfdngui* while it operates on a Windows-based computer. Shown is an example of the execution of the program on the modified IEEE 34-bus test system. Simultaneous minimization of the Fcost and Ploss (Fcost_Ploss) is selected as the objective function, and GSA is selected as the solution method. The OPF was performed for the time = 10 h which determines load level, wind speed, ambient temperature, and solar irradiance on the daily diagrams of forecasted values.

As can be seen from Figure 9.14, the *opfdngui* is organized in following components:

1. Panel TEST SYSTEM. It consists of one pop-up menu and two buttons. Pop-up menu enables to selection of a test system such as the IEEE 31-bus, the IEEE 33-bus, the modified IEEE 34-bus, and the IEEE 69-bus distribution networks. When clicking on the button Single-Line Diagram, the selected test system will be displayed. The button System Data opens the input file of the selected test system. The input file contains branch data, bus data, generator data, transformer data, Var compensation data, as well as the data of daily load diagram and forecasted values of wind speed, ambient temperature, and solar irradiance for the selected test system. The data are given in matrix form. The user can change the data and save those modifications.
2. Panel OPTIONS. It consists of two pop-up menus. In the first pop-up menu, the user performs selection of an objective function among six offered OPF objectives, such as minimization of Fcost, minimization of real Ploss, voltage profile improvement (VD), simultaneous minimization of Fcost_Ploss, simultaneous minimization of Fcost and VD, and simultaneous minimization of Fcost, real Plosses, and VD. Pop-up menu Method enables to choose a metaheuristic optimization method to solve the OPF problem. The user can select

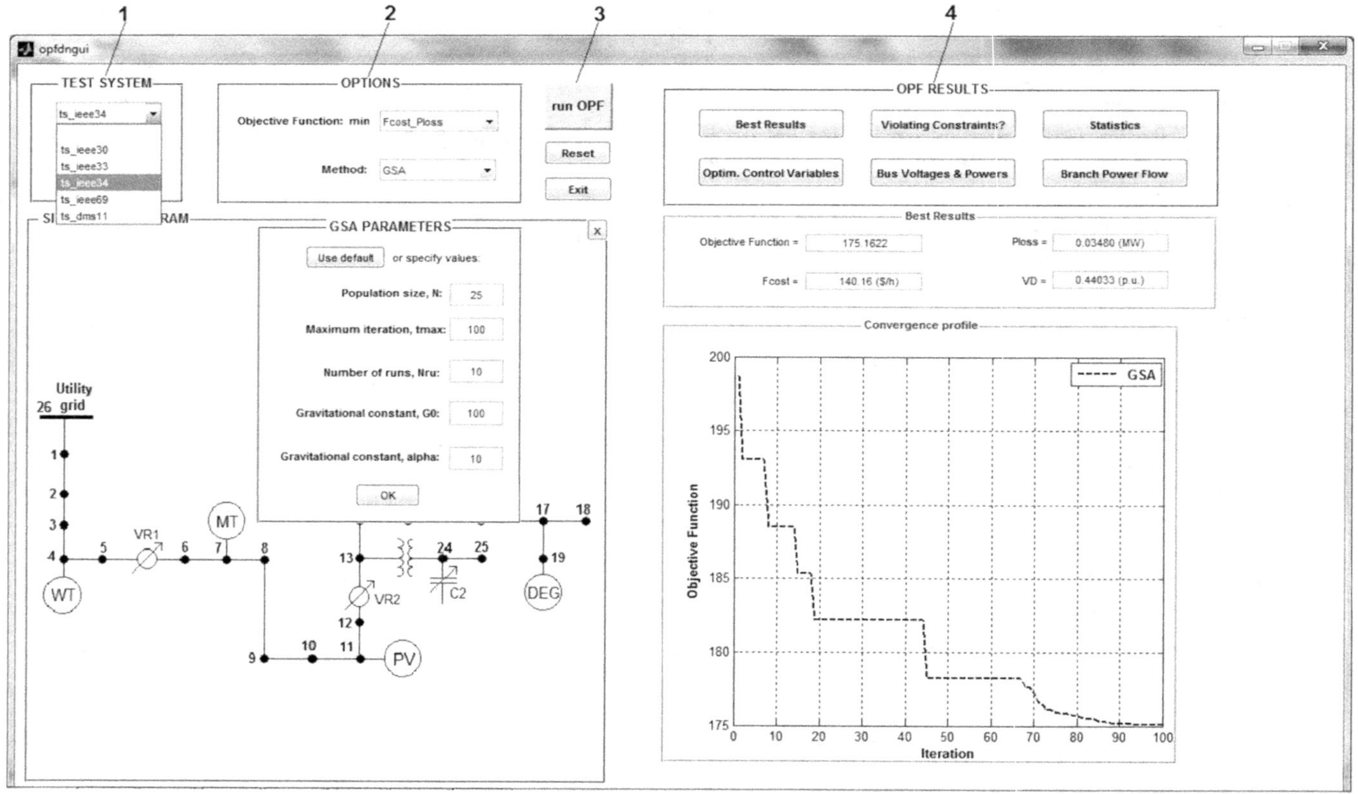

Figure 9.14 Main window of the opfdngui

one of 12 methods, such as particle swarm optimization (PSO), gravitational search algorithm (GSA), hybrid PSO and GSA (PSOGSA), artificial bee colony algorithm (ABC), wind driven optimization (WDO), grey wolf optimizer (GWO), teaching learning based optimization (TLBO), etc. When clicking on the optimization method, the panel to settings of the algorithm parameters opens. The user can used the default parameters by clicking on button Use default, or to enter the desired value for each of algorithm parameters.

3. After selection of appropriate options on the panels TEST SYSTEM and OPTIONS, the user can start the program by clicking the button run OPF.
4. Panel OPF RESULTS. It manages the access to the different segments of the OPF results obtained from the program. The best results, that is, objective function values in the last iteration and convergence profile of the optimization algorithm, will be displayed immediately after execution of the program (Figure 9.14). Besides, the program generates other important results, also. When clicking the button Optimum Control Variables the obtained optimal values of control variables such as generator active power outputs, generator voltages, transformer tap settings, and shunt Var compensations will be displayed, as shown in Figure 9.15.

The button Bus Voltages & Powers enables to display bus voltages and bus powers in the system under optimum control variables (Figure 9.16).

The button Branch Power Flow enables to display branch power flows and losses in the system under optimum control variables (Figure 9.17).

The button Violating Constraints is used to check if operating constraints are violated (Figure 9.18), as well as which operating constraints are violated.

Because of the stochastic nature of metaheuristic optimization methods, a very important segment of the OPF program is statistical evaluation of the results. When the user clicks on the button Statistics, a statistical evaluation of the optimal results which are obtained using the selected optimization method in specified number of program execution will be displayed (Figure 9.19). These statistical indicators are number of program execution, the minimum value of the objective function, the maximum value of the objective function, the mean value of the objective function, the standard deviation of the objective function, and the mean time of execution of the program. On the basis of these statistical indicators, a relevant comparison of different metaheuristic optimization methods can be achieved.

9.8 Conclusion

OPFs in distribution networks are achieved by the proper coordination of the operation of DG units, voltage regulators, shunt Var compensators, and other regulation resources. This formulation includes the DG units with renewable and nonrenewable energy sources. The OPF is very important tool for achieving economic, secure, and reliable operation of distribution networks.

To achieve a more realistic approach to the problem of optimizing distribution networks, many random disturbances or uncertain factors, such as the variation of load, the change in network configuration, and especially the stochastic nature of the

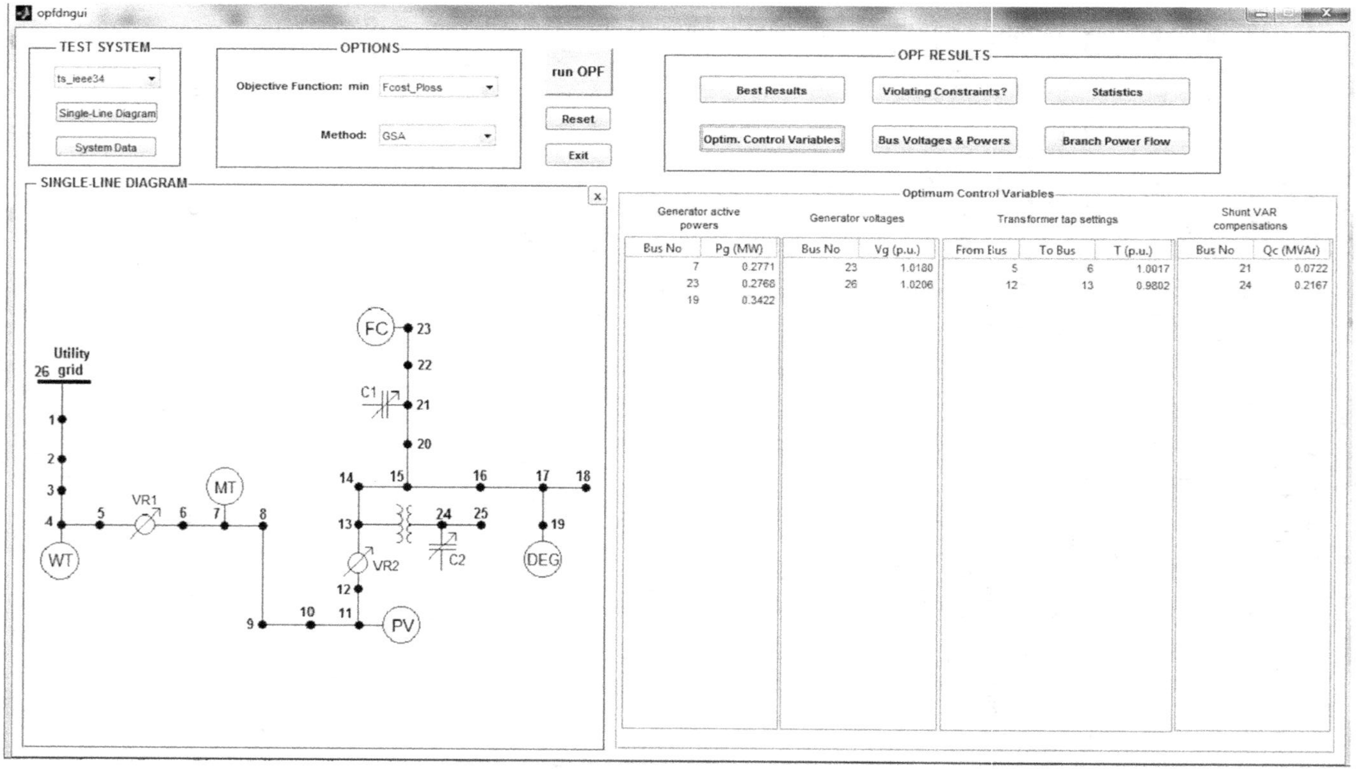

Figure 9.15 Display of the results: optimum control variables

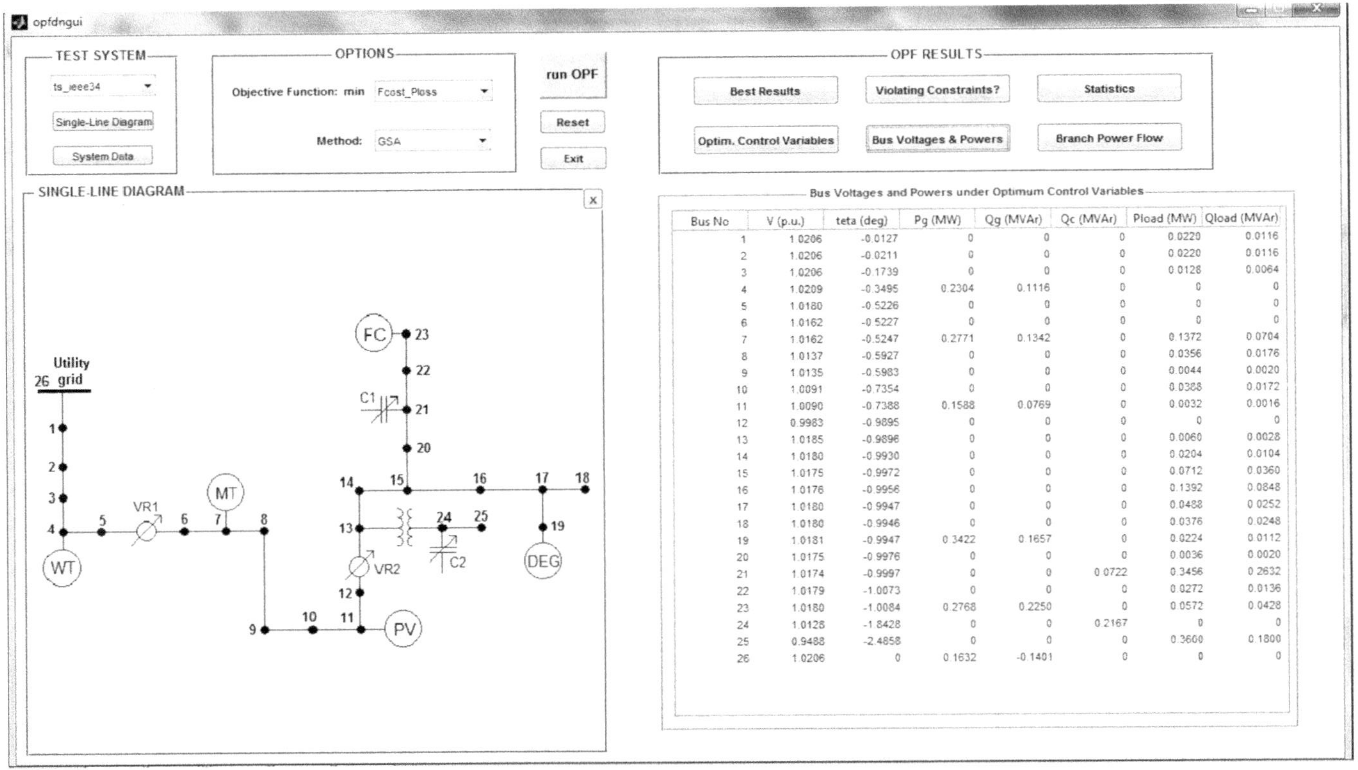

Figure 9.16 Display of the results: bus voltages and powers

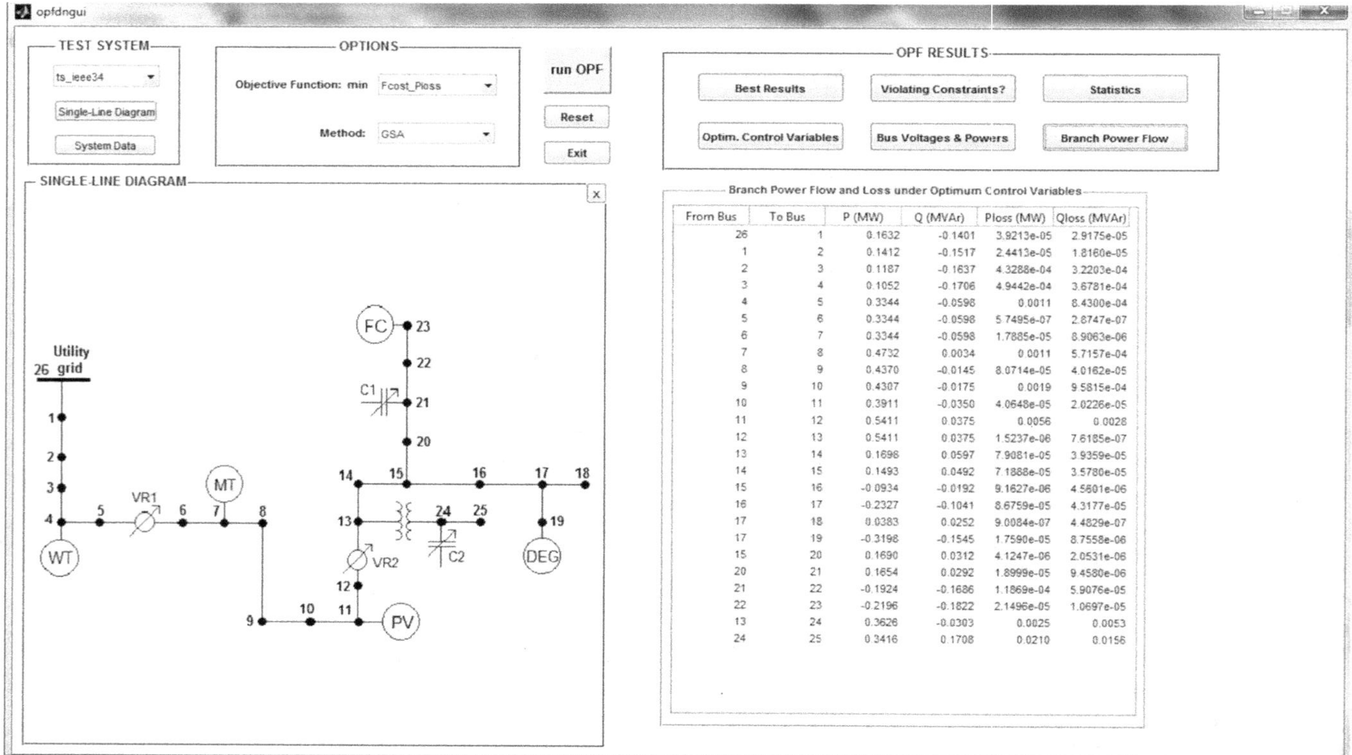

Figure 9.17   Display of the results: branch power flow and loss

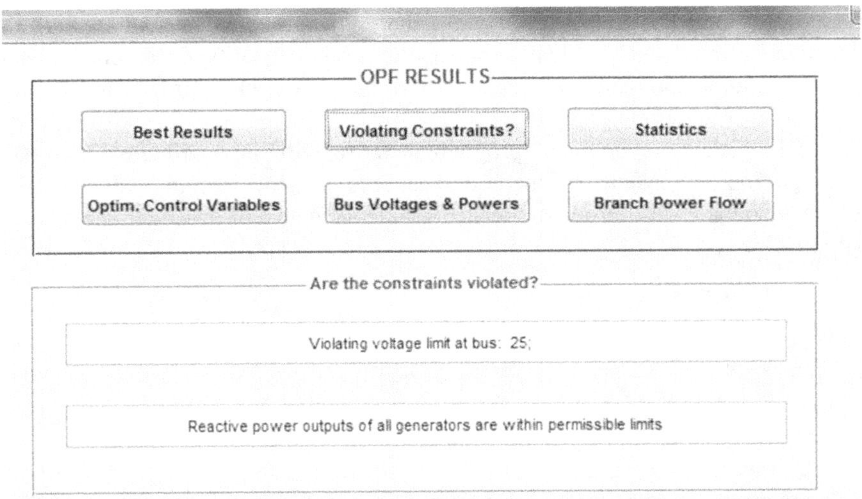

Figure 9.18 Display of the results: checking of the operating constraints

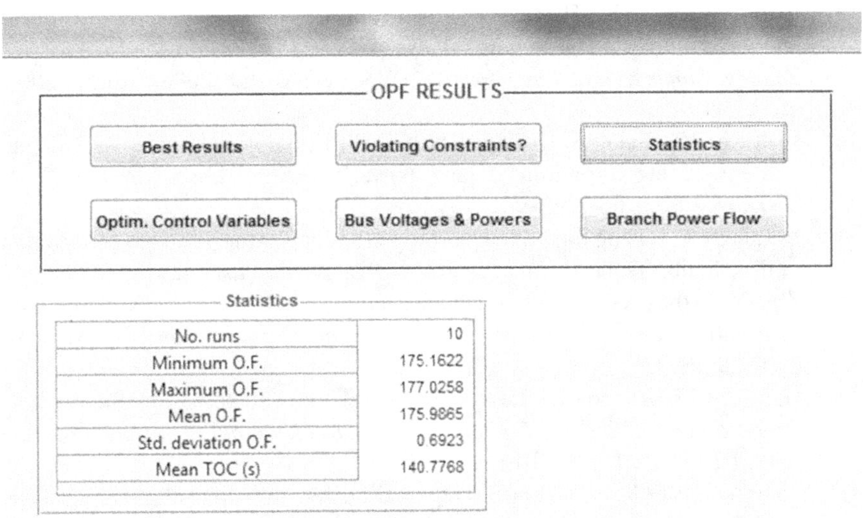

Figure 9.19 Statistical evaluations of the results

output power of the DG with renewable energy such as WT and PV units, must be taken in consideration. This implies transformation of the deterministic OPF in a probabilistic OPF approach. The OPF solution can serve as a useful decision-making supporting tool for distribution network operators and help to find out how the different uncertainties in input random variables affect the statistical indicators of the depending variables which defines operating state of the distribution network.

From a mathematical point of view, the OPF is a complex nonlinear multi-objective optimization problem with equality and inequality constraints. Metaheuristic optimization methods, such as GA and GSA, have proved to be very effective in solving this problem.

Additionally, the *opfdngui* program to solve OPF in distribution networks using different metaheuristic methods has been presented in this chapter. The software can be useful for researchers and students in area of distribution networks optimization.

References

[1] Safigianni A.S., Koutroumpezis G.N., Poulios V.C. 'Mixed distributed generation technologies in a medium voltage network'. *Electric Power Systems Research*. 2013;**96**:75–80

[2] Moghaddas-Tafreshi S.M., Mashhour E. 'Distributed generation modeling for power flow studies and a three-phase unbalanced power flow solution for radial distribution systems considering distributed generation'. *Electric Power Systems Research*. 2009;**79**:680–6

[3] Li S., Tomsovic K., Hiyama T. 'Load following functions using distributed energy resources'. *Proceedings of the IEEE/PES 2000 Power Engineering Society Summer Meeting*; Seattle, WA, USA, Jul 2000. IEEE; 2002. pp. 1756–61

[4] Li X., Li Y., Zhang S. 'Analysis of probabilistic optimal power flow taking account of the variation of load power'. *IEEE Transaction on Power Systems*. 2008;**23**(3):992–9

[5] Borghetti A. 'Using mixed integer programming for the volt/var optimization in distribution feeders'. *Electric Power Systems Research*. 2013;**98**:39–50

[6] Zhu Y., Tomsovic K. 'Optimal distribution power flow for systems with distributed energy resources'. *Electrical Power Energy Systems*. 2007;**29**:260–7

[7] Gan L., Li N., Topcu U., Low S.H. 'Optimal power flow in tree networks'. *Proceedings of 52nd IEEE Conference on Decision and Control*; Florence, Italy, Dec 2013. IEEE; 2014. pp. 2313–8

[8] Sheng W., Liu K.Y., Cheng S. 'Optimal power flow algorithm and analysis in distribution system considering distributed generation'. *IET Generation, Transmission & Distribution*. 2014;**8**(2):261–72

[9] Radosavljević J., Jevtić M., Klimenta D., Arsić N. 'Optimal power flow for distribution networks with distributed generation'. *Serbian Journal of Electrical Engineering*. 2015;**12**(2):145–70

[10] Shu J., Guan R., Wu L. 'Optimal power flow in distribution network considering spatial electro-thermal coupling effect'. *IET Generation, Transmission & Distribution*. 2017;**11**(5):1162–9

[11] Niknam T., 'An approach based on particle swarm optimization for optimal operation of distribution network considering distributed generators'.

Proceedings of 32nd Annual Conference on IEEE Industrial Electronics; Paris, France, Nov 2006. IEEE; 2007. pp. 633–7

[12] Niknam T., Meyimand H.Z., Mojarrad H.D. 'A novel multi-objective fuzzy adaptive chaotic PSO algorithm for optimal operation management of distribution network with regard to fuel cell power plants'. *European Transactions on Electrical Power*. 2011;**21**:1954–83

[13] Niknam T., Meyimand H.Z., Nayeripour M. 'A practical algorithm for optimal operation management of distribution network including fuel cell power plants'. *Renewable Energy*. 2010;**35**:1696–714

[14] Kaprál D., Braciník P., Roch M., Höger M. 'Optimization of distribution network operation based on data from smart metering systems'. *Electrical Engineering*. 2017; doi: 10.1007/s00202-017-0628-x

[15] Ghanegaonkar S.P., Pande V.N. 'Optimal hourly scheduling of distributed generation and capacitors for minimisation of energy loss and reduction in capacitors switching operations'. *IET Generation, Transmission & Distribution*. 2017;**11**(9):2244–50

[16] Radosavljević J., Jevtić M., Arsić N., Klimenta D. 'Optimal power flow for distribution networks using gravitational search algorithm'. *Electrical Engineering*. 2014;**96**:335–45

[17] Wang S., Wang N., Wang Q., Yang Y., Huang G. 'Reactive power optimization in distribution network by using improved gravitational search algorithm'. *Proceedings of 5th international conference on Electric Utility Deregulation and Restructuring and Power Technologies*; Changsha, China, Nov 2015. IEEE; 2015. pp. 178–82

[18] Shirmohammadi D., Hong H.W., Semlyen A., Luo G.X. 'A compensation-based power flow method for weakly meshed distribution and transmission networks'. *IEEE Transactions on Power Systems*. 1988;**3**(2):753–62

[19] Khushalani S., Solanki J.M., Shulz N.N. 'Development of three-phase unbalanced power flow using PV and PQ models for distributed generation and study of the impact of DG models'. *IEEE Transactions on Power Systems*. 2007;**22**(3):1019–25

[20] http://www.marind.ca/medias/pdf/equipement_industriel/hipower/HJW410T6.pdf [Accessed Aug 2017]

[21] Azmy A.M., Erlich I. 'Online optimal management of PEM fuel cells using neural networks'. *IEEE Transactions on Power Delivery*. 2005;**20**(2):1051–8

[22] Barbir F., Gomez T. 'Efficiency and economics of proton exchange membrane (PEM) fuel cells'. *International Journal of Hydrogen Energy*. 1997;**22**(10/11):1027–37

[23] Campanari S., Macchi E. 'Technical and tariff scenarios effect on microturbine trigenerative applications'. *Journal of Engineering for Gas Turbines and Power*. 2004;**126**:581–9

[24] Atwa Y.M., El-Saadany E.F., Salama M.M.A., Seethapathy R., Assam M., Counti S. 'Adequacy evaluation of distribution system including wind/solar DG during different modes of operation'. *IEEE Transactions on Power Systems*. 2001;**26**(4):1945–52

[25] https://en.wind-turbine-models.com/turbines/272-vestas-v-44 [Accessed Aug 2017]
[26] Tiwari G.N., Dubey S. *Fundamentals of photovoltaic modules and their applications*. Cambridge: RSC Energy Series; 2010. p. 422
[27] TÜV SÜD America (USA). Design qualification Type approval of commercial PV Modules. IEC 6121510.6 Performance at NOCT with IEC 60904-3 reference solar spectral irradiance distribution [online]. 2005. Available from http://www.tuvamerica.com/services/photovoltaics/TUV%20SUD%20test%20program%20IEC_61215_61646.pdf [Accessed 15 OCt 2017]
[28] Lee K.Y., El-Sharkawi M.A. *Modern heuristic optimization techniques with applications to power systems*. Cambridge: IEEE Power Engineering Society (02TP160); 2002. p. 221
[29] Rashedi E., Nezamabadi-pour H., Saryazdi S. 'GSA: a gravitational search algorithm'. *Information Sciences.* 2009;**179**:2232–48
[30] Caramia P., Carpinelli G., Varilone P. 'Point estimate schemes for probabilistic three-phase load flow'. *Electric Power Systems Research.* 2010;**80**:168–75
[31] Villanueva D., Pazos J.L., Feijo A. 'Probabilistic load flow including wind power generation', *IEEE Transactions on Power Systems*. 2011; **26**(3):1659–67
[32] Kaplanis S., Kaplani E. 'A model to predict expected mean and stochastic hourly global solar radiation $I(h;n_j)$ values'. *Renewable Energy.* 2007;**32**:1414–25
[33] Wan Nika W.B., Ibrahimb M.Z., Samoa K.B., Muzathika A.M. 'Monthly mean hourly global solar radiation estimation'. *Solar Energy.* 2012;**86**:379–87
[34] Thevenard D., Pelland S. 'Estimating the uncertainty in long-term photovoltaic yield predictions'. *Solar Energy.* 2013;**91**:432–45
[35] Jo B.K., Han J.H., Guo Q., Jang G. 'Probabilistic optimal power flow analysis with undetermined loads'. *Journal of International Council on Electrical Engineering*. 2012;**2**(3)321–5
[36] Conti S., Raiti S. 'Probabilistic load flow using Monte Carlo techniques for distribution network with photovoltaic generators'. *Solar Energy.* 2007;**81**:1473–81
[37] Morales J.M., Ruiz J.P. 'Point estimate schemes to solve the probabilistic power flow'. *IEEE Transactions on Power Systems*. 2007;**22**(4):1594–601
[38] Zhang P., Lee S.T. 'Probabilistic load flow computation using the method of combined cumulants and Gram–Charlier expansion'. *IEEE Transactions on Power Systems*. 2004;**19**(1):676–82
[39] Ruiz-Rodriguez F.J., Hernandez J.C., Jurado F. 'Probabilistic load flow for photovoltaic distributed generation using the Cornish–Fisher expansion'. *Electric Power Systems Research*. 2012;**89**:129–38
[40] Verbič G., Canizares C.A. 'Probabilistic optimal power flow in electricity markets based on a two-point estimate method'. *IEEE Transactions on Power Systems*. 2006;**21**(4):1883–93

[41] Distribution Test Feeders. Available https://ewh.ieee.org/soc/pes/dsacom/testfeeders/ [Accessed Aug 2017]

[42] Mwakabuta N., Sekar A. 'Comparative study of the IEEE 34 node test feeder under practical simplifications'. *39th North American Power Symposium*; Las Cruces, NM, USA, Oct 2007. IEEE; 2007. pp. 484–91

[43] Radosavljević J. *opfdngui—Matlab GUI based software for OPF in distribution networks*. 2017. Available at https://www.researchgate.net/profile/Jordan_Radosavljevic2/contributions

Chapter 10

Optimal Volt/Var control in distribution networks

10.1 Introduction

One of the most important parameters of the power quality is the voltage magnitude. At each point of the distribution network, the voltage must be within the permissible limits. Electrical devices are designed to have optimum operating characteristics at a certain, usually nominal voltage value. Deviation of the voltage beyond the allowable limits can cause irregular operation, reduction of the efficiency, and in extreme cases, the malfunction of electrical equipment. The voltage limits are standardized. The allowable voltage deviations (VDs) are defined in percentages of the nominal voltage value. Depending on law and local practice, the VDs of $\pm 5\%$ or $\pm 10\%$ may be considered acceptable for the medium-voltage (MV) and low-voltage (LV) distribution networks. In order to maintain voltage within allowable limits under different load levels, various types of devices are used to voltage control. In conventional distribution networks, the voltage control is performed by changing the transmission ratio of the transformers and by compensating for the reactive power. Accordingly, the basic resources for voltage control in the distribution networks are:

- Underload tap-changing transformers (ULTCTs); usually, this is the substation transformers (HV/MV) which connect the MV distribution network with a transmission (subtransmission) supply network;
- Off-voltage tap-changing transformers (OVTCTs); usually, this is distribution transformers (MV/LL) which connect the main MV distribution network with local LV distribution networks;
- Voltage regulators (VR) which are essentially transformers with tap-changers to adjust the voltage along the feeder, so as to compensate for the voltage drop over distance;
- Reactive power compensation devices: capacitors with unchangeable capacities (reactive power) and static Var compensators (SVCs) that have the capability to change reactive power.

The ULTCT can be controlled locally by automatic VR, or remotely by supervisory control and data acquisition system [1,2]. The tap ratio control scheme for ULTCT is needed to achieve appropriate change of voltage on the regulated bus according to load change. In addition to this basic requirement, when defining the control scheme for ULTCT, different factors should be consider, such as unbalanced load diversity on multiple feeders [3], the geometrical dimensions of the distribution network [4], the tap-changing operating times [5], etc.

In addition to the regulating transformers, the control of the voltage in the distribution networks is accomplished to a large extent by regulating the reactive power flows, i.e., compensating the reactive power by using shunt capacitor banks and/or SVCs. In this case, the aim of Volt/Var control is to find an optimal combination of dispatch for capacitor banks, SVC reactive power output, and ULTCT setting at substation to improve the voltage profile in whole distribution network [6–11].

Distributed generation (DG) can have a significant influence on the power flow and voltage conditions in a distribution network. Adequate voltage control requires the consideration of the DG units influence on the voltage profile. The impact of the DG units on the distribution network voltage depends on their size, location, and type [12–14].

The ability of a DG to participate in the voltage-control process in a distribution network is highly dependent on the connection type and operation mode. There are three possible typical cases [15]:

1. The DG is located near a large load, and it is designated to supply this load, with constant active and reactive power output. In this case, the impact of the DG is brought down to decreasing of the total power load at the connection point. The bus at which the DG is connected can be represented as a bus with constant active and reactive power, i.e. PQ bus, whereas the DG can be modeled as a negative power load.
2. The DG injects active power with an approximately constant power factor. The voltage at the DG connection point increases together with the increase of the active power of the DG. In this case too, the bus at which the DG is connected can be represented as a PQ node, whereas the DG can be modeled as a negative power load.
3. The DG injects power with specific voltage at the connection point. This DG controls the reactive power injection at the connection point, therefore controlling the voltage as well. The bus at which such a DG is connected must be treated as a PV bus. Their participation in the voltage control of the distribution network can be significant.

For distribution networks with the supply ULTCT and integrated DG which can control reactive power output, the optimal Volt/Var control implies the control scheme for coordination of the ULTCT actions with the reactive exchanges between the DG units and network [16–19].

Volt/Var control in distribution networks which include the ULTCT, shunt capacitor banks, SVCs, and DG units can be treated as an optimization problem with constraints. Solving this problem implies determining the control variables values in order to achieve the set optimization criteria. Depending on the available regulation resources, the control variables can be the tap-changer settings of ULTCTs, OVTCTs, and VRs; the voltage magnitudes at PV buses; the reactive powers of capacitors and SVCs. Dependent variables, which define an operation state of the distribution network, are the voltages at load buses. The objectives of optimal Volt/Var control can be different: minimizing VD [20,21], minimizing total power loss [22–30], minimizing operating costs [31–35], minimizing total number of switching operations [36,37], minimizing total harmonic distortion [38–40], and improving voltage stability [41].

Mathematically, the optimal Volt/Var control can be described as a nonlinear large-scale static optimization problem with constraints. Metaheuristic optimization methods such as genetic algorithm (GA) [7,9,22,24,30,37–40], particle swarm optimization (PSO) [11,33–35], gravitational search algorithm (GSA) [28], and evolutionary programing [29] have proven to be very successful in solving the Volt/Var problem considering different objective functions.

This chapter deals two methodologies for optimal voltage control. The first methodology related to determining the optimal transformer tap settings in radial (rural) distribution networks is based on the space and time decomposition of the voltage-control problem. Thereafter, the Volt/Var control is formulated as a nonlinear complex optimization problem with constraints and solved by using metaheuristic optimization methods.

10.2 Decomposition of the voltage-control problem

The distribution networks often have a very large number of nodes and branches (several hundreds), so it is also a large number of control variables to be considered. In addition, the specifics of distribution networks are the radial configuration and high r/x ratio. These specifics must be taken into account to create an efficient optimization model for optimal voltage control.

Taking into account the radial structure of the distribution network and the operation mode of transformers, a space and time decomposition of the voltage-control problem can be achieved [1,2]. The voltage-control activity performed by HV/MV (or MV/MV) ULTCTs influences voltages in the whole area "behind" (MV network, MV/LV transformers, and consequently their LV networks). At the same time, one ULTCT does not influence voltages in areas corresponding to other ULCTCs. Similarly, the voltage-control performed by one MV/LV OVTCT influences voltages in corresponding small areas of the distribution network—the individual LV networks only, i.e., the control activities on the MV/LV transformer do not have any influence on voltages in LV networks which belong other MV/LV transformers. This disposition and influences of voltage-control devices enable the space decomposition of the voltage-control problem in radial distribution networks. On the other hand, control action speeds of both transformer types differ to a great extent. Control actions of ULTCTs are "very fast," with a rate in minutes (hours). Changes of tap positions of OVTCTs are "very slow." These are usually performed a few times per year. Such speeds of transformer control actions enable the time decomposition of the voltage-control problem [1].

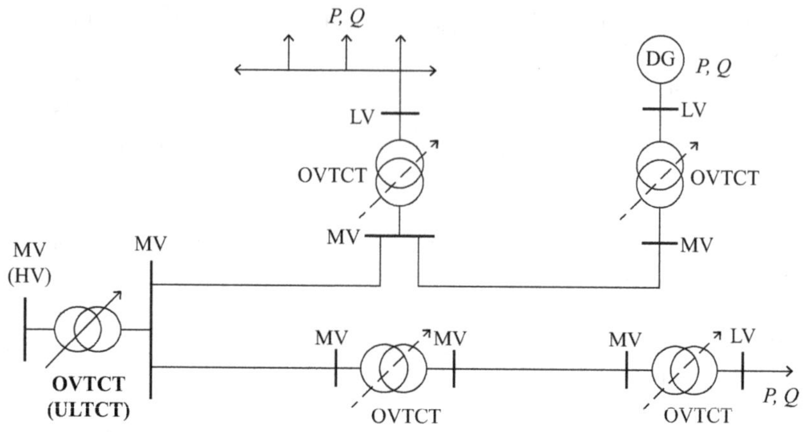

Figure 10.1 Radial distribution network

10.2.1 Seasonal control of voltage

Figure 10.1 shows a typical radial rural distribution network. It consists of a MV bus—the feed point of the network, OVTCT MV/MV supply transformer, an MV network and several OVTCT MV/LV transformers, and their belonging LV networks or LV feeders. Consumers and DGs are modeled as PQ buses.

Based on the space decomposition of the voltage-control problem in radial distribution networks, the optimal seasonal control of voltage can be carried out in two stages. The first stage consists of determining the optimal tap-changer positions of the MV/LV transformers, while the second one implies determining the optimal tap-changer position of the power supply MV/MV transformer [42].

10.2.1.1 The first stage of the optimal seasonal control of voltage

As already mentioned, due to the radial configuration of the distribution network, voltage control on the MV/LV transformers influences the voltage loads (as well as the DGs that are modeled as "negative loads") connected to the transformer, but it does not have any influence on the voltage loads and DGs connected to the other transformers. This enables an independent determination of the optimal tap-changer positions for each MV/LV distribution transformer. For the given power of the LV load, the voltage at the MV bus of the MV/LV transformer is mainly constant in regard to the tap-changer position of the transformer. Accordingly, the optimal tap-changer position can be determined as the function of the load power (i.e., the DG) on the LV side of the transformer, the voltage on the MV bus of the transformer, and the desired voltage value on the LV side of the transformer. The issue of choosing the desired voltage value can be regarded as the optimization criterion. However, the optimal operation of the system is implicitly determined by the nominal voltage values. The aim of the seasonal control of voltage is to keep load voltages as approximate to nominal values as possible.

Figure 10.2 shows the part of the distribution network consisting of the MV/LV transformer, line, and load (i.e., the DG). The parameters and variables in this figure are given in (p.u.), where $\underline{Z}_T = R_T + jX_T$ is the transformer impedance; $\underline{Z}_\ell = R_\ell + jX_\ell$ is the line impedance; $\underline{V}_1$ is the voltage at MV bus; $\underline{V}_2$ is the voltage at the line end; $\underline{S} = P + jQ$ is the complex injected power; $T = 1 + n \cdot \Delta T$ is the transmission ratio of transformer; n is the tap-changer position; and Δt is the step between tap-changer positions (p.u.).

The initial values of the bus voltages can be obtained by the power-flow calculation for a given power of loads and DG units, and transmission ratio of all transformers set to nominal value ($T = 1$). Assuming that the voltage magnitude V_1 is approximately constant, the relation between the voltages V_1 and V_2 can be written as follows:

$$\left(\frac{V_1}{T}\right)^2 = \left(V_2 + \frac{P \cdot R + Q \cdot X}{V_2}\right)^2 + \left(\frac{P \cdot X - Q \cdot R}{V_2}\right)^2 \tag{10.1}$$

where $R = R_T + R_\ell$ and $X = X_T + X_\ell$.

The transmission ratio T is calculated from (10.1), for the specified voltage value V_2:

$$T = \frac{V_1}{\sqrt{(V_2 + ((P \cdot R + Q \cdot X)/V_2))^2 + ((P \cdot X - Q \cdot R)/V_2)^2}} \tag{10.2}$$

It can now be calculated the optimal tap-changer position of the transformer:

$$n = \frac{T - 1}{\Delta t} \tag{10.3}$$

The real tap-changer position (n_r) is determined as the closest integer value from the range of the defined discrete values, based on the obtained value n according to (10.3). At most MV/LV distribution transformers, this range is $n_r = (-2, -1, 0, 1, 2)$. If the obtained value for n is greater than the maximum value from the defined range, the maximum value from the range, i.e., $n_r = n_{\max}$ is adopted as valid. In case the obtained value is smaller than the minimum value from the defined range, the minimum value $n_r = n_{\min}$ is adopted as valid for the tap-changer position. Since the real (adopted) position of the tap-changer n_r is different from the calculated value n, the real value of the voltage V_2 is obtained by square root equation (10.1) at V_2 or by the power-flow calculation in the whole distribution network, whereas the value of the transmission ratio is $T = 1 + \Delta t \cdot n_r$.

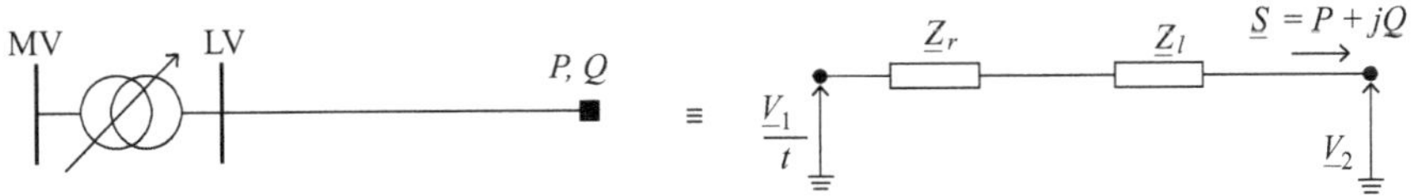

Figure 10.2 Segment of distribution network

Since the power of load and the DGs power output are not constant during the day, month, or year, it is necessary to determine their representative values over the given period. One of the possible approaches is the medium power model. Depending on the shape, annual power diagrams of loads, and DGs, they can be divided into seasons, usually two with the same or different durations. The seasonal power represents the mean value of the power segments which account for a season. In a general case, different power diagrams can be divided into different seasons. However, considering the fact that the seasonal voltage control implies a simultaneous coordination of the tap-changer positions of the supply transformer and all MV/LV transformers, it is necessary that the seasons are the same (coinciding) for the whole system. It is possible to meet this demand only in case the following assumptions can be applied:

1. All loads have approximately the same normalized annual power diagrams;
2. DGs connected in the distribution network have similar annual power diagrams.

In order to apply the described procedure for determining the optimal tap-changer position of the MV/LV OVTCT, it is necessary to transform the LV network into the equivalent line model, as shown in Figure 10.2. The principle of equal voltage sags and the principle of equal power losses in the real (LV network) and equivalent (line model) systems are used for determining the equivalent line model impedance and equivalent injection power at the end of the equivalent line model. Figure 10.3 shows the radial LV line with N nodes and sections as well as the equivalent line model. The impedance of section i, the injection power, and the voltage at the node i are marked with $\underline{Z}_i$, $\underline{S}_i$, and $\underline{V}_i$, respectively.

The equivalent line model impedance $\underline{Z}_{el}$ is determined according to the principle of equal voltage sags:

$$\underline{Z}_{el} = \frac{\underline{Z}_1 \cdot (\underline{S}_1^*/\underline{V}_1^*) + (\underline{Z}_1 + \underline{Z}_2) \cdot (\underline{S}_2^*/\underline{V}_2^*) + \cdots + (\underline{Z}_1 + \underline{Z}_2 + \cdots + \underline{Z}_N) \cdot (\underline{S}_N^*/\underline{V}_N^*)}{\sum_{i=1}^{i=N} (\underline{S}_i^*/\underline{V}_i^*)} \tag{10.4}$$

The voltage at the end of the equivalent line model is

$$\underline{V}_{el} = \underline{V}_0 - \underline{Z}_{el} \cdot \sum_{i=1}^{i=N} \frac{\underline{S}_i^*}{\underline{V}_i^*} \tag{10.5}$$

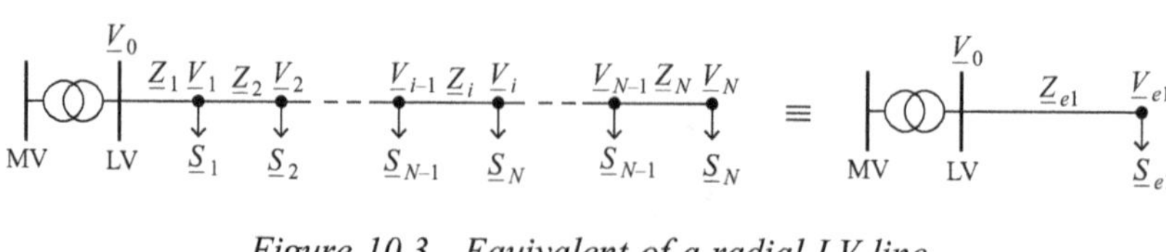

Figure 10.3 Equivalent of a radial LV line

whereas the injection power at the end of the equivalent line model is

$$\underline{S}_{e1} = \underline{V}_{e1} \cdot \sum_{i=1}^{i=N} \frac{\underline{S}_i}{\underline{V}_i} \tag{10.6}$$

The node voltages in per unit system are close to 1, and therefore (10.4) and (10.6) can be simplified:

$$\underline{Z}_{e1} = \frac{\underline{Z}_1 \cdot \underline{S}_1 + (\underline{Z}_1 + \underline{Z}_2) \cdot \underline{S}_2 + \cdots + (\underline{Z}_1 + \underline{Z}_2 + \cdots + \underline{Z}_N) \cdot \underline{S}_N}{\sum_{i=1}^{i=N} \underline{S}_i} \tag{10.7}$$

$$\underline{S}_{e1} = \sum_{i=1}^{i=N} \underline{S}_i \tag{10.8}$$

The equivalent line model of M radial LV lines, connected to the LV buses of the MV/LV transformers (Figure 10.4), can be determined based on the equal power loss principle:

$$\underline{Z}_e = \frac{\underline{Z}_{e1} \cdot \left|\underline{S}_{e1}^*/\underline{V}_{e1}^*\right|^2 + \underline{Z}_{e2} \cdot \left|\underline{S}_{e2}^*/\underline{V}_{e2}^*\right|^2 + \cdots + \underline{Z}_{eM} \cdot \left|\underline{S}_{eM}^*/\underline{V}_{eM}^*\right|^2}{\left|\sum_{j=1}^{j=M} \left(\underline{S}_{ej}^*/\underline{V}_{ej}^*\right)\right|^2} \tag{10.9}$$

The previous equation can be simplified:

$$\underline{Z}_e = \frac{\underline{Z}_{e1} \cdot \left|\underline{S}_{e1}\right|^2 + \underline{Z}_{e2} \cdot \left|\underline{S}_{e2}\right|^2 + \cdots + \underline{Z}_{eM} \cdot \left|\underline{S}_{M}\right|^2}{\left|\sum_{j=1}^{j=M} \underline{S}_{ej}\right|^2} \tag{10.10}$$

Therefore, the injection power at the end of the line model is now:

$$\underline{S}_e = \sum_{j=1}^{j=M} \underline{S}_{ej} \tag{10.11}$$

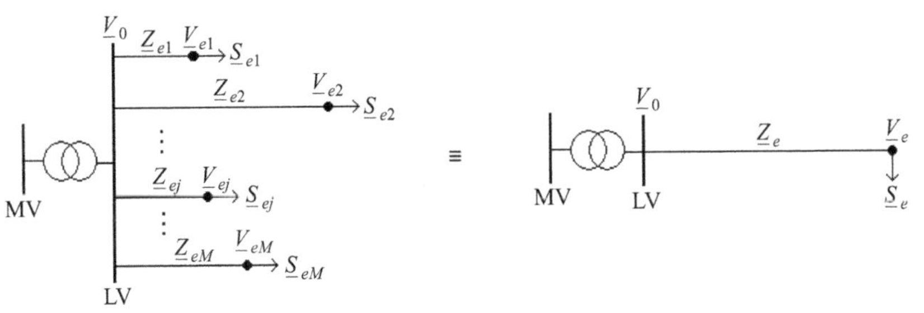

Figure 10.4 Equivalent of several radial LV lines connected to same bus

where $\underline{Z}_i$ is the impedance of the ith section of the radial LV line, $i = 1, 2, \ldots, N$, $\underline{S}_i$ is the injection power at the node i of the radial LV line, $\underline{V}_i$ is the voltage at the node i of the radial LV line, $\underline{Z}_{ej}$ is the equivalent model impedance of the jth radial LV line, $j = 1, 2, \ldots, M$, $\underline{S}_{ej}$ is the equivalent model injection power of the jth radial LV line, $\underline{V}_{ej}$ is the equivalent model voltage of the jth radial LV line, $\underline{Z}_e$ is the equivalent model impedance of the M radial LV lines (LV network), and $\underline{S}_e$ is the equivalent model injection power of the M radial LV lines.

10.2.1.2 The second stage of the optimal seasonal control of voltage

Determining the tap-changer position of the OVTCT MV/MV supply transformer located at root of the MV network represents the second stage of the procedure for solving the problem of the optimal seasonal control of voltage. Actually, this is an optimization problem with objective to minimize the VD at load buses in the downstream network. The objective function can be expressed as follows:

$$ObjFun = VD = \sum_{i=1}^{NL} \left| V_i - V_i^{ref} \right| \tag{10.12}$$

where V_i is the voltage at the bus i of the distribution network, V_i^{ref} is the reference (nominal) voltage value, and NL is the number of load buses in the distribution network.

Carrying out this optimization task is brought down to the power-flow calculation for all possible tap-changer positions of the supply transformer. During each power-flow calculation, the tap-changer positions of the MV/LV transformers are determined for the current tap-changer position of the supply transformer according to the procedure described in Section 10.2.1.1. At the same time, the value of the objective function (10.12) is calculated. The minimum value of the objective function implicates the optimal solution, i.e., the optimal seasonal control plan of the distribution transformers in the given distribution network. The adopted optimization criterion determines the control plan. The flowchart of the proposed procedure is given in Figure 10.5. The following procedure should be performed:

1. On regular basis, once a year, based on the updated load diagrams from the current year.
2. Extraordinarily, after significant changes in the distribution network that have occurred, such as adding a new DG unit, network reconstruction by building a new line, transformer, etc.

10.2.1.3 Optimal voltage control with ULTCTs

The optimal voltage control in the operation planning mode is performed for distribution networks with OVTCTs only, considering the mean power values over a given period. However, for distribution networks with ULTCT (Figure 10.1), the voltage control implies determining the optimal tap-changer settings of the ULTCT

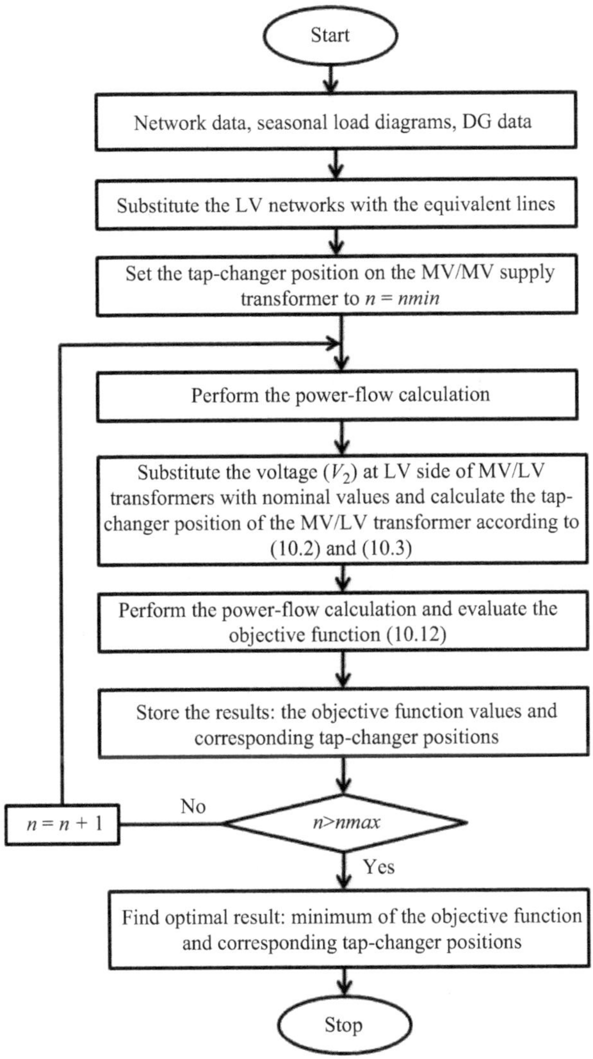

Figure 10.5 Flow chart of optimal seasonal voltage control

in accordance with the actual power of loads. Optimal real-time voltage control with ULTCT is based on the space and the time decomposition of the voltage-control problem. The space decomposition enables a solution of the distribution voltage-control problem for the MV network of each supply ULTCT (substation) separately. The time decomposition enables to solve the problem in the operation planning mode (determining the optimal seasonal voltage-control plan for OVTCTs in considered distribution network) and the real-time mode (determining the tap-changer position on the ULTCT). The objective function is defined by (10.12).

The procedure for optimal voltage control in the real time by using the ULTCT can be described as follows:

Step 1: Define the distribution network configuration, lines data, transformers data, loads data, and DG units data. Use the per unit system;
Step 2: Perform the algorithm for optimal seasonal control of voltage (Figure 10.5) and determine the optimal tap-changer positions for OVTCTs;
Step 3: Set the voltage value at the root bus (HV bus of ULTCT) and update the actual power of loads in the distribution network;
Step 4: Set the initial tap-changer position of ULTCT on $n = nmin$;
Step 5: Run the power-flow program based on backward/forward method and evaluate the objective function [Eq. (10.12)];
Step 6: Store the results: the objective function values and corresponding tap-changer positions of ULTCT;
Step 7: Set the tap-changer position of ULTCT on $n = n + 1$;
Step 8: Repeat steps 5–8 until $n < nmax$; Stop.

10.2.1.4 Simulation results

The procedure for optimal seasonal control of voltage is applied on a real distribution network located in the east of Serbia in the area of Stara Planina [43]. Figure 10.6 shows the distribution network containing 11 DGs. Those DGs are microhydro power plants (MHPs). The MV network is supplied through an OVTCT $(35 \pm 2 \cdot 2.5\%)/10$ kV/kV. The MV network supplies eight MV/LV OVTCTs $(10 \pm 2 \cdot 2.5\%)/0.4$ kV/kV. The data regarding the lines, transformers, and loads are given in Table 10.1. Figure 10.7 shows the annual power diagram of the MHPs.

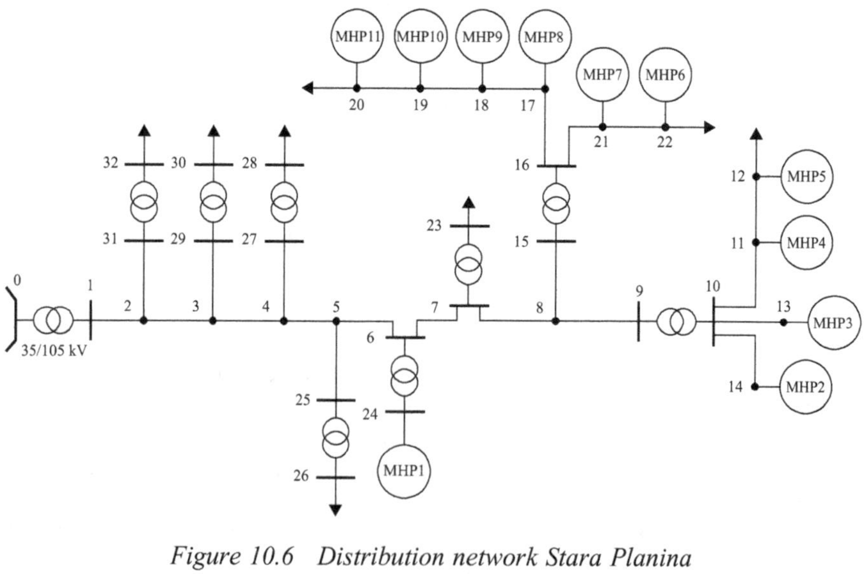

Figure 10.6 Distribution network Stara Planina

Table 10.1 Data on the distribution network Stara Planina; Sbase = 1 MV A

Send. node	Rec. node	R (p.u.)	X (p.u.)	Load at Rec. node	
				P_L (p.u.)	Q_L (p.u.)
0	1	0.0021	0.0149		
1	2	0.0083	0.0068	0	0
2	3	0.0066	0.0054	0	0
3	4	0.0072	0.0060	0	0
4	5	0.0027	0.0022	0	0
5	6	0.0145	0.0119	0	0
6	7	0.0010	0.0009	0	0
7	8	0.0376	0.0077	0	0
8	9	0.0517	0.0106	0	0
9	10	0.2000	0.3464	0	0
10	11	10.5262	3.2725	0	0
11	12	5.2631	1.6362	0.0074	0.0036
10	13	2.6316	0.8181	0	0
10	14	0.2256	0.0701	0	0
8	15	0.0451	0.0093	0	0
15	16	0.2000	0.3464	0	0
16	17	3.1579	0.9818	0	0
17	18	0.5263	0.1636	0	0
18	19	1.5789	0.4909	0	0
19	20	0.8271	0.2571	0.0014	0.0007
16	21	1.8045	0.5610	0	0
21	22	2.3308	0.7246	0.0008	0.0004
7	23	0.2000	0.3464	0.0028	0.0014
6	24	0.2000	0.3464	0	0
5	25	0.0199	0.0117	0	0
25	26	0.2000	0.3464	0.0096	0.0041
4	27	0.0024	0.00075	0	0
27	28	0.2000	0.3464	0.0126	0.0054
3	29	0.0042	0.0013	0	0
29	30	0.2000	0.3464	0.0178	0.0076
2	31	0.0120	0.0037	0	0
31	32	0.2000	0.3464	0.0099	0.0042

Two typical periods can be identified on the power diagrams of the MHPs: the period of the higher power (January–June) and the period of the lower power (July–December). The mean values of the MHP powers have been taken into consideration during the two seasons. It is assumed that the loads are constant, due to the fact the rural area has slight change in load power during the year. The voltages V_2 [Eq. (10.2)], V_{ref} [Eq. (10.12)], and root node voltage are specified to 1 (p.u.).

The impedances and injection powers of the equivalent line models of the LV networks, which belong to transformers 9–10 and 15–16, are shown in Table 10.2. The optimal seasonal voltage-control plan is given in Table 10.3. Figure 10.8 shows the voltage profiles for the period of higher MHPs powers (January–June). The upper and lower voltage limits in the MV and LV networks are specified according to the engineering standards on 1.05 and 0.95 (p.u.), respectively. Figure 10.8

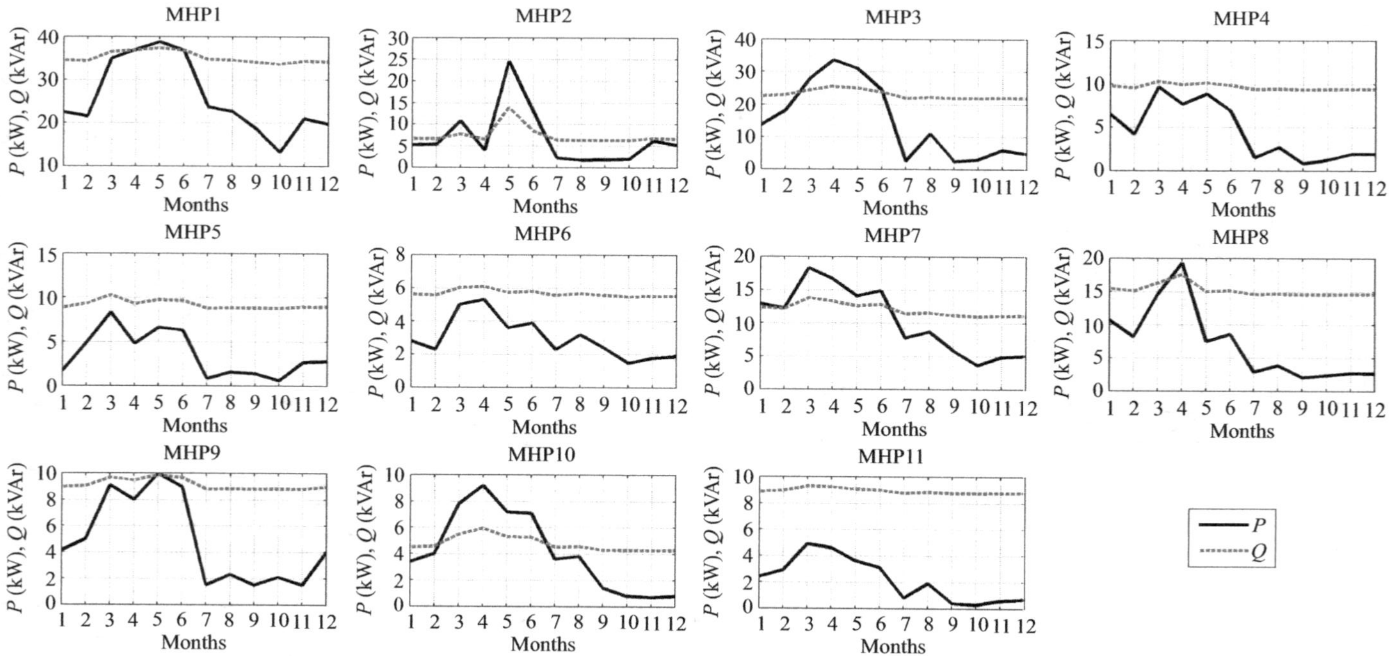

Figure 10.7 Annual power diagrams of MHPs

Table 10.2 Equivalent model impedances and injection powers of the LV networks

LV network		January–June		July–December	
TF	**Nodes**	$\underline{Z}_e$ **(p.u.)**	$\underline{S}_e$ **(p.u.)**	$\underline{Z}_e$ **(p.u.)**	$\underline{S}_e$ **(p.u.)**
9–10	11,12,13,14	$1.8899+j0.4162$	$-0.0408+j0.0243$	$5.6586+j1.2793$	$-0.004+j0.0193$
15–16	17,18,19, 20,21,22	$1.9149+j0.5245$	$-0.0456+j0.0283$	$2.0187+j0.5035$	$-0.0136+j0.0231$

Table 10.3 Optimal seasonal voltage-control plan

	Season: January–June		
Transformer	9–10	15–16	Other transformers
Tap-changer position, n_r	2	2	0
	Season: July–December		
Transformer	9–10	Other transformers	
Tap-changer position, n_r	−2	0	

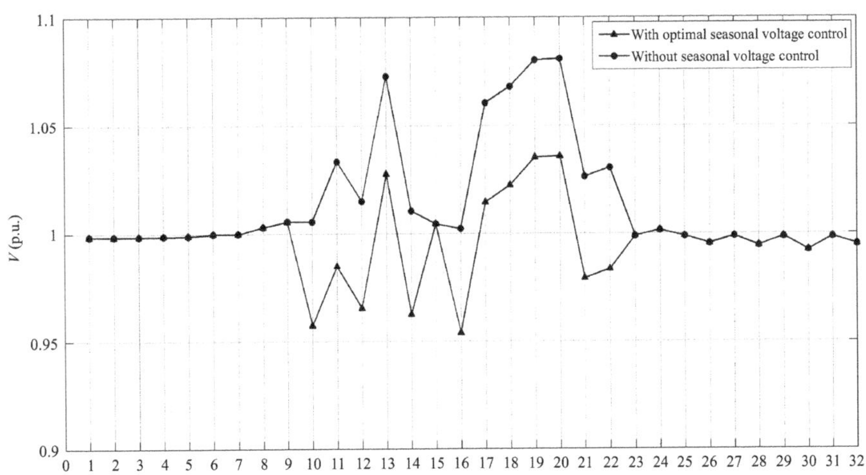

Figure 10.8 Voltage profiles in the cases without voltage control (the tap-changer positions of all transformers are n = 0) and with optimal seasonal control of voltage

clearly shows that the correction of the voltage ratio at transformers 9–10 and 15–16 in accordance with the optimal seasonal control of voltage enables reducing the voltages at nodes 13, and 17–20 within the allowable limits (0.95–1.05 p.u.). This is the maximum improvement of the voltage profile, which implicates a limited domain of the seasonal voltage control.

10.3 Optimal Volt/Var control using metaheuristic optimization

Optimum Volt/Var control implies the coordination of control variables of distribution networks in real time with aims to minimize the VD and power loss. This is possible for automated distribution networks with remote control and coordination of the system components. According to the available regulation devices, the control variables can be discrete (tap positions, number of capacitors) or continuous (voltage magnitudes and reactive power outputs). Dependent variables, which define an operation state of the distribution network, are the voltages at load buses. The objective function in this optimization problem can take different forms such as deviation of the voltage from the reference value, the costs of consumers due to VD from the nominal value, the voltage regulation costs (DG, capacitors, power losses, and reactive power), power losses in the network, etc. Some objective functions can be in collision, which means that satisfying one objective can have the opposite effect on the other objective function. For example, minimizing VDs can be at the expense of increasing power losses and vice versa. A compromise can be the choice of a multiobjective function that includes several individual objective functions that are weighted by appropriate coefficients in a way that depends on the goal set and the degree of importance of individual members of that function. The solution of the optimum voltage regulation must satisfy the operating constraints regarding to the voltage limits at buses and the maximum loadability of the branches in the distribution network.

10.3.1 Problem formulation

Optimal values of control variables imply values of dependent variables which have met the optimization criteria. This definition of the optimal Volt/Var control can be considered as one of the forms of the optimal power-flow problem in distribution networks. For a given network configuration and load level, each combination of control variable values defines a certain stationary state of the distribution network.

Generally, the optimal Volt/Var control problem can be formulated as follows:

$$\min F(\mathbf{x}, \mathbf{y}) \tag{10.13}$$

Subject to

$$g(\mathbf{x}, \mathbf{y}) = 0 \tag{10.14}$$

$$h(\mathbf{x}, \mathbf{y}) \leq 0 \tag{10.15}$$

$$\mathbf{x} \in \mathbf{X} \tag{10.16}$$

where F is the objective function; the objective function F can take different forms. Main objective in the optimal Volt/Var control is the voltage profile improvement that is minimization of the load bus *VD*s [Eq. (10.12)]. However, starting from the premise that the problem of voltage control is solved when the voltage values in all nodes of the network are within the permissible limits of min–max, the problem of optimal Volt/Var control can be viewed in a wider context. For example, minimizing active power losses or simultaneous minimization of VD and power losses can be adopted as objectives in the optimal Volt/Var control.

x is the vector of control variables, which generally consisting of following elements:

- The voltage magnitude of the root node V_0; actually this is specified voltage on secondary side of the supply ULTCT located at root of the network. Based on V_0 and known (constant) primary voltage, the tap-changer position of ULTCT is determined.
- The voltage magnitude of DG units modeled as PV buses, V_{DG}.
- The tap settings of regulating distribution transformers, T.
- The reactive power of shunt Var compensators, Q_C.

Hence, **x** can be expressed as

$$\mathbf{x} = [V_0, V_{G1} \ldots V_{GNG}, T_1 \ldots T_{NT}, Q_{C1} \ldots Q_{CNC}] \tag{10.17}$$

where *NG* is the number of DGs modeled as PV bus, *NT* is the number of regulating distribution transformers, and *NC* is the number of Var compensators in distribution network.

y is the vector of dependent variables consisting of load bus voltages V_L, DGs reactive power outputs Q_G, and branch loadings S_l. Accordingly, vector **y** can be expressed as

$$\mathbf{y} = [V_{L1} \ldots V_{LNL}, Q_{G1} \ldots Q_{GNG}, S_{l1} \ldots S_{lNB}] \tag{10.18}$$

where *NL* and *NB* are number of load buses and network branches, respectively.

The equality constraints (10.14) are the typical nonlinear power-flow equations.

$$\underline{J}_i = \underline{I}_{Li} + \underline{I}_{Ci} - \underline{I}_{DGi} + \underline{Y}_i^o \cdot \underline{V}_i + \sum_{\substack{\ell \in \alpha_{\ell i} \\ \ell \neq i}} \underline{J}_\ell; i = NB, NB-1, \ldots, 1; \tag{10.19}$$

$$\underline{V}_i = \underline{V}_m - \underline{Z}_i \underline{J}_i \tag{10.20}$$

where $\underline{J}_i$ is the current in branch i between nodes m and i (see Figure 5.7 in Chapter 5), $\underline{I}_{Li}$ is the complex current of load at bus i; $\underline{I}_{Ci}$ is the complex current of a Var compensator at bus i; $\underline{I}_{DGi}$ is the current injections at node i corresponding to DG; $\underline{J}_\ell$ is the complex current in branch ℓ emanating from bus i; $\underline{Y}_i^o$ is the shunt admittance at bus i; $\alpha_{\ell i}$ is the set of buses connected with bus i; $\underline{V}_i$ is the complex voltage at bus i, $\underline{V}_m$ is the complex voltage at bus m, and $\underline{Z}_i$ is the series impedance of branch i.

Inequality constraints (10.15) are the functional operating constraints, such as load bus voltage magnitude limits, DG reactive power output limits, and branch flow limits.

$$V_{Li}^{\min} \le V_{Li} \le V_{Li}^{\max}, \quad i = 1, \ldots, NL \tag{10.21}$$

$$Q_{Gi}^{\min} \le Q_{Gi} \le Q_{Gi}^{\max} \quad i = 1, \ldots, NG \tag{10.22}$$

$$S_{li} \le S_{li}^{\max} \quad i = 1, \ldots, NB \tag{10.23}$$

The constraints related to voltage at load buses (10.21) and power flow through network branches (10.23) are also known as the security constraints.

Constraints (10.16) define the feasibility region of the problem control variables, such as root bus voltage limits, DG bus voltage magnitude limits, transformer tap setting limits, and shunt Var compensation limits. Actually, those define the space of possible solutions for the Volt/Var control problem.

$$V_0^{\min} \le V_0 \le V_0^{\max} \tag{10.24}$$

$$V_{Gi}^{\min} \le V_{Gi} \le V_{Gi}^{\max}, \quad i = 1, \ldots, NG \tag{10.25}$$

$$T_i^{\min} \le T_i \le T_i^{\max}, \quad i = 1, \ldots, NT \tag{10.26}$$

$$Q_{Ci}^{\min} \le Q_{Ci} \le Q_{Ci}^{\max}, \quad i = 1, \ldots, NC \tag{10.27}$$

It is worth mentioning that the control variables ($\mathbf{x}$) are self-constrained. The hard inequality constraints of the dependent variables consisting of load bus voltages V_L are restricted by adding them as the quadratic penalty terms to the objective function. The new expanded objective function to be minimized becomes:

$$F_p(\mathbf{x}, \mathbf{y}) = F(\mathbf{x}, \mathbf{y}) + \lambda_V \sum_{i=1}^{NL} \left(V_{Li} - V_{Li}^{\lim}\right)^2 \tag{10.28}$$

where λ_V is the penalty factor. $V_{Li}^{\lim}$ is defined as follows:

$$V_{Li}^{\lim} = \begin{cases} V_{Li}^{\min}, & V_{Li} < V_{Li}^{\min} \\ V_{Li}^{\max}, & V_{Li} > V_{Li}^{\max} \\ V_{Li}, & V_{Li}^{\min} \le V_{Li} \le V_{Li}^{\max} \end{cases} \tag{10.29}$$

10.3.2 Solution method

In the population-based metaheuristic optimization algorithms, the population is defined by a set of individuals (agents) which represent potential solutions of the optimization problem. The number of agents (N) is named as the size of the population. In general, an agent can be represented as vector whose elements are the values of the control variables of the optimization problem. The number of control variables (n) is the search space dimension of the optimization problem.

The essence of metaheuristic methods is iterative correction of the solution, i.e., generating a new population by applying algorithmic operators with stochastic search mechanism on agents from the current population. The way in which the algorithmic operators are defined constitutes the essence of a particular metaheuristic optimization method. The efficiency and performance of metaheuristic optimization methods are dependent on the proper setting of the corresponding algorithmic parameters.

In recent years, various population-based metaheuristic optimization methods, such as GA, PSO, GSA, etc., have been suggested for solving the different power system problems.

The main performances of metaheuristics are fast search of large solution spaces, ability to find global solutions, and avoiding local optimum. Their main advantage compared to the classical (exact) optimization methods is that they are not limited with requirements for differentiability, nonconvexity, and continuity of the objective function or types of control variables. Moreover, these methods can be used for practical optimization problems taking into account various types of objective functions and constraints.

The basic elements of metaheuristic optimization methods can be defined as follows:

Agent, $\mathbf{x}(t)$: It is a candidate solution represented by an n-dimensional vector, where n is the number of control variables. At time (iteration) t, the ith agent $\mathbf{x}_i(t)$ can be described as $\mathbf{x}_i(t) = \left[x_i^1(t), \ldots, x_i^d(t), \ldots, x_i^n(t)\right]$, where $x_i^d(t)$ is the position of the ith agent with respect to the dth dimension, i.e., the value of the dth control variable in the ith candidate solution at iteration t.

Population, $\mathbf{POP}(t)$: It is a set of N agents at time (iteration) t, i.e., $\mathbf{POP}(t) = [\mathbf{x}_1(t), \ldots, \mathbf{x}_N(t)]^T$.

Space of possible solutions, $\mathbf{X}$: It is an n-dimensional solution space which is defined by lower and upper limits of control variables.

Fitness is a direct metric of the performance of the individual population member (agent). The fitness of each agent of the population is calculated from the value of the function being optimized.

General structure of metaheuristic optimization methods can be represented as follows:

Initialization

1. Defining the objective function $F(\mathbf{x}_i)$ and the space of possible solutions $\mathbf{X}$.
2. Generate initial population of N agents: $\mathbf{POP}(1) = [\mathbf{x}_1(1), \ldots, \mathbf{x}_i(1), \ldots, \mathbf{x}_N(1)]^T \subseteq \mathbf{X}$.

   Usually, the initial positions of each agent are randomly selected between minimum and maximum values of the control variables.

   Set the iteration counter: $t = 1$.

Iterative procedure

3. Calculate the fitness value $F(\mathbf{x}_i(t))$ for each agent $\mathbf{x}_i(t)$, $i = 1, \ldots, N$ in the current population $\mathbf{POP}(t)$.

4. Generate new population $\mathbf{POP}(t+1) = [\mathbf{x}_1(t+1), \ldots, \mathbf{x}_i(t+1), \ldots, \mathbf{x}_N(t+1)]^T \subseteq \mathbf{X}$ by applying the algorithmic operators on search agents from the current population $\mathbf{POP}(t)$.
5. Repeat the iterative procedure until the stop criteria is reached.
6. The optimal solution $\mathbf{x}^*$ is determined.

End

10.3.2.1 Implementation for optimal Volt/Var control

A general procedure of applying metaheuristic population-based optimization methods to solve the optimal Volt/Var control can be described in following steps:

Step 1: Define the distribution network configuration, lines data, transformers data, shunt Var compensators data, loads data, and DG units data. Use the per unit system.

Step 2: Specify the control variables $\mathbf{x}$ [Eq. (10.17)] and their lower and upper limits [Eq. (10.24)–(10.27)].

Specify the dependent variables $\mathbf{y}$ [Eq. (10.18)], and their lower and upper limits [Eqs. (10.21)–(10.23)].

Specify the objective function to be optimized $F(\mathbf{x}, \mathbf{y})$ [Eq. (10.13)], and the penalty factor (λ_V) to form the expanded objective function $F_p(\mathbf{x}, \mathbf{y})$ [Eq. (10.28)].

Step 3: Initialize algorithm parameters, such as the population size (N), maximum iteration number ($t_{\max}$), and other parameters depending of selected optimization method.

Step 4: Generate an initial random population of N agents:

$$\mathbf{POP}(0) = \begin{bmatrix} \mathbf{x}_1(0) \\ \mathbf{x}_2(0) \\ \vdots \\ \mathbf{x}_i(0) \\ \vdots \\ \mathbf{x}_N(0) \end{bmatrix} = \begin{bmatrix} x_1^1(0), & \ldots, & x_1^d(0), & \ldots, & x_1^n(0) \\ x_2^1(0), & \ldots, & x_2^d(0), & \ldots, & x_2^n(0) \\ & & \vdots & & \\ x_i^1(0), & \ldots, & x_i^d(0), & \ldots, & x_i^n(0) \\ & & \vdots & & \\ x_N^1(0), & \ldots, & x_N^d(0), & \ldots, & x_N^n(0) \end{bmatrix} \tag{10.30}$$

The initial positions of each agent are randomly selected between minimum and maximum values of the control variables. Therefore, the initial value of the dth control variable in the ith particle (candidate solution) is obtained as follows:

$$x_i^d(0) = x_{i,\min}^d + r \cdot \left(x_{i,\max}^d - x_{i,\min}^d \right) \tag{10.31}$$

where r is a random number in [0, 1], and $x_{i,\min}^d$ and $x_{i,\max}^d$ are the lower and upper limits of the dth control variable, respectively. According to (6.51), the ith agent can be written as follows:

$$\mathbf{x}_i = \left[V_0^1, V_{G1i}^{\cdots}, \ldots, V_{GNGi}, T_{T1i}, \ldots, T_{NTi}^d, Q_{C1i}^{.d+1}, \ldots, Q_{CNCi}^n\right] \tag{10.32}$$

where N is the number of particles (population size), and n is the number of control variables.

Step 5: Run the power-flow program based on backward/forward sweep method for each agent $\mathbf{x}_i(t)$ from the current population $\mathbf{POP}(t)$ and calculate the corresponding values of the objective function.

Step 6: Update the position of each agent by applying algorithmic operators of the selected optimization method.

Step 7: Repeat steps 5–6 until the stop criteria is reached, that is, maximum number of iterations $t_{\max}$.

Step 8: Return best solution obtained in the last iteration; Stop.

10.3.3 Simulation results

The proposed approach was implemented on the IEEE 31-bus test system shown in Figure 10.9. The data of the network including nominal powers of loads are given in Table 11.1 within Chapter 11, and also can be found in [44]. The network is fed through the ULTCT that maintains the voltage at the busbars 0 on the specified value of V_0. The DG connected at node 8 has constant active power output of 5 p.u., whereas operate in PV mode with reactive power in the range of −4 to 4 (p.u.). The reactive power compensator connected to the node 25 has the capability to generate reactive power in the range of 0–1.35 (p.u.). The limits of voltage V_0 at the secondary bus of ULTCT was set as follows: 0.97 (p.u.) $\leq V_0 \leq$ 1.05 (p.u.).

The task of optimal Volt/Var control in this case study is to determine the optimal voltage value of V_0, i.e., tap-changer position of the ULTCT, the optimal voltage at DG bus, i.e., the reactive power injection of the DG, and the optimum reactive power of the compensator. For solving this problem, it is assumed that all

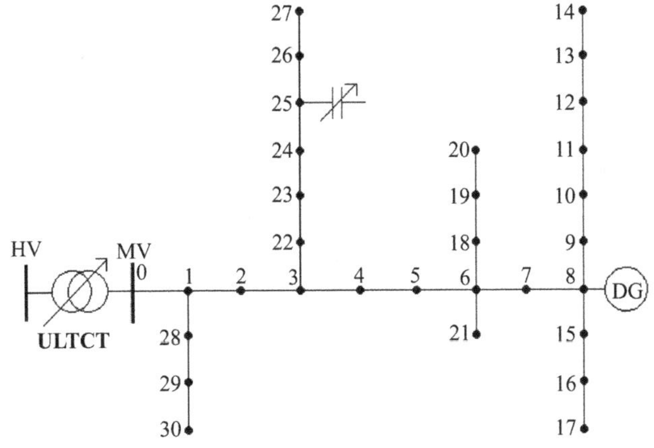

Figure 10.9 Single line diagram of the IEEE 31-bus system

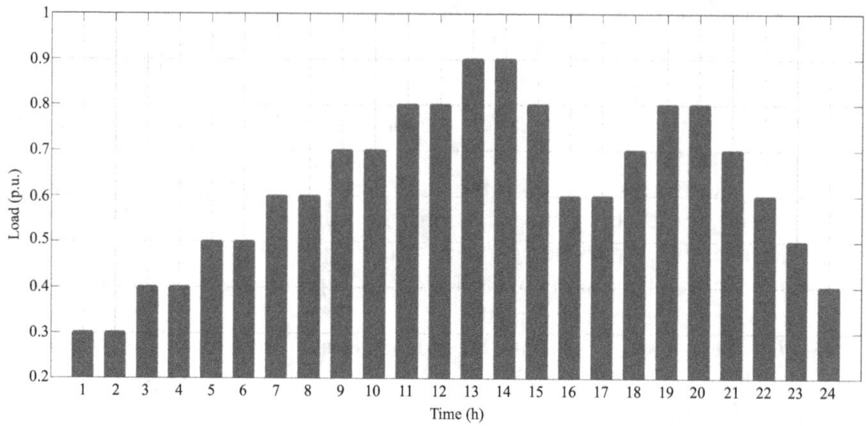

Figure 10.10 Daily load diagram

consumers in the distribution network follow a daily load diagram shown in Figure 10.10.

The optimal Volt/Var control problem was solved for three different objectives, as follows:

Case 1: minimization of VD at load buses

$$\min_{\mathbf{x}} F(\mathbf{x}, \mathbf{y}) = \min_{\mathbf{x}} VD(\mathbf{x}, \mathbf{y}) = \min_{\mathbf{x}} \sum_{i=1}^{NL} \left| V_i - V_i^{ref} \right| \tag{10.33}$$

Case 2: minimization of total power loss in the network

$$\min_{\mathbf{x}} F(\mathbf{x}, \mathbf{y}) = \min_{\mathbf{x}} Ploss(\mathbf{x}, \mathbf{y}) = \min_{\mathbf{x}} \sum_{i=1}^{NB} R_i \cdot J_i^2 \tag{10.34}$$

Case 3: simultaneous minimization of the VD and power loss

$$\min_{\mathbf{x}} F(\mathbf{x}, \mathbf{y}) = \min_{\mathbf{x}} \{w_V VD(\mathbf{x}, \mathbf{y}) + w_P Ploss(\mathbf{x}, \mathbf{y})\} \tag{10.35}$$

where w_V and w_P are the weighting factors.

The results shown were obtained by using the *opfdngui* software [44] described in Chapter 9. Figure 10.11 presents an example program run for Case 1, with load level of $L = 0.8$, and PSO as the solution method.

Complete results are given in Figures 10.12–10.15 in the form of daily diagrams of optimal values of control variables.

When scenario 1 is applied, there are high values of the reactive power of the compensator (Q_C) and a wide range of reactive power changes of DG (Q_{DG}), which are in function of maintaining the voltages at buses 25 and 8 on specified values to minimize the VD in the rest of network. Voltage control is directly related to the

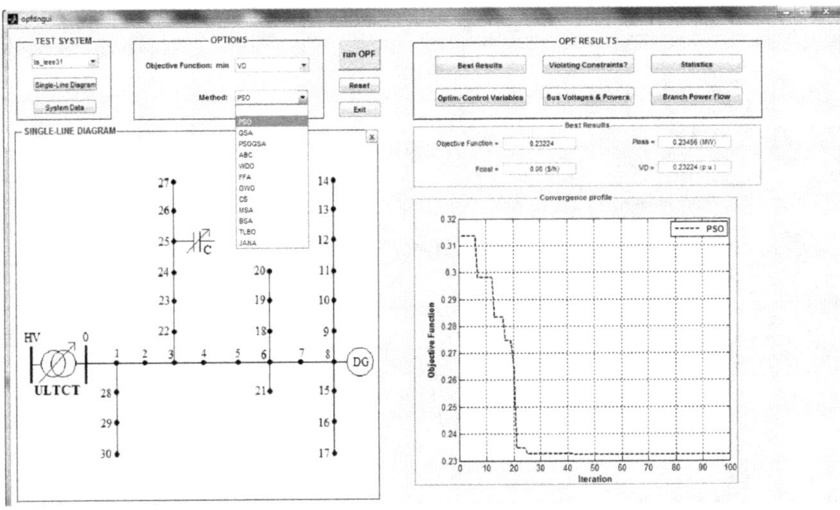

Figure 10.11 Main window of the opfdngui program used for optimal Volt/Var control

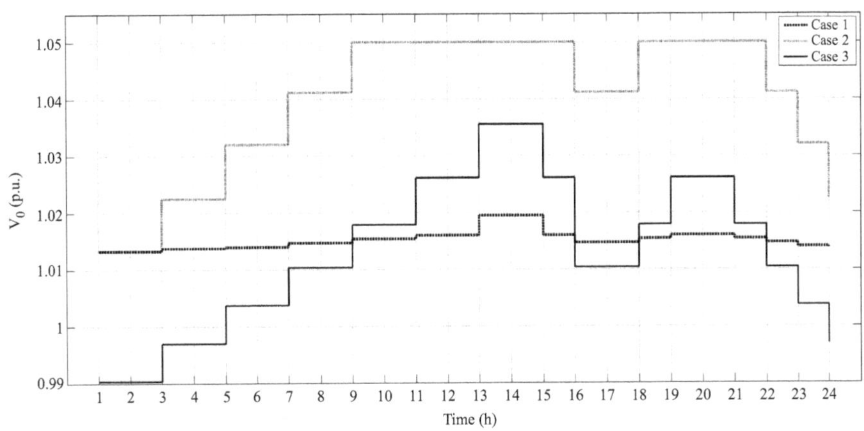

Figure 10.12 Optimal voltage at the root bus, V_0

compensation of reactive power. In Case 2, high values for V_0 and V_{DG} are obtained throughout the load range. This is because the loads were modeled with constant active and reactive power (PQ model). Since power losses are proportional to the square of the branch current, for constant load power, minimizing power losses is achieved by increasing the bus voltage.

Figure 10.16 shows the voltage profiles of the network for three load levels ($L = 0.3$, $L = 0.6$, and $L = 0.9$) for all three cases. As can be seen from Figure 10.16, Case 3 can be considered as a compromise solution, as it also reduces power losses and VDs at the same time.

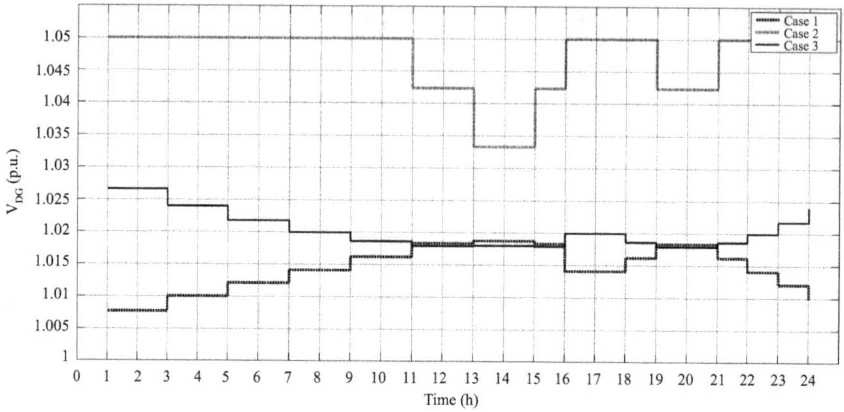

Figure 10.13 Optimal voltage at the DG bus, V_{DG}

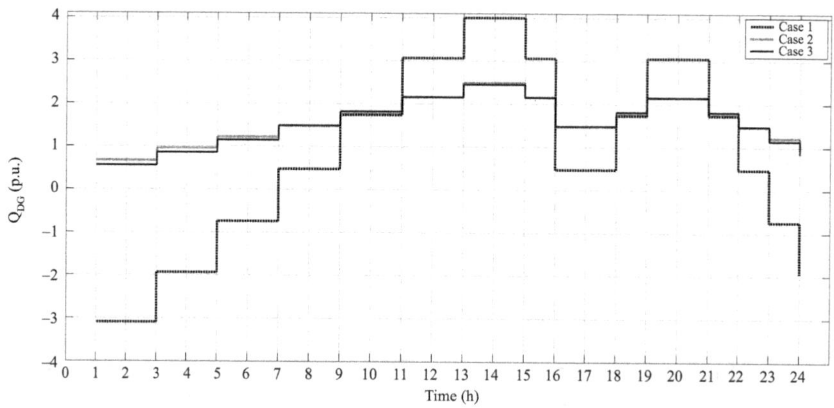

Figure 10.14 Reactive power output of the DG, Q_{DG}

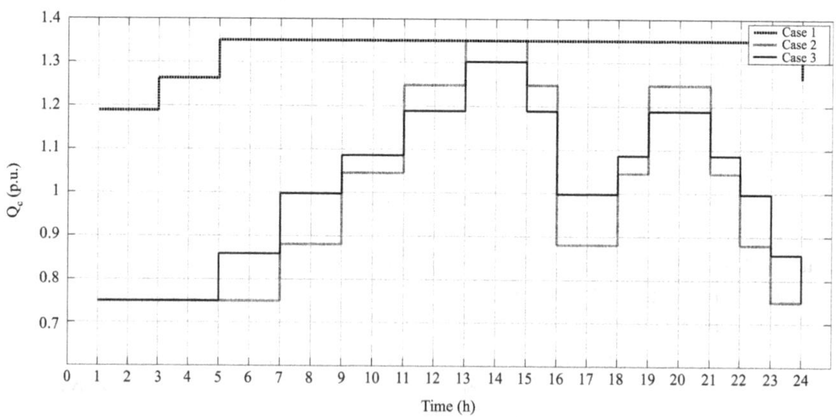

Figure 10.15 Optimal reactive power of the compensator at bus 25, Q_C

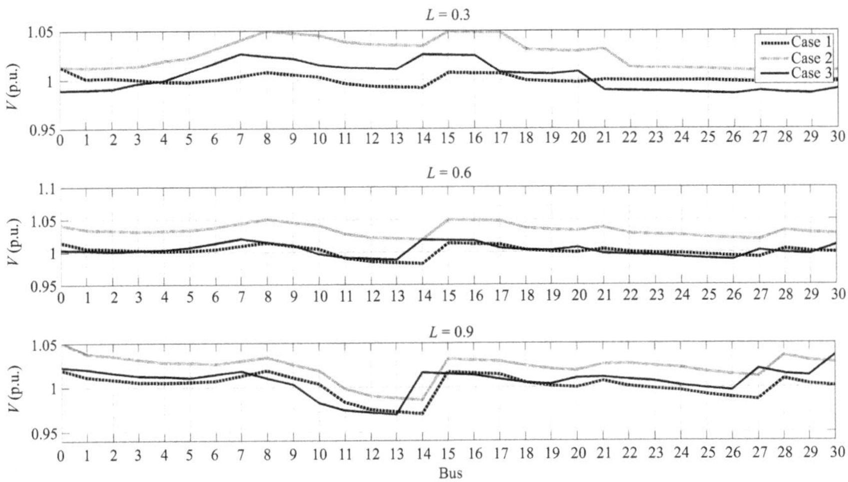

Figure 10.16 Voltage profile

10.4 Conclusion

This chapter deals two methodologies for optimal voltage control: (i) the methodology for optimal seasonal voltage control based on the space and time decomposition of the voltage-control problem and (ii) the methodology for optimal Volt/Var control in distribution networks based on metaheuristic optimization methods.

The main characteristics of the procedure for determining the optimal seasonal voltage-control plan in rural distribution networks with DGs are:

- Spatial decomposition principle of the voltage-control problem;
- Bringing down the LV network to the equivalent line with the aim to apply the optimal setting of the transfer ratio of MV/LV distribution transformer;
- Additional improvement of the network voltage profile by correcting the voltage at PV buses.

The case studies have shown the practical application of the described procedure in the distribution networks with a larger number of DGs modeled as PQ and PV buses.

The application of the procedure on a real 33-node distribution network with DGs modeled as PQ nodes has shown the following:

- The outstanding variation of the generated power during the year and a significant space dispersion of the MHPs caused the voltage variation out of the standard limits.
- The optimal model of the seasonal voltage control at transformers enables the control of the voltage values within the standard limits, except for some outlying nodes.

Simulation results presented in the second part of this chapter shows that the metaheuristic optimization methods are very suitable for solving the problem of optimal Volt/Var control in active distribution networks.

References

[1] Strezoski V.S., Katic N.A., Janjic D.S. 'Voltage control integrated in distribution management system'. *Electric Power Systems Research.* 2001;**60**:85–97

[2] Švenda G., Simendić Z., Strezocki V. 'Advanced voltage control integrated in DMS'. *Electrical Power and Energy Systems*. 2012;**43**:333–43

[3] Choi J.H., Kim J.C. 'The online voltage control of ULTC transformer for distribution voltage regulation'. *Electrical Power and Energy Systems*. 2001;**23**:91–8

[4] Saied M.M. 'The global voltage regulation: a suggested measure for the supply quality in distribution networks'. *Electrical Power and Energy Systems*. 2001;**23**:427–34

[5] Choi J.H., Moon S.I. 'The dead band control of LTC transformer at distribution substation'. *IEEE Transactions on Power Systems*. 2009;**24**(1): 319–26

[6] Liu Y., Zhang P., Qiu X. 'Optimal Volt/Var control in distribution systems.' *Electrical Power and Energy Systems*. 2002;**24**:271–6

[7] Hu Z., Wang X., Chen H., Taylor G.A. 'Volt/Var control in distribution systems using a time-interval based approach'. *IEE Proceedings – Generation, Transmission and Distribution*. 2003;**155**(5):548–54

[8] Kim G.W., Lee K.Y. 'Coordination control of ULTC transformer and STATCOM based on an artificial neural network'. *IEEE Transactions on Power Systems*. 2005;**20**(2):580–6

[9] Park J.Y., Nam S.R., Park J.K. 'Control of a ULTC considering the dispatch schedule of capacitors in a distribution system'. *IEEE Transactions on Power Systems*. 2007;**22**(2):755–61

[10] Liang R.H., Chen Y.K., Chen Y.T. 'Volt/Var control in a distribution system by a fuzzy optimization approach'. *Electrical Power and Energy Systems*. 2011;**33**:278–87

[11] Mokgonyana L., Zhang J., Zhang L., Xia X. 'Coordinated two-stage volt/var management in distribution networks'. *Electric Power System Research*. 2016;**141**:157–64

[12] Kojovic L. 'Impact of DG on voltage regulation'. *Proceedings of IEEE Power Engineering Society Summer Meeting*; Chicago, USA, Jul 2002. IEEE; 2005. pp. 97–102

[13] Kashem M.A., Ledwich G.L. 'Multiple distributed generators for distribution feeder voltage support'. *IEEE Transactions on Energy Conversion*. 2005;**20**(3):676–84

[14] Mollen M.H.J., Sannino A. 'Voltage control with inverter-based distributed generation'. *IEEE Transactions on Power Delivery*. 2005;**20**(1): 519–20

[15] Cheng C.S., Shirmohammadi D. 'A three-phase power flow method for real-time distribution system analysis'. *IEEE Transactions on Power Systems*. 1995;**10**:671–9

[16] Kim T.E., Kim J.E. 'Voltage regulation coordination of distributed generation system in distribution system'. *Proceedings of Power Engineering Society Summer Meeting*; Vancouver, Canada, Jul 2001. IEEE; 2002. pp. 480–4

[17] Bignucolo F., Caldon R., Prandoni V. 'Radial MV networks voltage regulation with distribution management system coordinated controller'. *Electric Power Systems Research*. 2008;**78**:634–45

[18] Kim M., Hara R., Kita H. 'Design of the optimal ULTC parameters in distribution system with distributed generators'. *IEEE Transactions on Power Systems*. 2009;**24**(1):297–305

[19] Robbins B.A., Dominguez-Garcia A.D. 'Optimal reactive power dispatch for voltage regulation in unbalanced distribution systems'. *IEEE Transactions on Power Systems*. 2016;**31**(4):2903–13

[20] Vaiwan F.A., Sannino A., Daalder J. 'Voltage control with on-load tap changers in medium voltage feeders in presence of distributed generation'. *Electric Power System Research*. 2007;**77**:1314–22

[21] Castro J.R., Saad M., Lefebvre S., Asber D., Lenoir L. 'Optimal voltage control in distribution network in the presence of DGs'. *Electrical Power and Energy Systems*. 2016;**78**:239–47

[22] Niknam T., Ranjbar A.M., Shirani A.R. 'Impact of distributed generation on Volt/Var control in distribution networks'. *Proceedings of IEEE Bologna PowerTech Conference*; Bologna, Italy, Jun 2003. IEEE; 2003. pp. 1–7

[23] Vaiwan F.A., Karlsson D. 'Combined local and remote voltage and reactive power control in the presence of induction machine distributed generation'. *IEEE Transactions on Power Systems*. 2007;**22**(4):2003–12

[24] Hong Y.Y., Luo Y.F. 'Optimal VAR control considering wind farms using probabilistic load-flow and Gray-based genetic algorithm'. *IEEE Transactions on Power Delivery*. 2009;**24**(3):1441–9

[25] Liu M.B., Canizares C.A., Huang W. 'Reactive power and voltage control in distribution systems with limited switching operations'. *IEEE Transactions on Power Systems*. 2009;**24**(2):889–99

[26] Mohapatra A., Bijwe P.R., Panigrahi B.K. 'An efficient hybrid approach for Volt/Var control in distribution systems'. *IEEE Transactions on Power Delivery*. 2014;**29**(4):1780–8

[27] Kolenc M., Papič I., Blažič B. 'Coordinated reactive power control to achieve minimal operating costs'. *Electrical Power and Energy Systems*. 2014;**63**:1000–7

[28] Wang S., Wang N., Wang Q., Yang Y., Huang G. 'Reactive power optimization in distribution network by using improved gravitational search algorithm'. *Proceedings of 5th International Conference on Electric Utility Deregulation and Restructuring and Power Technologies*; Changsha, China, Nov 2015. IEEE; 2015. pp. 178–82

[29] Madureira A.G., Lopes J.A.P. 'Coordinated voltage support in distribution networks with distributed generation and microgrids'. *IET Renewable Power Generation*. 2009;**3**(4):439–54

[30] Senjyu T., Miyazato Y., Yona A., Urasaki N., Funabashi T. 'Optimal distribution voltage control and coordination with distributed generation'. *IEEE Transactions on Power Delivery*. 2008;**23**(2):1236–42

[31] Villacci D., Bontempi G., Vaccaro A. 'An adaptive local learning-based methodology for voltage regulation in distribution networks with dispersed generation'. *IEEE Transactions on Power Systems*. 2006;**21**(3):1131–40

[32] Galdi V., Vaccaro A., Villacci D. 'Voltage regulation in MV networks with dispersed generations by a neural-based multiobjective methodology'. *Electric Power System Research*. 2008;**78**:785–93

[33] Niknam T., Firouzi B.B., Ostadi A. 'A new fuzzy adaptive particle swarm optimization for daily Volt/Var control in distribution networks considering distributed generators'. *Applied Energy*. 2010;**87**:1919–28

[34] Cheng S., Chen M.Y. 'Multi-objective reactive power optimization strategy for distribution system with penetration of distributed generation'. *Electrical Power and Energy Systems*. 2014;**62**:221–8

[35] Chen S., Hu W., Su C., Zhang X., Chen Z. 'Optimal reactive power and voltage control in distribution networks with distributed generators by fuzzy adaptive hybrid particle swarm optimization method'. *IET Generation, Transmission & Distribution*. 2015;**9**(11):1096–103

[36] Daratha N., Das B., Sharma J. 'Robust voltage regulation in unbalanced radial distribution system under uncertainty of distributed generation and loads'. *Electrical Power and Energy Systems*. 2015;**73**:516–27

[37] Zhang L., Tang W., Liang J., Cong P., Cai Y. 'Coordinated day-ahead reactive power dispatch in distribution network based on real power forecast errors'. *IEEE Transactions on Power Systems*. 2016;**31**(3):2472–80

[38] Ulinuha A., Masoum M.A.S., Islam S. 'Hybrid genetic-fuzzy algorithm for volt/var/total harmonic distortion control of distribution systems with high penetration of non-linear loads'. *IET Generation, Transmission & Distribution*. 2011;**5**(4):425–39

[39] Jashfar S., Esmaeili S. 'Volt/var/THD control in distribution networks considering reactive power capability of solar energy conversion'. *Electrical Power and Energy Systems*. 2014;**60**:221–33

[40] Jashfar S., Hosseini-Biyouki M.M., Esmaeili S. 'A stochastic programming to Volt/VAR/total harmonic distortion control in distribution networks including wind turbines'. *Electric Power Components and Systems*. 2015;**43**(7):733–46

[41] Seeling B.H.T., de Souza L.J., Prada R.B. 'Voltage control in heavily loaded networks'. *International Transactions on Electrical Energy Systems*. 2015;**25**:3275–87

[42] Radosavljević J., Jevtić M., Klimenta D. 'Optimal seasonal voltage control in rural distribution networks with distributed generators'. *Journal of Electrical Engineering*. 2010;**61**(6):321–31

[43] Radosavljević J., Jevtić M., Klimenta D. 'The analysis of a rural distribution network with distributed generation in catchment area of Stara Planina'. *Facta Universitatis, Series: Electronics and Energetics*. 2009;**22**(1):71–89

[44] Radosavljević J. *opfdngui—Matlab GUI based software for OPF in distribution networks*. 2017. Available at https://www.researchgate.net/profile/Jordan_Radosavljevic2/contributions

Chapter 11

Optimal placement and sizing of distributed generation in distribution networks

11.1 Introduction

Distributed generation (DG) consists of various electrical energy sources of relatively small powers integrated into a local distribution network. Over the past decades, the DG units have been widely used due to a continuous consumption increase, deregulation, and liberalization of the electrical power market, as well as the limitation of pollutant emission. Generally speaking, the integration of DG units in a distribution network has positive technical implications. The DG units can affect the power loss reduction, improving the voltage quality, increasing the reliability and efficiency of consumer supplying by electric power, and pollutant emission reduction (by installation of renewable energy sources).

The degree and quality of an effect a DG has on the distribution network performances depend on its power, type, and location. Optimal placement and sizing of a DG unit aim to maximize the positive and minimize the negative effects it has on the distribution network operation. Accordingly, selection of the objective function depends on the desired objective, which is usually the following:

- minimizing power and energy losses;
- improving the network voltage profile, that is, minimization of voltage deviations;
- increasing the availability of system elements;
- increasing the reliability of power supply to consumer units;
- minimizing the levels of short-circuit currents;
- minimizing harmonic distortions;
- minimizing total costs pertaining to the operation of a network (installation of equipment, purchase of power, power losses, power supply interruptions); and
- minimizing investment, procurement and costs relating to DG connecting, etc.

Based on the possibility of determining locations and sizes, DGs may be classified into two groups. Locations and sizes in the instance of the first group are to the largest extent determined by geographic, hydrological, and meteorological factors (small hydroelectric power plants, wind generators). In this case, optimization of the connection point to the existing distribution network is possible only if a certain set of buses to which a DG may be connected is specified. Optimal solutions may be looked for during the stage of planning of an entirely new or reconstruction of the existing network in a specified (geographically and meteorologically) permissible area. The second group includes those DGs with predictable power outputs (micro turbines, fuel cells, diesel generators, and similar), which may be connected to any point in a distribution network.

Several approaches to resolving the problem of selecting an optimal location and size of DGs are represented in scientific literature, starting from analytical techniques, through methods of linear and quadratic programing, all the way to recently rather popular metaheuristic optimization methods based on mimicking natural processes and patterns.

An analytical approach to determining optimal locations of DGs in networks with radial and loop topology, which does not include computations of power flows, is presented in [1]. For simple radial systems, the procedure is based on the assumption that the feeder's parameters are constant (with the same cross-section in all sections), whereas load is uniformly distributed along the feeder. According to this assumption, an analytic function of change in the current is derived along the feeder with the change in the distance from the feed point. The optimal point of connection of a DG with a defined power is determined based on the condition of minimum power losses. In large networks with looped topology, the procedure is based on modifying the bus admittance matrix when adding DGs to network buses and examining the values of the objective function after the modification that has occurred. Assessing the favorability of buses for connecting a DG in terms of improved reliability of load supply, reduction of losses and an improved voltage profile of the network in [2] is carried out by means of the voltage sensitivity index and loss sensitivity index (*LSI*) of network buses. According to [3], when characteristics of costs per unit of DG power are known, optimal location and power of a DG in a distribution network are determined according to the optimal power flow principle. The method is based on computing bus location marginal costs. The aim of such optimization is to generate the maximum profit for DG owners, since in deregulated energy markets, the price of energy is largest in peak (maximum) load periods; therefore, engaging a DG in said period generates the largest profit for the owner. Recently, some interesting analytical approaches proposed to solve the problem of optimal siting and sizing of DG in radial networks. In [4], the authors formulated a loss sensitivity factor for the distribution networks and employed it for determination of the optimum size and location of DG with aims to minimize total power loss. Elsaiah *et al.* [5] proposed an analytical method for placement and sizing of DG based on a new formulation for the power flow problem, which is noniterative, direct and involves no convergence issues even for systems with high r/x branch ratios. Naik *et al.* [6] determine optimal siting and sizing of DG using suitable analytical expressions which are based on change in active and reactive components of branch currents caused by the DG placement. Similar approach was proposed by Viral and Khatod [7]. They developed analytical expressions which are based on a minimizing the loss associated with the active and reactive component of branch currents by placing the DG at various locations.

The methodology based on application of linear programing presented in [8] solves the problem of optimal locations for the purpose of maximizing sizes of DGs which may be connected while not impeding technical constraints pertaining to permissible values of bus voltages and limit values of currents/powers of feeders and transformers. An optimization methodology for establishing optimal locations and sizes of DGs is developed in [9] by applying the sequential quadratic programing algorithm with the purpose of reducing losses of power and costs of DGs in a distribution network, while respecting voltage constraints.

In recent years, most authors use some of metaheuristic population-based methods for solving the problem of optimal placement and sizing of DG units.

Genetic algorithm (GA). In a large number of scientific papers, the GA is used to solve the problem of optimal DG placement and size. A procedure based on the GA for establishing an optimal DG location is proposed in [10], while taking into consideration all technical constraints such as feeder capacity limits, the feeder voltage profile and permissible voltage, and level of short-circuit current deviation in the network buses. Total network reconstruction costs are taken as an objective function to be minimized, which would have the same effect on the loss reduction as it is the case with connecting a DG. A methodology for determining optimal DG allocation and sizing is presented in [11] with the purpose of maximizing the profit/cost ratio, where the profit is equivalent to the power loss reduction in the network, and costs pertain to the investment of constructing (purchasing) and connecting a DG while guaranteeing an acceptable level of reliability and of the network voltage profile. The GA combined with methods to evaluate DG impacts on reliability, power losses, and voltage profile was used to solve the optimization task. Solving the optimal DG placement problem in [12] is achieved by combining the fuzzy technique and the GA. The optimization objective selected was to minimize power losses in addition to technical constraints pertaining to either the number or size of a DG and deviation of the bus voltage. The objective function with constraints is transformed into the multiobjective function and modeled by a fuzzy set. Encompassing several objective functions during the process of determining optimal DG placement and size, such as minimization of network reconstruction costs, energy purchase, energy losses, and voltage interruptions, is achieved in [13], where a form of the GA is used as a method of optimization. The problem of optimal DG placement and size is observed in [14] in the context of optimal power flow. Application of the GA is proposed to solve the problem. The method presented in [15] employs the GA to simultaneously establish optimal DG and reactive power compensators location and size with the purpose of increasing the reliability and reducing power losses in a distribution network. The authors [16] showed that load models can significantly affect the optimal location and sizing of DGs in distribution networks. A multiobjective performance index (IMO) for the performance calculation of distribution systems for DG size and location planning with load model considers is defined. GA is used to find the best solution. Based on simulation results, the authors [17] concluded that the GA method for optimal placement and sizing of the DGs in radial and networked distribution networks gives better results than an analytical approach. The authors in [18,19] proposed a combined

GA/particle swarm optimization (PSO) for optimal location and sizing of DG in distribution systems. The problem is defined as multiobjective optimization with several indices such as network power losses, voltage deviation index, voltage stability index, and reliability index. In [20], the authors used a MATLAB® realization of GA for optimal allocation of different types of DGs in order to minimization of cumulative average daily active power losses. The authors [21] evaluated the impact of micro-DG on the overall process of low voltage network planning using GA. Some of the authors [22,23] devoted special attention on the problem of optimal DG placement in context of power quality improvement. GA-based tabu search (TS) method is proposed in [24] for determination of optimal location of DG units as independent private sector in a distribution system. The non-dominated sorting GA (NSGAII) approach is proposed in [25] for solving the optimal sitting of DG units in power system as a probabilistic multiobjective optimization problem with objectives such as total imposed costs, total network losses, customer outage costs as well as absorbed private investments. The authors [26] considered optimal placement of DG as a tool for improving the distribution system reliability. The GA is used for solving the nonlinear optimization problem. A combination of analytical and GA methods is proposed in [27] for determination of optimal location and size of DG units to minimize the distribution system power loss. Wu *et al.* [28] proposed cloud theory adapted GA to determine the optimal location and the capacity of DG units in the distribution networks considering different aspects such as energy loss, voltage quality and stability, and line loadability.

PSO. The authors [29] used a local PSO variant to solve the problem of optimal placement of DG units, taking into account load variations in the distribution network. In [30,31], PSO is applied for optimal sitting and sizing of DG to voltage profile improvement, loss reduction, and total harmonic distortion reduction in distribution networks. Prommee and Ongsakul proposed an adaptive PSO [32] and an improved reinitialized social structure PSO [33] for solving optimal multiple DG placement in a distribution network in order to minimize the real power losses. The authors [34] introduced a hybrid method, which employs discrete PSO and optimal power flow to achieve optimal site and size of DG units. PSO was used in [35] to determine the optimal size and location of different types of DGs by minimizing the system power loss considering the exact loss formula as objective function. Siano and Mokryani [36] proposed a hybrid optimization method for optimal allocation of wind turbines in radial distribution networks. This method used PSO to select the optimal sizes of wind turbines while the market-based OPF is used to determine the optimal number of wind turbines at each candidate bus. Ameli *et al.* [37] applied multiobjective PSO to find the optimal location and size of DGs as well as their optimal-generated electricity prices in a competitive market, whereas Zeinalzadeh *et al.* [38] proposed a new application of multiobjective PSO with the aim of determining optimal location and size of DGs and shunt capacitor banks simultaneously with the consideration of load uncertainty in distribution systems. Reference [39] propose a long-term scheduling for optimal allocation and sizing of different types of DG units in the distribution networks in order to minimize power losses. PSO is used to solve the optimization problem. Nasri *et al.* [40] introduced an optimal DG planning methodology based on PSO algorithm in order to minimize the system energy losses. The planning problem is formulated for dispatchable and nondispatchable DG units in different scenarios. An improved multiobjective PSO with a preference strategy has been proposed and successfully applied to determine the optimal siting and sizing of DG units into the distribution system [41]. Bohre *et al.* [42] introduced a novel multiobjective function for optimal sizing and placement of DGs considering different indices such as active and reactive power loss indices, voltage deviation index, reliability index, and shift factor indices. PSO was used to optimize the multiobjective function. Recently, Kumawat *et al.* [43] proposed a PSO-based method for determining the optimal size and location of DGs in distribution network to minimize annual energy loss, considering time-varying characteristics of electrical load demand.

Gravitational search algorithm (GSA). Usually, the main objective in optimal location and sizing of DG units is to minimize active power loss and improve voltage profile of distribution system. Mistry *et al.* [44] use GSA to achieve this objective. The authors in [45,46] employed GSA along with PSO to find best location and size of multi-DG units in distribution system considering different aspects such as the total real power losses, the voltage profile, the mega volt ampere intake by the grid, the DG quantity, and the greenhouse gases emission. An improved GSA by introducing a chaotic gravitational constant is applied in [47] for determining the optimal placement and sizing of DG in a radial distribution system to minimize the total losses, total voltage harmonic distortion, and voltage deviation. One more application of GSA to the optimal allocation and sizing of multi-DG in distribution system for minimizing power loss is given in [48]. The authors concluded that the GSA has better performances in comparison to evolutionary programing. Kundu and Chakrabarti [49] used GSA for optimal placement and sizing of DG units in order to improve nodal pricing and system voltage profiles. Niknam *et al.* [50] proposed θ-self-adaptive GSA to achieve the optimal places for fuel cell power plants (FCPPs) in distribution networks and daily optimal active powers of distribution substation and FCPPs.

In addition to GA, PSO, and GSA, also in recent literature, other population-based metaheuristic algorithms have been proposed to solve the problem of optimal siting and sizing of DG units. Nayak *et al.* [51] proposed differential evolution (DE) algorithm for optimal placement and sizing of DG in radial distribution system to minimize the total real power loss and improve the voltage profile within the framework of system operation and security constraints. The authors [52,53] applied backtracking search algorithm for optimal placement and sizing DGs. Ant colony optimization (ACO) [54], artificial bee colony (ABC) [55], and a hybrid ACO–ABC algorithm [56] have been applied to solve the problem of optimal placement and sizing of DGs. Sultana *et al.* [57] proposed gray wolf optimizer for multiple DG allocation in the distribution system. Hassan *et al.* [58] hybridized GA with firefly algorithm to solve DG placement problem. Cuckoo search algorithm is proposed in [59] for optimal location and sizing of DG to minimize the total power losses and improve voltage stability within a radial distribution network. Bacterial foraging optimization algorithm [60], teaching–learning-based optimization [61], imperialistic competitive algorithm [62], and some other metaheuristic algorithms are also applied to the solution of this problem.

A higher penetration of a DG requires a somewhat altered distribution network planning strategy. In that respect, an optimization model has been developed in [63,64] which, amongst other things, takes into consideration an optimal location and size of a DG in a network. An optimal solution is employed to minimize investment and operating costs. A method for optimal planning of the location and size of microhydropower resources in a catchment area is presented in [65] with the purpose of minimizing investment costs per unit of generated electrical power. This method enables selecting an optimal turbine type for each site, setting its optimal head, flow, and power output.

The problem of optimal DG positioning and size is regarded in [66] from the aspect of mobile reactive compensators, which are to ensure maximum voltage stability in a distribution network. A quantitative voltage stability index of load buses is proposed. An optimal DG location and size is established based on the mentioned index. The TS algorithm is applied in [67] for concurrent determination of locations and sizes of DGs, as well as of reactive power sources, whereas a multiobjective function of costs is employed and it consists of power losses, feeder load, and total reactive power capacity of a distribution network.

The algorithms for optimal DG placement and sizing are important supporting tools for the distribution network operators (DNO) since it provide a guideline for proper strategic planning regarding DG penetration. Within the current liberalized energy market, the DG installation is usually ruled by economic incentives related to private investment plans, or by environmental and geographical constraints and criteria. Nevertheless, the prioritization regarding the critical installation nodes for DG could allow the DNO to promote DG installation at critical nodes to satisfy the specifications defined by the optimal DG placement planning. Although the DNO might be unable to canalize the funds of private investments, proper planning about the prioritization of such investment actions should be performed, and be implemented, whenever possible [29].

In order to determine the optimal location and size of a DG, it is necessary to either find or detect the relation between the power injected by a DG at a given point and change in the size of the value which needs to be optimized. The simplest (most trivial) manner is successive placement of DGs at each bus and computation of the given value, whereas various sizes and types of DGs must be taken into consideration, as well as different load levels throughout the observed period. When everything is taken into account, along with the fact that real distribution networks contain up to several thousands of buses, the number of possible combinations that should be taken into consideration becomes immense. However, this number may be significantly reduced by taking into account certain information on the configuration and manner in which the network operates. In addition, a preliminary ranking list of favorable locations may be compiled by applying the procedure for computing the bus location coefficients (*BLCs*).

A procedure for determining optimal DG locations and sizes in radial balanced symmetrical three-phase distribution networks with the objective of minimizing active power and energy losses is presented in this chapter [68,69]. This is a two-step procedure. Step 1 includes computing *BLCs* based on the change in active power losses with the change in injecting active and reactive power in the network buses. In this manner, a preliminary ranking list of favorable locations (buses) is compiled for connecting a DG. By taking the initial *M* buses from the preliminary ranking list out of the total *NB* buses in the network, the number of combinations to be taken into consideration is significantly decreased. Step 2 includes selecting optimal DG locations and sizes from the narrowed set of possible solutions by means of the selected search technique. To that end, two search techniques are used:

- Partial search of variants;
- GA.

11.2 Preliminary locations of DG

The procedure is based on identifying the change in the total power loss of a distributed network caused by the change in the bus injection power. The measure of favorability to connect a DG is expressed by *BLCs*.

Active and reactive power losses of a network constitute an inevitable consequence of the active and reactive energy flow, the powers of which are P and Q, respectively. The power flows, and thereby system power losses are dependent on voltage. Accordingly, the change in power loss, P_{loss}, caused by the change in the injection, P and Q may be expressed in the following manner [2]:

$$\begin{bmatrix} \dfrac{\partial P_{gub}}{\partial P} \\ \dfrac{\partial P_{gub}}{\partial Q} \end{bmatrix} = \mathbf{J}^{-1} \cdot \begin{bmatrix} \dfrac{\partial P_{gub}}{\partial \theta} \\ \dfrac{\partial P_{gub}}{\partial V} \end{bmatrix} \tag{11.1}$$

where $\mathbf{J}$ is the Jacobian matrix,

$$\mathbf{J} = \begin{bmatrix} \dfrac{\partial \Delta P}{\partial \theta} & \dfrac{\partial \Delta P}{\partial V} \\ \dfrac{\partial \Delta Q}{\partial \theta} & \dfrac{\partial \Delta Q}{\partial V} \end{bmatrix} \tag{11.2}$$

Total power losses may be determined as a sum of injected active powers in all network buses:

$$P_{gub} = \sum_{i=1}^{NB} P_i = \sum_{i=1}^{NB} \sum_{j \in \alpha_i} V_i V_j \left[G_{ij} \cos\left(\theta_i - \theta_j\right) + B_{ij} \sin\left(\theta_i - \theta_j\right) \right]; \tag{11.3}$$

The derivative of power losses with respect to the angle and magnitude of the voltage at bus i may be determined from (11.3) as follows:

$$\frac{\partial P_{gub}}{\partial \theta_i} = \sum_{i=1}^{NB} \sum_{j \in \alpha_i} V_i V_j \left[-G_{ij} \sin\left(\theta_i - \theta_j\right) + B_{ij} \cos\left(\theta_i - \theta_j\right) \right] \tag{11.4}$$

$$\frac{\partial P_{gub}}{\partial V_i} = \sum_{i=1}^{NB} \sum_{j \in \alpha_i} V_j \left[G_{ij} \cos\left(\theta_i - \theta_j\right) + B_{ij} \sin\left(\theta_i - \theta_j\right) \right] \tag{11.5}$$

where P_i and Q_i are active and reactive injecting powers at the bus i; V_i, θ_i are the magnitude and angle of the voltage at bus i; V_j, θ_j are the magnitude and angle of the voltage at bus j; α_i is the set of buses connected with the bus i; B_{ij}, G_{ij} are the susceptance and conductance of the network bus admittance matrix element $\left(\underline{Y}_{ij} = G_{ij} + jB_{ij}\right)$, and NB is the total number of buses.

The hybrid form of the equations of power debalans in network buses are as follows:

$$\Delta P_i = P_i - V_i^2 G_{ii} - V_i \sum_{\substack{j=1 \\ j \neq i}}^{NB} V_j \left[G_{ij} \cos\left(\theta_i - \theta_j\right) + B_{ij} \sin\left(\theta_i - \theta_j\right) \right] \tag{11.6}$$

$$\Delta Q_i = Q_i + V_i^2 B_{ii} - V_i \sum_{\substack{j=1 \\ i \neq j}}^{NB} V_j \left[G_{ij} \sin\left(\theta_i - \theta_j\right) - B_{ij} \cos\left(\theta_i - \theta_j\right) \right] \tag{11.7}$$

Elements of the Jacobian matrix are computed by partial derivatives of (11.6) and (11.7) with respect to the angle and magnitude of the bus voltages:

$$\frac{\partial \Delta P_i}{\partial \theta_i} = V_i \sum_{\substack{j=1 \\ j \neq i}}^{NB} V_j \left[G_{ij} \sin\left(\theta_i - \theta_j\right) - B_{ij} \cos\left(\theta_i - \theta_j\right) \right] \tag{11.8}$$

$$\frac{\partial \Delta P_i}{\partial \theta_j} = -V_i V_j \left[G_{ij} \sin\left(\theta_i - \theta_j\right) - B_{ij} \cos\left(\theta_i - \theta_j\right) \right], \quad j \neq i \tag{11.9}$$

$$\frac{\partial \Delta P_i}{\partial V_i} = -2V_i G_{ii} - \sum_{\substack{j=1 \\ j \neq i}}^{N} V_j \left[G_{ij} \cos\left(\theta_i - \theta_j\right) + B_{ij} \sin\left(\theta_i - \theta_j\right) \right] \tag{11.10}$$

$$\frac{\partial \Delta P_i}{\partial V_j} = -V_i \left[G_{ij} \cos\left(\theta_i - \theta_j\right) + B_{ij} \sin\left(\theta_i - \theta_j\right) \right], \quad j \neq i \tag{11.11}$$

$$\frac{\partial \Delta Q_i}{\partial \theta_i} = -V_i \sum_{\substack{j=1 \\ i \neq j}}^{N} V_j \left[G_{ij} \cos\left(\theta_i - \theta_j\right) + B_{ij} \sin\left(\theta_i - \theta_j\right) \right] \tag{11.12}$$

$$\frac{\partial \Delta Q_i}{\partial \theta_j} = V_i V_j \left[G_{ij} \cos\left(\theta_i - \theta_j\right) + B_{ij} \sin\left(\theta_i - \theta_j\right) \right], \quad j \neq i \tag{11.13}$$

$$\frac{\partial \Delta Q_i}{\partial V_i} = 2V_i B_{ii} - \sum_{\substack{j=1 \\ i \neq j}}^{N} V_j \left[G_{ij} \sin\left(\theta_i - \theta_j\right) - B_{ij} \cos\left(\theta_i - \theta_j\right) \right] \tag{11.14}$$

$$\frac{\partial \Delta Q_i}{\partial V_j} = -V_i \left[G_{ij} \sin\left(\theta_i - \theta_j\right) - B_{ij} \cos\left(\theta_i - \theta_j\right) \right], \quad j \neq i \tag{11.15}$$

Values of the magnitudes and angles of the bus voltages are determined by computing the power flows. This enables computing the Jacobian matrix and derivative of the power loss with respect to the bus voltage magnitude and phase. Equation (11.1) enables determining the changes in active power losses in a distribution network caused by the change in injection power in the nodes. As a measure of favorability for DG connection, values of the *LSI* are taken as the *BLC*s, which are, according to [2], determined in the following manner:

$$BLC_i = LSI_i = w \frac{\partial P_{gub}}{\partial P_i} + (1 - w) \frac{\partial P_{gub}}{\partial Q_i} \tag{11.16}$$

In expression (11.16), w is the weight factor dependent on the network r/x ratio. The ratio of resistance and reactance (r/x) of feeders and transformers in distribution networks varies broadly. The approach adopted herein implies that in distribution networks with a vast range of the r/x ratio, the weight factor w is computed for each bus individually. Due to the radial topology of the network and relatively small powers, the change in injection power at a given bus has the largest effect on power losses at the correspondent branch. Given the manner of enumeration of buses and branches in radial distribution network, computation of weight factors is rather simple:

$$w_i = 1 - \frac{1}{(r_i/x_i) + 1} \tag{11.17}$$

where w_i is the weight factor for the bus i; r_i and x_i are resistance and reactance of the ith branch, respectively.

The need for such an approach is particularly justified for lateral branches of a radial distribution network, since ratios of resistances and reactances of distribution transformers differ significantly from the r/x ratio of medium voltage feeders to which they are connected, particularly when smaller distribution transformers are considered in a network. If a distribution network is homogenous (one voltage level), the feeders have approximately the same parameters; therefore, the r/x ratio is within narrow limits. In that instance, a unique weight factor is employed for the entire system and it is determined as a mean value of weight factors of all buses computed according to expression (11.17).

When power is successively injected in two buses with different *BLC* values, the relation between the *BLC* values and power losses in a network depends on the ratio between the DG power and power load. In other words, *BLC* values as a criterion for assessment of favorability of a bus for DG connection should be construed while taking into account the ratio between the DG power and the total network load power. This ratio is important since it defines the power distribution in a network and consequentially losses as well. In general, if DG power is smaller than 50% of the power load, buses with larger *BLC* values are more suitable for connecting, and vice versa: when DG power is larger than 50% of the power load, buses with smaller *BLC* values are considered more suitable for DG connecting.

On the basis of *BLC* values, a set of preliminary buses is defined as a starting point for determining an optimal location or several optimal locations for DG connecting for the purpose of minimizing power and energy losses in a distribution network. To that end, a certain optimization procedure or a search technique is applied. The algorithm for creating preliminary DG locations based on the *BLC*s may be summarized in the following six steps:

1. Defining the topology (configuration) of the network and parameters of elements. The network is defined by enumerating buses by branches, i.e., buses on the longest branch are marked first, followed by marking of buses on other branches, starting from the branch most distant from the root bus.
2. Defining the power load in a given distribution network. It is difficult to determine the exact values, but in the practice, minimum and maximum power of each bus over a certain period may be familiar. On this occasion, power load is determined by multiplying rated values (P_{ni}, Q_{ni}) with the load coefficient (k_{load}), the value of which ranges between 0 and 1,

$$P_{Pi} = k_{load} \cdot P_{ni} \tag{11.18}$$

$$Q_{Pi} = k_{load} \cdot Q_{ni} \tag{11.19}$$

3. Preliminary computation of power flows and bus voltages.
4. Using the magnitudes and angles of network bus voltages computed in step 3, computing the elements of Jacobian matrix [Eqs. (11.8)–(11.15)], and derivatives of the power loss function [Eqs. (11.4) and (11.5)].
5. Computing the *BLC* values for all network buses (11.16).
6. Selection of preliminary DG locations based on the *BLC* values.

If there are DGs in a specific distribution network, when determining *BLC* buses, these are treated as negative constant power loads (*PQ*).

11.3 Partial search of variants

Variant search includes computation of the selected objective function for each combination of the DG size, type, and location. The optimal solution is obtained by comparing the values of objective function. To evaluate a variant, the power flow computation is performed. As the number of possible combinations may be rather large, they are reduced by taking into account certain information on real operating of the network. Accordingly, partial search algorithms appear as a separate group of methods. In such a manner, better or worse local optima are found with a smaller or larger extent of efficiency. The goal is to have them as close to the global optimum as possible, whereas it is insisted that this is performed in as short a time as possible, that is, to search for as many number of combinations in as efficiently formulated manner as possible. This procedure directly implies (includes)

- selection of a favorable set of buses for searching;
- adequate DG modeling;
- taking into account various levels of load over a given period;
- defining the objective function;
- power flow computation;
- taking into account the operating constraints.

11.3.1 Optimal DG placement by using partial search of variants

The basic idea is to narrow the starting set of possible solutions by employing the results of previous research while starting from characteristics of the specific network, and thereby obtain a solution faster. Criteria for narrowing the set may vary. For instance, it was demonstrated that connecting a DG close to the root bus has a minimal impact on the loss reduction and voltage improvement, that is, the best effects in radial networks are achieved by placing a DG in peripheral parts (buses), or on the basis of *BLCs* determined according to the algorithm referred to in Section 11.2.

The number of preliminary locations may be selected arbitrarily. A larger number ensures reliability of the optimal solution, but the number of combinations and time of computation is consequentially increased. On the other hand, a smaller number of preliminary locations not only imply a smaller number of search variants but also a larger probability that the optimal solution found has a local character. In this manner, the number of search variants is reduced significantly. For other practical reasons, sometimes a specific set of buses is given to which a DG may be connected. Therefore, taken as a starting point herein are M_p buses ($M_p < NB$), which constitute a preliminary set of buses for DG connecting. A single combination of locations (buses) for connecting the given number of DGs constitutes a single variant. Thereby, it is adopted that only one DG may be connected to a single bus. If the number of preliminary locations is M_p, and the number of DGs for which optimal locations are to be established is N_{DG}, then the number of variants to be explored is equal to the number of variations without repetition:

$$N_{\text{var}} = N_{DG}!\binom{M_p}{N_{DG}} = M_p(M_p - 1)(M_p - 2)\ldots(M_p - N_{DG} + 1) \quad (11.20)$$

For each possible variant, it is necessary to perform at least one power flow computation in order to determine its quality according to the adopted optimization criterion. Therefore, the efficiency of the power flow computation method is very important. A DG must be adequately modeled based on the output power and voltage regulation ability. The DGs are modeled as *PQ* and *PV* buses herein. The relevance of the DG placement and sizing results is directly related to the accuracy of the model and level of load over the observed period. The term period employed herein implies a single year (8,760 h). The annual load diagram is approximated by a step curve as shown in Figure 11.1. The larger the number of segments in the step curve, the more accurate the approximation.

The symbols in Figure 11.1 have the following meaning: $S_{mean,j}$ is the mean value of power of a typical daily load diagram in the segment j of the annual load diagram; ΔT_j is the duration of the segment j of the annual load diagram; N_{NP} is the total number of annual load diagram segments.

It is assumed that the form of the annual load diagram is the same for all consumer units in the considered distribution network. It means that it is necessary to define only one diagram in which the size is expressed in relative units ranging from 0 to 1. The actual annual load diagram of a consumer unit is obtained by multiplying the rated or maximum power of the consumer unit with the relative diagram.

The optimal solution (optimal DG locations) depends on the adopted optimization criterion. Most commonly used optimization criterion is the effect of a DG on the loss reduction. In order to encompass the effect of the change in load during the course of a year, total annual active energy loss in the distribution network is taken as objective function. Therefore, the aim to be achieved by optimization is minimizing annual active energy loss:

$$\min \Delta W_v = \min \sum_{j=1}^{j=N_{NP}} P^j_{loss,v} \cdot \Delta T_j \tag{11.21}$$

where ΔW_v denotes total annual active energy loss in the network for the vth variant of DG locations, $v = 1,\ldots,N_{var}$; $P^j_{loss,v}$ is the power loss in the network in the segment j of the annual load diagram for the vth variant of DG locations; ΔT_j is the duration of the segment j in the annual loss diagram; N_{NP} is the total number of annual load diagram segments.

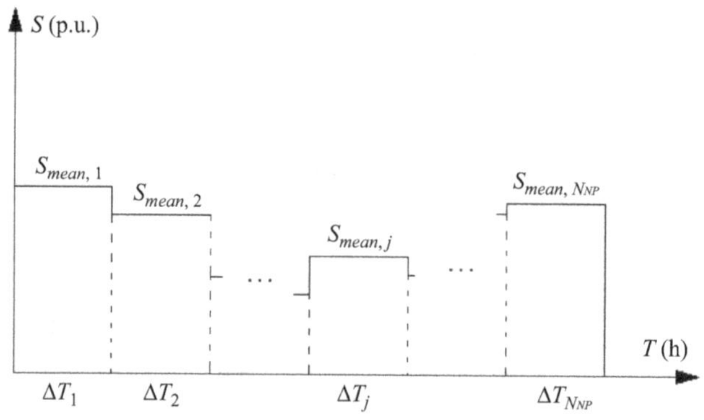

Figure 11.1 Mean values of load over a period (year)

In addition, operational constraints must be satisfied:

$$V_{i\min} \le V_i^j \le V_{i\max} \tag{11.22}$$

$$J_i^j \le J_{i\max} \tag{11.23}$$

where V_i^j is the voltage at bus i for the jth load level; $V_{i\min}$, $V_{i\max}$ are minimum and maximum permissible values of the voltage at bus i, respectively; J_i^j is the current in branch i for the j load level; $J_{i\max}$ is the maximum permissible root mean square (RMS) current of branch i of the network.

The algorithm for determining optimal DG locations is quite simple and it consists of several steps as follows:

1. Defining the network configuration, data on parameters of network elements, and on loads;
2. Defining the annual load diagram parameters: $S_{mean,j}$ and ΔT_j, $j = 1,\ldots,N_{NP}$ (according to Figure 11.1);
3. Defining preliminary locations (buses)—based on $BLCs$, BLC_i, $i = 1,\ldots,M_p$;
4. Defining the number N_{DG}, type, and size of DGs to be allocated;
5. Defining possible DG location variants, $v = 1,\ldots,N_{var}$;
6. Computing the objective function ΔW_v (11.21) and examining the constraints (11.22) and (11.23). If any of the constraints are violated, the current variant v is excluded;
7. Comparing the criterion function values obtained and determining the optimal combination of locations for connecting DGs of the defined types and sizes.

11.3.2 Optimal DG sizing by using partial search of variants

This problem may be characterized as both operation-related and planning-related. An operation-related problem refers to establishing optimal distribution of power among several DGs, that is, establishing the optimal DG operation in a given stationary network mode aimed at loss reduction. A planning-related problem pertains to selection of installed power of DGs to be connected to predefined locations, which are not necessarily optimal. Both cases refer to the selection of optimal variant of operation of DGs connected to given network buses. A single variant is a single possible solution that constitutes a combination of DG output powers from the defined DG power ranges. Various DGs may operate within various power ranges. Owing to the combinatorial nature of the problem, it is convenient to represent the mentioned range as a segment function in the following form:

$$P_{DGk} = a \cdot \Delta P_{DG} \tag{11.24}$$

where a is a coefficient, ($a = 1,2,\ldots,N_s$); N_s is the total number of segments within the DG power range; ΔP_{DG} is a power change step within the defined range.

In a general case, DGs may operate within various power ranges. However, all of them have the same number of segments, the width of which depends on the width of the range of the specific DG. If N_{DG} is the number of DG locations, and if the power range of all DGs is divided into N_s segments, then the number of possible solutions is equal to the number of variations with the repetition $N^p_{\text{var}} = N_s^{N_{DG}}$. As an illustration, if $N_{DG} = 3$ and $N_s = 50$, then the number of variants to be examined is $N^p_{\text{var}} = 125,000$. If we assume that a computer requires 0.1 s to examine one solution variant for a network of several dozens of buses, then it would take about 3.5 h to examine all 125,000 possible variants in this example. Therefore, it is necessary to resort to a drastic decrease in the number of variants to be examined. The approach applied herein includes simultaneous repetition of search. For instance, if the initial DG power range from the previous example is divided into ten segments, the number of possible solutions is 1,000. The first search results in a rough estimate of optimal DG size. Then, in the following step, new DG size ranges are estimated close to the optima obtained in the previous step. The search process is repeated on the newly formed size ranges with the same number of segments. The entire cycle is repeated several times. In this manner, a satisfactory solution is obtained with a drastic decrease in the number of variants examined. Figure 11.2 shows the described approach of reducing the number of variants in the partial search process.

The objective is to minimize power losses of a network for the given load power level. The process of partial search of variants is performed for any level of consumption over a given period. The result is optimal DG size for all load levels over a given period. For each DG size variant from within the defined ranges, constraints (11.22) and (11.23) are examined. Violation of any constraint means eliminating the current variant. The criterion function constitutes power losses of a network, $P^j_{loss,v}$, where v denotes (is the number of) the variant and j is the segment over the period (year). The algorithm for optimal DG sizes at the given locations by applying partial search of variants includes the following steps:

1. Defining the network configuration, data on parameters of network elements, and on loads;
2. Defining the load diagram parameters in a given period: $S_{mean,j}$ and ΔT_j, $j = 1, \ldots, N_{NP}$;
3. Defining locations and DG power ranges;
4. Defining the number of N_s segments for the DG size range;

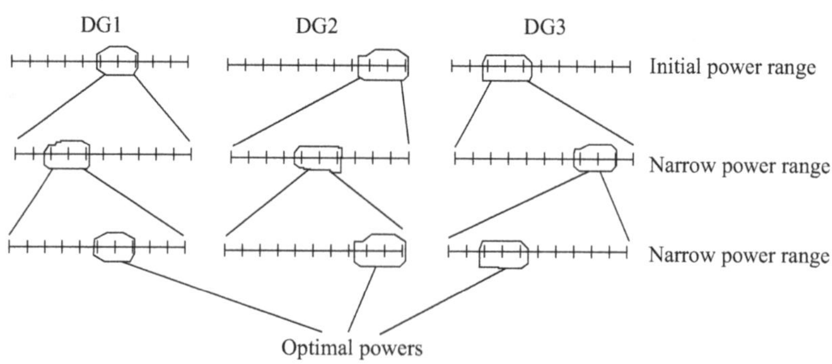

Figure 11.2 Reducing the number of variants by narrowing the DG power range

5. Defining the number of repetitions so as to narrow the range;
6. Dividing the size ranges of all DGs into a uniform number of segments N_s;
7. Defining possible variants of DG sizes—solutions, $v = 1, \ldots, N_{\text{var}}$;
8. Computing power losses, $P^j_{loss,v}$, for various variants of DG sizes (v) for a given level of load power (j);
9. Verifying whether conditions (11.22) and (11.23) have been met. If any of the given conditions have been breached, the current variant is excluded;
10. Comparing the obtained values of power losses and establishing optimal DG sizes;
11. Narrowing the DG size range close to optimal values and returning to step 6.

The algorithm is performed for any power load level $j = 1, \ldots, N_{NP}$.

11.4 Genetic algorithm

Distribution network optimization involves solving the problem of optimal locations and DG powers. Distribution networks of real dimensions have up to few hundreds of nodes. The nature of variables whose values need to be optimized is twofold. In addition to continuous variables such as voltage and power, there are discrete and integer variables such as locations (nodes), and number of DG units. For this case, the application of classical optimization methods based on the differentiation of an objective function can be difficult. Therefore, the so-called "evolutionary optimization techniques" are used to overcome this problem. They are based on searching for a set of possible solutions using the rules of natural evolution. Due to the large number of possible solutions, the problems of optimizing the distribution network with DG can be considered as combinatorial optimization problems.

One of the well-established evolutionary methods that is very effective in solving these combinatorial optimization problems is the GA. The GA as an optimization method was created by Holland [70]. He elaborated an idea how to transform the characteristics of natural evolution into a computer program.

The GA operate with a population of individuals. Each individual constitutes a potential solution to a given optimization problem. An individual can be described as a set of variable conditions, the values of which are being optimized. The quality of an individual is quantified over the value of the fitness function or the function of goodness. A population of individuals constitutes a set of solutions to the given optimization problem. One generation has a population with a certain number of individuals with better or worse fitness function values. Through a series of GA operations, a new population is obtained and its individuals are created by the individuals from the previous population according to the natural evolution principles: the selection, crossover, and mutation.

The basic steps of the GA are:

Step 1. Representation of individuals: All data (variables) that make an individual are written in a string. A string is composed of substrings. Each substring represents a binary encoded variable on which the process of optimization is carried out.

Step 2. Initialization: Individuals with random strings are generated that set up the initial population.

Step 3. Fitness function calculation: It is used to rate the quality of an individual and it represents an equivalent of the function that should be optimized, that is, objective function.

Step 4. Selection: During the selection process, the individuals that will participate in the reproduction (parents) are selected. The point of the selection is to store and transfer good individuals to the next generation.

Step 5. Crossover: The way in which coded column parts (substrings) are crossed over actually makes a GA. Crossover is an exchange process of column parts between two individuals, that is, "parents." One or two new individuals engender by the crossover, that is, a "child." The possibility of inheriting the first parent's characteristics by a child is introduced during this process.

Step 6. Mutation: Mutation is a way to give a new piece of information to an individual. Mutation represents an accidental bit variation of an individual, generally with a constant probability for each bit within a population.

Step 7. Ending conditions: The GA is an iterative process which ends when a maximum number of generations is achieved or when another criterion is fulfilled, such as a minimum offset from the best fitness value and medium fitness value of all individuals in a current population. Otherwise, return to step 3.

The main advantages of GA over the classical optimization methods are:

- The objective function to be optimized is entirely optional, that is, there are no special requirements such as continuity, differentiability, etc.;
- Reliability of the results obtained may be increased by simple repetition of the process;
- They solve any problem that may be presented as optimization-related, regardless of whether the state variables are real numbers, bits, or characters;
- Availability of program support. There are ready-made software packages that may be applied to solving concrete problems. For instance, the global optimum toolbox in the MATLAB software package.

The main weakness of GA are as follows:

- Parameters exert a considerable impact on efficiency. There is no universal rule for parameter setting;
- It is impossible to obtain a solution that is 100% reliable.

11.4.1 Optimal DG placement and sizing by using GA

In this optimization problem, the total energy losses of distribution network is considered as an objective function, whereas the control variables are locations (i.e., indexes of connecting buses) and power outputs of the DG units. Therefore, a potential solution of the problem can be represented by a vector (string) consisting a combination of locations and powers of DG units as shown in Figure 11.3.

Both DG types, *PQ* and *PV* are considered. The total energy losses are computed based on the system power losses and time during which the mentioned losses occur. In order to obtain as realistic a model as possible, the load curve is approximated by a step curve with N_{NP} levels, as shown in Figure 11.1. An objective function of energy losses, in a general case, may be formulated in the following manner:

$$\min \Delta W = \min \sum_{j=1}^{j=N_{NP}} P^j_{gub} \cdot \Delta T_j \tag{11.25}$$

where P^j_{loss} is the power loss at the jth load level; ΔT_j is the duration of the jth load level.

At the same time, the following operating constraints must be satisfied:

1. Power flow equations

$$f^j\left(x_j, P^j_{DGi}\right) = 0 \tag{11.26}$$

where P^j_{DGi} is the power of DG at the bus i in case the jth load level; x_j is the state variable (bus voltages) for the jth load level.
2. *The bus voltages and branch currents must be within the permissible limits*:

$$V_{i\min} \le V_i^j \le V_{i\max} \tag{11.27}$$

$$J_i^j \le J_{i\max} \tag{11.28}$$

where V_i^j is the voltage at bus i for the load level j; $V_{i\min}, V_{i\max}$ are minimum and maximum permissible values of voltage at the bus i; J_i^j is current of the ith branch for the load level j; $J_{i\max}$ is the maximally permissible current of the branch i.
3. *Constraints pertaining to the size or number of DGs that may be connected to certain distribution network buses*

$$P^j_{DGi\min} \le P^j_{DGi} \le P^j_{DGi\max} \tag{11.29}$$

where P^j_{DGi} is the size of the DG at the bus i for the jth load level; $P^j_{DGi\min}$, $P^j_{DGi\max}$ are minimum and maximum sizes of a DG that may be connected to the bus i for the jth load level.

1		2			N_{DG}	
Bus	P_{DG}	Bus	P_{DG}	. . .	Bus	P_{DG}

Figure 11.3 An individual (possible solution) for optimal location and size of DG

A general flowchart which briefly describe the application of GA in determining the optimal locations and size of DGs in distribution networks is presented in Figure 11.4. The MATLAB program realization of the GA within the *toolbox/gads* module is applied herein.

Starting from the assumption that the total number of DGs to be connected to a distribution network is N_{DG}, a single set of buses is associated to each DG of the defined size type and range. Classification of any bus into a set depends on the given or adopted criterion of favorability for connecting the DG to that bus. Criteria

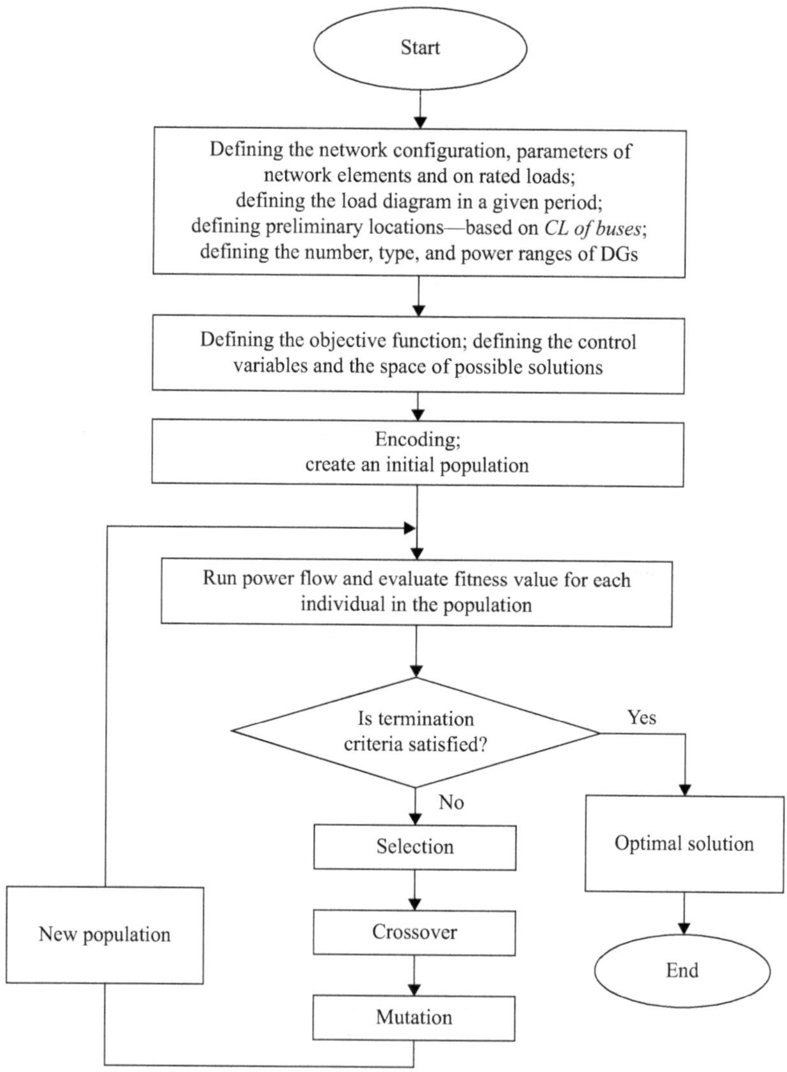

Figure 11.4. General flow chart of DG placement and sizing using GA

may vary, as already mentioned above. The *BLC* values determined in the manner described in Section 11.2 have been taken herein as a criterion for defining the set of preliminary buses.

11.5 Simulation results

Testing of the algorithms presented in Sections 11.2–11.4 was performed on two distribution networks. The first one is the IEEE 31-bus system, for which data was taken from [71]. The other test network is an actual distribution network from the energy distribution company (EDC) in the Zajecar city area, Serbia.

11.5.1 IEEE 31-bus system

The network is shown in Figure 11.5. The data on the network elements, data and load data are given in Table 11.1. In all the computations that follow, the root bus voltage is constant and is equal to 1 p.u.

11.5.1.1 Preliminary locations—*BLC*s

The *BLC*s, adopted as a criteria for determining the preliminary locations of DGs in the network, were computed for two levels of power load: 80% and 40% of the rated power. The results are shown in Figure 11.6. Although the *BLC*s were obtained for two different levels of power load, the shape of the *BLC* diagram in Figure 11.6 is the same in both cases. Somewhat higher values are obtained at a lower level of power load. In both instances, the maximum *BLC* value is at bus 14, whereas the minimum value is at bus 28. The order of buses according to *BLC*

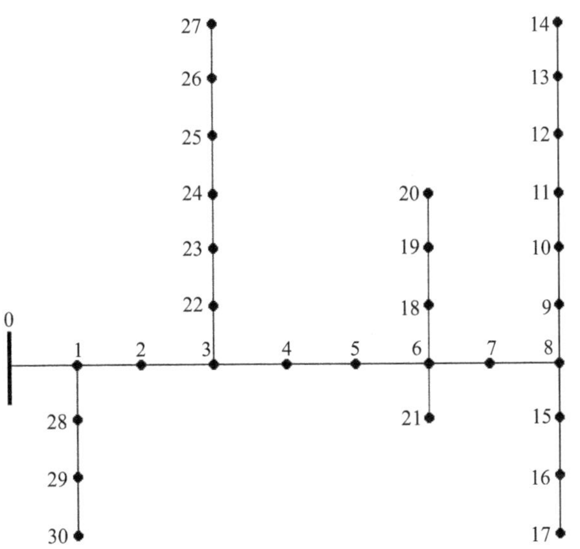

Figure 11.5 IEEE 31-bus test system

Table 11.1 Data for IEEE 31-bus test system

Branch	V_n (kV)	R (Ω)	X (Ω)	R (p.u.)	X (p.u.)	P_p (kW)	Q_P (kVAr)	P_p (p.u.)	Q_P (p.u.)
0–1	23	0.5096	1.7030	0.000963	0.003219	–	–	–	–
1–2	23	0.2191	0.0118	0.000414	0.000022	522	174	0.522	0.174
2–3	23	0.3485	0.3446	0.000659	0.000651	–	–	–	–
3–4	23	1.1750	1.0214	0.002221	0.001931	936	312	0.936	0.312
4–5	23	0.5530	0.4806	0.001045	0.000908	–	–	–	–
5–6	23	1.6625	0.9365	0.003143	0.001770	–	–	–	–
6–7	23	1.3506	0.7608	0.002553	0.001438	–	–	–	–
7–8	23	1.3506	0.7608	0.002553	0.001438	–	–	–	–
8–9	23	1.3259	0.7469	0.002506	0.001412	189	63	0.189	0.063
9–10	23	1.3259	0.7469	0.002506	0.001412	–	–	–	–
10–11	23	3.9709	2.2369	0.007506	0.004229	336	112	0.336	0.112
11–12	23	1.8549	1.0449	0.003506	0.001975	657	219	0.657	0.219
12–13	23	0.7557	0.4257	0.001429	0.000805	783	261	0.783	0.261
13–14	23	1.5389	0.8669	0.002909	0.001639	729	243	0.729	0.243
8–15	23	0.4752	0.4131	0.000898	0.000781	477	159	0.477	0.159
15–16	23	0.7282	0.4102	0.001377	0.000775	549	183	0.549	0.183
16–17	23	1.3053	0.7353	0.002467	0.001390	477	159	0.477	0.159
6–18	23	0.4838	0.4206	0.000915	0.000795	432	144	0.432	0.144
18–19	23	1.5898	1.3818	0.003005	0.002612	672	224	0.672	0.224
19–20	23	1.5389	0.8669	0.002909	0.001639	495	165	0.495	0.165
6–21	23	0.6048	0.5257	0.001143	0.000994	207	69	0.207	0.069
3–22	23	0.5639	0.5575	0.001066	0.001054	522	174	0.522	0.174
22–23	23	0.3432	0.3393	0.000649	0.000641	1,917	639	1.917	0.639
23–24	23	0.5728	0.4979	0.001083	0.000941	–	–	–	–
24–25	23	1.4602	1.2692	0.002760	0.002399	1,116	372	1.116	0.372
25–26	23	1.0627	0.9237	0.002009	0.001746	549	183	0.549	0.183
26–27	23	1.5114	0.8514	0.002857	0.001609	792	264	0.792	0.264
1–28	23	0.4659	0.0251	0.000881	0.000047	882	294	0.882	0.294
28–29	23	1.6351	0.9211	0.003091	0.001741	882	294	0.882	0.294
29–30	23	1.1143	0.6277	0.002106	0.001187	882	294	0.882	0.294

values is obviously not dependent on the level of power load for which these are computed. The form or trend in the diagram shown in Figure 11.6 clearly indicates that the buses at peripheral parts of the network have higher *BLC* values. There is a clear pattern in terms of increase in the *BLC* values for buses which are part of certain feeders. Thus, the trend of increase in the *BLC* values for the main feeder from bus 2 to bus 14 is approximately of a linear character. A similar pattern in terms of increase in the *BLC* values is observed in auxiliary feeders between buses 15–17, 18–20, or 22–27. It may be concluded that the network configuration has a crucial impact on *BLC* values of buses.

Power and energy losses were computed at successive connecting of DGs of various sizes at various load levels for the purpose of observing the relation between *BLC* values and suitability of buses for connecting DGs in terms of power loss reduction in a network. The results are shown in Figures 11.7 and 11.8.

When the diagrams in Figures 11.6 and 11.7 are compared, a relation between the *BLC* values and power losses in the network becomes evident. When DG power is injected in buses with the largest *BLC* values, the loss values are minimal. It is

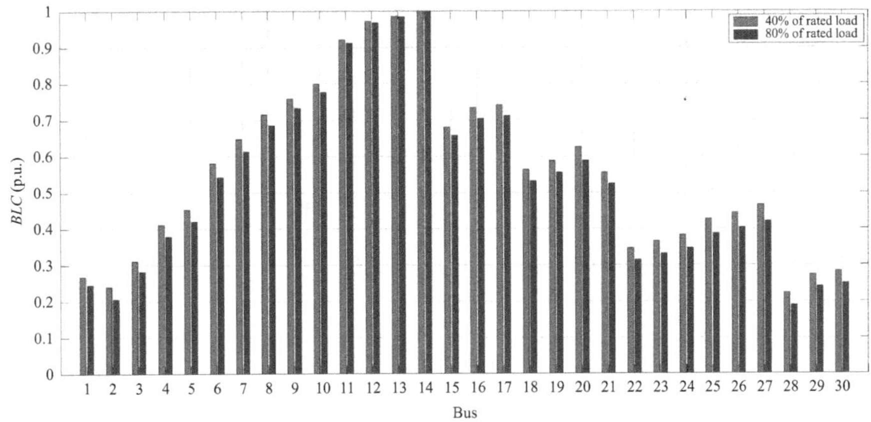

Figure 11.6 BLC values for the IEEE 31-bus test system for two load levels

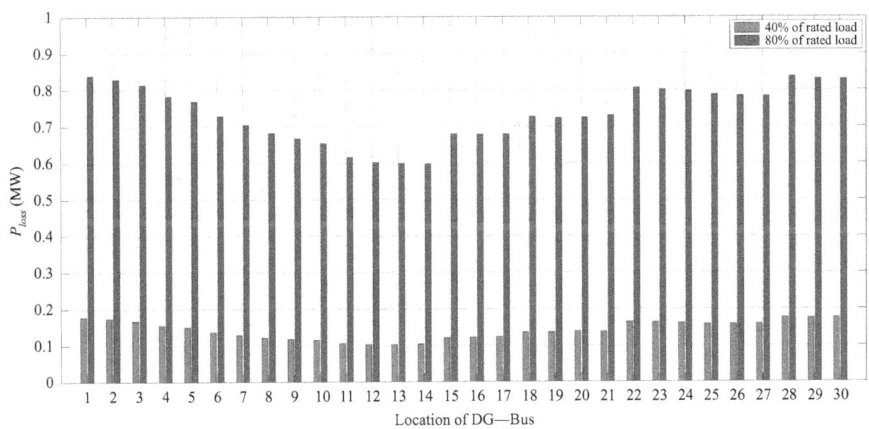

Figure 11.7 Active power losses in the IEEE 31-bus network at successive connecting of a 1-MW DG to buses for two load levels

apparent that the trend of increase in *BLC* values follows the trend of power loss reduction. However, annual energy losses of a network at successive connecting of DGs of various sizes are given in Figure 11.8. Three typical cases were selected: 3 MW, which is 30% of the total rated power load, 7.5 MW, which is 50% of the total rated power load, and 15 MW, which accounts for 100% of the total rated power load of the network. The results obtained clearly indicate that the optimal location of DGs depends on the ratio between the DG size and distribution network load. For a DG of a smaller size, optimal locations are in peripheral parts of the network. As the size of a DG increases, the optimal location is shifted toward the network root. When the diagrams in Figures 11.8 and 11.6 are compared, for the 7.5 and 15-MW DGs which account for 50% and 100% of the network power load,

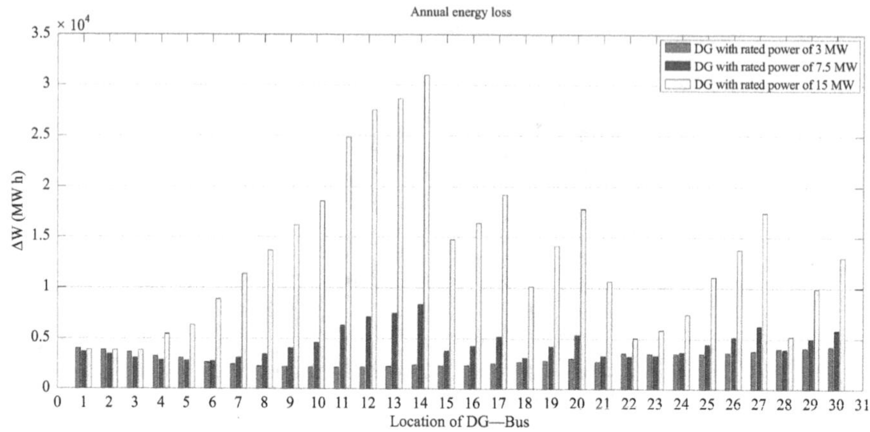

Figure 11.8 Active energy losses in the IEEE 31 bus network at successive connecting of DGs of various sizes to buses at the annual load levels given in Table 11.2

the trend of change in energy losses follows the trend of *BLC* bus values, that is, the diagram shapes become equal to the increase in the DG size. It means that the *BLC* values of buses, as a criterion for forming preliminary locations, must be taken conditionally. In that respect, a tentative framework is given herein for defining preliminary locations based on the *BLC* values depending on the ratio between the DG size and a distribution network power load:

- If the DG size is smaller than 30% of the total network power load, buses with larger *BLC* values are more suitable for connecting a DG compared with buses with smaller *BLC* values. The initial M_n buses are selected for preliminary locations, sorted according to the descending order of *BLC* values starting from the bus with the maximum *BLC* value.
- If the DG size is larger than 70% of the total network power load, buses with smaller *BLC* values are more suitable for connecting a DG compared with buses with larger *BLC* values. Preliminary locations now are M_n buses sorted according to the ascending order of *BLC* values starting from the bus with the minimum *BLC* value.
- In instances of DGs the sizes of which fall between 30% and 70% of the power load, *BLC* values may not be employed as clear criteria for assessing favorability of buses for connecting a DG.

11.5.1.2 Optimal DG placement and sizing by using partial search of variants

The test is performed for DG sizes smaller than 30% of the total power load. Accordingly, ten buses with the largest *BLC* values determined in Section 11.5.1.1 were taken as preliminary optimal locations. These are the following buses: 14, 13, 12, 11, 10, 9, 17, 16, 8, and 15. A 1-year period is observed. The data on power load levels and the duration thereof are given in Table 11.2. Three different scenarios (tests) were considered in this case.

Test B1: Optimal location of DGs of various sizes. The procedure of determining an optimal location is carried out for three values of the rated power of the DG: 3, 2, and 1 MW, whereby the DG was modeled as PQ bus with the unit power factor. The results obtained are given in Figure 11.9 in the form of a diagram of annual energy losses. According to the results shown in Figure 11.9, optimal buses vary for various values of installed power of DGs. For a 3-MW DG, optimal is bus 11, whereas for a 2-MW DG optimal is bus 12, and for a 1-MW DG the bus in question is 13. As the power of a DG increases, the optimal bus is shifted toward the network root. A 1-MW DG covers the power load at bus 13 and partially at bus 14. In this manner, preceding feeders are unloaded and losses are reduced. As the DG power consumption increases, its capacity to supply more load downstream is increased concurrently, thus unloading the feeders upstream (toward the network root). That is the underlying reason for shifting the optimal point in this specific instance. The same conclusion may be drawn when the diagram in Figure 3.8(b) is observed.

Test B2: Optimal locations of several DGs with specified sizes. In the instance of a number of DGs of various sizes and types that must not be connected to a single bus, optimal combination of buses should be determined so as to meet the specified optimization criterion—minimization of energy losses. Since the number

Table 11.2 Annual load levels and their durations

Load level (p.u.)	1.0	0.8	0.6	0.4
Duration (h)	1,215	1,190	1,515	4,840

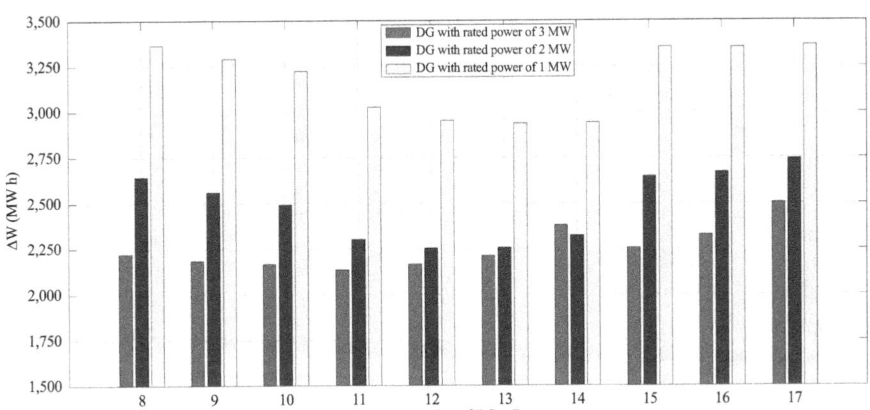

Figure 11.9 Annual energy losses in IEEE 31-bus system with DGs of different sizes located at preliminary buses

of preliminary locations (buses) is 10, and the number of DGs to be allocated is 3, the total number of possible combinations of buses for connecting DGs is 720. Data on the DGs is given in Table 11.3, which includes the optimization results presented in the last two columns.

Optimal locations for three DGs were established in three examples different by the DG operating mode. The results differ significantly at first sight. However, if the network topology is regarded, it may be noted that in all three examples, the optimal locations are concentrated around bus 8 as a root of that specific part of the network. The results vary due to different manners of modeling and operation of DGs. In Examples 1 and 2, all DGs are represented by the *PQ* model with a constant power factor, whereas in Example 3, DGs are represented by a *PV* model with a variable power factor in the given range. Minimum annual energy losses occur in Example 2, when DGs inject constant active and reactive power. The specified voltage value at all buses is a rated value (1 p.u.). In Example 3, DGs inject reactive power in the positive and negative direction with the purpose of achieving and maintaining the specified voltage value. It means that during the optimization process, attention must be paid to the values of the voltage magnitude specified at *PV* buses, in accordance with the value of the root bus voltage (network power supply transformer), consumption rate levels, active DG power, and possible range of the change in reactive power of DG. It, however, falls in the domain of exploitation, and therefore the optimization of the specified values of the voltage magnitude at *PV* buses is specifically dealt with in the chapter on optimal power flows and optimal voltage regulation. If the results of the Test B1 and Test B2 are compared, it is noticeable that optimal locations closer to the network root correspond to larger DG sizes. It quite certainly depends on the network configuration, level and distribution of load on network buses, and the ratio between the power load and the DG size. In the mentioned test examples, DG sizes are constant and selected arbitrarily. In practice, it is necessary to take into account the variability of the DG

Table 11.3 Data on the DGs and the results—optimal locations

	DG	Type	P_{DG} (MW)	$tg\varphi$	Results	
					Optimal locations of DG	Annual energy loss ΔW (MW h)
Example 1	DG1	*PQ*	3	0	8	2,206
	DG2	*PQ*	2	0	12	
	DG3	*PQ*	1	0	16	
Example 2	DG1	*PQ*	3	0.4	8	1,820
	DG2	*PQ*	2	0.4	15	
	DG3	*PQ*	1	0.4	13	
Example 3	DG1	*PV*	3	−0.4 to 0.4	15	1,991
	DG2	*PV*	2	−0.4 to 0.4	8	
	DG3	*PV*	1	−0.4 to 0.4	11	

size over a certain period, a year for instance. In order to obtain relevant results, the most plausible power to operate a DG or the power to operate a DG for the longest time over the considered period should be taken as a starting point when establishing the optimal location. The mentioned power depends on the rated power, technology of DG, and type of the primary energy source.

Test B3: Optimal power outputs of DGs at the specified locations. This test example includes determining optimal power outputs of DGs connected at predefined buses. The aim of this task is optimal operation of DGs for a given level of power load. Optimal DG power outputs are to be determined at their simultaneous operation with the purpose of minimizing power losses of the distribution network. The test was conducted with the input data given in Table 11.4. Optimal DG locations were selected according to the results of Test B2 presented in Table 11.3. In the procedure of successive reduction of variants, the DG power ranges were divided into ten equal segments. The results are shown in Table 11.5.

The obtained values in Table 11.5 constitute local optima simply because the specified ranges of possible DG powers in the procedure of determining optimal power outputs were divided into a finite number of segments (see Figure 11.2).

Reduction of annual energy losses of the network when DGs operate at optimal power outputs relative to operating with rated power outputs is as follows:

Example 1: 2,206 – 1,357 MW h = 849 MW h
Example 2: 1,820 – 978 MW h = 842 MW h
Example 3: 1,991 – 1,134 MW h = 857 MW h

Energy saving is apparent when a DG operates at optimal power output. However, the underlying motive (aim, interest) of distribution system owners is to generate as much energy as possible and supply it to the network, that is, to consumers. From the point of individual (private) owner of a distribution system, energy losses of a distribution network are not too relevant and therefore, optimization of a DG size becomes senseless. Nevertheless, if an owner of a distribution system is a distribution company operating a distribution network, it must pay attention to various technical and financial aspects of connecting and operating DGs. In this case,

Table 11.4 Data on DGs for which optimal power outputs need to be determined

	DG	Type	P_{DG} (MW)	$tg\varphi$
Example 1	DG1 at bus 8	*PQ*	0–3	0
	DG2 at bus 12	*PQ*	0–2	0
	DG3 at bus 16	*PQ*	0–1	0
Example 2	DG1 at bus 8	PQ	0–3	0.4
	DG2 at bus 15	*PQ*	0–2	0.4
	DG3 at bus 13	*PQ*	0–1	0.4
Example 3	DG1 at bus 15	*PV*	0–3	−0.4 to 0.4
	DG2 at bus 8	*PV*	0–2	−0.4 to 0.4
	DG3 at bus 11	*PV*	0–1	−0.4 to 0.4

Table 11.5 Optimal power outputs of DGs

Example 1	Load level		1.0	0.8	0.6	0.4
	DG1 at bus 8	P (MW)	3	2.4	1.8	1.2
	DG2 at bus 12	P (MW)	2	2	1.5	1
	DG3 at bus 16	P (MW)	1	1	0.7	0.5
	Active power loss, P_{loss} (MW)		0.4334	0.2618	0.1436	0.0623
	Annual energy loss, ΔW (MW h)		1,357			
Example 2	Load level		1.0	0.8	0.6	0.4
	DG1 at bus 8	P (MW)	3	3	2	1.2
	DG2 at bus 15	P (MW)	2	1.4	1	0.6
	DG3 at bus 13	P (MW)	1	1	1	0.9
	Active power loss, P_{loss} (MW)		0.3279	0.1898	0.0994	0.0419
	Annual energy loss, ΔW (MW h)		978			
Example 3	Load level		1.0	0.8	0.6	0.4
	DG1 at bus 15	P (MW)	2.9	3	2.7	1.6
	DG2 at bus 8	P (MW)	1.3	1	0.2	0
	DG3 at bus 11	P (MW)	1	1	1	1
	Active power loss, P_{loss} (MW)		0.3857	0.2275	0.1123	0.0463
	Annual energy loss, ΔW (MW h)		1,134			

the optimization of the DG size becomes very significant. In addition, optimal operation of DGs also includes coordinating the manner in which DGs operate along with other regulation resources in a distribution network.

11.5.1.3 Optimal DG placement and sizing by using GA

Test C1: Optimal DG location and size. An optimal DG location and size was determined for the possible DG size range between 0 and 5 MW and for the list of preliminary buses (8, 9, 10, 11, 12, 13, 14, 15, 16, and 17) according to the criterion of minimizing annual active energy loses. The test was conducted for two DG models:

- PQ model with $tg\varphi = 0$;
- PV model with $tg\phi = (-0.4 \text{ to } 0.4)$.

The MATLAB realization of the GA within the additional *gads* module was used to solve the problem, with parameters and options given in Table 11.6. The results are shown in Table 11.7. Applying the GA ensures certain approximation to the global optimum; however, owing to the stochastic nature of the solutions, these are nevertheless local optima. For the purpose of illustration, values of two local optima are given, one of which was selected as a global optimum and final solution (in bold). The global optimal solution is best solution obtained from ten consecutive runs of GA. Figures 11.10 and 11.11 contain the best and mean values of the fitness function during the GA runs for these two examples.

Test C2: Optimal locations of several DGs with specified sizes. Optimal DG locations were determined for Example 2 in Table 11.3. Owing to the stochastic nature of the GA, the solutions obtained are local optima. For that reason, the procedure was conducted several times. Table 11.8 presents the best results of ten

Table 11.6 GA parameters and options

Generation	50
Population	20
Initial population	Feasible population
Scaling function	Top, quantity: 0.4
Selection	Tournament, tournament K: 4
Crossover	Heuristic, ratio: 1.2
Mutation	Adaptive feasible
Ending conditions	Max generation: 50
	Termination tolerance on the function value: 10^{-6}

Table 11.7 Optimal DG location and size obtained by using GA

DG	Optimal location—Bus	Optimal power (MW)	Annual energy loss ΔW (MW h)
PQ	**8**	**4.0736**	**2,072.8**
	11	2.8383	2,127.4
PV	8	3.5251	1,717.8
	8	**3.5392**	**1,715.5**

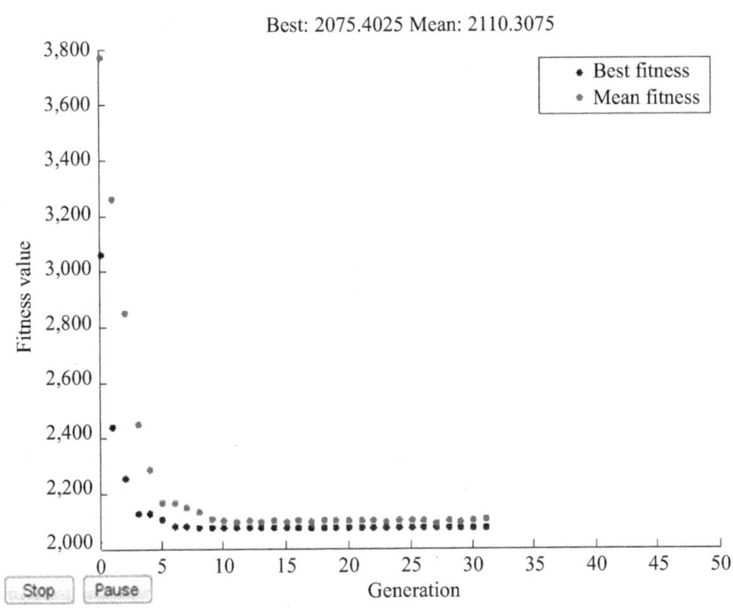

Figure 11.10 Convergence of GA in the case of a PQ-type DG

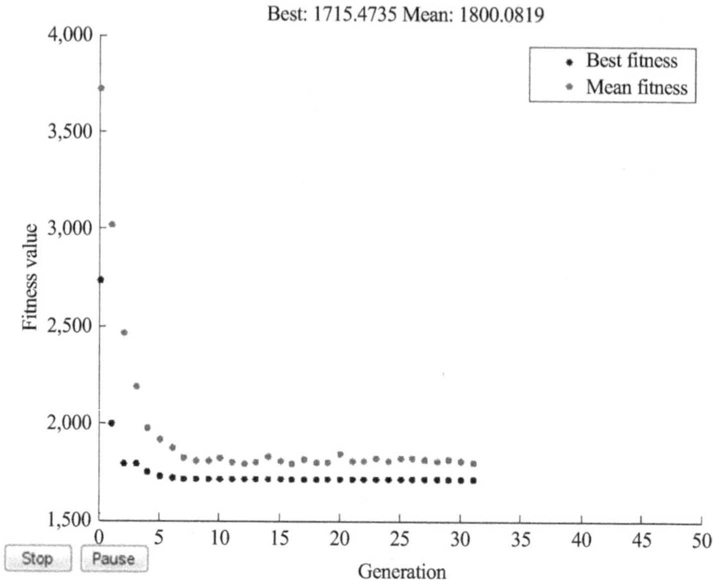

Figure 11.11 Convergence of GA in the case of a PV-type DG

Table 11.8 Optimal locations of DG using GA

Locations—Buses			Objective function ΔW (MW h)
DG1	**DG2**	**DG3**	
8	15	13	1,819.7

consecutive runs of GA. Since a unique optimal solution was determined by applying the partial search of variants in Test B2, it may be said with certainty that the local optimum specified in Table 11.8 is concurrently a global optimum.

Test C3: Optimal power outputs of DGs at the specified locations. In this case, the GA was used to determine the optimal power outputs of DGs for Examples 1 and 3 as given in Table 11.4. The results are given in Table 11.9. When the results shown in Table 11.9 are compared with the corresponding results from Table 11.5, it is noticeable that they are rather similar. However, in neither instance can it be said that unique optimal solutions have been achieved (global optimum). The accuracy in the case using partial search of variants depends on the number of DG power range segments considered. The larger the number of segments, the more accurate the obtained solution. In the instance when the GA is applied, the reason for the above-mentioned lies in the stochastic nature of the process. In addition to said reasons, there is often more than one global optimum, that is, there are a larger number of solutions—combinations of variables (DG size) for which the minimum of the criterion function is achieved.

Table 11.9 Optimal power outputs of DGs obtained by using GA

Example 1	Load level		1.0	0.8	0.6	0.4
	DG1 at bus 8	P (MW)	3	2.4137	1.8083	1.2060
	DG2 at bus 12	P (MW)	2	1.9838	1.4822	0.9876
	DG3 at bus 16	P (MW)	1	0.9786	0.722	0.4702
	Active power loss, P_{loss} (MW)		0.4334	0.2618	0.1436	0.0623
	Annual energy loss, ΔW (MW h)			1,357		
Example 3	Load level		1.0	0.8	0.6	0.4
	DG1 at bus 15	P (MW)	2.9019	3	2.9	1.6059
	DG2 at bus 8	P (MW)	1.3211	0.9373	0.0504	0.0161
	DG3 at bus 11	P (MW)	0.9931	1	1	1
	Active power loss, P_{loss} (MW)		0.3851	0.2274	0.1120	0.0463
	Annual energy loss, ΔW (MW h)			1,132.3		

11.5.2 Distribution network Zajecar

The single line diagram of the test network Zajecar is shown in Figure 11.12. This is a part of a 10-kV distribution network in area of city Zajecar in Serbia. There is already one DG in the network. It is an mini hydropower (MHP) plant at bus 14, which has two synchronous generators, the rated power of which is 160 kVA each and $\cos\varphi = 0.7$. In that area, the conditions are favorable for construction of microhydropower and MHP plants, therefore the results below may be of use when selecting an optimal connection point, as well as in considering the effect of MHPs on the distribution network operation.

The network components parameters data, and the loads data are given in Table 11.10. Rated powers of consumer units at the network buses were estimated according to the rated powers of the associated transformers, whereas it was adopted that the power factor is 0.9. Active transformer resistances and reactances were computed for a higher voltage level, that is, in 0–1 transformers for the voltage level of 35 kV, and in other transformers for the voltage level of 10 kV. Transforming the variables and parameters into per unit system was carried out for the base power of 1 MVA and base voltages

- 35 kV for the 0–1 transformer;
- 10 kV for other distribution transformers and feeders;
- 6 kV for the MHP.

11.5.2.1 Preliminary locations—*BLCs*

The procedure of determining the *BLCs* was performed for two values of load level and power output of DG in bus 14, obtained by random selection

- 45.43% of the rated power and
- 95.72% of the rated power.

Figure 11.13 shows the *BLC* values in both cases. The largest values are at low-voltage buses of MV/LV transformers to which the loads are connected. The reason

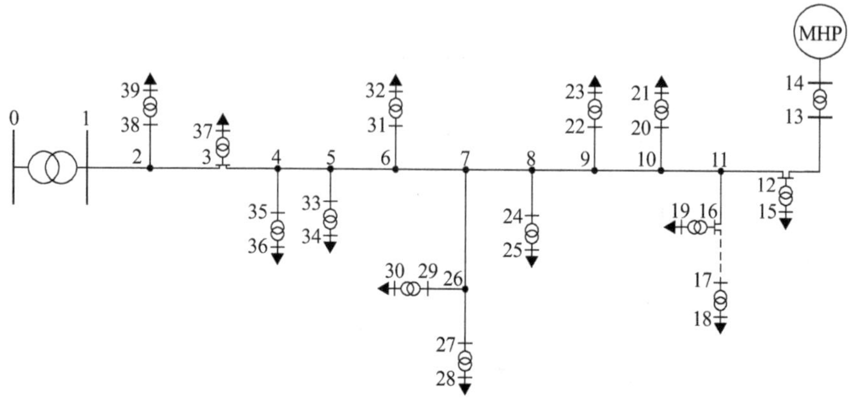

Figure 11.12 Distribution network Zajecar

is a large transformer impedance relative to MV feeders, as well as the fact that they are located on the periphery of the radial network. The *BLC* diagram form is approximately the same for both the levels of the load power. It means that the network configuration and values of parameters of elements (resistance and reactance) have a decisive effect on the *BLC* values. There is already a DG (MHP) at bus 14, therefore the *BLC* value of that bus is markedly different from other buses. This is yet another indirect indicator that *BLC* values can be used to assess the favorability of buses for connecting a DG thereto (a bus to which a DG is already connected may not be included in the list of preliminary locations for connecting new DGs, which is clearly indicated by the *BLC* value of the bus in question).

For the purpose of identifying a relation between the *BLCs* and favorability of buses for connecting DGs in terms of power loss reduction in the network, power losses when 100 kW DGs are successively connected were computed for all levels of power load in the same manner as the *BLC* values. It was assumed that DGs operate at a unit power factor. Figure 11.14 shows the values of network losses when DG power is injected in certain network buses. When the *BLC* diagrams in Figure 11.13 are compared with the loss diagrams in Figure 11.14, the relation between the *BLC* values and power losses in the network is clearly visible. Minimum loss values are detected when DG power is injected in buses with large *BLC* values. Minimum power losses for both the load power levels correspond to the case of connecting DGs to buses 15, 18, 19, and 25. These buses may be selected as preliminary locations in the process of determining optimal locations of DGs.

11.5.2.2 Optimal DG placement and sizing

Buses with the largest *BLC* values, determined in Section 11.5.2.1, were taken as preliminary locations. These buses are as follows: 15, 18, 19, 21, 23, 25, 28, 30, 32, 34, 36, 37, and 39. A 1-year period is observed. The annual power load diagram has four levels relative to the rated power of distribution transformers (1; 0.8; 0.6; 0.4) with the power factor of 0.9. The output power of the DG at bus 14 is changed

Table 11.10 Branch data and load data for the test network Zajecar

Branch	V_n (kV)	R (Ω)	X (Ω)	R (p.u.)	X (p.u.)	P_L (kW)	Q_L (kVAr)	P_L (p.u.)	Q_L (p.u.)
0–1	35/10.5	1.2630	9.1000	0.00103	0.00743	–	–	–	–
1–2	10	0.10108	0.05967	0.0010108	0.0005967	–	–	–	–
2–3	10	0.10703	0.06318	0.0010703	0.0006318	–	–	–	–
3–4	10	0.27350	0.16146	0.0027350	0.0016146	–	–	–	–
4–5	10	0.42810	0.25272	0.0042810	0.0025272	–	–	–	–
5–6	10	0.10703	0.06318	0.0010703	0.0006318	–	–	–	–
6–7	10	0.40433	0.23868	0.0040433	0.0023868	–	–	–	–
7–8	10	0.82649	0.48789	0.0082649	0.0048789	–	–	–	–
8–9	10	0.29135	0.17200	0.0029135	0.0017200	–	–	–	–
9–10	10	0.17838	0.10530	0.0017838	0.0010530	–	–	–	–
10–11	10	0.05946	0.03510	0.0005946	0.0003510	–	–	–	–
11–12	10	0.07286	0.03981	0.0007286	0.0003981	–	–	–	–
12–13	10	0.17989	0.10299	0.0017989	0.0010299	–	–	–	–
13–14	10/6	3.4300	14.000	0.03430	0.1400	–	–	–	–
12–15	10/0.4	3.4370	13.570	0.03437	0.1357	360	170	0.36	0.17
11–16	10	0.19212	0.08326	0.0019212	0.0008326	–	–	–	–
16–17	10	0.16217	0.02542	0.0016217	0.0002542	–	–	–	–
16–19	10/0.4	6.8800	15.326	0.0688	0.1533	225	109	0.225	0.109
17–18	10/0.4	11.719	22.080	0.1172	0.2208	144	70	0.144	0.070
10–20	10	0.03557	0.02106	0.0003557	0.0002106	–	–	–	–
20–21	10/0.4	20.000	34.6400	0.2000	0.3464	90	43	0.090	0.043
9–22	10	0.11892	0.07020	0.0011892	0.0007020	–	–	–	–
22–23	10/0.4	20.000	34.6400	0.2000	0.3464	90	43	0.090	0.043
8–24	10	0.14865	0.08775	0.0014865	0.0008775	–	–	–	–
24–25	10/0.4	3.4370	13.570	0.03437	0.1357	–	–	–	–
7–26	10	1.05211	0.33460	0.0105211	0.0033460	–	–	–	–
26–27	10	0.06014	0.01865	0.0006014	0.0001865	–	–	–	–
27–28	10/0.4	20.000	34.640	0.2000	0.3464	90	43	0.090	0.043
26–29	10	0.76848	0.33304	0.0076848	0.003304	–	–	–	–
29–30	10/0.4	48.000	64.000	0.4800	0.6400	90	43	0.090	0.043
6–31	10	0.72168	0.22380	0.0072168	0.002238	–	–	–	–
31–32	10/0.4	48.000	64.000	0.4800	0.6400	90	43	0.090	0.043
5–33	10	0.10703	0.06318	0.0010703	0.0006318	–	–	–	–
33–34	10/0.4	48.000	64.000	0.4800	0.6400	90	43	0.090	0.043
4–35	10	0.48112	0.14920	0.0048112	0.0014920	–	–	–	–
35–36	10/0.4	20.000	34.640	0.2000	0.3464	90	43	0.090	0.043
2–38	10	0.20811	0.12285	0.0020811	0.0012285	–	–	–	–
3–37	10/0.4	3.4370	13.570	0.03437	0.1357	360	170	0.360	0.170
38–39	10/0.4	3.4370	13.570	0.03437	0.1357	360	170	0.360	0.170

according to the same pattern as load (relative to its rated power). It is represented by the model of constant power as a negative load with the rated power factor of 0.7. All load levels have the same duration of 2,190 h.

Test D1: Optimal location of DG with various rated powers. The procedure of determining an optimal location was carried out for the following DG rated power values: 300, 200, and 100 kW, whereas DGs were modeled as PQ buses with the unit power factor. Since only one DG is treated, the process is rather simple and it includes determining annual power losses of the network when DGs of the specified size are successively connected to the defined preliminary buses.

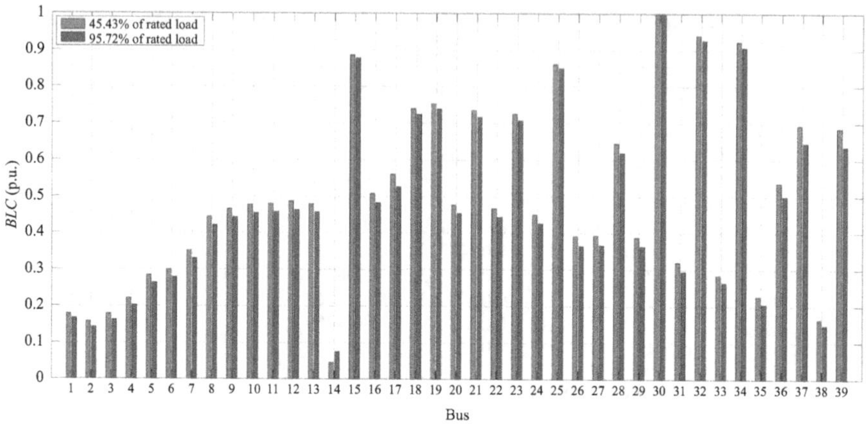

Figure 11.13 BLCs for test network Zajecar

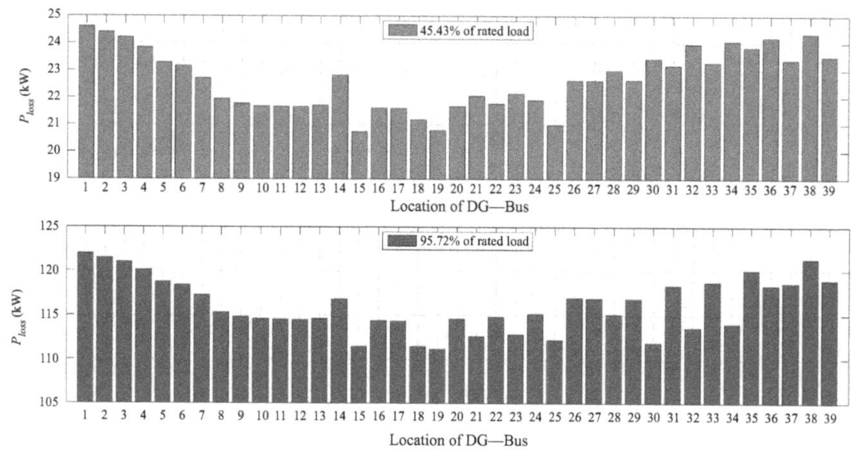

Figure 11.14 Active power losses in the distribution network Zajecar with DG of 100 kW in different buses

Figure 11.15 includes the results—annual power losses of the network for certain values of DG rated power. Minimum annual power losses for 300 and 200 kW DGs are observed at bus 15. Accordingly, bus 15 is the optimal bus for connecting DGs of the mentioned installed power. If a 100-kW DG is to be installed, the optimal bus would be bus 19, although the difference compared to bus 15 in that instance is practically negligible, therefore it would be no mistake if also bus 15 was selected as optimal. According to the results in Figure 11.13, bus 15 is not the one with the largest *BLC* value, but bus 30. Therefore, *BLC* values of buses should not be taken as absolute a priori indicators of optimality, but solely as a guideline or a good initial guess for a starting point in the further process of final establishing of

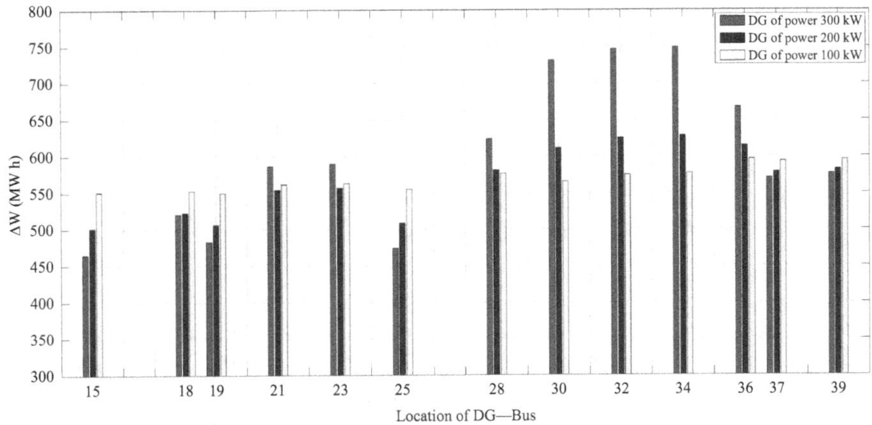

Figure 11.15 Annual active energy losses for various DG sizes in preliminary buses

Table 11.11 Data on the DGs for Test D2

DG	**Type**	**P_{DG} (kW)**	**$tg\varphi$**
DG1	*PV*	300	−0.4 to 0.4
DG2	*PQ*	200	0
DG3	*PQ*	100	0

Table 11.12 Results—optimal DG locations

Partial search of variants		**GA**		**ΔW (MW h)**
DG	**Optimal location**	**DG**	**Optimal location**	
DG1	15	DG1	15	326
DG2	25	DG2	25	
DG3	19	DG3	19	

an optimal location or locations for DG connecting. As in the case of the IEEE 31 test system, in this example, optimal buses are also located on the periphery, that is, further away from the network root relative to other preliminary locations, which is the result of the relation between the load power and DGs.

Test D2: Optimal locations for a number of DGs with specified rated power. Table 11.12 contains the results of optimization for the DG on which the data is given in Table 11.11. This table also contains the results of applying both procedures—partial search of variants and GA application. The results are identical.

Given the number of preliminary locations and the number of DGs, there are 1,716 possible variants—combinations examined when performing partial search of variants. Since all possible solution variants are encompassed according to the defined criteria, the obtained solution constitutes a global optimum.

Given a small number of variables of the same nature (integers), when applying a GA, the solution is obtained rather quickly, after only nine generations. In addition, a global solution is obtained in spite of the stochastic character of the GA and owing to the aforementioned nature of variables and relatively narrow search set (preliminary locations). This was confirmed by several repeated GA runs, whereas identical results as those specified in Table 11.12 are obtained in each run. Figure 11.16 shows the convergence characteristic of GA for this test case.

Test D3: Optimal power outputs of DGs at specified locations. The test was performed for the data given in Table 11.13. The results of partial search of variants are given in Table 11.14. Table 11.15 presents the solutions obtained by applying the GA.

Since maximum powers (rated powers) were obtained as optimal power outputs of DGs for all load levels (excluding 0.4), it is evident that the objective

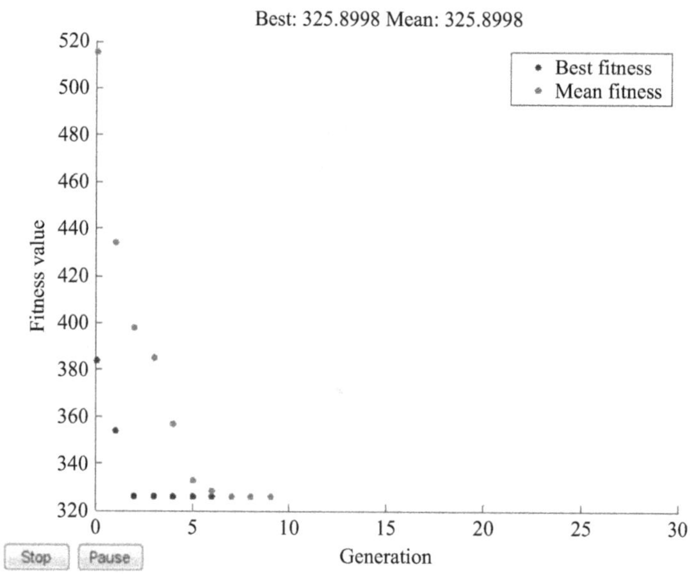

Figure 11.16 Convergence characteristic of GA in Test D2

Table 11.13 Data on the DGs for Test D3

DG	Type	P_{DG} (kW)	$tg\varphi$
DG1 at bus 15	*PV*	0–300	−0.4–0.4
DG2 at bus 25	*PQ*	0–200	0
DG3 at bus 19	*PQ*	0–100	0

Table 11.14 Optimal results for Test D3 obtained by applying partial search of variants

Load level	Optimal power of DG1 (kW)	Optimal power of DG2 (kW)	Optimal power of DG3 (kW)	Power loss, P_{loss} (kW)
1.0	300	200	100	76.6
0.8	300	200	100	42.5
0.6	300	200	100	20.5
0.4	210	180	100	8.5

Table 11.15 Optimal results for Test D3 obtained by applying GA

Load level	Optimal power of DG1 (kW)	Optimal power of DG2 (kW)	Optimal power of DG3 (kW)	Power loss, P_{loss} (kW)
1.0	300	200	100	76.6
0.8	300	200	100	42.5
0.6	300	200	100	20.5
0.4	202.3	184	100	8.5

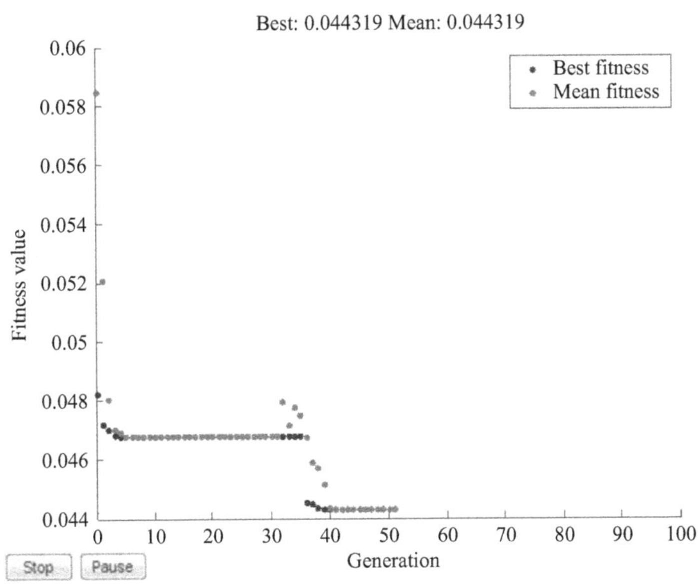

Figure 11.17 Convergence characteristic of GA in Test D3 for the load level of 0.8

function is decreasing and that in order to achieve its minimum, it is necessary to increase the DG sizes at given locations. However, optimization includes taking into account realistic conditions, which in turn imply that DGs are of limited power capacities. Accordingly, in terms of the defined DG size range, the results

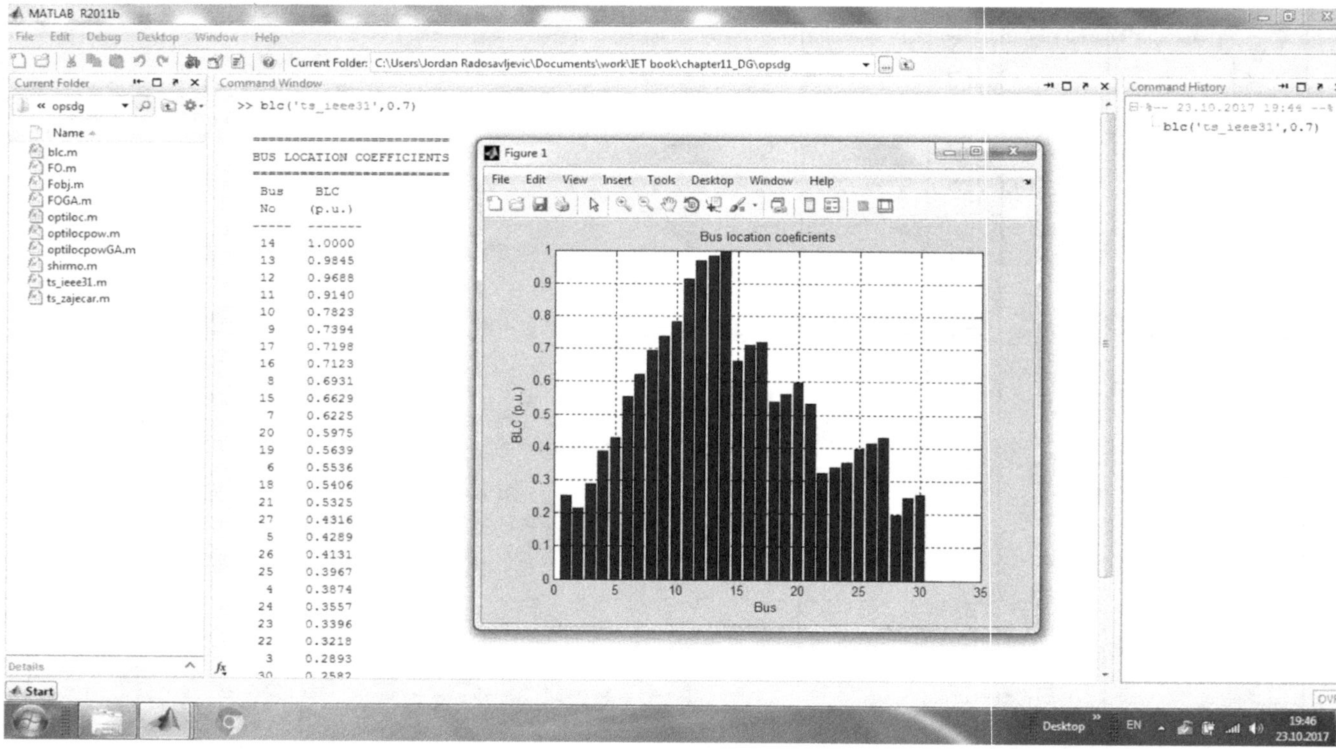

Figure 11.18 Main window of the opsfg program package

obtained constitute global optima for all levels of power load other than for 0.4.

When partial search of variants was applied, DG size ranges were divided into ten equal segments. For all levels of power load other than for 0.4, already during the first search, the results obtained were the same as those specified in Table 11.14. However, the procedure was conducted once again on the narrowed DG power ranges close to the previously obtained optima (for DG1 200–300 kW, for DG2 100–200 kW, and for DG3 the same range as in the first search was used). The results obtained are identical.

When GA was applied in this test example, the optimal solution was obtained after 50 generations, as shown in Figure 11.17. This procedure is several times longer than in the example from Test D2. The reason for this is that the variables are not integers and have a much larger range of possible values and thereby a much larger number of possible solutions. If these results are compared with the results in Table 11.14, it is apparent that global optima were obtained for all levels of power load other than for 0.4 when the obtained solutions have a character of a local optimum relative to the defined DG power ranges.

11.6 Educational program package *opsdg*

Based on the methods described above, the *opsdg* program package was developed. The *opsdg* is an educational program package of MATLAB m-files for determining Optimal Placement and Sizing of DG units in distribution networks. The program package is available at [72]. The simulation results presented in Sections 11.5 are obtained by the *opsdg* program package on a 2.20-GHz PC with 3.0 GB RAM. An example of usage of the program package is shown in Figure 11.18. The heads of the m functions are given in Table 11.16.

Table 11.16 Content of the opsdg program package

blc.m	`function [BLC,indexBLC]=blc(testsystem,kload)`
	`%blc - program for Bus Location Coefficients (BLC) calculation`
	`%-----------------------------------------------`
	`%Input variables:`
	`%testsystem - test system for which is BLC calculated`
	`%kload - load level`
	`%Output variables:`
	`%BLC - Bus Location Coefficients based on Loss Sensitivity Index -LSI`
	`%indexBLC - bus index from highest to lowest BLC value`
	`...`

optiloc.m	`function [opti_loc_DG,deltaWmin]=optiloc(testsystem)`
	`% optiloc - program for optimal locations of DG using partial search of variants`
	`% for maximum 3 DG units`
	`%-----------------------------------------------`
	`%Input variables:`

	%testsystem - test system ('ts_ieee31' or 'ts_zajecar') %Output variables: %opti_loc_DG - optimal locations (bus indexes) %deltaWmin - minimum of the objective function, i.e., minimum value of annual energy losses ...
optilocpow.m	function [opti_loc_DG,opti_pow_DG,deltaWmin]=optilocpow(testsystem) % optilocpow - program for optimal power of DG using GA %--- %Input variables: % testsystem - test system ('ts_ieee31' or 'ts_zajecar') %Output variables: %opti_loc_DG - optimal locations (bus indexes) determined by program optiloc.m %opti_pow_DG - optimal powers of DG units determined by GA %deltaWmin - minimum of the objective function, i.e., minimum value of annual energy losses ...
optilocpowGA.m	function [opti_loc_DG,opti_pow_DG,deltaWmin]=optilocpowGA (testsystem) % optilocpowGA - program for location and power of DG using GA %--- %Input variables: % testsystem - test system ('ts_ieee31' or 'ts_zajecar') %Output variables: %opti_loc_DG - optimal locations (bus indexes) %opti_pow_DG - optimal powers of DG units %deltaWmin - minimum of the objective function, i.e., minimum value of annual energy losses ...
FO.m	function [deltaW]=FO(x,testsystem,opti_loc_DG) %FO - Calculate the objective function for optilocpow.m %Input variable: %x - vector of control variables ...
Fobj.m	function [sum_ploss]=Fobj(branch,bus,kload) %Fobj - Calculate the objective function for optilocpow.m ...
FOGA.m	function [deltaW]=FOGA(x,testsystem) %FOGA - Calculate the objective function for optilocpowGA.m ...
ts_ieee31.m	function [branch,bus,loadlevel,dg]=ts_ieee31 %Data on the IEEE 31-bus distribution test system. ...
ts_zajecar.m	function [branch,bus,loadlevel,dg]=ts_zajecar %Data on the test system Zajecar. ...
shirmo.m	function [V,theta]=shirmo(testsystem,kload) %shirmo - program for power flow computation; it is used in the program blc.m in

11.7 Conclusion

Procedure for optimal siting and sizing of DG units requires two levels. The first level aims to reduce the solution space by introducing a list of preliminary locations based on the sensitivity of power losses with power injections into the network buses. This significantly reduces the number of possible variants of the solution, which are searched in the second stage of the procedure.

Two search techniques were used for searching: partial search of variants and metaheuristic optimization methods—GA.

Simulation results on two test networks confirm the efficiency and robustness of this approach for both search techniques. However, it should be noted that the partial search technique is suitable for small distribution networks, while the GA also provides high-quality solutions for real systems with a large number of buses and potential locations for DG units.

References

[1] Wang C., Nehrir M.H. 'Analytical approaches for optimal placement of distributed generation sources in power systems'. *IEEE Transactions on Power Systems*. 2004;**19**(4):2068–76

[2] Popović D.H., Greatbanks J.A., Begović M., Pregelj A. 'Placement of distributed generators and reclosers for distribution network security and reliability'. *Electrical Power and Energy Systems*. 2005;**27**:398–408

[3] Gautam D., Mithulananthan N. 'Optimal DG placement in deregulated electricity market'. *Electric Power Systems Research*. 2007;**77**:1627–36

[4] Gözel T., Hocaoglu M.H. 'An analytical method for the sizing and siting of distributed generators in radial systems'. *Electric Power Systems Research*. 2009;**79**:912–8

[5] Elsaiah S., Benidris M., Mitra J. 'Analytical approach for placement and sizing of distributed generation on distribution systems'. *IET Generation, Transmission & Distribution*. 2014;**8**(6):1039–49

[6] Naik S.N.G., Khatod D.K., Sharma M.P. 'Analytical approach for optimal siting and sizing of distributed generation in radial distribution networks'. *IET Generation, Transmission & Distribution*. 2015;**9**(3):209–20

[7] Viral R., Khatod D.K. 'An analytical approach for sizing and siting of DGs in balanced radial distribution networks for loss minimization'. *Electrical Power and Energy Systems*. 2015;**67**:191–201

[8] Keane A., O'Malley M. 'Optimal allocation of embedded generation on distribution networks'. *IEEE Transactions on Power Systems*. 2005;**20**(3):1640–6

[9] Le A.D.T., Kashem M.A., Negnevitsky M., Ledwich G. 'Optimal distributed generation parameters for reducing losses with economic consideration'. *Proceedings of IEEE Power Engineering Society General Meeting*; Tampa, FL, USA, Jun 2007. IEEE; 2007. pp. 1–8

[10] Celli G., Pilo F. 'Optimal distributed generation allocation in MV distribution networks'. *Proceedings of 22nd IEEE Power Engineering Society International Conference on Power Industry Computer Applications*; Sydney, Australia, May 2001. IEEE; 2002. pp. 81–6

[11] Borges C.L.T., Falcao D.M. 'Optimal distributed generation allocation for reliability, losses and voltage improvement'. *Electrical Power and Energy Systems*. 2006;**28**:413–20

[12] Kim K.H., Lee Y.J., Rhee S.B., Lee S.K., You S.K. 'Dispersed generator placement using fuzzy-GA in distribution systems'. *Proceedings of Power Engineering Society Summer Meeting*; Chicago, IL, USA, Jul 2002. IEEE; 2005. pp. 1148–53

[13] Celli G., Ghiani E., Mocci S., Pilo F. 'A multiobjective evolutionary algorithm for the sizing and siting of distributed generation'. *IEEE Transaction on Power Systems*. 2005;**20**(2):750–7

[14] Harrison G.P., Piccolo A., Siano P., Wallace A.R. 'Hybrid GA and OPF evaluation of network capacity for distributed generation connections'. *Electric Power Systems Research*. 2008;**78**:392–8

[15] Khoshbakht A.F., Raoofat M. 'Optimal allocation of DGs and RCSs to improve distribution network reliability and network energy loss, *Proceedings of 2nd International Power and Energy Conference*; Johor, Malaysia, Dec 2008. IEEE; 2009. pp. 1586–91

[16] Singh D., Singh D., Verma K.S. 'Multiobjective optimization for dg planning with load models'. *IEEE Transactions on Power Systems*. 2009;**24**(1):427–36

[17] Singh R.K., Goswami S.K. 'Optimum siting and sizing of distributed generations in radial and networked systems'. *Electric Power Components and Systems*. 2009;**37**:127–45

[18] Moradi M.H., Abedini M. 'A combination of genetic algorithm and particle swarm optimization for optimal DG location and sizing in distribution systems'. *Electrical Power and Energy Systems*. 2012;**34**:66–74

[19] Bohre A.K., Agnihotri G., Dubey M. 'Optimal sizing and sitting of DG with load models using soft computing techniques in practical distribution system'. *IET Generation, Transmission & Distribution*. 2016;**10**(11):2606–21

[20] Prenc R., Škrlec D., Komen V. 'Distributed generation allocation based on average daily load and power production curves'. *Electrical Power and Energy Systems*. 2013;**53**:612–22

[21] Mendoza J.E., Lopez M.E., Fingerhuth S.C., Pena H.E., Salinas C.A. 'Low voltage distribution planning considering micro distributed generation'. *Electric Power Systems Research*. 2013;**103**:233–40

[22] Biswas S., Goswami S.K., Chatterjee A. 'Optimal distributed generation placement in shunt capacitor compensated distribution systems considering voltage sag and harmonics distortions'. *IET Generation, Transmission & Distribution*. 2014;**8**(5):783–97

[23] Savić A., Đurišić Ž. 'Optimal sizing and location of SVC devices for improvement of voltage profile in distribution network with dispersed photovoltaic and wind power plants'. *Applied Energy*. 2014;**134**:114–24

[24] Mohammadi M., Nafar M. 'Optimal placement of multitypes DG as independent private sector under pool/hybrid power market using GA-based tabu search method'. *Electrical Power and Energy Systems*. 2013;**51**:43–53

[25] Dehghanian P., Hosseini S.H., Moeini-Aghtaie M., Arabali A. 'Optimal siting of DG units in power systems from a probabilistic multi-objective optimization perspective'. *Electrical Power and Energy Systems*. 2013;**51**: 14–26

[26] Dezaki H.H., Agheli A., Vahidi B., Abyaneh H.A. 'Optimized placement of connecting the distributed generations work stand alone to improve the distribution systems reliability'. *Journal of Electrical Engineering*. 2013; **64**(2):76–83

[27] Vatani M., Alkaran D.S., Sanjari M.J., Gharehpetian G.B. 'Multiple distributed generation units allocation in distribution network for loss reduction based on a combination of analytical and genetic algorithm methods'. *IET Generation, Transmission & Distribution*. 2016;**10**(1):66–72

[28] Wu C., Lou Y., Lou P., Xiao H. 'DG location and capacity optimization considering several objectives with cloud theory adapted GA'. *International Transactions on Electrical Energy Systems*. 2014;**24**:1076–88

[29] Gkaidatzis P.A., Bouhouras A.S., Doukas D.I., Sgouras K.I., Labridis D.P. 'Load variations impact on optimal DG placement problem concerning energy loss reduction'. *Electric Power Systems Research*. 2017;**152**:36–47

[30] Alinejad-Beromi Y., Sedighizadeh M., Sadighi M. 'A particle swarm optimization for sitting and sizing of distributed generation in distribution network to improve voltage profile and reduce THD and losses'. *Proceedings of 43rd International Universities Power Engineering Conference*; Padova, Italy, Sep 2008. IEEE; 2008. pp. 1–5

[31] Amanifar O. 'Optimal distributed generation placement and sizing for loss and THD reduction and voltage profile improvement in distribution systems using Particle Swarm Optimization and sensitivity analysis'. *Proceedings of 16th Conference on Electrical Power Distribution Networks*; Bandar Abbas, Iran, Apr 2011. IEEE; 2011. pp. 1

[32] Prommee W., Ongsakul W. 'Optimal multi-distributed generation placement by adaptive weight particle swarm optimization'. *Proceedings of International Conference on Control, Automation and Systems*; Seoul, Korea, Oct 2008. IEEE; 2008. pp. 1663–8

[33] Prommee W., Ongsakul W. 'Optimal multiple distributed generation placement in microgrid system by improved reinitialized social structures particle swarm optimization'. *European Transactions on Electrical Power*. 2011;**21**:489–504

[34] Gomez-Gonzalez M., Lopez A., Jurado F. 'Optimization of distributed generation systems using a new discrete PSO and OPF'. *Electric Power Systems Research*. 2012;**84**: 174–80

[35] Kansal S., Kumar V., Tyagi B. 'Optimal placement of different type of DG sources in distribution networks'. *Electrical Power and Energy Systems*. 2013;**53**:752–60

[36] Siano P., Mokryani G. 'Assessing wind turbines placement in a distribution market environment by using particle swarm optimization'. *IEEE Transactions on Power Systems*. 2013;**28**(4):3852–64

[37] Ameli A., Bahrabi S., Khazaeli F., Haghifam M.R. 'A multiobjective particle swarm optimization for sizing and placement of DGs from DG owner's and distribution company's viewpoints'. *IEEE Transactions on Power Delivery*. 2014;**29**(4):1831–40

[38] Zeinalzadeh A., Mohammadi Y., Moradi M.H. 'Optimal multi objective placement and sizing of multiple DGs and shunt capacitor banks simultaneously considering load uncertainty via MOPSO approach'. *Electrical Power and Energy Systems*. 2015;**67**:336–49

[39] Karimyan P., Gharehpetian., Abedi M., Gavili A. 'Long term scheduling for optimal allocation and sizing of DG unit considering load variations and DG type'. *Electrical Power and Energy Systems*. 2014;**54**:277–87

[40] Nasri A., Golshan M.E.H., Najed S.M.S. 'Optimal planning of dispatchable and non-dispatchable distributed generation units for minimizing distribution system's energy loss using particle swarm optimization'. *International Transactions on Electrical Energy Systems*. 2014;**24**:504–19

[41] Cheng S., Chen M.Y., Fleming P.J. 'Improved multi-objective particle swarm optimization with preference strategy for optimal DG integration into the distribution system'. *Neurocomputing*. 2015;**148**:23–9

[42] Bohre A.K., Agnihotri G., Dubey M. 'Optimal sizing and sitting of DG with load models using soft computing techniques in practical distribution system'. *IET Generation, Transmission & Distribution*. 2016;**10**(11): 2606–21

[43] Kumawat M., Gupta N., Jain N., Bansal R.C. 'Swarm-intelligence-based optimal planning of distributed generators in distribution network for minimizing energy loss'. *Electric Power Components and Systems*, 2017;**0**(0):1–12

[44] Mistry K., Bhavsar V., Roy R. 'GSA based optimal capacity and location determination of distributed generation in radial distribution system for loss minimization'. *Proceedings of 11th International Conference on Environment and Electrical Engineering*; Venice, Italy, May 2012. IEEE; 2012. pp. 1–6

[45] Tan W.S., Hassan M.Y., Rahman H.A., Abdullah P., Hussin F. 'Multi-distributed generation planning using hybrid particle swarm optimisation-gravitational search algorithm including voltage rise issue'. *IET Generation, Transmission & Distribution*. 2013;**7**(9):929–42

[46] Sharma D., Yadov N.K., Gunjan V.K., Bala A. 'Impact of distributed generation on voltage profile using different optimization techniques'. *Proceedings of International Conference on Control, Computing, Communication and Materials*; Allahabad, India, Oct 2016. IEEE; 2017. pp. 1–6

[47] Kadir A.F.A., Mohamed A., Shareef H., Ibrahim A.A., Khatib T., Elmenreich W. 'An improved gravitational search algorithm for optimal placement and sizing of renewable distributed generation units in a distribution system for power quality enhancement'. *Journal of Renewable and Sustainable Energy*. 2014;**6**(3):1–17

[48] Rasdi M.R.H.M., Musirin I., Hamid Z.A., Haris H.C.M. 'Gravitational search algorithm application in optimal allocation and sizing of multi distributed generation'. *Proceedings of 8th International Power Engineering and Optimization Conference*; Langkawi, Malaysia, Mar 2014. IEEE; 2014. pp. 364–8

[49] Kundu B., Chakrabarti A. 'OPF governed determination of location and size of distribution generators using gravitational search algorithm'. *Proceedings of 2nd International Conference on Control, Instrumentation, Energy & Communication*; Kolkata, India, Jan 2016. IEEE; 2016. pp. 446–50

[50] Niknam T., Bornapour M., Gheisari A. 'Combined heat, power and hydrogen production optimal planning of fuel cell power plants in distribution networks'. *Energy Conversion and Management*. 2013;**66**:11–25

[51] Nayak M.R., Dash S.K., Rout P.K. 'Optimal placement and sizing of distributed generation in radial distribution system using differential evolution algorithm', in Panigrahi B.K., Das S., Suganthan P.N., Nanda P.K. (eds.) *Swarm, Evolutionary, and Memetic Computing*. Berlin: Springer 2012. pp. 133–42

[52] Fadel. W, Kilic U., Taskin S. 'Placement of Dg, Cb, and Tcsc in radial distribution system for power loss minimization using back-tracking search algorithm'. *Electrical Engineering*. 2016; doi: 10.1007/s00202-016-0448-4

[53] El-Fergany A. 'Optimal allocation of multi-type distributed generators using backtracking search optimization algorithm'. *Electrical Power and Energy Systems*. 2015;**64**:1197–205

[54] Falaghi H., Haghifam M.R. 'ACO based algorithm for distributed generation sources allocation and sizing in distribution networks'. *Proceedings of IEEE Lausanne Power Technology*; Lausanne, Switzerland, Jul 2007. IEEE; 2008. pp. 555–60

[55] Abu-Mouti F.S., El-Hawary M.E. 'Optimal distributed generation allocation and sizing in distribution systems via artificial bee colony algorithm'. *IEEE Transactions on Power Delivery*. 2011;**26**(4):2090–101

[56] Kefayat M., Ara A.L., Niaki S.A.N. 'A hybrid of ant colony optimization and artificial bee colony algorithm for probabilistic optimal placement and sizing of distributed energy resources'. *Energy Conversion and Management*. 2015;**92**:149–61

[57] Sultana U., Khairuddin A.B., Mokhtar A.S., Zareen N., Sultana B. 'Grey wolf optimizer based placement and sizing of multiple distributed generation in the distribution system'. *Energy*. 2016;**111**:525–36

[58] Hassan M.Y., Tan W., Majid M., Rahman H. 'A hybrid genetic-firefly algorithm for siting and sizing of distributed generation'. *Journal of Engineering Technology*. 2012;**2**:36–45

[59] Tan W., Hassan M.Y., Majid M., Rahman H. 'Allocation and sizing of DG using cuckoo search algorithm'. *Proceedings of 2012 IEEE international conference on Power and Energy*; Kota Kinabalu, Malaysia, Dec 2012. IEEE; 2013. pp. 133–8

[60] Devi S., Geethanjali M. 'Application of modified bacterial foraging optimization algorithm for optimal placement and sizing of distributed generation'. *Expert Systems with Applications*. 2014;**41**:2772–81

[61] García J.A.M., Mena A.J.G. 'Optimal distributed generation location and size using a modified teaching learning based optimization algorithm'. *Electrical Power and Energy Systems*. 2013;**50**:65–75

[62] Poornazaryan B., Karimyan P., Gharehpetian G.B., Abedi M. 'Optimal allocation and sizing of DG units considering voltage stability, losses and load variations'. *Electrical Power and Energy Systems*. 2016;**79**:42–52

[63] El-Khattam W., Hegazy Y.G., Salama M.M.A. 'An integrated distributed generation optimization model for distribution system planning'. *IEEE Transactions on Power Systems*. 2005;**20**(2):1158–65

[64] Mori H., Iimura Y. 'An improved tabu search approach to distribution network expansion planning under new environment'. *Proceedings of International Conference on Power System Technology*; Singapore, Singapore, Nov 2004. IEEE; 2005. pp. 981–6

[65] Roy S. 'Optimal planning of generating units over micro-hydro resources within a catchment area'. *IEEE Transactions on Energy Conversion*. 2005;**20**(1):231–6

[66] Yuan Y., Qian K., Zhou C. 'The optimal location and penetration level of distributed generation'. *Proceedings of 2nd International Universities Power Engineering Conference*; Brighton, UK, Sep 2007. IEEE; 2008. pp. 917–23

[67] Golshan M.E.H., Arefifar S.A. 'Optimal allocation of distributed generation and reactive sources considering tap positions of voltage regulators as control variables'. *European Transactions on Electrical Power*. 2007;**17**(3): 219–39

[68] Radosavljević J., Jevtić M., Klimenta D. 'Determining optimal location of generators in radial distribution networks'. *Elektroprivreda*. 2006;**59**(1): 63–71

[69] Radosavljević J., Jevtić M. 'Optimal location and sizing of distributed generators in radial distribution networks using genetic algorithm'. *Proceedings of 28th Conference JUKO CIGRE*; Vrnjačka Banja, Serbia, Sep 2007. pp. 1–9

[70] Holland J.H. *Adaptation in natural and artificial systems*. Cambridge: MIT Press; 1975

[71] Santoso N.I., Tan O.T. 'Neural-net based real-time control of capacitors installed on distribution systems'. *IEEE Transactions on Power Delivery*. 1990;**5**(1):266–72

[72] Radosavljević J. *Opsdg – program package of Matlab files for optimal placement and sizing of DG units*. 2017. Available at https://www.researchgate.net/profile/Jordan_Radosavljevic2/contributions

Chapter 12
Optimal energy and operation management of microgrids

12.1 Introduction

In recent two decades, the distributed generation (DG) of energy has constant growth. Integration of DG units with renewable energy such as photovoltaic (PV) systems and small wind turbines (WTs) is dependent on geographical and meteorological conditions. Moreover, the natures of their output powers are both time-varying and difficult to predict. On the other hand, the DG units with non-renewable energy, such as fuel cells (FCs), gas-fired microturbines (MTs), diesel generators (DEs), and energy-storage devices, can be connected to any point in the distribution network and have predictive output power. Depending on the DG unit technologies, the primary energy of DG units may be injected into the distribution network via either a synchronous or asynchronous electric machine that is directly connected to the grid, a combination of an electric machine and a power electronic interface, or only via a power electronic interface.

In general, DG units, if properly planned and controlled, enable improvement of power quality, power-loss reduction, enhancement of reliability, and reduce emissions. A good way to realize the maximum potential of DG is coordinated operation and control together with storage devices and controllable loads. This brings about the concept of microgrids (MGs) [1–7].

Accordingly, an MG can be defined as local distribution network comprising various DG units, controllable loads, and energy-storage devices, which can operate either interconnected or isolated from the main distribution grid as a controlled entity. The MGs represent the core of smart distribution networks within modern power systems.

In order to achieve the full benefits from the operation of MGs, such as better economics and reduction dependence on the main grid, it is important that the integration of the DG units into MG, and their relation with the main distribution network upstream, will contribute to optimize the general operation of the system [1]. In order to achieve this goal, it is necessary to use the appropriate control system architecture. A usual control system architecture of MG comprises three hierarchical control levels. As shown in Figure 12.1, the different control levels are [1–5]

1. Local controllers (LCs);
2. MG system central controller (MGCC); and
3. Distribution management system (DMS).

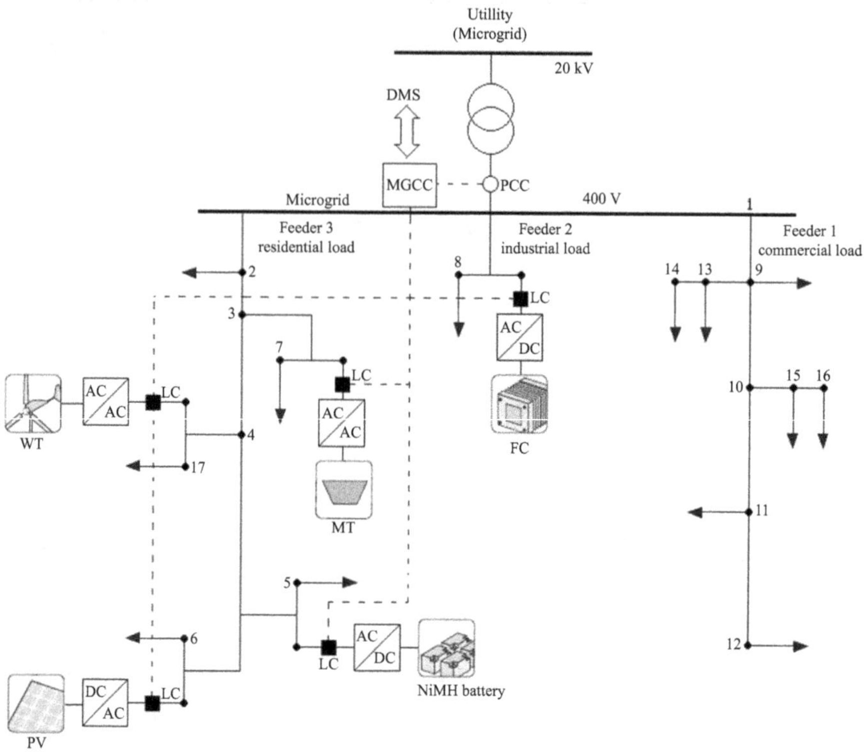

Figure 12.1 Typical low-voltage microgrid

The LC takes advantage of the power electronic interface of the DG units and can be enhanced with various degrees of intelligence. LCs are responsible to control components of an MG and have to be adapted to each type of the component such as DG units (PV, WT, FC, MT, etc.), storage devices, and controllable loads. It uses local information to control the voltage and frequency in case isolated operating mode of MG. When the MG is connected to the main distribution grid, LCs follow the demands from the MGCC and perform local optimization of the DG power production.

The MGCC acts as an interface between the MG and the rest of power system. Its functions can range from monitoring the actual active and reactive power of DG units to full responsibility for the maximization of the MG's value and the optimization of its operation by sending control signal settings to the DG units, storage devices, and controllable loads.

Conventional approaches to DMSs need to be enhanced with new features related to the operation MGs connected to the main distribution grid. MGs connected on the feeders of DMS should ideally look like concentrated loads. In a network where more than one MGs exist, MGCCs communicate with DMS to ensure a harmony, reliable, and safe operation mode of the MGs.

In grid connected mode of operation, decision on local generation is based on maximization of the MG value, according to the availability of the primary energy sources and the energy prices. Depending on the decision-making scheme, the MGs control systems can be categorized as centralized and decentralized. Centralized control tends to maximize the local production according to market prices. For this to occur, there is a two-way communication between the MGCC and each LC. Conversely, in decentralized control, the LCs act in a smart fashion and communicate with each other to increase the revenue and the performance of the MG [3].

Therefore, the main optimization problem for the MGs connected to the main grid is the energy and operation management (EOM). Network limits, namely, capacity of the supply transformer interconnecting the MG with main grid or MG network congestion have to be of course respected [1].

In recent 10 years, many researchers have been focused in formulation of the EOM problem and developing efficient solution methodologies for optimal EOM. There are two general approaches for EOM of MG. Those are deterministic and probabilistic approaches. In the deterministic EOM of the MGs, it is assumed that the output powers of the PV and the WT, as well as load power and market prices, are equal to their forecast values. When some of the input variables are uncertain, the EOM problem becomes probabilistic. Because of the probabilistic nature of wind speed and solar irradiance, the power outputs of the WT and the PV units are random variables as well. Moreover, because of unexpected disturbance, forecast error, or load variation, it is very unlikely that the forecast of the load demand will be exactly correct. Likewise, in the open-access market, there are always errors in forecast values for market prices [8].

Recently, numerous methods have been proposed by researchers to solve both deterministic and probabilistic EOM of MGs.

The authors in [2] proposed a centralized MGs control and assessed the optimal operation of MG by optimizing both the active power of the DG units and power exchange with the upstream distribution network. Hatziargyriou *et al.* [9] studied the effect of the use of an MGCC to achieve coordinated operation of various DG units, storage devices, and controllable loads with regards to the potential economic benefits and the power losses avoided in the local network. A linear programing algorithm is used in [10] to optimize MG operation cost and battery charge states. Because of the uncertainty of renewable DG and load, batteries as energy storage devices play an important role in ensuring the safety and reliability of MG. Liu *et al.* [11] established an economic scheduling model of MG in grid-connected mode with the consideration of battery lifetime. The mesh adaptive direct search algorithm is used in [12] to minimize the cost function of the MG, taking into consideration the cost of the emissions nitrogen oxides (NO_x), sulfur dioxide (SO_2), and carbon dioxide (CO_2).

Several authors have proposed metaheuristic population-based optimization methods for solving the EOM of MGs.

In [13], a smart energy management system based on the matrix real-coded genetic algorithm (GA) to optimize the operation of MG is presented. Mohamed and Koivo [14] applied GA for solving the optimal EOM problem which is considered as a nonlinear constrained multiobjective optimization, where the objective function includes the costs of the emissions, NO_x, SO_2, and CO_2, start-up costs, as well as the operation and maintenance costs. A probabilistic approach for EOM based on GA and two-point-estimate method is proposed in [15] with objective to minimize cost and increase efficiency of an MG consisting of PV and wind generation, energy storage, controllable heating, ventilation, and air-conditioning load. The nondominated sorting GA II is applied for optimal planning of an islanded MG system [16] and for optimal EOM of a grid connected MG [17,18] considering the uncertainties of wind speed, solar radiation, and load demand.

An expert multiobjective adaptive modified particle swarm optimization (PSO) algorithm is proposed and implemented in [19] to solve the multioperation management problem in a typical MG with renewable energy sources accompanied by a backup MT/FCl/Battery hybrid power source to level the power mismatch or to store the surplus of energy when it's needed. The problem is formulated as a nonlinear constraint multiobjective optimization problem to minimize the total operating cost and the net emission simultaneously. A method based on optimal power flow and PSO algorithm is suggested in [20] to study two MGs, including wind farms. An improved PSO algorithm combined with Monte Carlo simulation is used in [21] to solve the dynamic economic dispatch of an MG system with both renewable and nonrenewable energy sources. Radosavljević *et al.* [22] presented an efficient algorithm based on PSO for EOM of an MG including different DG units and energy storage devices. Owing to the stochastic nature of energy produced from renewable sources, i.e., WTs and PV systems, as well as load uncertainties and market prices, a probabilistic approach in the EOM is introduced also. Nikmehr and Ravadanegh used PSO [23] and imperialist competitive algorithm [24] to solve the optimal power dispatch problem of interconnected MGs with minimum operating cost and compared the obtained results with Monte Carlo simulation method considering generated power and load uncertainties. Recently, Mohan *et al.* [25] proposed a stochastic weight trade-off PSO-based backward–forward sweep OPF method to obtain the online optimal schedules of distributed energy resources in MG considering renewable energy, grid power trade, and demand response. In [26], a heuristic algorithm for the energy management on stand-alone MGs, which avoids wastage of the existing renewable potential at each time interval, is proposed.

A probabilistic approach for the EOM of renewable MGs considering uncertainties in load demand, market prices, and the available electrical power of renewable DG units, such as WTs and PV systems, is proposed in [8]. The proposed framework consists of $2m$ point estimate method for covering the existing uncertainties in the MGs and a self-adaptive optimization algorithm based on the gravitational search algorithm (GSA) to determine the optimal energy management of MGs. Li *et al.* [27] applied an improved GSA based on opposition learning and elite strategy to solve a multiobjective EOM problem of MG. The multiobjective function consists of the fuel costs, operation and maintenance costs, and pollutant emissions of DG units.

The $2m + 1$ point estimation method along with an adaptive modified firefly optimization algorithm (AMFA) [28] and a self-adaptive charged system search [29] have been proposed to probabilistic optimal operation management of an MG with consideration of uncertainty in the market prices, the load demand, and the electric power generation of the PV and WT units. Mohammadi *et al.* [30] present a stochastic framework based on a scenario production technique, such that the uncertainty associated with the load forecast error, WT/PV power generation, and market price has been considered in the operation management of MGs. The optimization method based on AMFA is employed to solve each of the deterministic problems generated by the scenario production technique. Moradi *et al.* [31] proposed a hybrid optimization method that combines the quadratic programing and PSO algorithms to determine the optimum capacity of the DG units as well as the appropriate operational strategy for energy management in MG.

To schedule power in an MG, the dual-decomposition method was utilized in [32] to decompose the original problem into smaller subproblems solved by the LC of conventional generators, dispatchable loads, distributed storage units, and the renewable energy sources. Prodan and Zio [33] proposed a framework for reliable MG EOM based on receding horizon control to minimize a cost function by battery scheduling.

Rest of this chapter describes an efficient algorithm based on PSO for optimal deterministic and probabilistic EOM of MGs comprising different DG units, loads, and energy-storage devices.

12.2 Problem formulation of EOM

A typical low-voltage (LV) grid-connected MG comprising different types of DG sources, storage device, and consumers is shown in Figure 12.1. The MG is supplied from the utility (MV distribution grid) through an MV/LV transformer. The MGCC is located downstream of the transformer, and the LCs are installed close to the DGs. The MGCC optimizes the MG operation according to the open-market prices, bids received by the DG sources and forecast loads, and sends signals to the LCs of DG sources to be committed, and if applicable, to determine the level of their production. The optimization procedure implemented in MGCC depends on the market policy adopted in the MG operation. For example, two market policies are proposed in [2,9].

In the first policy, the MGCC aims to serve the total demand of the MG, using its local production, as much as possible, without exporting power to the utility. From the consumers' point of view, the MGCC minimizes the operational cost of the MG, taking into account open-market prices, demand, and DG bids.

In the second policy, the MG participates in the open market, buying and selling active power to the grid, probably via an aggregator or similar energy service provider. According to this policy, the MGCC tries to maximize the value of the MG, i.e., to maximize the corresponding revenues of the aggregator, by exchanging power with the grid.

In this work, the EOM problem is defined accordingly to first market policy. Therefore, the EOM problem solution in a typical MG aims to minimize the total operating cost of the MG through optimal adjustment of the DG's power generation, while satisfying various system-operating constraints. The total operating cost of the MG includes the DG bids and the market prices of power exchange between the MG and the utility. The mathematical model of such a problem can be expressed as follows.

12.2.1 Objective function

$$\min_{\mathbf{x}} F = \min_{\mathbf{x}} \sum_{t=1}^{NT} Cost^t(\mathbf{x}^t, y^t) = \min_{\mathbf{x}} \sum_{t=1}^{NT} \sum_{i=1}^{NG} \left[B_{Gi}\left(P^t_{Gi}\right) + MP^t \cdot P^t_{Grid} \right] \tag{12.1}$$

where $\mathbf{x}^t$ is the vector of control variables, including the active power of the generation and storage units within the MG, and can be expressed as

$$\mathbf{x}^t = \left[P^t_{G1}, P^t_{G2}, \ldots, P^t_{GNG}\right] \tag{12.2}$$

y^t is the dependent variable consisting of the active power that is bought (sold) from (to) the utility:

$$y^t = P^t_{Grid} \tag{12.3}$$

where NT is the total number of time intervals (hours) in considered period, NG is the total number of DG including storage units, P^t_{Gi} is the active power outputs of the ith DG at time t, $B_{Gi}\left(P^t_{Gi}\right)$ is the bid of the ith DG unit as a function of its active power at time t, P^t_{Grid} is the active power that is bought (sold) from (to) the utility at time t, and MP^t is the market price of power exchange between the MG and the utility at time t.

The cost function (12.1) is generally a nonlinear function because the DG bids are typically either piecewise linear or smooth quadratic function of P_{Gi} [32].

12.2.2 Constraints

12.2.2.1 Power balance

The total output power of DGs, energy-storage devices, and the utility must cover the total load demand in the MG for each time interval t. The real power loss in the MG is neglected. Accordingly, the power balance constraint can be represented as follows:

$$\sum_{i=1}^{NG} P^t_{Gi} + P^t_{Grid} = \sum_{D=1}^{ND} P^t_{L_D} \tag{12.4}$$

where P_{L_D} is the amount of the Dth load level, ND is the total number of load levels.

12.2.2.2 Real power generation capacity

For stable operation, the real power output of each unit in the MG, including the utility, is restricted by minimum and maximum power limits as follows:

$$P^t_{Gi,\min} \leq P^t_{Gi} \leq P^t_{Gi,\max} \tag{12.5}$$

$$P^t_{Grid,\min} \leq P^t_{Grid} \leq P^t_{Grid,\max} \tag{12.6}$$

where $P^t_{Gi,\min}$ and $P^t_{Grid,\min}$ are the minimum active powers of the ith DG, and the utility at time t; $P^t_{Gi,\max}$ and $P^t_{Grid,\max}$ are the maximum active powers of the ith DG, and the utility at time t.

12.2.2.3 Spinning reserve

The spinning reserve is necessary to maintain the system reliability, owing to power fluctuations in renewable energy and load fluctuations. To meet the spinning reservation, the following inequality constraint should be satisfied [21,28]:

$$\sum_{i=1}^{NG} P^t_{Gi,\max} + P^t_{Grid,\max} \geq \sum_{D=1}^{ND} P^t_{L_D} + R^t \tag{12.7}$$

where R^t is the scheduled spinning reserve at time t.

In an MG, the spinning reserve constraint is considered by adding an extra value to the total power demand, which should be supplied by the DG units. It is worth to note that the maximum power capacity (not the operating point) of the power sources is considered in the above equation.

12.2.2.4 Energy-storage limits

Since there are some limitations on charge and discharge rate of storage devices during each time interval, the following equation and constraints can be expressed for a typical battery [19]:

$$W_{ess,t} = W_{ess,t-1} + \eta_{\text{charge}} P_{\text{charge}} \Delta t - \frac{1}{\eta_{\text{discharge}}} P_{\text{discharge}} \Delta t \tag{12.8}$$

$$\begin{cases} W_{ess,\min} \leq W_{ess,t} \leq W_{ess,\max} \\ P_{\text{charge},t} \leq P_{\text{charge,max}}; P_{\text{discharge},t} \leq P_{\text{discharge,max}} \end{cases} \tag{12.9}$$

where $W_{ess,t}$ and $W_{ess,t-1}$ are the amount of energy storage inside the battery at hour t and $t-1$, respectively; $P_{\text{charge}}(P_{\text{discharge}})$ is the permitted rate of charge (discharge) during a definite period of time (Δt); $\eta_{\text{charge}}(\eta_{\text{discharge}})$ is the efficiency of the battery during charge (discharge) process; $W_{ess,\min}$ and $W_{ess,\max}$ are the lower and upper limits on amount of energy storage inside the battery; and $P_{\text{charg,max}}(P_{\text{discharge,max}})$ is the maximum rate of battery charge (discharge) during each time interval Δt.

12.2.2.5 Calculation of the active power from (to) the utility

To enforce the active power balance constraint given in (12.4), the active power from (to) the utility (MV distribution grid) is considered as a dependent variable. The value of grid power is calculated using the following equation:

$$P^t_{Grid} = \sum_{D=1}^{ND} P^t_{L_D} - \sum_{i=1}^{NG} P^t_{Gi} \tag{12.10}$$

The obtained P^t_{Grid} is checked as to whether it satisfies the constraint defined in (12.6) or not. Consequently, the variable $P^t_{Grid,\lim}$ is defined as

$$P^t_{Grid,\lim} = \begin{cases} P^t_{Grid,\max} & \text{if} \quad P^t_{Grid} > P^t_{Grid,\max} \\ P^t_{Grid,\min} & \text{if} \quad P^t_{Grid} < P^t_{Grid,\min} \\ P^t_{Grid} & \text{if} \quad P^t_{Grid,\min} \le P^t_{Grid} \le P^t_{Grid,\max} \end{cases} \tag{12.11}$$

It is worth mentioning that the control variables are self-constrained while the dependent variable, i.e., P^t_{Grid}, must be added to the objective function as a quadratic penalty term. In this term, a penalty factor multiplied with the square of the difference between the actual value and the limiting value of the dependent variable is added to the objective function, and any unfeasible solution obtained during the optimization process is ignored. The new expanded objective function to be minimized becomes:

$$\min_{\mathbf{x}} F_p = \min_{\mathbf{x}} \sum_{t=1}^{NT} Cost^t(\mathbf{x}^t, y^t) + \lambda_p \left(P^t_{Grid} - P^t_{Grid,\lim} \right)^2 \tag{12.12}$$

where λ_p is the penalty factor.

12.2.3 Distributed generation bids calculation

DG bids are considered quadratic, as presented in (12.13), according to the cost function of the units, if any, the feedback from the market prices, and the need for making some profit, necessary for the annual depreciation of the installation cost [9].

$$B_{Gi} = a_i \cdot (P_{Gi})^2 + b_i \cdot P_{Gi} + c_i \tag{12.13}$$

12.2.3.1 Microturbine and fuel cell

The MT and FC bids in (€/h) can be calculated as follows [12,34]:

$$B_G = C_{fuel} \frac{P_G}{\eta_G} + C_{inv} \tag{12.14}$$

where P_G is the electrical output power (kW) of DG, e.g., MT and FC, η_G is the DG electric efficiency, C_{fuel} is the price of fuel (natural gas) to supply the DG (€/kW h), C_{inv} is the hourly payback amount for the investment cost of the DG in (€/h).

C_{inv} is a function of the annual investment cost for depreciation (AC) in (€/kW-year), the annual production (AP) in (kW h/kW), and the nominal power of the DG in kW:

$$C_{inv} = AC \frac{P_{Gnom}}{AP} \tag{12.15}$$

$$AC = \frac{i(1+i)^n}{(1+i)^n - 1} \cdot IC \tag{12.16}$$

where i is the interest rate, n the depreciation period in years, and IC is the installation cost of the DG.

The efficiency of the MT increases with the increase of the output power [35]. The electric efficiency characteristic of the MT can be estimated as a quadratic function of its output power. The coefficients of this quadratic function can be obtained by fitting of the electrical efficiency curve of the MT based on manufacturers' data.

The FC efficiency is a nonlinear function of power level and can be expressed as follows (for a proton-exchange membrane fuel cell—PEM-FC) [36]:

$$\eta_{FC} = \frac{1}{2.964}\left[V_0 + \sqrt{V_0^2 - 4\left(V_0 V_{Pnom} - V_{Pnom}^2\right)\cdot \xi}\right] \tag{12.17}$$

where V_0 is the theoretical or thermodynamic FC potential (0.9 V), V_{Pnom} is the selected cell potential at nominal power (0.45–0.75 V), $\xi = P_{FC}/P_{nom}$ is the power level, and P_{nom} is the nominal power of the FC.

The coefficients of quadratic function (12.13) can be obtained by curve fitting, constructed according to (12.14) for a power range of P_G, e.g., $P_{G\,\min} \le P_G \le P_{G\,\max}$.

12.2.3.2 Diesel generator

The diesel-fuel-consumption data (l/h) at 25%, 50%, 70%, and 100% of the DE power rating (kW) are usually given by the manufacturer. Based on these data, the fuel-consumption characteristic of the DE can be estimated as a quadratic function of its active power output:

$$Fuel_{DE} = a_f P_{DE}^2 + b_f P_{DE} + c_f \tag{12.18}$$

where $Fuel_{DE}$ is the DE fuel consumption (l/h), P_{DE} is the DE power output (kW), and a_f, b_f, and c_f are coefficients of the fuel-consumption characteristic. Accordingly, the DE bids (€/h) can be expressed as

$$B_G = C_{fuel} Fuel_{DE} + C_{inv} \tag{12.19}$$

where C_{fuel} is the diesel-fuel price to supply the DE (€/l), and C_{inv} is the hourly payback amount for the investment cost of the DE (€/h) defined by (12.15).

12.2.3.3 Wind turbine and photovoltaic

The bid functions of WT and PV consider the annual investment cost for depreciation of equipment (AC) (€/kW) and the annual energy production per kW (AP) (kW h/kW):

$$B_G = \frac{AC}{AP} P_G \tag{12.20}$$

$$AC = \frac{i(1+i)^n}{(1+i)^n - 1} \cdot IC \tag{12.21}$$

where i is the interest rate, n the depreciation period in years, and IC is the installation cost of DGs, e.g., WT and PV.

Renewable source-based DGs, e.g., WT and PV cannot be regulated, and their output is determined by the availability of the primary source, i.e., wind or solar irradiance. To account for their production in the optimization functions, WT and PV forecasting is required.

To calculate the power output of a WT, two main factors must be known: the wind speed at a certain location and the power curve of the WT. According to [37,38], the power curve of a WT can be modeled by means of a function split into four different parts:

$$P_{WT} = \begin{cases} 0, & v \leq v_{ci} \\ \dfrac{v^2 - v_{ci}^2}{v_{nom}^2 - v_{ci}^2} \cdot P_{WT,nom}, & v_{ci} < v \leq v_{nom} \\ P_{nom}, & v_{nom} < v \leq v_{co} \\ 0, & v > v_{co} \end{cases} \tag{12.22}$$

where $P_{WT,nom}$, v_{nom}, v_{ci}, and v_{co} are nominal power, nominal wind speed, cut-in wind speed, and cut-out wind speed of the WT, respectively; P_{WT} is the power output of the WT; and v is the wind speed.

The power output of the PV module is dependent on the solar irradiance and ambient temperature of the site, as well as the characteristics of the module itself. The following equation can be used to calculate the power output of the PV module, P_{PV} [38,39]:

$$P_{PV} = P_{STC} \frac{Is}{1,000} [1 + \gamma(T_c - 25)] \tag{12.23}$$

where P_{STC} is PV module maximum power under standard test condition (STC) (W), Is is solar irradiance on the PV module surface (W/m^2), γ is the PV module temperature coefficient for power (°C^{-1}), T_c is PV cell (module) temperature (°C). The PV module temperature can be calculated as a function of solar irradiance and ambient temperature based on the module's nominal operating cell temperature (NOCT). The NOCT model equation is [39–41]

$$T_c = T_a + \frac{Is}{800} \cdot (T_{NOCT} - 20) \tag{12.24}$$

where T_a is the ambient temperature (°C), and T_{NOCT} is the nominal operating cell temperature (°C) of the module.

Bids coming from the batteries are defined by (12.20).

12.3 Solution method

This section describes applications of PSO algorithm for EOM of MGs.

12.3.1 Overview of PSO

PSO is basically developed through simulation of bird flocking in two-dimensional space [42]. Bird flocking optimizes a certain objective function. PSO uses a number of particles (search agents) that fly around in the search space to find best solution. Meanwhile, all the particles look at the best particle (best solution) in their paths. In other words, particles consider their own best solutions as well as the best solution found so far.

The basic elements of the PSO method are briefly stated and defined as follows [43]:

Particle, $\mathbf{x}(t)$: It is a candidate solution represented by an n-dimensional vector, where n is the number of control variables. At time (iteration) t, the ith particle $\mathbf{x}_i(t)$ can be described as $\mathbf{x}_i(t) = \left[x_i^1(t), \ldots, x_i^d(t), \ldots, x_i^n(t)\right]$, where $x_i^d(t)$ is the position of the ith particle with respect to the dth dimension, i.e., the value of the dth control variable in the ith candidate solution.

Population, POP(t): It is a set of N particles at time (iteration) t, i.e., $\mathbf{POP}(t) = [\mathbf{x}_1(t), \ldots, \mathbf{x}_N(t)]^T$ (where T denote transposition operator).

Swarm: It is an apparently disorganized population of moving particles that tend to cluster together while each particle seems to be moving in a random direction.

Particle velocity, $\mathbf{v}(t)$: It is the velocity of the moving particles represented by an n-dimensional vector. At time (iteration) t, the ith particle velocity $\mathrm{v}_i(t)$ can be described as $\mathbf{v}_i(t) = \left[v_i^1(t), \ldots, v_i^d(t), \ldots, v_i^n(t)\right]$, where $v_i^d(t)$ is the velocity component of the ith particle with respect to the dth dimension.

Individual best, $\mathbf{pbest}_i(t)$: It is individual best. As a particle moves through the search space, it compares its fitness value at the current position to the best fitness value it has ever attained at any time up to the current time. The best position that is associated with the best fitness encountered so far is called the individual best, $\mathbf{pbest}(t)$. For the ith particle, individual best can be expressed as $\mathbf{pbest}_i(t) = \left[pbest_i^1(t), \ldots, pbest_i^d(t), \ldots, pbest_i^n(t)\right]$.

Global best, $\mathbf{gbest}(t)$: It is the best position among all individual best position achieved so far. Hence, global best can be expressed as $\mathbf{gbest}(t) = \left[gbest^1(t), \ldots, gbest^d(t), \ldots, gbest^n(t)\right]$.

Fitness is a direct metric of the performance of the individual population member (agent). The fitness of each agent of the population is calculated from the value of the function being optimized.

Stopping criteria: These are the conditions under which the search process will terminate. It is often stated the search will terminate if the number of iterations t reaches the maximum allowable number $t_{\max}$.

In PSO, each agent tries to modify its position using the following information: the current position, the current velocity, the distance between the current position and **pbest**, and the distance between the current position and **gbest**. This modification can be represented by the concept of velocity. Velocity of each agent can be modified by the following equation:

$$\mathbf{v}_i(t+1) = w \cdot \mathbf{v}_i(t) + C_1 \cdot r_1 \cdot (\mathbf{pbest}_i(t) - \mathbf{x}_i(t)) + C_2 \cdot r_2 \cdot (\mathbf{gbest}(t) - \mathbf{x}_i(t)) \tag{12.25}$$

where C_1 and C_2 are positive constants defined as acceleration coefficients, r_1 and r_2 are uniformly distributed random numbers in [0,1], and w is a inertia weight. This w plays the role of balancing the global search and local search.

After updating the velocity, the current position of particle (searching point in the solution space) can be modified by the following equation:

$$\mathbf{x}_i(t+1) = \mathbf{x}_i(t) + \mathbf{v}_i(t+1) \tag{12.26}$$

The first term of (12.25), $w \cdot \mathbf{v}_i(t)$, provides exploration ability for PSO. For initial stages of the search process, large w to enhance the global exploration is recommended while, for last stages, the weight factor is reduced for better local search. The second term of (12.25), $C_1 \cdot r_1 \cdot (\mathbf{pbest}_i(t) - \mathbf{x}_i(t))$, represents the cognitive part of PSO where the particle changes its velocity based on its own thinking and memory. The third term of (12.25), $C_2 \cdot r_2 \cdot (\mathbf{gbest}(t) - \mathbf{x}_i(t))$, represents the social part of PSO where the particle changes its velocity based on the social–psychological adaptation of knowledge.

The PSO starts by randomly placing the particles in a problem space. In each iteration, the velocities of particles are calculated using (12.25). After defining the velocities, the positions of particles can be calculated as (12.26). The process of changing particles' positions will continue until an end criterion is met.

12.3.2 Application of PSO to EOM

The procedure for implementing the PSO algorithm to solve the optimal EOM problem can be summarized in the following steps:

Step 1: Define the MG system configuration, DG units data, storage units data, loads data, and control strategy.

Step 2: Specify the control variables **x** (12.2) and their lower and upper limits (12.5).

Specify the dependent variable y (12.3), and its lower and upper limit (12.6); Specify the objective function to be optimized $F(\mathbf{x},\mathbf{y})$ (12.1), and the penalty factor (λ_P) to form the expanded objective function $F_p(\mathbf{x},\mathbf{y})$ (12.12).

Step 3: Initialize PSO parameters, such as the population size (N), maximum iteration number ($t_{\max}$), inertia weight (w), cognitive constant (C_1), and social constant (C_2).

Step 4: Generate an initial random population of N particles.

$$\mathbf{POP}(0) = \begin{bmatrix} \mathbf{x}_1(0) \\ \mathbf{x}_2(0) \\ \vdots \\ \mathbf{x}_i(0) \\ \vdots \\ \mathbf{x}_N(0) \end{bmatrix} = \begin{bmatrix} x_1^1(0), & \ldots, & x_1^d(0), & \ldots, & x_1^n(0) \\ x_2^1(0), & \ldots, & x_2^d(0), & \ldots, & x_2^n(0) \\ & & \vdots & & \\ x_i^1(0), & \ldots, & x_i^d(0), & \ldots, & x_i^n(0) \\ & & \vdots & & \\ x_N^1(0), & \ldots, & x_N^d(0), & \ldots, & x_N^n(0) \end{bmatrix} \tag{12.27}$$

The initial positions of each particle are randomly selected between minimum and maximum values of the control variables. Therefore, the initial value of the dth control variable in the ith particle (candidate solution) is obtained as follows:

$$x_i^d(0) = x_{i,\min}^d + r \cdot \left(x_{i,\max}^d - x_{i,\min}^d\right) \tag{12.28}$$

where r is a random number in [0,1], $x_{i,\min}^d$ and $x_{i,\max}^d$ are the lower and upper limits of the dth control variable.

Initialize the velocity of particles by a zero matrix of dimension $N \times n$; where N is the number of particles (population size) and n is the number of control variables.

Step 5: Calculate the active power from (to) the utility for each particle agent in current population (12.10) and check the constraint (12.6).

Step 6: Calculate the fitness value for each particle using (12.1) and (12.12).

Step 7: Update $\mathbf{pbest}(t)$ for each particle and determine $\mathbf{gbest}(t)$.

Step 8: Update the velocity of all particles using (12.25).

Step 9: Update the position of each particle using (12.26).

Step 10: Repeat steps 5–9 until the stop criteria are reached, that is, max number of iterations $t_{\max}$.

Step 11: Return best solution obtained in the last iteration; stop.

12.4 Probabilistic EOM of MG

When some of the input variables are uncertain, the EOM problem becomes probabilistic. Because of the probabilistic nature of wind speed and solar irradiance, the power produced by the WT and PV units is also a random variable. Moreover, it is very unlikely that the forecast of the load demand will be exactly correct, owing to unexpected disturbance, forecast errors, or load variations. Likewise, in the open-access market, there are always errors in forecasted values for market prices. Such randomly occurring factors will be the main source of uncertainties. Every probabilistic formulation requires statistical characterization of the input random variables and a method for evaluating statistical features of the output variables [44].

12.4.1 Statistical characterization of the input random variables

12.4.1.1 Wind-speed modeling

The wind-speed probability density function (PDF) at a certain location, and consequently the WT power output (12.22), can generally be described by a Weibull distribution [45]:

$$f_v(v) = \frac{k}{C} \cdot \left(\frac{v}{C}\right)^{k-1} \cdot e^{-\left(\frac{v}{C}\right)^k} \tag{12.29}$$

The cumulative density function (CDF) for the Weibull distribution is

$$F_v(v) = 1 - \cdot e^{-(v/C)^k} \tag{12.30}$$

The CDF with its inverse has been utilized to calculate the wind speed:

$$v = C \cdot (-\ln(r))^{1/k} \tag{12.31}$$

where r are the random numbers uniformly distributed on [0,1]. Constants C and k are the scale and shape parameters of the Weibull distribution. Different methods can be used to calculate the Weibull parameters. Usually, parameters k and C are calculated, approximately, using mean wind speed v_m and standard deviation (STD) σ as follows [38]:

$$k = \left(\frac{\sigma}{v_m}\right)^{-1.086} \tag{12.32}$$

$$C = \frac{v_m}{\Gamma(1 + 1/k)} \tag{12.33}$$

where $\Gamma(x)$ is the gamma function defined as

$$\Gamma(x) = \int_0^\infty t^{x-1} e^{-t} dt \quad \text{for} \quad x > 0 \tag{12.34}$$

12.4.1.2 Solar irradiance, load demand, and market-price modeling

In this study, it was assumed that the solar irradiance (i.e., the PV power output), load demand, and market price have normal distribution function. The corresponding PDF of the variable x_i (i.e., the solar irradiance, load demand, and market price) is formulated as

$$f_{x_i}(x_i) = \frac{1}{\sigma\sqrt{2\pi}} \cdot e^{-\left((x_i-\mu)^2/2\sigma^2\right)} \tag{12.35}$$

The CDF for the normal distribution is

$$F_{x_i}(x_i) = \frac{1}{2}\left[1 + \mathrm{erf}\left(\frac{x_i - \mu}{\sqrt{2}\sigma}\right)\right] \tag{12.36}$$

The CDF with its inverse has been utilized to calculate the variable x_i:

$$x_i = \mu + \sqrt{2} \cdot \sigma \cdot \mathrm{erf}^{-1}(2r - 1) \tag{12.37}$$

where r is a random variable with uniform distribution in [0,1], μ is the mean value of the variable x_i, and σ is the STD of the variable x_i, erf and erf^{-1} are the error function and the inverse error function, respectively, defined as follows:

$$\mathrm{erf}(x) = \frac{2}{\sqrt{\pi}} \int_0^x e^{-t^2} dt \tag{12.38}$$

$$\mathrm{erf}^{-1}(x) = 1 - \mathrm{erf}(x) = \frac{2}{\sqrt{\pi}} \int_x^\infty e^{-t^2} dt \tag{12.39}$$

12.4.2 Statistical evaluation of the output variables

Generally, the probabilistic EOM can be expressed as

$$\mathbf{Y} = F(\mathbf{X}) \tag{12.40}$$

where $\mathbf{X}$ is the vector of input random variables and $\mathbf{Y}$ is the vector of output random variables.

In the technical literature, there are several methods for evaluating the statistical features of the output variables. These methods may be classified into the three main categories [46]: Monte Carlo simulation, analytical methods, and approximate methods.

Monte Carlo simulation randomly generates values for uncertain input variables, and these values are taken into account to solve a deterministic problem. The simulation is repeated many times (several hundred to several thousand), in order to determine with sufficient accuracy the statistical characteristics of the results, (i.e., $\mathbf{Y}$). The main drawback of the Monte Carlo method is the great number of simulations required to attain convergence.

Analytical methods are based on certain mathematical simplifications of the problem, in order to obtain analytically the statistical characteristics of random output variable Y based on the statistical characteristics of random input variables $\mathbf{X}$.

Approximate methods provide an approximate description of the statistical properties of output random variables. One of these methods has been explained and applied in this chapter. It is the point-estimate method. This method, like Monte Carlo, uses a deterministic approach to solving a probabilistic problem, but with a drastically lower number of simulations.

The point-estimate method can guarantee a great reduction of the computational efforts compared to the Monte Carlo simulation procedure. This method concentrates the statistical information provided by the first few central moments of the m input random variables on $K = 2$, 3, or 5 points for each variable, named concentrations. By using these points and function F, which relates input and output variables, statistical moments of the output variables can be obtained. To obtain these moments, function F has to be calculated $2m$, $2m+1$, or $4m+1$ times depending on the adopted scheme. In this work, $2m+1$ scheme was used.

The $2m+1$ scheme is more accurate than the $2m$ scheme, and performances of the $2m+1$ and $4m+1$ schemes are practically the same. The $2m+1$ scheme requires $2m+1$ evaluations of function F and uses only a two-point concentration for each input random variable [44].

12.4.3 Procedure for solving probabilistic EOM

Procedure for computing the moments of the output variables for the probabilistic EOM problem can be summarized by the following steps:

1. Determine the number m of input random variables.
2. Set the vector of the jth moment of the output variable equal to zero: $E(\mathbf{Y}^j) = 0$.
3. Set $t = 1$, $(t = 1,2,\ldots,m)$.
4. Determine the two standard locations:

$$\xi_{t,i} = \frac{\lambda_{t,3}}{2} + (-1)^{3-i} \cdot \sqrt{\lambda_{t,4} - \frac{3}{4}\lambda_{t,3}^2} \quad i = 1,2 \tag{12.41}$$

   where $\lambda_{t,3}$ is the skewness, and $\lambda_{t,4}$ is the kurtosis of the input random variable x_t.
5. Determine the two locations $x_{t,i}$:

$$x_{t,i} = \mu_{x_t} + \xi_{t,i} \cdot \sigma_{x_t} \quad i = 1,\ 2 \tag{12.42}$$

   where μ_{x_t} and σ_{x_t} are the mean and STD of x_t, respectively.
6. Run *the deterministic EOM algorithm* (described in Section 12.3.2) for both locations $x_{t,i}$ using the two input variable vectors:

$$\mathbf{X}_i = \left[\mu_{x1}, \mu_{x2}, \ldots, x_{t,i}, \ldots, \mu_{xm}\right] \quad i = 1,2 \tag{12.43}$$

   where μ_{xk} $(k = 1,2,\ldots,m$; and $k \neq t)$ is the mean value of the remaining random input variables.
7. Determine the weight factors:

$$w_{t,i} = \frac{(-1)^{3-i}}{\xi_{t,i}\left(\xi_{t,1} - \xi_{t,2}\right)} \quad i = 1,2 \tag{12.44}$$

8. Update $E(\mathbf{Y}^j)$:

$$E\left(\mathbf{Y}^j\right) = E\left(\mathbf{Y}^j\right) + \sum_{i=1}^{2} w_{t,i}[F(\mathbf{X}_i)]^j \tag{12.45}$$

9. Repeat steps 4–8 for $t = t + 1$ until the list of random input variables is exhausted.

10. Run *the deterministic EOM algorithm* using as input variable vector:

$$\mathbf{X}_{\mu} = [\mu_{x1}, \mu_{x2}, \ldots, \mu_{x,t}, \ldots, \mu_{xm}] \tag{12.46}$$

11. Determine the weight factor of EOM solution of step 10:

$$w_0 = 1 - \sum_{t=1}^{m} \frac{1}{\lambda_{t,4} - \lambda_{t,3}^2} \tag{12.47}$$

12. Update $E(\mathbf{Y}^j)$:

$$\begin{aligned} E(\mathbf{Y}^j) &= E(\mathbf{Y}^j) + w_0 [F(\mathbf{X}_{\mu})]^j \\ E(\mathbf{Y}^j) &= \sum_{t=1}^{m} \sum_{i=1}^{2} w_{t,i} [F(\mu_{x_1}, \mu_{x_1}, \ldots, x_{t,i}, \ldots, \mu_{x_m})]^j + w_0 [F(\mathbf{X}_{\mu})]^j \end{aligned} \tag{12.48}$$

Knowing statistical moments of the output random variable, the mean and STD can be computed:

$$\mu_Y = E(\mathbf{Y}) \tag{12.49}$$

$$\sigma_Y = \sqrt{E(\mathbf{Y}^2) - \mu_Y^2} \tag{12.50}$$

Based on statistical moments, it is possible to approximate the PDFs of the output random variables of interest using the Gram–Charlier series approach. For example, let us consider Y as any random variable with a continuous distribution, whose mean value is μ_Y and STD σ_Y. For the standardized variable $x = (Y - \mu_Y)/\sigma_Y$, its cumulative function and density function are denoted as $F(x)$ and $f(x)$, respectively. According to Gram–Charlier expansion, the cumulative and the density functions can be written as [47]

$$F(x) = \Phi(x) + \frac{c_1}{1!}\Phi'(x) + \frac{c_2}{2!}\Phi''(x) + \frac{c_3}{3!}\Phi'''(x) + \cdots \tag{12.51}$$

$$f(x) = \phi(x) + \frac{c_1}{1!}\phi'(x) + \frac{c_2}{2!}\phi''(x) + \frac{c_3}{3!}\phi'''(x) + \cdots \tag{12.52}$$

where $\Phi(x)$ and $\phi(x)$ represent the cumulative distribution function (CDF) and probabilistic density function (PDF) of normal distribution with μ=0 and σ=1. The constant coefficients c_k are calculated as follows:

$$\begin{aligned} c_1 &= c_2 = 0 \\ c_3 &= -\frac{E(Y^3) - 3\mu_Y E(Y^2) + 2\mu_Y^3}{\sigma_Y^3} \\ c_4 &= \frac{E(Y^4) - 3[E(Y^2)]^2 - 4\mu_Y E(Y^3) + 12\mu_Y^2 E(Y^2) - 6\mu_Y^4 + 3\sigma_Y^4}{\sigma_Y^4} - 3 \end{aligned} \tag{12.53}$$

Note: $E(Y^1)$, $E(Y^2)$, $E(Y^3)$, $E(Y^4)$, ... are the jth statistical moments of the output random variable Y, $E(Y^j), j = 1,2,3,4, \ldots$.

12.5 Simulation results

Two MGs were considered. The MG1 is a standard MG [8,19,28,30] used in this study to compare the obtained optimal solutions of the EOM problem with those referred to in literature. The MG2 is considered in order to show how can be determined the bids of the different DG units.

The proposed solution algorithm has been implemented in MATLAB® computing environment and run on a 2.20-GHz PC with a 3.0-GB RAM. The program *eom.m* developed to solving the deterministic EOM in both MGs was given at the end of this chapter.

12.5.1 Microgrid MG1

The MG1 is a typical LV grid-connected MG as shown in Figure 12.1. The MG1 consists of different types of DG sources such as gas-fired MT, PEM-FC, WT, PV, and storage device (NiMH-Battery). It is assumed that all DG sources produce active power at unity power factor. Furthermore, there is a power-exchange link between the utility and the MG1 to trade energy during a day based on decisions of the MGCC. The load demand within the MG1 for a typical day comprises one primarily residential area, one industrial feeder serving a small workshop and, one feeder with light commercial consumers, which is equivalent to a total energy demand of 1,695 kW h for that day. Table 12.1 offers the minimum and maximum production limits and the bid coefficients of the DG units in the MG1. The hourly forecast load demands, market energy prices, and WT and PV power productions are given in Table 12.2. The system data are taken from [19].

The proposed PSO-based approach is implemented to find the optimal solutions of the deterministic and probabilistic EOM problem of the MG1 for three cases, as follows:

- *Case 1*: It is supposed that both renewable energy sources (WT and PV) act at their available maximum power outputs during each hour of the day, and the rests of DGs, including MT, FC, battery, and the distribution grid (utility), can operate just within their power limits while satisfying the needed constraints. All DGs with related characteristics produce electricity within the MG1, and additional demand or surplus of energy inside the grid is exchanged with the utility from the point of common coupling (PCC).
- *Case 2*: In the second case, it is assumed that all the DGs including both renewable energy sources (WT and PV) and the utility can operate just within their power limits while satisfying the needed constraints.
- *Case 3*: In this case, it is supposed that the utility behaves as an unconstraint unit and exchanges energy with the MG1 without any limitation, while the rest of DGs act as in Case 2.

The above cases are considered to compare the results of the proposed method with those presented in [8,19].

Table 12.1 The power limits and coefficients of bid functions of the installed DG units

ID	Type	Min power (kW)	Max power (kW)	a_i (€ct/kW h^2)	b_i (€ct/kW h)	c_i (€ct/h)
1	MT	6	30	0	0.457	0
2	FC	3	30	0	0.294	0
3	PV	0	25	0	2.584	0
4	WT	0	15	0	1.073	0
5	Battery	−30	30	0	0.38	0
6	Utility	−30	30	0	–	0

Table 12.2 The forecast values of the load demand, market price, and WT and PV power production

Time (h)	Load (kW)	MP (€ct/kW h)	WT (kW)	PV (kW)
1	52	0.23	1.7850	0
2	50	0.19	1.7850	0
3	50	0.14	1.7850	0
4	51	0.12	1.7850	0
5	56	0.12	1.7850	0
6	63	0.20	0.9150	0
7	70	0.23	1.7850	0
8	75	0.38	1.3050	0.200
9	76	1.50	1.7850	3.750
10	80	4.00	3.0900	7.525
11	78	4.00	8.7750	10.450
12	74	4.00	10.410	11.950
13	72	1.50	3.9150	23.900
14	72	4.00	2.3700	21.050
15	76	2.00	1.7850	7.875
16	80	1.95	1.3050	4.225
17	85	0.60	1.7850	0.550
18	88	0.41	1.7850	0
19	90	0.35	1.3020	0
20	87	0.43	1.7850	0
21	78	1.17	1.3005	0
22	71	0.54	1.3005	0
23	65	0.30	0.9150	0
24	56	0.26	0.6150	0

12.5.1.1 Deterministic EOM

The best results obtained using the proposed PSO approach for Cases 1, 2, and 3 are given in Tables 12.3–12.5, respectively. By analyzing these results, it can be seen that all equality and inequality constraints are satisfied.

Table 12.3 Best solutions obtained deterministic EOM problem using PSO for Case 1

Time (h)	Power (kW)						Cost (€ct/h)
	PV	WT	MT	FC	Battery	Utility	
1	0	1.785	6	30	−15.785	30	14.379
2	0	1.785	6	30	−17.785	30	12.419
3	0	1.785	6	30	−17.785	30	10.919
4	0	1.785	6	30	−16.785	30	10.699
5	0	1.785	6	30	−11.785	30	12.599
6	0	0.915	6	30	−3.915	30	17.0561
7	0	1.785	6	30	2.215	30	21.219
8	0.2	1.305	6	30	29.223	8.272	27.7272
9	3.75	1.785	30	30	30	−19.535	16.2328
10	7.525	3.09	30	30	30	−20.615	−25.7698
11	10.45	8.775	28.775	30	30	−30	−50.2115
12	11.95	10.41	21.64	30	30	−30	−47.8418
13	23.9	3.915	14.185	30	30	−30	47.6609
14	21.05	2.37	18.58	30	30	−30	−34.3527
15	7.875	1.785	30	30	30	−23.66	8.8743
16	4.225	1.305	30	30	30	−15.53	15.9642
17	0.55	1.785	30	30	30	−7.335	32.8655
18	0	1.785	6	30	30	20.215	33.1655
19	0	1.302	6	30	22.698	30	32.0843
20	0	1.785	6	30	30	19.215	33.1398
21	0	1.3005	30	30	30	−13.3005	19.7639
22	0	1.3005	30	30	30	−20.3005	24.3632
23	0	0.915	6	30	−1.915	30	20.8161
24	0	0.615	6	30	−10.615	30	15.9882
						Total cost (€ct)	269.75999

The numerical results of Table 12.3 indicate that in the periods from 1 to 8 h and from 23 to 24 h, a large portion of the load is supplied by the FC and the utility, because the bids of corresponding units are lower in comparison with those of others during the examined period. In the periods from 9 to 17 h and from 21 to 22 h, the market prices are much higher, and the output power of MT is increased and surplus of energy is exported from MG1 to utility. The charging process of the battery is done at the hours of the day when the market prices are low (1–6 h and 23–24 h).

The results for Case 2 are shown in Table 12.4. It can be seen from Table 12.4 that the operating cost of the MG1 decreases greatly in comparison with Case 1 and demonstrates a reduction of 42.54% in related cost. The reason for this is significantly less participation by WT and PV, which have much higher bids than the other DGs. Similar to Case 1, the output power of FC has maximum value throughout the day, and the output power of MT has minimum value during periods of the day with low values of the market prices and maximum values during periods

Table 12.4 Best solutions obtained deterministic EOM problem using PSO for Case 2

Time (h)	Power (kW)						Cost (€ct/h)
	PV	WT	MT	FC	Battery	Utility	
1	0	0	6	30	−14	30	13.142
2	0	0	6	30	−16	30	11.182
3	0	0	6	30	−16	30	9.682
4	0	0	6	30	−15	30	9.462
5	0	0	6	30	−10	30	11.362
6	0	0	6	30	−3	30	16.422
7	0	0	6	30	4	30	19.982
8	0	0	6	30	30	9	26.382
9	0	1.785	30	30	30	−15.785	12.1678
10	7.525	3.09	30	30	30	−20.615	−25.7698
11	9.225	8.775	30	30	30	−30	−52.817
12	3.59	10.41	30	30	30	−30	−65.6235
13	0	3.915	30	30	30	−21.915	5.2583
14	9.63	2.37	30	30	30	−30	−58.6431
15	0	1.785	30	30	30	−15.785	4.2753
16	0	1.305	30	30	30	−11.305	13.2855
17	0	0	30	30	30	−5	30.93
18	0	0	6	30	30	22	31.982
19	0	0	6	30	24	30	31.182
20	0	0	6	30	30	21	31.992
21	0	1.3005	30	30	30	−13.3005	19.7639
22	0	0	30	30	30	−19	23.67
23	0	0	6	30	−1	30	20.182
24	0	0	6	30	−10	30	15.562
						Total cost (€ct)	155.01333

with high market prices. The battery-charging process corresponds with periods of low market prices. In periods with low market prices, the power is exchanging from the utility to the MG1.

The optimal operating points of the DGs for Case 3 are shown in Table 12.5. The results of Table 12.5 indicate that allocation of optimal powers to DGs regarding an unlimited power exchange situation ends in reductions of 74.73% and 56.02% in operation costs of the MG1 in comparison with Cases 1 and 2, respectively. In Case 3, the utility takes the lead in supplying the load inside the MG during the periods from 1 to 8 h, at 19 h and from 23 to 24 h, when the market price is low, while purchasing energy in bulk from the MG1 during the peak times. From an economical point of view, PV and WT start up when a shortage of power generation occurs inside the MG1 or there is a need to export more energy to the utility. Likewise, other DGs such as FC, MT, and battery adjust their generation set-points according to load levels during each hour of the day in an economical manner.

Table 12.5 Best solutions obtained deterministic EOM problem using PSO for Case 3

Time (h)	Power (kW)						Cost (€ct/h)
	PV	WT	MT	FC	Battery	Utility	
1	0	0	6	3	−30	73	9.014
2	0	0	6	3	−30	71	5.714
3	0	0	6	3	−30	71	2.164
4	0	0	6	3	−30	72	0.864
5	0	0	6	3	−30	77	1.464
6	0	0	6	3	−30	84	9.024
7	0	0	6	3	−30	91	13.154
8	0	0	6	30	−30	69	26.382
9	0	1.785	30	30	30	−15.785	12.1678
10	7.525	3.09	30	30	30	−20.615	−25.7698
11	10.45	8.775	30	30	30	−31.225	−54.5516
12	11.95	10.41	30	30	30	−38.36	−77.4613
13	0	3.915	30	30	30	−21.915	5.2583
14	21.05	2.37	30	30	30	−41.42	−74.8138
15	0	1.785	30	30	30	−15.785	4.2753
16	0	1.305	30	30	30	−11.305	13.2855
17	0	0	30	30	30	−5	30.93
18	0	0	6	30	30	22	31.982
19	0	0	6	30	−30	84	29.562
20	0	0	6	30	30	21	31.992
21	0	1.3005	30	30	30	−13.3005	19.7639
22	0	0	30	30	30	−19	23.67
23	0	0	6	30	−30	59	17.862
24	0	0	6	3	−30	77	12.244
						Total cost (€ct)	68.17626

Under the same system data, control variable limits, and constraints, the results obtained using the proposed PSO approach reported in this study are compared to some other algorithms reported in the literature as shown in Table 12.6. Bold font has been used to indicate the proposed algorithm and better performance. From Table 12.6, it can be seen that the proposed approach outperforms many techniques used to solve EOM problems, because the results obtained using PSO are either better than or comparable to those obtained using other techniques. This highlights its ability to find better quality solutions. The calculated STD of PSO for all three cases is equal to zero, which confirms the excellent robustness of the PSO. Moreover, Figure 12.2 shows the fast convergence characteristics of the proposed PSO for the best results of Cases 1–3.

12.5.1.2 Probabilistic EOM

The $2m+1$ point-estimate method is implemented along with the PSO to model the uncertainties in the output powers of the PV and WT units, the market prices and load demands, and to find the statistical moments, mean, and STD of the output

Table 12.6 Comparison of the simulation results for MG1

Case	Method	Best solution	Worst solution	Standard deviation
Case 1	GA [19]	277.7444	304.5889	13.4421
	PSO [19]	277.3237	303.3791	10.1821
	FSAPSO [19]	276.7867	291.7562	8.3301
	CPSO-T [19]	275.0455	286.5409	6.2341
	CPSO-L [19]	274.7438	281.1187	5.9697
	AMPSO-T [19]	274.5507	275.0905	0.3210
	AMPSO-L [19]	274.4317	274.7318	0.0921
	GSA [8]	275.5369	282.1743	2.9283
	SGSA [8]	269.7600	269.7600	0
	PSO	**269.75999**	**269.75999**	**0**
Case 2	GA [19]	162.9469	198.5134	24.5125
	PSO [19]	162.0083	180.2282	12.6034
	FSAPSO [19]	161.5561	175.5402	10.0025
	CPSO-T [19]	161.0580	165.3110	2.9971
	CPSO-L [19]	160.7708	163.5512	1.9660
	AMPSO-T [19]	159.9244	160.4091	0.3427
	AMPSO-L [19]	159.3628	159.6813	0.0963
	PSO	**155.01333**	**155.01333**	**0**
Case 3	GA [19]	91.3293	127.7625	13.4005
	PSO [19]	90.7629	112.8628	10.8689
	FSAPSO [19]	90.6919	108.7761	9.7874
	CPSO-T [19]	90.5545	102.1001	8.1639
	CPSO-L [19]	90.4833	100.8786	7.3505
	AMPSO-T [19]	89.9917	90.6221	0.4457
	AMPSO-L [19]	89.9720	90.0431	0.0921
	PSO	**68.17626**	**68.17626**	**0**

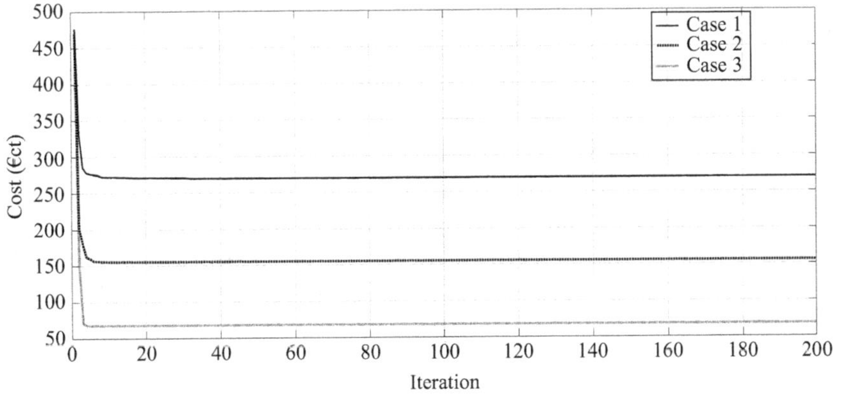

Figure 12.2 Convergence characteristics of PSO

random variables, i.e., the energy and operation cost of the MG1. It is assumed that the market price, load demand, and the output powers of the PV units have normal distribution with STD of 5%. The Weibull distribution is considered for the output power of the WT units, also with an STD of 5%. In this part, only Case 1 is considered. In this case, both renewable energy sources (WT and PV) act at their available maximum power outputs, and the rest of the DGs can operate just within their power limits while satisfying the needed constraints.

Table 12.7 shows the locations (concentrations) of the input random variables for the EOM problem. In order to determine the impact of individual input random variables on the statistical characteristics of the results, in the last row of Table 12.7, the corresponding expected costs are given. From these results, it is clear that the load demand and market price as input random variables have a higher impact on the STD of the expected cost than the output powers of PV and WT. The reason is that the output powers of PV and WT units are relatively small.

The expected power generation and expected cost for Case 1, evaluated by proposed method, are shown in Table 12.8. Comparing the results in Tables 12.3 and 12.8, it can be seen that, considering the effect of uncertainty, the expected cost has increased.

Figures 12.3 and 12.4 show the PDFs of the hourly expected costs for Case 1 obtained using the Gram–Charlier expansion. Moreover, Table 12.9 reports the statistical moments, mean, and STD of the total expected cost, and Figure 12.5 shows the PDF of the total expected cost for Case 1. These results prove that the presented method could find solution to the EOM problem under uncertain environments.

12.5.2 Microgrid MG2

The MG2 is an MG consisting of five different DG units, i.e., MT, FC, DE, WT, PV, and electric loads. Table 12.10 summarizes the data assumed for the DG units. The data were used to calculate the DG bids coefficients presented in Table 12.11. The 24-h load demand profile and the forecasted market energy prices are given in Table 12.12. Similarly, as in the MG1, it is assumed that all DG sources produce active power at unity power factor. There is a power-exchange link between the utility and the MG2 in order to trade energy during a day based on decisions of the MGCC.

The PSO is implemented to find optimal solutions of the EOM problem for the MG2 under a scenario of operation as follows: It is supposed that the renewable energy sources (WT and PV) act at their available maximum power outputs during each hour of the day, and the rest of DGs including MT, FC, and DE can operate just within their power limits while satisfying the needed constraints. All DGs with related characteristics produce electricity within the MG2, and additional demand or surplus of energy inside the grid is exchanged with the utility from the PCC. The utility behaves as an unconstraint unit and exchanges energy with the MG2 without any limitation.

Table 12.7 Locations of the input random variables for EOM and corresponding costs (according to Section 12.4.3)

T (h)	PV1 (kW)	PV2 (kW)	WT1 (kW)	WT2 (kW)	Load1 (kW)	Load2 (kW)	MP1 (€ct/kW h)	MP2 (€ct/kW h)	PVμ (kW)	WTμ (kW)	Loadμ (kW)	MPμ (€ct/kW h)
1	0	0	1.9100	1.6010	56.4525	47.2317	0.2502	0.2094	0	1.7889	51.9198	0.2300
2	0	0	1.8969	1.6066	54.4831	45.3188	0.2062	0.1741	0	1.7835	50.0021	0.1903
3	0	0	1.9127	1.5523	54.0711	45.7446	0.1524	0.1279	0	1.7840	49.9262	0.1402
4	0	0	1.9214	1.5520	55.3704	46.7322	0.1309	0.1094	0	1.7883	51.0369	0.1199
5	0	0	1.9115	1.5868	60.9813	51.0615	0.1297	0.1092	0	1.7857	55.9733	0.1195
6	0	0	0.9847	0.8134	68.5998	57.6901	0.2175	0.1830	0	0.9177	63.0725	0.1999
7	0	0	1.9087	1.5847	75.8243	63.9493	0.2510	0.2098	0	1.7846	70.0102	0.2298
8	0.2171	0.1823	1.3965	1.1675	81.6577	68.8102	0.4118	0.3467	0.1995	1.3057	75.1199	0.3796
9	4.0607	3.4274	1.9130	1.5653	82.3238	69.0709	1.6286	1.3654	3.7407	1.7847	76.1712	1.4937
10	8.1221	6.9365	3.3016	2.7502	86.7500	73.9082	4.3292	3.6443	7.5270	3.0867	80.2363	3.9991
11	11.3162	9.5398	9.4131	7.7950	84.8142	71.6062	4.3354	3.6443	10.4460	8.7809	78.0237	4.0046
12	13.0312	10.8307	11.1047	9.3827	80.4433	67.0277	4.3551	3.6340	11.9342	10.3968	73.8199	3.9957
13	26.0343	21.7962	4.1911	3.4923	78.1132	65.8875	1.6354	1.3764	23.8946	3.9201	72.0903	1.5041
14	22.9668	19.1433	2.5247	2.1291	78.5574	66.0522	4.3565	3.6568	21.0404	2.3702	72.1915	4.0060
15	8.5812	7.1641	1.9118	1.5600	82.3312	68.9979	2.1708	1.8374	7.8860	1.7840	75.9745	2.0009
16	4.6020	3.8538	1.3975	1.1477	86.5766	73.1214	2.1110	1.7784	4.2166	1.3049	79.7009	1.9490
17	0.6013	0.4995	1.9138	1.5590	92.2610	78.3103	0.6502	0.5460	0.5504	1.7838	85.2089	0.6007
18	0	0	1.9218	1.5585	94.8496	80.4597	0.4439	0.3751	0	1.7863	87.7317	0.4090
19	0	0	1.3827	1.1851	97.6795	82.0252	0.3780	0.3188	0	1.3021	89.9397	0.3491
20	0	0	1.9139	1.5802	94.6771	79.8851	0.4675	0.3942	0	1.7884	87.0910	0.4302
21	0	0	1.3900	1.1608	84.5054	71.3368	1.2761	1.0706	0	1.3008	77.8435	1.1693
22	0	0	1.3927	1.1515	77.3392	64.6437	0.5882	0.4942	0	1.3041	71.0001	0.5398
23	0	0	0.9821	0.8105	70.4745	59.6663	0.3252	0.2740	0	0.9171	64.9475	0.2994
24	0	0	0.6554	0.5499	60.5972	51.2704	0.2821	0.2376	0	0.6145	56.0443	0.2603
Total cost (€ct)	282.8	256.6	271.03	267.88	407.02	153.36	222.58	316.49	269.94			

Table 12.8 Expected power generation and cost evaluated by 2m + 1 PEM and PSO for Case 1

Time (h)	Power (kW)						Expected cost (€ct/h)
	PV	WT	MT	FC	Battery	Utility	
1	0	1.7889	5.9995	29.9974	−15.8162	29.9974	14.3665
2	0	1.7835	5.9996	29.9981	−17.7828	29.9981	12.4211
3	0	1.7840	6.0000	29.9999	−17.8124	29.9999	10.9108
4	0	1.7883	6.0006	30.0029	−16.7719	30.0029	10.7056
5	0	1.7857	6.0007	30.0035	−11.7961	30.0034	12.5918
6	0	0.9177	6.0000	30.0001	−3.8880	30.0001	17.0665
7	0	1.7846	6.0003	30.0014	2.2190	30.0014	21.2195
8	0.1995	1.3057	6.0002	30.0009	30.8156	6.6866	27.6104
9	3.7407	1.7847	29.9974	29.9974	29.9974	−19.5233	16.2544
10	7.5270	3.0867	29.9991	29.9991	29.9991	−20.6270	−25.8194
11	10.4460	8.7809	27.8642	30.0011	29.9973	−29.0836	−46.9688
12	11.9342	10.3968	21.6750	29.9965	29.9585	−29.9997	−47.8402
13	23.8946	3.9201	14.1913	30.0033	30.0033	−30.0032	47.6631
14	21.0404	2.3702	18.5848	30.0004	30.0004	−30.0004	−34.3574
15	7.8860	1.7840	29.8973	29.9994	29.9994	−23.5601	9.0267
16	4.2166	1.3049	30.0005	30.0005	30.0005	−15.5468	15.9319
17	0.5504	1.7838	30.0007	30.0007	30.0007	−7.3183	32.8760
18	0	1.7863	6.0003	30.0013	28.2537	21.8562	33.1067
19	0	1.3021	6.0644	29.9984	22.6144	29.9984	32.0729
20	0	1.7884	10.0222	29.9988	29.9988	15.2255	33.1136
21	0	1.3008	30.0029	30.0029	30.0029	−13.3602	19.7009
22	0	1.3041	29.9985	29.9985	29.9985	−20.3003	24.3642
23	0	0.9171	6.0003	30.0013	−1.9355	30.0013	20.8035
24	0	0.6145	5.9998	29.9989	−10.6000	29.9989	15.9959
					Total expected cost (€ct)		272.8162

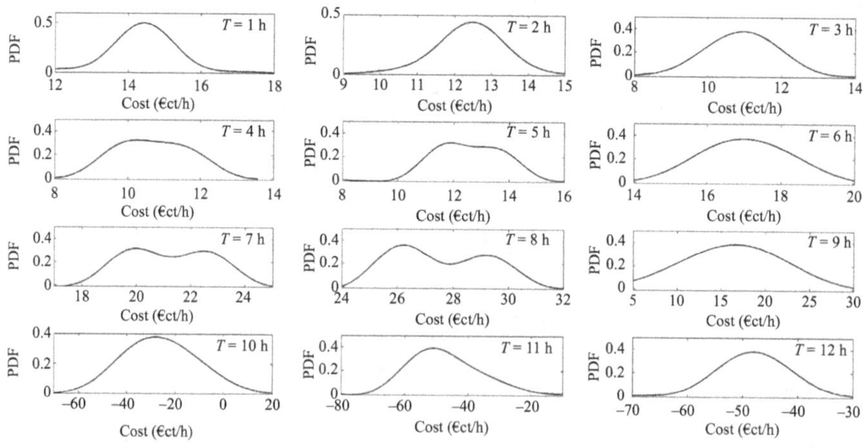

Figure 12.3 PDFs of hourly expected cost (T = 1–12 h) for Case 1

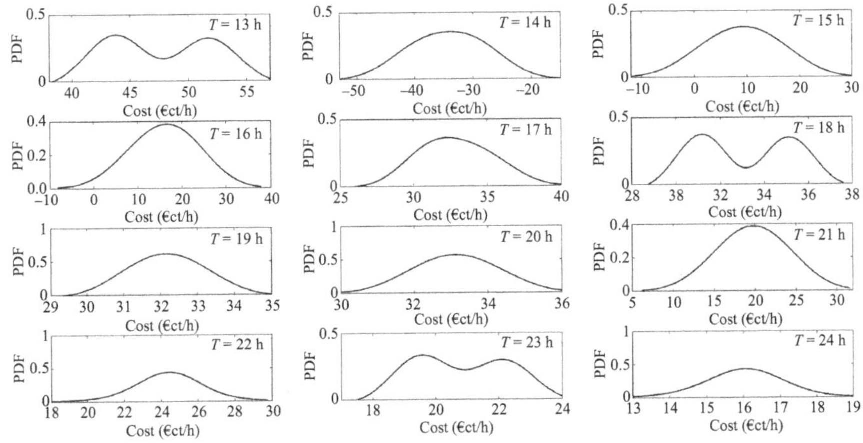

Figure 12.4 PDFs of hourly expected cost (T = 13–24 h) for Case 1

Table 12.9 Statistical moments of total expected cost for Case 1

$E(Y)$ (€ct)	$E(Y^2)$ (€ct)^2	$E(Y^3)$ (€ct)^3	$E(Y^4)$ (€ct)^4	μ_Y (€ct)	σ_Y (€ct)
272.8162	1.8111e+004	3.6247e+004	2.7502e+007	272.8162	90.8210

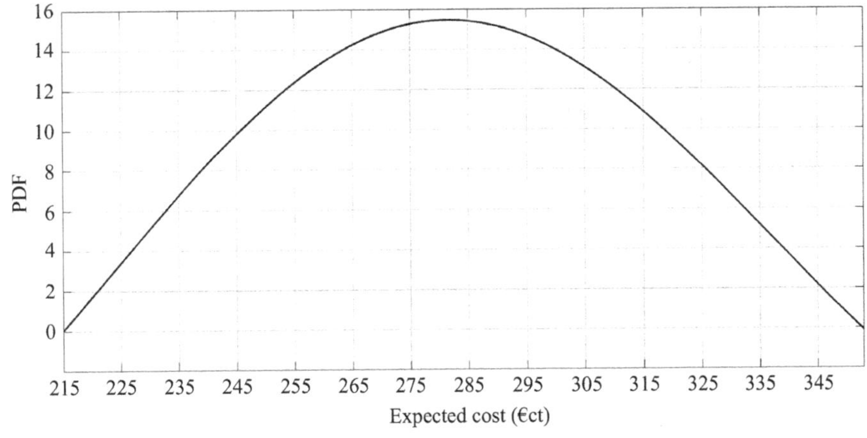

Figure 12.5 PDF of total expected cost for Case 1

In this scenario, the influence of variations in the market price of energy exchanged with utility on the total operating cost of the MG2 is discussed under the following three cases:

Case A: The market price of energy is low;
Case B: The market price of energy is in medium range;
Case C: The market price of energy reaches its real value.

Table 12.10 DG units data for MG2

MT	Tip: CGT301-302; $P_{nom} = 300$ kW; Efficiency characteristic [35]: P_{MT} = [90 120 150 180 210 240 270 300] kW; η_{MT} = [33.6 35.4 37.5 39.0 39.7 40.2 40.5 40.0]% C_{fuel} = 1.5 €ct/kW h; IC = 1,200 €/kW; i=7%; n=10 year; AP=7,884 kW h/kW
FC	Tip: PEM-FC; $P_{nom} = 300$ kW; V_0=0.9 V; V_{Pnom}=0.6 V; C_{fuel} = 1.5 €ct/kW h; IC = 4,000 €/kW; i=7%; n=10 year; AP=7,884 kW h/kW
DE	Tip: MDJW 410 T6; $P_{nom} = 369$ kW; Fuel consumption characteristic [48]: P_{DE} = [92.5 184.5 276.75 369.0] kW; $Fuel_{DE}$ = [29.5 51.9 74.8 100.9] L/h C_{fuel} = 0.6 €/L; IC = 500 €/kW; i=7%; n=10 year; AP=4,380 kW h/kW
WT	Tip: Vestas V44/600 kW; P_{nom} = 600 kW; v_{ci}=4 m/s; v_{nom}=16 m/s; v_{co}=20 m/s; IC = 1,600 €/kW; i=7%; n=10 year; AP=3,504 kW h/kW; Figure 12.6 shows the wind speed data used to calculate the power generated by the WT
PV	Tip: SW 250 mono; P_{nom} = 250 kW; P_{STC} = 250 W; $\gamma = -0.0045$ °C^{-1}; T_{NOCT} = 46 °C; IC = 6,000 €/kW; i=7%; n=20 year; AP=1,300 kW h/kW; Figure 12.6 shows the ambient temperature and solar irradiance data used to calculate the output power generated by the PV modules

Table 12.11 Power limits and calculated coefficients of bid functions of the installed DG units in MG2

ID	Type	Min power (kW)	Max power (kW)	a_i (€ct/kW h^2)	b_i (€ct/kW h)	c_i (€ct/h)
1	MT	90	300	0.0018	2.63	811
2	FC	30	300	0.0062	1.62	2,204
3	DE	0	369	0.0065	12.41	1,177
4	WT	0	600	0	6.50	0
5	PV	0	250	0	43.57	0

The simulation results associated with Case A, Case B, and Case C are shown in Tables 12.13–12.15, respectively.

In the Case A, given the low market price of electricity, the utility takes the lead in supplying the load inside the MG2, particularly during the periods with low and mean load levels.

In the Case B, the market price of energy is in medium range and significantly higher in relation to Case A. Most of the load portion is supplied from MT and FC. During the periods with mean load levels, corresponding with mean market price (5.4 €ct/kW h), the surplus of energy is exported from the MG to the utility.

In the third case, the market price of energy is increased to its real market value. The MT and FC are employed with their maximum power in periods with mean and high load level. Moreover, the surplus of energy is exported from the MG2 to the utility in most period of the day.

Table 12.12 The forecast load demands and market prices for MG2

Time (h)	Load (kW)	MP (€ct/kW h)		
		Case A	Case B	Case C
1	480	1.4	2.8	3.5
2	480	1.4	2.8	3.5
3	600	1.4	2.8	3.5
4	600	1.4	2.8	3.5
5	720	2.8	5.6	7
6	720	2.8	5.6	7
7	840	2.8	5.6	7
8	840	2.8	5.6	7
9	960	2.8	5.6	7
10	960	2.8	5.6	7
11	1,080	4.2	11.2	12.6
12	1,080	4.2	11.2	12.6
13	1,200	4.2	11.2	12.6
14	1,200	4.2	11.2	12.6
15	1,080	4.2	11.2	12.6
16	840	2.8	5.6	7
17	840	2.8	5.6	7
18	960	2.8	5.6	7
19	1,080	4.2	11.2	12.6
20	1,080	4.2	11.2	12.6
21	960	2.8	5.6	7
22	840	2.8	5.6	7
23	720	2.8	5.6	7
24	600	1.4	2.8	3.5

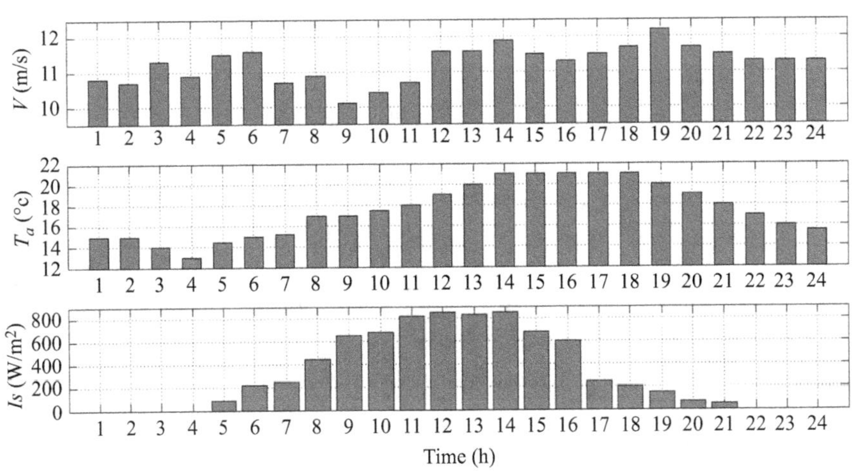

Figure 12.6 Wind speed data for MG2

Table 12.13 *Best solutions for Case A in MG2*

Time (h)	Power (kW)						Cost (€/h)
	MT	FC	DE	WT	PV	Utility	
1	90	30	0	252	0	108	62.85
2	90	30	0	246	0	114	62.57
3	90	30	0	279	0	201	65.94
4	90	30	0	257	0	223	64.80
5	90	95	0	291	23	221	81.75
6	90	95	0	296	56	183	95.19
7	90	95	0	246	63	346	99.65
8	90	95	0	257	109	289	118.88
9	90	95	0	215	153	407	138.53
10	90	95	0	230	159	386	141.51
11	300	208	0	246	187	139	160.74
12	300	208	0	296	192	84	163.84
13	300	208	0	296	187	209	166.98
14	300	208	0	314	190	188	168.54
15	300	208	0	291	156	125	149.67
16	90	95	0	279	140	236	132.09
17	90	95	0	291	61	303	100.63
18	90	95	0	302	49	423	99.57
19	300	208	0	332	38	202	103.92
20	300	208	0	302	18	252	95.46
21	90	95	0	291	13	471	84.20
22	90	95	0	279	0	376	75.20
23	90	95	0	279	0	256	71.84
24	90	30	0	279	0	201	65.94
						Total cost (€/h)	2,570.29

Table 12.14 *Best solutions for Case B in MG2*

Time (h)	Power (kW)						Cost (€/h)
	MT	FC	DE	WT	PV	Utility	
1	90	95	0	252		43	64.10
2	90	95	0	246	0	49	63.90
3	90	95	0	279	0	136	68.48
4	90	95	0	257	0	158	67.66
5	300	300	0	291	0	−194	80.04
6	300	300	0	296	23	−232	92.41
7	300	300	0	246	56	−69	101.44
8	300	300	0	257	63	−126	119.07
9	300	300	0	215	109	−8	142.02
10	300	300	0	230	153	−29	144.41
11	300	300	0	246	159	47	164.55
12	300	300	0	296	187	−8	163.79
13	300	300	0	296	192	117	175.66
14	300	300	0	314	187	96	175.78
15	300	300	0	291	190	33	152.52
16	300	300	0	279	156	−179	130.81
17	300	300	0	291	140	−112	101.22
18	300	300	0	302	61	8	103.52
19	300	300	0	332	49	110	112.17
20	300	300	0	302	38	160	107.19
21	300	300	0	291	18	57	89.51
22	300	300	0	279	13	−39	77.82
23	300	300	0	279	0	−159	71.10
24	90	95	0	279	0	136	68.48
						Total cost (€/h)	2,637.65

Table 12.15 Best solutions for Case C in MG2

Time (h)	Power (kW)						Cost (€/h)
	MT	FC	DE	WT	PV	Utility	
1	242	152	0	252	0	−165	63.79
2	242	152	0	246	0	−160	63.63
3	242	152	0	279	0	−73	68.82
4	242	152	0	257	0	−50	68.15
5	300	300	0	291	23	−194	77.33
6	300	300	0	296	56	−232	89.16
7	300	300	0	246	63	−69	100.47
8	300	300	0	257	109	−126	117.30
9	300	300	0	215	153	−8	141.91
10	300	300	0	230	159	−29	144.00
11	300	300	15	246	187	32	165.19
12	300	300	15	296	192	−23	163.66
13	300	300	15	296	187	102	177.28
14	300	300	15	314	190	81	177.12
15	300	300	15	291	156	19	152.97
16	300	300	0	279	140	−179	128.30
17	300	300	0	291	61	−112	99.65
18	300	300	0	302	49	8	103.64
19	300	300	15	332	38	96	113.70
20	300	300	15	302	18	145	109.41
21	300	300	0	291	13	57	90.30
22	300	300	0	279	0	−39	77.27
23	300	300	0	279	0	−159	68.87
24	242	152	0	279	0	−73	68.82
					Total cost (€/h)		2,630.75

12.5.3 MATLAB program eom *used for deterministic EOM*

```
function []=eom(MG,Case)
T=0;
while T<24
   T=T+1;
   [GlobBestCos,Fbest,gbest]=PSO(MG,Case,T);
   FBEST(T)=Fbest;
   LBEST(T,:)=gbest;
   GLOBESTCOST(T,:)=GlobBestCos;
end
PLOTGLBEST=sum(GLOBESTCOST);
plot(PLOTGLBEST,'-k','linewidth',2);
title(['\fontsize{12}\bf EOM']);
xlabel('\fontsize{12}\bf Iteration');
ylabel('\fontsize{12}\bf Total Cost');
legend('\fontsize{10}\bf PSO',1);
format short

switch MG
   case 'MG1'
       [DGs,Ppv,Pwt,Pload,Bgrid,Pgridmin,Pgridmax]=MG1;
       Bgen=DGs(:,4);
       if Case == 'Case1'
```

```
            for k=1:24
                LBEST(k,4)=Pload(k)-sum(LBEST(k,1:3))-Ppv(k)-Pwt(k);
            end
            Cg1=LBEST(:,1)*Bgen(1);
            Cg2=LBEST(:,2)*Bgen(2);
            Cg3=Ppv*Bgen(3);
            Cg4=Pwt*Bgen(4);
            Cg5=LBEST(:,3)*Bgen(5);
            Cg6=LBEST(:,4).*Bgrid';
            Cost_h=Cg1+Cg2+Cg3'+Cg4'+Cg5+Cg6;
            GBEST(:,1)=LBEST(:,1);
            GBEST(:,2)=LBEST(:,2);
            GBEST(:,3)=Ppv;
            GBEST(:,4)=Pwt;
            GBEST(:,5)=LBEST(:,3);
            GBEST(:,6)=LBEST(:,4);
        else
            for k=1:24
                LBEST(k,6)=Pload(k)-sum(LBEST(k,1:5));
            end
            Cg1=LBEST(:,1)*Bgen(1);
            Cg2=LBEST(:,2)*Bgen(2);
            Cg3=LBEST(:,3)*Bgen(3);
            Cg4=LBEST(:,4)*Bgen(4);
            Cg5=LBEST(:,5)*Bgen(5);
            Cg6=LBEST(:,6).*Bgrid';
            Cost_h=Cg1+Cg2+Cg3+Cg4+Cg5+Cg6;
            GBEST=LBEST;
        end
        Cost24=sum(Cost_h);
        Time=[1:24]';
disp(' ')
disp('R E S U L T S: ')
disp(' ')
disp('     Time      MT       FC         PV       WT     Battery    Utility   Cost_h')
disp('     (h)      (kW)     (kW)       (kW)     (kW)     (kW)        (kW)   (Ect/h)')
disp('-----------------------------------------------------------------------------')
disp([Time GBEST Cost_h]);
fprintf('Total cost = %7.5f (Ect/day)\n',Cost24);


    case 'MG2'
        [DGs,Ppv,Pwt,Pload,Bgrid,Pgridmin,Pgridmax]=MG2;
        a=DGs(:,4);b=DGs(:,5);c=DGs(:,6); %B(Pg)=a*Pg^2+b*Pg+c
        for k=1:24
            LBEST(k,4)=Pload(k)-sum(LBEST(k,1:3))-Ppv(k)-Pwt(k);
        end
        Cg1=a(1)*LBEST(:,1).^2+b(1)*LBEST(:,1)+c(1);
        Cg2=a(2)*LBEST(:,2).^2+b(2)*LBEST(:,2)+c(2);
        Cg3=a(3)*LBEST(:,3).^2+b(3)*LBEST(:,3)+c(3);
        Cg4=Pwt*b(4);
        Cg5=Ppv*b(5);
        if Case == 'CaseA'
            Cg6=LBEST(:,4).*Bgrid(:,1);
        elseif Case == 'CaseB'
            Cg6=LBEST(:,4).*Bgrid(:,2);
        elseif Case == 'CaseC'
            Cg6=LBEST(:,4).*Bgrid(:,3);
        end
        GBEST(:,1)=LBEST(:,1);
        GBEST(:,2)=LBEST(:,2);
        GBEST(:,3)=LBEST(:,3);
        GBEST(:,4)=Pwt;
        GBEST(:,5)=Ppv;
        GBEST(:,6)=LBEST(:,4);
        Cost_h=(Cg1+Cg2+Cg3+Cg4'+Cg5'+Cg6)/100; %(Euro/h)
```

```
      Cost24=sum(Cost_h);
      Time=[1:24]';
disp(' ')
disp('R E S U L T S: ')
disp(' ')
disp('     Time      MT       FC          DE        WT        PV      Utility     Cost_h')
disp('     (h)      (kW)     (kW)        (kW)      (kW)      (kW)      (kW)       (E/h)' )
disp('  - - - - - - - - - - - - - - - - - - - - - - - - - - - - - - - - - - - - - - - -')
disp([Time GBEST Cost_h]);
fprintf('Total cost = %7.5f (E/day)\n',Cost24);
end

%==============================================================================
% MGs data
%==============================================================================
function [DGs,Ppv,Pwt,Pload,Bgrid,Pgridmin,Pgridmax]=MG1
%DGs - DG units data: Pgmin, Pgmax - DG power limits; Bgen - DG bids (Ect/kWh)
%Ppv - forecasted power of PV generator
%Pwt - forecasted power of WT generator
%Pload - forecasted power of load
%Bgrid - forecasted market energy price
%Pgridmin, Pgridmax - power limits from/to main grid

% | DG |Pgmin|Pgmax| Bgen |StarShut|TipGen|
DGs=[1    6    30    0.457   0.96    %MT
     2    3    30    0.294   1.65    %PAFC
     3    0    25    2.584   0       %PV
     4    0    15    1.073   0       %WT
     5  -30    30    0.38    0];     %Batt

Ppv=[0,0,0,0,0,0,0,0.008,0.150,0.301,0.418,0.478,0.956,0.842,0.315,0.169,...
    0.022,0,0,0,0,0,0,0]*25; % kW

Pwt=[0.119,0.119,0.119,0.119,0.119,0.061,0.119,0.087,0.119,0.206,0.585,0.694,...
    0.261,0.158,0.119,0.087,0.119,0.119,0.0868,0.119,0.0867,0.0867,0.061,
0.041]*15;%kW

Pload=[52,50,50,51,56,63,70,75,76,80,78,74,72,72,76,80,85,88,90,87,78,71,65,56];
%kW

Bgrid=[0.23,0.19,0.14,0.12,0.12,0.20,0.23,0.38,1.50,4.0,4.0,4.0,1.50,4.0,2.0,...
      1.95,0.60,0.41,0.35,0.43,1.17,0.54,0.30,0.26]; %(Ect/kWh)

Pgridmin=-30; %kW
Pgridmax=30; %kW
return
%- - - - - - - - - - - - - - - - - - - - - - - - - - - - - - - - - - - - - - - -
function [DGs,Ppv,Pwt,Pload,Bgrid,Pgridmin,Pgridmax]=MG2
%DGs - DG units data: Pgmin, Pgmax - DG power limits; a,b,c - DG bids coefficients
%Ppv - forecasted power of PV generator
%Pwt - forecasted power of WT generator
%Pload - forecasted power of load
%Bgrid - forecasted market energy price
%Pgridmin, Pgridmax - power limits from/to main grid

  %| DG |Pgmin|Pgmax|    a     |  b    |  c    | TipGen |
DGs=[1    90    300   0.0018    2.63    811      %MT
     2    30    300   0.0062    1.62    2204     %PAFC
     3     0    369   0.0065    12.41   1177     %DE
     4     0    600   0         6.50    0        %WT
     5     0    250   0         43.57   0];      %PV

Ppv =[0,0,0,0,23.2670,55.7054,62.9711,109.1461,152.9023,158.8310,186.8729,...
     191.8211,186.9808,189.9086,156.1535,139.5375,61.3398,49.4375,37.5211,...
     17.7933,12.8023,0,0,0]; %kW
```

```
Pwt =[251.6,246.225,279.225,257.025,290.625,296.4,246.225,257.025,215.025,...
     230.4,246.225,296.4,296.4,314.025,290.625,279.225,290.625,302.225,...
     332.1,302.225,290.625,279.225,279.225,279.225]; %kW

Pload=[0.4,0.4,0.5,0.5,0.6,0.6,0.7,0.7,0.8,0.8,0.9,0.9,1.0,1.0,0.9,...
      0.7,0.7,0.8,0.9,0.9,0.8,0.7,0.6,0.5]*1200; %kW

BgridA=[1.4,1.4,1.4,1.4,2.8,2.8,2.8,2.8,2.8,2.8,4.2,4.2,4.2,4.2,4.2,2.8,...
       2.8,2.8,4.2,4.2,2.8,2.8,2.8,1.4]; %(Ect/kWh)
BgridB=[2.8,2.8,2.8,2.8,5.6,5.6,5.6,5.6,5.6,5.6,11.2,11.2,11.2,11.2,11.2,...
       5.6,5.6,5.6,11.2,11.2,5.6,5.6,5.6,2.8]; %(Ect/kWh)
BgridC=[3.5,3.5,3.5,3.5,7,7,7,7,7,7,12.6,12.6,12.6,12.6,12.6,7,7,7,12.6,...
       12.6,7,7,7,3.5]; %(Ect/kWh)
Bgrid=[BgridA' BgridB' BgridC'];

Pgridmin=-2000;  Pgridmax=2000; %kW
return

%================================================================================
% Particle Swarm Optimization (PSO) algorithm
%================================================================================
function [GlobBestCos,Fbest,gbest] = PSO(MG,Case,T)

%Defining parameters of PSO:
%- - - - - - - - - - - - - - - - - - - - - - - - - - - - - - - - - - - - - - -
N=50;                  %population size (number of particles N)
tmax=200;              %maximum number of iterations
C1=2;                  %cognitive constant C1
C2=2;                  %social   constant C2
wmax=0.9;              %max inertia weight in w, wmax
wmin=0.4;              %min inertia weight in w, wmin
type_opt='min';        %type of optimization: 'min' or 'max'

%Generating the initial random population.
%- - - - - - - - - - - - - - - - - - - - - - - - - - - - - - - - - - - - - - -
[numbervar,lowerbound,upperbound]=bounds(MG,Case,T);
if size(upperbound,2)==1
   X=rand(N,numbervar).*(upperbound-lowerbound)+lowerbound;
end
if size(upperbound,2)>1
   for i=1:numbervar
       high=upperbound(i);
       low=lowerbound(i);
       X(:,i)=rand(N,1).*(high-low)+low;
   end
end

%Initializing velocity of particles
%- - - - - - - - - - - - - - - - - - - -
V=zeros(N,numbervar);
% ITERATIVE PROCEDURE OF PSO
% - - - - - - - - - - - - - - - - - - - - - - - - - - - - - - - - - - - - - - - - -
for t=1:tmax
   for i=1:N
       x=X(i,:,t);  %position of particle i at iteration t

       % Calculate the fitness value for each particle
       %- - - - - - - - - - - - - - - - - - - - - - - - -
       fitness(i,t)=Fobj(x,MG,Case,T);

       %Determination of pbest (individual best) and gbest (global best)
       %- - - - - - - - - - - - - - - - - - - - - - - - - - - - - - - - - - - -
       switch type_opt
           case 'min'
               [pbest_fitness,index]=min(fitness(i,:));
               pbest=X(i,:,index);
```

```
            if(i==1 && t==1)
               gbest=pbest;
               gbest_fitness=pbest_fitness;
            elseif fitness(i,t)<gbest_fitness
               gbest_fitness=fitness(i,t);
               gbest=X(i,:,t);
            end
         case 'max'
            [pbest_fitness,index]=max(fitness(i,:));
            pbest=X(i,:,index);

            if(i==1 && t==1)
               gbest=pbest;
               gbest_fitness=pbest_fitness;
            elseif fitness(i,t)>gbest_fitness
               gbest_fitness=fitness(i,t);
               gbest=X(i,:,t);
            end
         otherwise
            error('fitness_type mismatch')
      end

      %Updating the inertia weight
      %- - - - - - - - - - - - - - - -
      w(t)=wmin+((tmax - t)*(wmax - wmin))/(tmax-1);

      %Updating the velocity (V) of each particle
      %- - - - - - - - - - - - - - - - - - - - - - - - -
      V(i,:,(t+1))=w(t)*V(i,:,t)+C1*rand(1,numbervar).*(pbest-X(i,:,t))...
         +C2*rand(1,numbervar).*(gbest-X(i,:,t));

      %Updating the position (X) of each particle
      %- - - - - - - - - - - - - - - - - - - - - - - - -
      X(i,:,(t+1))= X(i,:,t) + V(i,:,(t+1));

      %Checking the limits
      %- - - - - - - - - - - -
      for k=1:length(lowerbound)
         if upperbound(k)<X(i,k,(t+1))
            X(i,k,(t+1))=upperbound(k);
         elseif lowerbound(k)>X(i,k,(t+1))
            X(i,k,(t+1))=lowerbound(k);
         end
      end

   end
   GlobBestCos(t)= gbest_fitness;
end                                  %end of PSO iteration  procedure
Fbest=GlobBestCos(tmax);


%============================================================================
% Calculation of the objective function
%============================================================================
function [Fobj]=Fobj(L,MG,Case,T)
switch MG
   case 'MG1'
      [DGs,Ppv,Pwt,Pload,Bgrid,Pgridmin,Pgridmax]=MG1;
      [ngen,nkolg]=size(DGs);
      Bg=DGs(:,4);
      if Case=='Case1'
         Pg(1)=L(1);Pg(2)=L(2);Pg(3)=Ppv(T);Pg(4)=Pwt(T);Pg(5)=L(3);
         Pgrid=Pload(T)-sum(Pg);
         if Pgrid<Pgridmin
            Pglim=Pgridmin;
         elseif Pgrid>Pgridmax
```

```
            Pglim=Pgridmax;
         else
            Pglim=Pgrid;
         end
         Pf_Pgrid=10^6*(Pgrid-Pglim)^2;
      elseif Case=='Case2'
         Pg(1)=L(1);Pg(2)=L(2);Pg(3)=L(3);Pg(4)=L(4);Pg(5)=L(5);
         Pgrid=Pload(T)-sum(Pg);
         if Pgrid<Pgridmin
            Pglim=Pgridmin;
         elseif Pgrid>Pgridmax
            Pglim=Pgridmax;
         else
            Pglim=Pgrid;
         end
         Pf_Pgrid=10^6*(Pgrid-Pglim)^2;
      elseif Case=='Case3'
         Pg(1)=L(1);Pg(2)=L(2);Pg(3)=L(3);Pg(4)=L(4);Pg(5)=L(5);
         Pgrid=Pload(T)-sum(Pg);
      end
      CostG=0;
      for k=1:ngen
         CostG=CostG+Bg(k)*Pg(k);
      end
      Cost=CostG+Bgrid(T)*Pgrid;
      if Case=='Case1'
         Fobj=Cost+Pf_Pgrid;
      elseif Case=='Case2'
         Fobj=Cost+Pf_Pgrid;
      elseif Case=='Case3'
         Fobj=Cost;
      end

   case 'MG2'
      [DGs,Ppv,Pwt,Pload,Bgrid,Pgridmin,Pgridmax]=MG2;
      [ngen,nkolg]=size(DGs);
      a=DGs(:,4);b=DGs(:,5);c=DGs(:,6); % B(Pg)=a*Pg^2+b*Pg+c
      Pg(1)=L(1);Pg(2)=L(2);Pg(3)=L(3);Pg(4)=Pwt(T);Pg(5)=Ppv(T);
      Pgrid=Pload(T)-sum(Pg);
      %-----------------------------------------------------
      if Pgrid<Pgridmin
         Pglim=Pgridmin;
      elseif Pgrid>Pgridmax
         Pglim=Pgridmax;
      else
         Pglim=Pgrid;
      end
      Pf_Pgrid=10^6*(Pgrid-Pglim)^2;
      %-----------------------------------------
      CostG=0;
      for k=1:ngen
         CostG=CostG+a(k)*Pg(k)^2+b(k)*Pg(k)+c(k);
      end
      if Case == 'CaseA'
         Cost=CostG+Bgrid(T,1)*Pgrid;
      elseif Case == 'CaseB'
         Cost=CostG+Bgrid(T,2)*Pgrid;
      elseif Case == 'CaseC'
         Cost=CostG+Bgrid(T,3)*Pgrid;
      end

      Fobj=Cost+Pf_Pgrid;
end
return
```

```
%=====================================================================================
% Defining the space of possible  solutions
%=====================================================================================
function [numbervar,lowerbound,upperbound]=bounds(MG,Case,T)
switch MG
    case 'MG1'
        [DGs,Ppv,Pwt,Pload,Bgrid,Pgridmin,Pgridmax]=MG1;
        [ngen,nkolg]=size(DGs);
        Pgmin=DGs(:,2); Pgmax=DGs(:,3);
        if Case=='Case1'
            numbervar=3;
            lowerbound=[Pgmin(1),Pgmin(2),Pgmin(5)];
            upperbound=[Pgmax(1),Pgmax(2),Pgmax(5)];
        elseif Case=='Case2'
            numbervar=5;
            lowerbound=[Pgmin(1),Pgmin(2),Pgmin(3),Pgmin(4),Pgmin(5)];
            upperbound=[Pgmax(1),Pgmax(2),Ppv(T)+eps,Pwt(T),Pgmax(5)];
        elseif Case=='Case3'
            numbervar=5;
            lowerbound=[Pgmin(1),Pgmin(2),Pgmin(3),Pgmin(4),Pgmin(5)];
            upperbound=[Pgmax(1),Pgmax(2),Ppv(T)+eps,Pwt(T),Pgmax(5)];
        end
    case 'MG2'
        [DGs,Ppv,Pwt,Pload,Bgrid,Pgridmin,Pgridmax]=MG2;
        Pgmin=DGs(:,2); Pgmax=DGs(:,3);
        numbervar=3;
        lowerbound=[Pgmin(1),Pgmin(2),Pgmin(3)];
        upperbound=[Pgmax(1),Pgmax(2),Pgmax(3)];
end
return
```

An example of execution of the *eom* program was displayed in Figure 12.7.

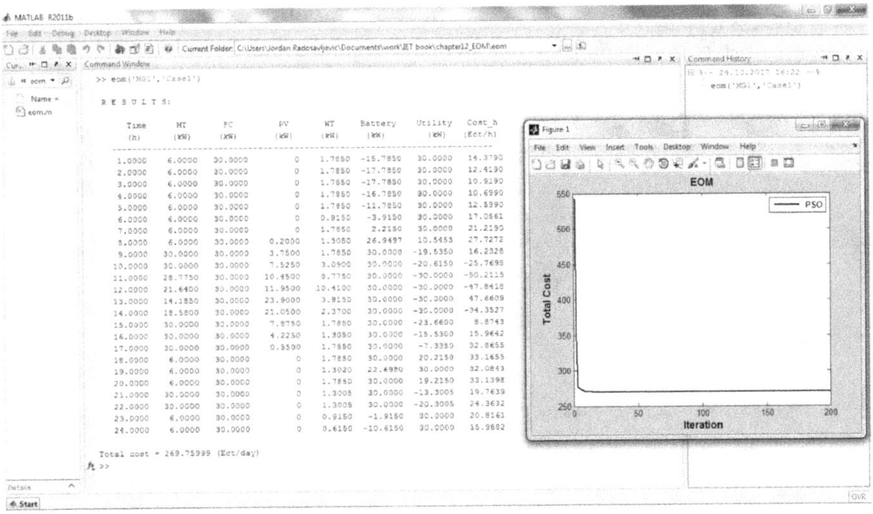

Figure 12.7 Example execution of the eom program

12.6 Conclusion

In this chapter, an efficient PSO-based approach has been proposed and successfully applied to solve EOM problem in an MG. The Weibull and normal distributions are employed to model the input random variables, namely, the output power of the WT and PV units, the load demand, and the market price. The $2m+1$ point estimate method and the Gram–Charlier expansion theory are used to obtain the statistical moments and the PDFs of the EOM results.

The proposed approach has been tested and investigated on two grid-connected MGs including different DG units and energy storage. The simulation results show the efficiency of the proposed approach to solve both deterministic and probabilistic EOM problems under different operational scenarios of the MGs. Moreover, the results obtained using the proposed PSO algorithm are either better or comparable to those obtained using other techniques reported in the literature. As such, it can serve as a useful decision-making supporting tool for MG operators and help to find out how the input random variables affect the statistical characteristics of the EOM results.

References

[1] Hatziargyriou N. *MICROGRIDS—large scale integration of micro-generation to low voltage grids*. 2006. Available at http://microgrids.eu/micro2000/presentations/16.pdf

[2] Tsikalakis A.G., Hatziargyriou N.D. 'Centralized control for optimizing microgrids operation'. *IEEE Transactions on Energy Conversion*. 2008;**23**(1):241–8

[3] Ustun T.S., Ozansoy C., Zayegh A. 'Recent developments in microgrids and example cases around the world—a review'. *Renewable and Sustainable Energy Reviews*. 2011;**15**:4030–41

[4] Jiayi H., Chuanwen J., Rong X. 'A review on distributed energy resources and MicroGrid'. *Renewable and Sustainable Energy Reviews*. 2008;**12**:2472–83

[5] Lopes J.A.P., Moreira C.L., Madureira A.G. 'Defining control strategies for microgrids islanded operation'. *IEEE Transactions on Power Systems*. 2006;**21**(2):916–24

[6] Basak P., Chowdhury S., Halder nee Dey S., Chowdhury S.P. 'A literature review on integration of distributed energy resources in the perspective of control, protection and stability of microgrid'. *Renewable and Sustainable Energy Reviews*. 2012;**16**:5545–56

[7] Planas E., Gil de Muro A., Andreu J., Kortabarria I., Martinez del Algria I. 'General aspects, hierarchical controls and droop methods in microgrids: a review'. *Renewable and Sustainable Energy Reviews*. 2013;**17**:147–59

[8] Niknam T., Golestaneh F., Malekpour A. 'Probabilistic energy and operation management of a microgrid containing wind/photovoltaic/fuel cell

generation and energy storage devices based on point-estimate method and self-adaptive gravitational search algorithm'. *Energy*. 2012;**43**:427–37

[9] Hatziargyriou N.D., Anastasiadis A.G., Tsikalakis A.G., Vasiljevska J. 'Quantification of economic, environmental and operational benefits due to significant penetration of microgrids in a typical LV and MV Greek network'. *European Transaction on Electrical Power*. 2011;**21**(2):1217–37

[10] Chakraborty S., Weiss M.D., Simoes M.G. 'Distributed intelligent energy management system for a single-phase high frequency AC microgrid'. *IEEE Transactions on Industrial Electronics*. 2007;**54**:97–109

[11] Liu C., Wang X., Wu X., Guo J. 'Economic scheduling model of microgrid considering the lifetime of batteries'. *IET Generation, Transmission & Distribution*. 2017;**11**(3):759–67

[12] Mohamed F.A., Koivo H.N. 'System modeling and online optimal management of MicroGrid using Mesh Adaptive Direct Search'. *International Journal of Electrical Power & Energy Systems*. 2010;**32**:398–407

[13] Chen C., Duan S., Cai T., Liu B., Hu G. 'Smart energy management system for optimal microgrid economic operation'. *IET Renewable Power Generation*. 2011;**5**:258–67

[14] Mohamed F.A., Koivo H.N. 'Online management genetic algorithms of microgrid for residential application'. *Energy Conversion and Management*. 2012;**64**:562–8

[15] Arabali A., Ghofrani M., Etazadi-Amoli M., Fadali S., Baghzouz Y. 'Genetic-algorithm-based optimization approach for energy management'. *IEEE Transactions on Power Delivery*. 2013;**28**(1):162–70

[16] Guo L., Wang N., Lu H., Li X., Wang C. 'Multi-objective stochastic optimal planning method for stand-alone microgrid system'. *IET Generation, Transmission & Distribution*. 2014;**8**(7):1263–73

[17] Kuznetsova E., Ruiz C., Li Y.-F., Zio E. 'Analysis of robust optimization for decentralized microgrid energy management under uncertainty'. *International Journal of Electrical Power & Energy Systems*. 2015;**64**:815–32

[18] Sanseverino E.R., Di Silvestre M.L., Ippolito M.G., De Paola A., Le L.G. 'An execution, monitoring and replanning approach for optimal energy management in microgrids'. *Energy*. 2011;**36**:3429–36

[19] Moghaddam A.A., Seifi A., Niknam T., Pahlavani M.R.A. 'Multi-objective operation management of a renewable MG (micro-grid) with back-up micro-turbine/fuel cell/battery hybrid power source'. *Energy*. 2011;**36**: 6490–507

[20] Sortome E., El-Sharkawi M.A. 'Optimal power flow for a system of microgrids with controllable loads and battery storage'. *IEEE/PES Power Systems Conference and Exposition*; Seattle, WA, USA, Mar 2009. IEEE; 2009. pp. 1–5

[21] Wu H., Liu X., Ding M. 'Dynamic economic dispatch of a microgrid: mathematical models and solution algorithm'. *International Journal of Electrical Power & Energy Systems*. 2014;**63**:336–46

[22] Radosavljević J., Jevtić M., Klimenta D. 'Energy and operation management of a microgrid using particle swarm optimization'. *Engineering Optimization.* 2016;**48**(5):811–30
[23] Nikmehr N., Ravadanegh S.N. 'A study on optimal power sharing in interconnected microgrids under uncertainty'. *International Transactions on Electrical Energy Systems.* 2016;**26**:208–32
[24] Nikmehr N., Ravadanegh S.N. 'Optimal operation of distributed generations in micro-grids under uncertainties in load and renewable power generation using heuristic algorithm'. *IET Renewable Power Generation.* 2015;**9**(8):982–90
[25] Mohan V., Singh J.G., Ongsakul W., Suresh M.P.R. 'Performance enhancement of online energy scheduling in a radial utility distribution microgrid'. *International Journal of Electrical Power & Energy Systems.* 2016;**79**:98–107
[26] Tomoiaga B., Chindric M., Sumper A., Marzband M. 'The optimization of microgrids operation through a heuristic energy management algorithm'. *Advanced Engineering Forum.* 2013;**8–9**:185–94
[27] Li P., Xu W., Zhou Z., Li R. 'Optimized operation of microgrid based on gravitational search algorithm'. *Proceedings of International Conference on Electrical Machines and Systems*; Busan, Korea, Oct 2013. IEEE; 2013. pp. 338–42
[28] Mohammadi S., Mozafari B., Solimani S., Nikman T. 'An adaptive modified firefly optimization algorithm based on Hong's point estimate method to optimal operation management in a microgrid with consideration of uncertainties'. *Energy.* 2013;**51**:339–48
[29] Niknam T., Golestaneh F., Shafiei M. 'Probabilistic energy management of a renewable microgrid with hydrogen storage using self-adaptive charge search algorithm'. *Energy.* 2013;**49**:252–67
[30] Mohammadi S., Soleymani S., Mozafari B. 'Scenario-based stochastic operation management of microgrid including wind, photovoltaic, micro-turbine, fuel cell and energy storage devices'. *International Journal of Electrical Power & Energy Systems.* 2014;**54**:525–35
[31] Moradi M., Eskandari M., Showkati H. 'A hybrid method for simultaneous optimization of DG capacity and operational strategy in microgrids utilizing renewable energy resources'. *International Journal of Electrical Power & Energy Systems.* 2014;**56**:241–58
[32] Zhang Y., Gatsis N., Giannakis G.B. 'Robust energy management for microgrids with high-penetration renewables'. *IEEE Transactions on Sustainable Energy.* 2013;**4**(4):944–53
[33] Prodan I., Zio E. 'A model predictive control framework for reliable microgrid energy management'. *International Journal of Electrical Power & Energy Systems.* 2014;**61**:399–409
[34] Azmy A.M., Erlich I. 'Online optimal management of PEM fuel cells using neural networks'. *IEEE Transactions on Power Delivery.* 2005;**20**(2):1051–8

[35] Campanari S., Macchi E. 'Technical and tariff scenarios effect on microturbine trigenerative applications'. *Journal of Engineering for Gas Turbines and Power*. 2004;**126**:581–9

[36] Barbir F., Gomez T. 'Efficiency and economics of proton exchange membrane (PEM) fuel cells'. *International Journal of Hydrogen Energy*. 1997; **22**(10/11):1027–37

[37] Villanueva D., Pazos J.L., Feijo A. 'Probabilistic load flow including wind power generation'. *IEEE Transactions on Power Systems*. 2011;**26**(3):1659–67

[38] Atwa Y.M., El-Saadany E.F., Salama M.M.A., Seethapathy R., Assam M., Conti S. 'Adequacy evaluation of distribution system including wind/solar DG during different modes of operation'. *IEEE Transactions on Power Systems*. 2011;**26**(4):1945–52

[39] Tiwari G.N., Dubey S. *Fundamentals of photovoltaic modules and their applications*. London: RSC Energy Series Cambridge; 2010. p. 423

[40] TÜV SÜD America (USA). Design qualification Type approval of commercial PV Modules. IEC 6121510.6 Performance at NOCT with IEC 60904-3 reference solar spectral irradiance distribution [online]. 2005. Available from http://www.tuvamerica.com/services/photovoltaics/TUV%20SUD%20test%20program%20IEC_61215_61646.pdf [Accessed 15 Oct 2017]

[41] Oh J. *Building applied and back insulated photovoltaic modules: thermal models (Master thesis)*. Phoenix: Arizona State University, 2010

[42] Kennedy J., Eberhart R.C. 'Particle swarm optimization'. *Proceedings of IEEE International Conference on Neural Networks*; Perth, Australia, 1995. IEEE; 1995. pp. 1942–8

[43] Abido M.A. 'Optimal power flow using particle swarm optimization'. *International Journal of Electrical Power & Energy Systems*. 2002;**24**: 563–71

[44] Caramia P., Carpinelli G., Varilone P. 'Point estimate schemes for probabilistic three-phase load flow'. *Electric Power Systems Research*. 2010;**80**:168–75

[45] Villanueva D., Pazos J.L., Feijo A. 'Probabilistic load flow including wind power generation'. *IEEE Transactions on Power Systems*. 2011;**26**(3): 1659–67

[46] Morales J.M., Ruiz J.P. 'Point estimate schemes to solve the probabilistic power flow'. *IEEE Transactions on Power Systems*. 2007;**22**(4):1594–601

[47] Zhang P., Lee S.T. 'Probabilistic load flow computation using the method of combined cumulants and Gram–Charlier expansion'. *IEEE Transactions on Power Systems*. 2004;**19**(1):676–82

[48] Hardy Diesel Generators. Model: HDJW 225 T6 [online]. 2012. Available from http://www.hardydiesel.com/john-deere-generators/dl/john-deere-generator-hdjw-225-t6.pdf [Accessed 15 Sep Sep 2017]

Chapter 13

Optimal coordination of directional overcurrent relays

13.1 Introduction

Generally, a power system contains many sections and each section must be protected against overcurrents. In order to ensure optimum-supply reliability, it is important to obtain selectivity, i.e., to switch off the part of the power system where a fault occurs and to avoid switching off the parts of the system where no fault occurs. In case a first overcurrent device fails (although an overcurrent occurs, the overcurrent device does not interrupt this fault current), a second overcurrent device will detect the overcurrent and interrupt the fault current. Due to this back-up principle, a safe-system situation remains even in case an overcurrent device fails.

When considering a radial system with one feeding point and one directional power flow, it is rather easy to obtain the desired selectivity using overcurrent relays having appropriate time delays. For radial system with multisource points and for system with closed ring topology when there is bidirectional power flow, it is insufficient to use overcurrent relays having an appropriate delay to obtain the desired selectivity property. In these cases, the directional overcurrent relays (DOCRs) must be used, which are able to detect the polarity of the overcurrent.

The DOCRs are commonly used for the protection of subtransmission and distribution system or as a secondary protection of transmission system [1–7]. The coordination of DOCRs in power systems consists of selecting their suitable settings such that their fundamental protective function is met under the requirements of sensitivity, selectivity, reliability, and speed [8]. These requirements can be translated into conditions such as [7,8]

1. a relay must get sufficient chance to operate, if a fault occurs within the relay jurisdiction;
2. if a primary relay fails, a backup relay should operate;
3. the sequence of primary/backup (P/B) relays operations should be with sufficient coordination margins, without excessive time delays.

The goal in coordination of DOCRs is to find the appropriate time dial setting (TDS) and pickup current setting (PCS) of relays, considering various constraints, such as coordination and boundary limits [9,10].

The coordination of DOCRs has been formulated as an optimization problem and solved with different optimization methods, including conventional and heuristic techniques.

In [2,5], optimum coordination has been obtained considering the independent optimization of relay settings. In [10], an analytic approach to optimal coordination of DOCRs was proposed. It has been shown that the conventional optimization methods such as linear programing (LP) [1,6,11], sequential quadratic programing (SQP) [12], quadratically constrained quadratic programing [13], gradient-based method [14], and binary integer programing [15] can be applied to optimize DOCR settings. However, conventional optimization methods have a drawback, some time they may get trapped to local minima and fail to give global optimal solution. In addition, as the size of system increases, the convergence is also weak [16]. It becomes essential to develop optimization methods able to overcome these drawbacks and handle such difficulties.

Recently, many population-based metaheuristic methods have been used to solve complex constrained optimization problems. Generally, achieving optimal or near optimal solution for a specific problem will require multiple trials as well as appropriate tuning of associated parameters. A wide variety of population-based techniques such as differential evolution (DE) [17], modified differential evolution (MDE) [3], adaptive differential evolution (ADE) [18], informative differential evolution (IDE) [16], opposition-based chaotic differential evolution (OCDE) [19], genetic algorithm (GA) [20], nondominated sorting GA-II [7], biogeography-based optimization (BBO) [21], artificial bee colony [22], teaching–learning-based optimization (TLBO) [23], particle swarm optimization (PSO) [9,24], seeker optimization algorithm (SOA) [25], and gravitational search algorithm (GSA) [26] have been applied in solving the optimal coordination of DOCRs. The authors [27] confirmed efficiency of different metaheuristic methods including GA, PSO, DE, SOA, and harmony search algorithm for solving the protection coordination problem of DOCRs.

Combination of several optimization algorithms in hybrid algorithm allows them to exploit the strength of each algorithm. In [4], the optimal coordination of DOCRs is formulated as a non-LP (NLP) problem, and hybrid GA–NLP approach is proposed to find the optimum solution. In [21], a hybridization of BBO with LP is proposed for optimal coordination of DOCRs to improve the convergence speed by reducing the required number of iterations and, hence, reduces the CPU time. In [28], a hybrid PSO–LP approach is used to find the optimal TDS and PS values of each DOCR, making use of the advantages of PSO and LP technique and, at the same time, overcoming their drawbacks.

The current chapter describes application of a hybrid algorithm of GSA with local search technique based on SQP in solving the DOCRs coordination problem. In this approach, GSA is used as a global optimizer and SQP as a local optimizer to fine-tune the solution [29].

13.2 Problem formulation

The protection zone of the DOCR is a function of the measured fault current. The relay will start while the measured fault current is greater than the predetermined setting of pickup current. In the coordination of the DOCRs, each relay must offer primary protection of its line (where the relay was installed) and backup protection of all the adjacent lines.

This study takes into account the same assumptions that have been considered in other studies reported in the literature related to optimal coordination of DOCRs. The assumptions are as follows:

- Each protection zone corresponds with one of the lines.
- DOCRs are placed at both ends of each line.
- Primary relay should trip for a fault in its protection zone (line) only.
- Only if the primary relay is fail to clear the fault, the backup relay operates after a time delay.
- Fault currents for different locations and corresponding P/B relay pairs are predetermined. Usually, near-end, far-end, and middle-point fault currents are used, providing coordination for most of the fault situations.
- Networks topologies are fixed; the transient configuration of the system is neglected during relay coordination; thermal limits of lines are not considered.
- DOCR coordination problem is to determine the sequences of relay operations for each predetermined fault location so that faulted line is isolated with sufficient coordination margins and without excessive time delays.

Figure 13.1 shows a simple system with the relay pairs for a near-end fault (F1) and far-end fault (F2). The P/B relay pairs for faults F1 and F2 are R1/R5, R1/R4, and R2/R8.

Obviously, when the network topology is changed, or when the distributed generation (DG) units are added at one or more buses in the network, the short-circuit currents seen by the relay are changed. If the same settings of the basic system are used, it is possible that relays see fault currents smaller than their pickup currents at certain fault locations. This will cause some cases of miscoordination between primary and backup relays. For that, it is necessary to calculate new optimal settings of DOCRs for the system with changed topology and/or added new DG units.

The optimal coordination problem of DOCRs using optimization technique consists of minimizing an objective function (OF) subject to certain coordination criteria and limits on problem variables [3].

13.2.1 Objective function

The relay, which is supposed to operate first to clear the fault, is called the primary relay. Conventionally, OF in coordination studies is constituted as the summation

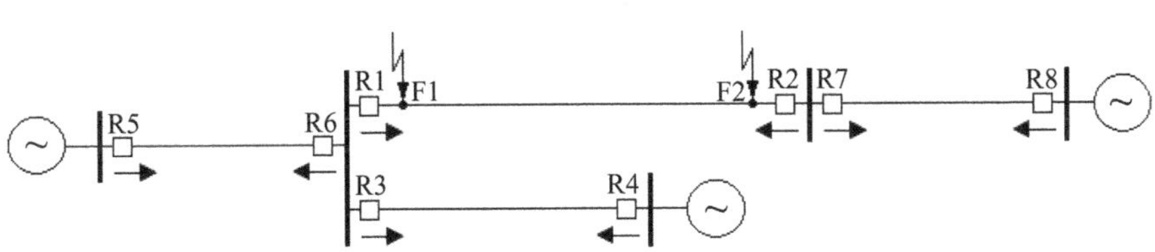

Figure 13.1 Simple network with primary/backup relay pairs

of operating times (OTs) of all primary relays. Therefore, the OF may be defined as follows [15]:

$$\min_{\mathrm{PCS}_i,\mathrm{TDS}_i} \mathrm{OF} = \sum_{i=1}^{m} \sum_{k} T_{ik} \tag{13.1}$$

where m is the number of relays in the network; PCS_i and TDS_i are the pickup current and time dial settings of the ith relay, respectively; and T_{ik} is the operation time of the ith relay for a fault at k.

Concerning the operating characteristics of DOCR, the most common and widely used time–current curves, according to IEC Standard 60255-151, are considered as follows:

$$T_{ik} = \frac{\mathrm{TDS}_i \times \alpha}{(I_{ik}/\mathrm{PCS}_i)^n - 1} \tag{13.2}$$

where I_{ik} is the fault current seen by the ith relay for a fault at k; α and n are constant values, which demonstrate the relay characteristic according to Table 13.1, illustrated in Figure 13.2.

With each relay, one current transformer (CT) is used to reduce the level of the current so that relay can withstand it. The primary rating of CT is known in the problem. Therefore, the fault current seen by the relay (I_{ik}) is obtained as a ratio of the fault current at CT primary terminals ($I_{f,ik}$) and the primary rating of the CT, (CT_{rating}), which is as follows:

$$I_{ik} = \frac{I_{f,ik}}{CT_{rating}} \tag{13.3}$$

13.2.2 Limits of the settings

The bounds of the settings can be expressed by

$$\mathrm{TDS}_i^{\min} \le \mathrm{TDS}_i \le \mathrm{TDS}_i^{\max}, \quad i = 1, \ldots, m \tag{13.4}$$

$$\mathrm{PCS}_i^{\min} \le \mathrm{PCS}_i \le \mathrm{PCS}_i^{\max}, \quad i = 1, \ldots, m \tag{13.5}$$

where $\mathrm{TDS}_i^{\min}$ and $\mathrm{TDS}_i^{\max}$ are the lower and upper limits of TDS of the ith relay, respectively; $\mathrm{PCS}_i^{\min}$ and $\mathrm{PCS}_i^{\max}$ are the lower and upper limits of PCS of the ith relay, respectively.

Table 13.1 Characteristics of the overcurrent relays (IEC standard)

Characteristic type	α	n
Long time inverse	120	1
Extremely inverse	80	2
Very inverse	13.5	1
Normally inverse	0.14	0.02

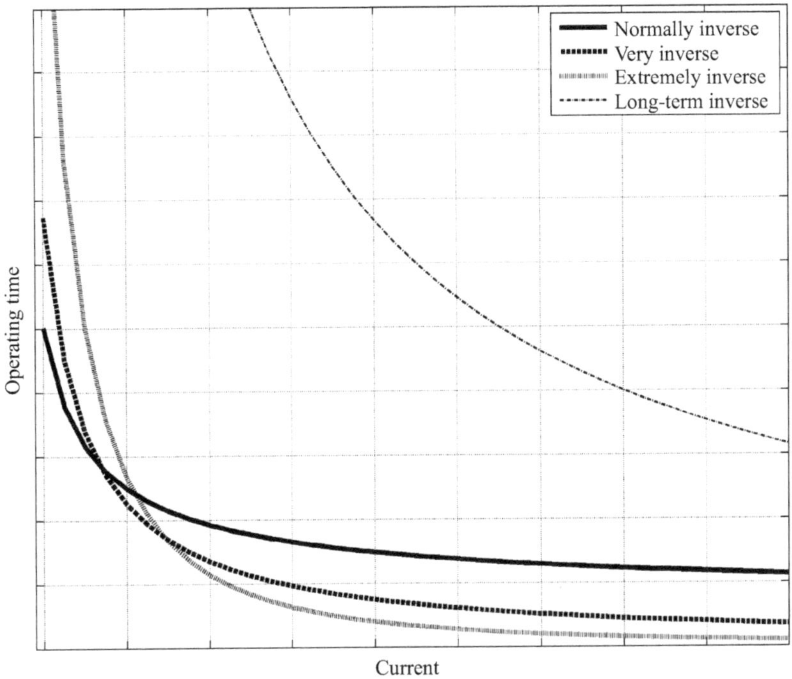

Figure 13.2 Typical inverse time–current characteristics of DOCRs

The PCS_i should be less than the minimum fault current and, at the same time, greater than the maximum load current seen by each relay, including a safety margin, which depends on the relay technology and CT errors [10,15].

13.2.3 Limits of relay operation time

Minimum and maximum OTs can be established for each relay by

$$T_{ik}^{\min} \leq T_{ik} \leq T_{ik}^{\max}, \quad i = 1, \ldots, m \tag{13.6}$$

where $T_{ik}^{\min}$ and $T_{ik}^{\max}$ are the minimum and maximum operation times of the ith relay for a fault k, respectively. $T_{ik}^{\min}$ depends on the relay manufacturer, whereas $T_{ik}^{\max}$ depends on the critical clearing time required to prevent equipment damage and preserve system stability [21]. The acceptable range of the primary relay OT is between values of 0.1 and 1.1 s [13].

13.2.4 Coordination criteria

The backup relay operates only when the corresponding primary relay fails. To ensure relay coordination, the OT of the backup relay must be greater than the corresponding primary relay for all the faults. Coordination time interval (CTI) can be defined as follows:

$$\mathrm{CTI} = T_{jk} - T_{ik}, \quad i = 1, \ldots, m \tag{13.7}$$

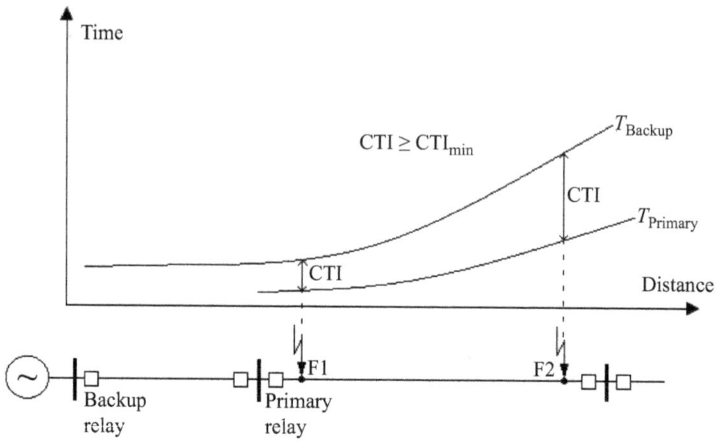

Figure 13.3 Coordination between primary and backup relays

where T_{jk} is the operation time of the jth backup relay for a fault at k inside the zone protected by the ith primary relay.

Therefore, the CTI constraint can be stated as

$$\mathrm{CTI} \geq \mathrm{CTI}_{\min} \tag{13.8}$$

where $\mathrm{CTI}_{\min}$ is a minimum coordination time interval. The typical value of $\mathrm{CTI}_{\min}$ is 0.2–0.3 s [18].

This constraint must be satisfied for all P/B pairs. Usually, near-end, far-end, and middle-point fault currents are used, providing coordination for most of the faults situations. Figure 13.3 shows an example of the CTI constraints for coordination between the primary and backup relays in case the near- and far-end faults (F1 and F2).

13.2.5 Modification of objective function for minimization of CTI

Although there is a minimum CTI between the primary and backup relays is desirable for proper selectivity, much delayed operation of backup relays is not preferable from effective relay coordination point of view. In order to optimize the CTI between primary and backup relays, expression for OF is modified, as follows [7,23]:

$$\min_{\mathrm{PCS}_i,\mathrm{TDS}_i} \mathrm{MOF} = \alpha_1 \sum_{i=1}^{m} \sum_{k} T_{ik}^2 + \alpha_2 \sum_{p=1}^{m_p} \left[\Delta T_{mbp} - \beta\left(\Delta T_{mbp} - \left|\Delta T_{mbp}\right|\right)\right]^2 \tag{13.9}$$

$$\Delta T_{mbp} = T_{jk} - T_{ik} - \mathrm{CTI} \tag{13.10}$$

where ΔT_{mbp} is OT difference with CTI between pth relay pair; m_p is the number of P/B relay pairs; p represents each P/B relay pair and varies from 1 to m_p; α_1 and α_2

are used to control the weighting of first and second terms of the modified OF (MOF) and β is used to consider the miscoordination. By increasing β, miscoordination is decreased, but OTs of the relays increased. Therefore, the suitable β is fit to a value by which the miscoordination is omitted.

13.3 Solution method

As can be seen above, the coordination between DOCRs is formulated as a nonlinear and highly constrained optimization problem in which two settings, namely, TDS and PCS of each relay are considered as control variables. The main objective is to minimize the sum of OTs of all the primary relays, which are expected to operate in order to clear the faults of their corresponding zones. To solve this problem, a hybrid GSA–SQP algorithm is proposed in this study.

13.3.1 Overview of GSA

The GSA is a metaheuristic population-based optimization method developed by Rashedi *et al.* [30]. In GSA, the search agents are a collection of masses, which interact with each other based on the Newtonian gravity and the laws of motion. In this algorithm, agents are considered as objects, and their performances are measured by their masses. All these objects attract each other by the gravity force, and this force causes a global movement of all objects toward the objects with heavier masses. The position of the mass corresponds to the solution of the problem, and its gravitational and inertial masses are determined using a fitness function. In other words, each mass presents a solution. The algorithm is navigated by properly adjusting the gravitational and inertial masses. By lapse of time, the masses will be attracted by the heaviest mass that it represents as an optimum solution in the search space.

The GSA could be considered as an isolated system of masses. It is like a small artificial world of masses obeying the Newtonian laws of gravitation and motion.

In a system with N agents (masses), the position and velocity of the ith agent is defined by

$$\mathbf{x}_i = \left(x_i^1, \ldots, x_i^d, \ldots, x_i^n\right) \quad \text{for } i = 1, 2, \ldots, N \tag{13.11}$$

$$\mathbf{v}_i = \left(v_i^1, \ldots, v_i^d, \ldots, v_i^n\right) \quad \text{for } i = 1, 2, \ldots, N \tag{13.12}$$

where n is the search-space dimension of the problem, x_i^d defines the position of the ith agent in the dth dimension, and v_i^d is the velocity of the ith agent in the dth dimension ($d = 1, 2, 3, \ldots, n$).

After evaluating the current population fitness, based on OF of the optimization problem, the mass of each agent is calculated as follows [30]:

$$M_i(t) = \frac{m_i(t)}{\sum_{j=1}^{N} m_j(t)} \tag{13.13}$$

where

$$m_i(t) = \frac{fit_i(t) - worst(t)}{best(t) - worst(t)} \tag{13.14}$$

where $fit_i(t)$ represent the fitness value of the agent i at time (iteration) t; $best(t)$ and $worst(t)$ is the best and worst fitness of all agents, respectively, and defined as follows (for a minimization problem):

$$best(t) = \min_{j \in \{1,\ldots,N\}} fit_j(t) \tag{13.15}$$

$$worst(t) = \max_{j \in \{1,\ldots,N\}} fit_j(t) \tag{13.16}$$

According to Newton gravitation theory, the total force that acts on the ith agent in the dth dimension at t time is specified as follows:

$$\mathbf{F}_i(t) = \sum_{j \in Kbest, j \neq i} r \cdot G(t) \frac{M_j(t) \cdot M_i(t)}{R_{i,j}(t) + \varepsilon} (\mathbf{x}_j(t) - \mathbf{x}_i(t)) \tag{13.17}$$

where r is a random number within the interval [0,1]; $G(t)$ is gravitational constant at iteration t; $M_i(t)$ and $M_j(t)$ are masses of agents i and j; ε is a small constant; $R_{ij}(t)$ is the Euclidian distance between the two agents i and j, given by the following equation:

$$R_{ij}(t) = \|\mathbf{x}_i(t), \mathbf{x}_i(t)\|_2 \tag{13.18}$$

Kbest is the set of first K agents with the best fitness value and biggest mass, which is a function of time, initialized to K_0 at the beginning and decreased with time. In such a way, at the beginning, all agents apply the force, and as time passes, *Kbest* is decreased linearly and at the end, there will be just one agent applying force to the others.

By the law of motion, the acceleration of the ith agent at t time in the kth dimension is given by following equation:

$$\mathbf{a}_i(t) = \frac{\mathbf{F}_i(t)}{M_i(t)} \tag{13.19}$$

The searching strategy on this notion can be defined to find the next velocity and next position of an agent. Next velocity of an agent is defined as a function of its current velocity added to its current acceleration. Hence, the next velocity and next position of an agent can be computed as follows:

$$\mathbf{v}_i(t+1) = r_i \cdot \mathbf{v}_i(t) + \mathbf{a}_i(t) \tag{13.20}$$

$$\mathbf{x}_i(t+1) = \mathbf{x}_i(t) + \mathbf{v}_i(t+1) \tag{13.21}$$

where r_i is a uniform random variable in the interval [0, 1]. This random number is utilized to give a randomized characteristic to the search.

It must be pointed out that the gravitational constant $G(t)$ given in (13.17) is important in determining the performance of GSA. It is initialized at the beginning and will be reduced with time to control the search accuracy. In other words, the gravitational constant is a function of the initial value G_0 and time t:

$$G(t) = G_0 \cdot e^{-\alpha(t/t_{\max})} \tag{13.22}$$

where α is a user-specified constant, t is the current iteration, and $t_{\max}$ is the maximum iteration number. The parameters of maximum iteration $t_{\max}$, population size N, initial gravitational constant G_0, and constant α control the performance of GSA.

13.3.2 Overview of SQP

The SQP method seems to be the best NLP method for constrained optimization problems. SQP is a direct method in which the constraints are handled in an explicit manner. It outperforms every other NLP method in terms of efficiency, accuracy, and percentage of successful solutions over a large number of test problems [4,31]. The method closely resembles Newton's method for constrained optimization, just as is done for unconstrained optimization. At each iteration, an approximation is made of the Hessian of the Lagrangian function using Broyden–Fletcher–Goldfarb–Shanno (BFGS) quasi-Newton updating method. The result of the approximation is then used to generate a quadratic programing subproblem, the solution of which is used to form a search direction for a line-search procedure. Since the OF to be minimized is nonconvex, SQP ensures a local minimum for an initial solution.

Each iteration in the SQP algorithm involves three major steps [4,31]: (i) solve the QP subproblem and obtain the search direction; (ii) line search and merit function calculation; and (iii) updating of the Hessian matrix of the Lagrangian function, using BFGS formula.

13.3.3 Hybrid GSA–SQP algorithm

GSA as a probabilistic-based multipoint search technique has a drawback of, sometimes, converging to the values that may not be optimum. On the other hand, GSA searches a large solution space with excellent convergence providing approaching to the global optimum. NLP methods, such as SQP, being single-point search methods, have a drawback of being trapped in local optimum point, if the initial choice is nearer to the local optimum. NLP method gives global optimum solution, if proper initial choice is made [4].

To use of the advantage of GSA and SQP methods, and at the same time to overcome the drawbacks of these methods, a hybrid GSA with SQP is proposed in [29]. The SQP routine is incorporated in GSA as a local search mechanism to improve the convergence. First, the GSA algorithm is executed normally, and the best fitness per generation is selected in each iteration. From that best fitness, the corresponding agent is set as the initial values of variables to the SQP method. Then, the SQP routine is executed depending on the adopted probability of local search α_{LS}, providing improvement of the best fitness obtained from GSA in the current iteration. In this way, the hybrid GSA–SQP algorithm gives the global optimum solution.

13.3.4 Implementation of hybrid GSA–SQP algorithm

Proposed hybrid GSA–SQP approach has been applied to solve the DOCRs coordination problem. The control variables of the DOCRs coordination problem

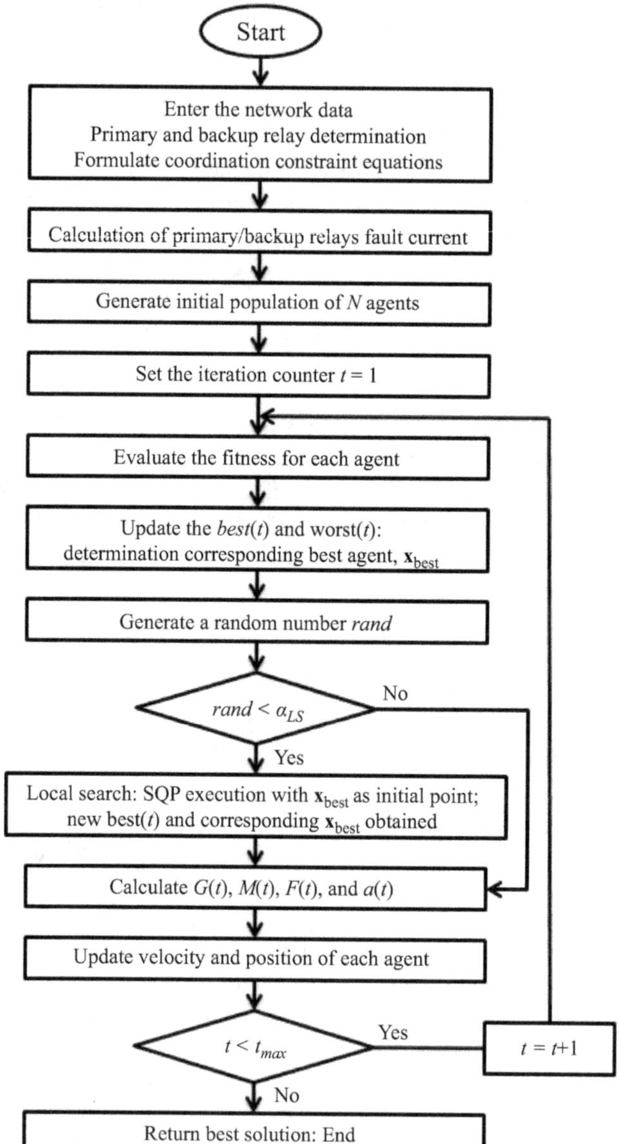

Figure 13.4 General flowchart of the hybrid GSA–SQP for optimal coordination of DOCRs

constitute the individual position of several agents that represent a complete solution set. In a system with N agents, the position of the ith agent is defined by

$$\mathbf{x}_i = \left[\mathrm{TDS}_i^1, \ldots, \mathrm{TDS}_i^m, \mathrm{PCS}_i^1 \ldots, \mathrm{PCS}_i^m\right] \tag{13.23}$$

where $i = 1, 2, \ldots, N$, and m is the number of relays in the network.

Figure 13.4 illustrates the general flowchart for the hybrid GSA–SQP algorithm used to calculate the optimal settings of the DOCRs.

13.4 Simulation results

The proposed GSA, SQP, and hybrid GSA–SQP algorithms are simulated over three different test systems with 6, 24, and 38 DOCRs. These test systems are widely used as benchmarks in the power-system field for solving the DOCRs coordination problem.

The algorithm parameters used for the simulation are adopted as follows: α is set to 20 and G_0 is set to 100; probability of local search α_{LS} is set to 0.05. The population size (N) and maximum iteration number ($t_{\max}$) are set to 50 and 200, respectively, for all case studies. The weighting factors in (13.9) are selected as follows: α_1 is set to 2, α_2 is set to 15, and β is set to 200 for test system 1, 500 for test system 2, and 1,000 for test system 3.

13.4.1 Test system 1

This is the IEEE 3-bus test system, which is shown in Figure 13.5. As can be seen from Figure 13.5, the IEEE 3-bus system contains one synchronous generator, three lines, and six DOCRs with the normally inverse characteristics ($\alpha = 0.14$ and $n = 0.02$). The objective is to coordinate the settings of all the six relays responding to clear all near- and far-end faults. Accordingly, there are 12 decision variables in this problem (two for each relay), i.e., TDS^1–TDS^6 and PCS^1–PCS^6. The minimum and maximum limits of TDS are 0.05 and 1.1, respectively. The range of PCS is

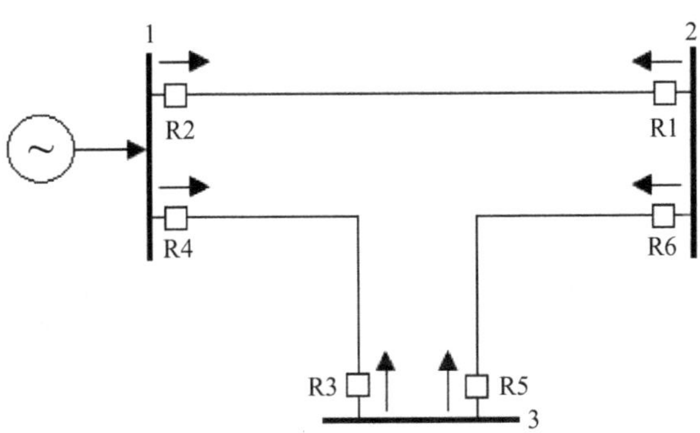

Figure 13.5 Test system 1: IEEE 3-bus system

Table 13.2 Fault currents seen by relays for test system 1 [2,19]

Relay	Fault current		Line
	Near-end	Far-end	
R1	9.46/2.06	14.08/2.06	1–2
R2	29.91/2.06	100.63/2.06	1–2
R3	8.81/2.23	12.07/2.23	1–3
R4	37.68/2.23	136.23/2.23	1–3
R5	17.93/0.80	25.90/0.80	2–3
R6	14.35/0.80	19.20/0.80	2–3

Table 13.3 Fault currents for primary/backup relays in test system 1 [2,19]

Primary relay	Fault current		Backup Relay	Fault current	
	Near-end	Far-end		Near-end	Far-end
R1	9.46/2.06	14.08/2.06	R5	9.46/0.80	14.08/0.80
R3	8.81/2.23	12.07/2.23	R6	8.81/0.80	12.07/0.80
R5	17.93/0.80	25.90/0.80	R4	17.93/2.23	25.90/2.23
R6	14.35/0.80	19.20/0.80	R2	14.35/2.06	19.02/2.06

Table 13.4 Optimal TDS and PCS for test system 1

Relay	GSA		SQP		Hybrid GSA–SQP			
	TDS	PCS	TDS	PCS	TDS	PCS	OT_{near_end}	OT_{far_end}
R1	0.05000	1.28160	0.05000	1.25	0.05000	1.25	0.26549	0.20253
R2	0.21202	1.31711	0.19766	1.50	0.19764	1.50	0.62549	0.38351
R3	0.05000	1.25000	0.05000	1.25	0.05000	1.25	0.30066	0.23533
R4	0.21253	1.46965	0.20903	1.50	0.20903	1.50	0.58970	0.38027
R5	0.18727	1.40548	0.18120	1.50	0.18120	1.50	0.45649	0.40035
R6	0.18095	1.49486	0.18069	1.50	0.18067	1.50	0.49712	0.44361
OF (s)	4.82566		4.78081		4.78057			

from 1.25 to 1.50, respectively. The CTI_{min} is set to its typical value of 0.3 s. Other data, including the CT ratings and the flowing fault currents for all of P/B relays that were calculated for near- and far-end faults are given in Tables 13.2 and 13.3.

Table 13.4 shows the optimal settings of TDS and PCS obtained by the proposed hybrid GSA–SQP. The OTs of all primary relays for near- and far-end faults are given in the in the last two columns of Table 13.4. The OTs are in an acceptable range of 0.1–1.1 s. The total OT of the primary relays is 4.78057 s. As can be seen from Table 13.5, the DOCRs operate with no miscoordination pairs.

Table 13.5 CTI for test system 1

Fault	Relay		CTI
	Backup	Primary	
Near-end	5	1	0.30000
	6	3	0.30000
	4	5	0.30000
	2	6	0.39008
Far-end	5	1	0.33624
	6	3	0.32118
	2	6	0.40043
	4	5	0.30000

Table 13.6 Comparison of the simulation results for test system 1

Algorithm	OF (s)	Note
DE [2]	4.8421	
MDE5 [2]	4.7806	
RST2 [2]	4.8354	
GA [2]	5.0761	
SOMGA [2]	4.7898	
ADE [18]	4.7636	Miscoordination
TLBO [23]	5.3349	
DE [19]	4.8422	
OCDE2 [19]	4.7806	
Hybrid GSA–SQP	4.78057	

For the purpose of comparison, the same problem was solved using only GSA and only SQP method. For the SQP solver, initial choice of variables was taken as the lower limit of the control variables. It can be seen from Table 13.4 that the hybrid GSA–SQP has better performance than the conventional GSA and SQP. The results of the proposed hybrid GSA–SQP algorithm are compared with various algorithms reported in recent years. The comparison is shown in Table 13.6. The algorithm reported in [18] resulted in lesser minimum OF than the proposed GSA–SQP approach, but with miscoordination. It is clear that the proposed GSA–SQP algorithm results in the best settings, compared with the other algorithms.

13.4.2 Test system 2

The 9-bus system shown in Figure 13.6 is considered as test system 2. There are 24 digital DOCRs with the normally inverse characteristics ($\alpha = 0.14$ and $n = 0.02$). The range of TDS is from 0.01 to 1.0 and for PCS from 0.5 to 2.5. The CTI_{min} is set to the value of 0.2 s. The CT ratio for each relay is 500/1.

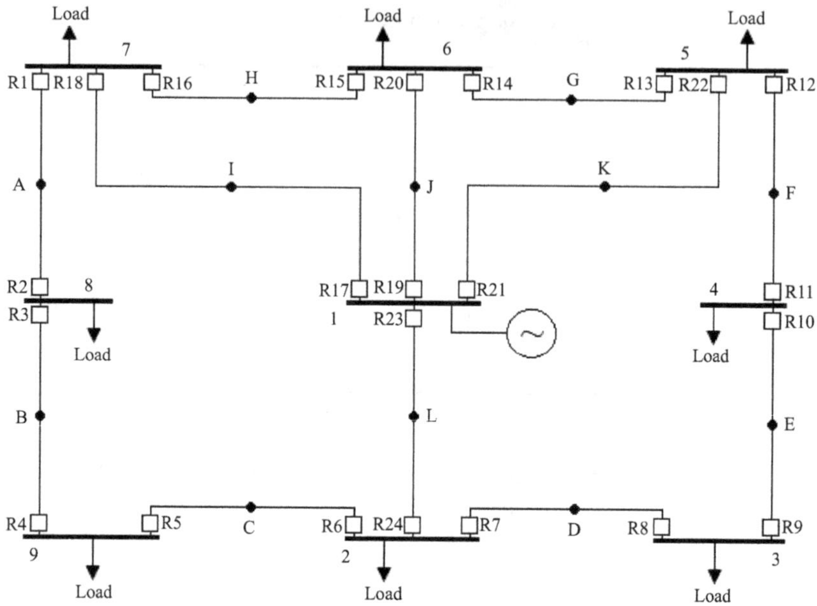

Figure 13.6 Test system 2: 9-bus system

Table 13.7 Fault currents for test system 2 [16]

Location	Primary relay	Backup relay	$I_{f,primary}$ (A)	$I_{f,backup}$ (A)	Location	Primary relay	Backup relay	$I_{f,primary}$ (A)	$I_{f,backup}$ (A)
A	R1	R15	24,779	9,150	G	R13	R11	16,087	3,088
	R1	R17	24,779	15,632		R13	R21	16,087	13,000
	R2	R4	8,327	8,327		R14	R16	18,213	6,285
B	R3	R1	16,390	16,930		R14	R19	18,213	11,934
	R4	R6	14,671	14,671	H	R15	R13	18,218	6,285
C	R5	R3	9,454	9,454		R15	R19	18,218	11,935
	R6	R8	23,280	4,777		R16	R2	16,087	3,088
	R6	R23	23,280	18,507		R16	R17	16,087	13,000
D	R7	R5	23,280	4,777	I	R18	R2	8,161	2,426
	R7	R23	23,280	18,507		R18	R15	8,161	5,736
	R8	R10	9,454	9,454	J	R20	R13	9,286	4,644
E	R9	R7	15,304	15,304		R20	R16	9,286	4,644
	R10	R12	16,490	16,490	K	R22	R11	8,161	2,426
F	R11	R9	8,326	8,327		R22	R14	8,161	5,736
	R12	R14	24,779	9,150	L	R24	R5	6,149	3,075
	R12	R21	24,779	15,631		R24	R8	6,149	3,075

Short-circuit faults are generated on the middle of each line. Fault points are marked as A to L. Fault currents seen in each P/B relays are listed in Table 13.7. There are 48 variables existing in the optimization problem. Those are TDS^1–TDS^{24} and PCS^1–PCS^{24}.

Table 13.8 Results for test system 2

Relays	IDE [16]		GSA		SQP		Hybrid GSA–SQP		
	TDS	PCS	TDS	PCS	TDS	PCS	TDS	PCS	$OT_{primary}$
R1	0.6352	0.5382	0.6192	1.3782	0.3040	2.4259	0.2765	2.5000	0.6288
R2	0.1895	1.3645	0.2211	1.4531	0.1959	1.1581	0.0918	2.5000	0.3323
R3	0.4605	0.7755	0.5547	1.0784	0.2301	2.2583	0.1975	2.5000	0.5235
R4	0.3814	0.5000	0.3483	1.0006	0.2244	1.8590	0.1470	2.5000	0.4076
R5	0.3974	0.5970	0.3521	1.1756	0.1831	1.8752	0.1385	2.5000	0.4696
R6	0.5101	0.5000	0.5128	0.6301	0.2721	2.4934	0.2191	2.5000	0.5093
R7	0.5123	0.5560	0.3667	2.1981	0.2471	2.5000	0.2209	2.5000	0.5135
R8	0.2403	1.6853	0.3984	0.6316	0.1934	1.9260	0.1377	2.5000	0.4667
R9	0.3418	0.9210	0.3539	1.0966	0.1841	2.1560	0.1475	2.5000	0.4021
R10	0.4740	0.5001	0.3444	1.7680	0.2300	2.5000	0.1967	2.5000	0.5200
R11	0.2012	1.6954	0.2347	1.5582	0.1303	1.9619	0.0923	2.5000	0.3344
R12	0.4190	1.6966	0.5083	1.0618	0.3069	2.4758	0.2723	2.5000	0.6193
R13	0.5414	0.5011	0.5640	0.6511	0.2488	1.7959	0.1903	2.5000	0.5081
R14	0.3947	1.3848	0.4328	1.2291	0.4519	0.6029	0.2377	2.5000	0.6046
R15	0.5640	0.6420	0.5579	1.1417	0.3011	1.7832	0.2404	2.5000	0.6115
R16	0.2634	2.1479	0.4537	0.7078	0.2387	2.1922	0.1887	2.5000	0.5038
R17	0.4808	2.1229	0.8032	1.7951	0.5458	0.5206	0.5001	0.5419	–
R18	0.1462	2.4547	0.1359	1.2924	0.0499	1.7434	0.0715	1.9786	0.2323
R19	0.4627	2.4999	0.7782	1.1285	0.3350	2.1148	0.3152	1.6934	–
R20	0.2454	0.5016	0.0341	1.7836	0.0717	2.0845	0.0637	2.2012	0.2046
R21	0.9327	0.5851	0.6259	1.6575	0.5190	0.6396	0.4822	0.5967	–
R22	0.1667	1.8427	0.1389	0.8907	0.0417	2.4976	0.0719	1.9017	0.2290
R23	0.9995	0.5005	0.7280	0.8679	0.3827	2.3378	0.4238	0.6822	–
R24	0.1481	1.4887	0.1113	0.8264	0.0516	2.3987	0.1013	0.9638	0.2714
OF (s)	14.0448		14.7384		10.2030		8.8923		

The proposed hybrid GSA–SQP algorithm is applied for solution of the DOCRs coordination problem. The optimized TDS and PCS as well as OF value are given in Table 13.8. The OTs of all primary relays are in an acceptable range of 0.1–1.1 s. The optimized values of decision variables (TDS and PCS) and OF are compared with results obtained using GSA, SQP, and IDE algorithm [16]. Table 13.9 shows the corresponding values of CTI. The results show that total OT of primary DOCRs is reduced using proposed hybrid GSA–SQP as compared to IDE [16], and with original GSA and SQP. The optimal results give no mis-coordination as can be seen from Table 13.9. Moreover, the CTI improves using hybrid GSA–SQP algorithm, because the sum of CTI values is reduced as compared with other algorithms.

Figure 13.7 shows the comparative convergence profiles of GSA and proposed hybrid GSA–SQP for test system 2. It may be observed from this figure that the hybrid GSA–SQP explores the global optima thanks to exploration and exploitation of GSA in the first part of the iterative process and local search ability of SQP in subsequent iterations.

13.4.3 Test system 3

To evaluate the efficiency of the proposed approach in solving a larger power system, the IEEE 30-bus system is considered as test system 3. The system can be

Table 13.9 CTI for test system 2

Relays		IDE [16]	GSA	SQP	Hybrid GSA–SQP
Backup	**Primary**				
R15	R1	0.200	0.2017	0.2000	0.2000
R17	R1	0.279	0.7448	0.2110	0.2000
R4	R2	0.218	0.2235	0.2000	0.2000
R1	R3	0.200	0.2122	0.2000	0.2000
R6	R4	0.212	0.2017	0.2000	0.2000
R3	R5	0.200	0.4543	0.2000	0.2000
R8	R6	0.200	0.2001	0.2000	0.2000
R23	R6	0.800	0.5085	0.3114	0.2043
R5	R7	0.201	0.3367	0.2000	0.2000
R23	R7	0.782	0.4922	0.3688	0.2000
R10	R8	0.202	0.2005	0.2006	0.2000
R7	R9	0.200	0.2295	0.2005	0.2000
R12	R10	0.200	0.2004	0.2000	0.2000
R9	R11	0.200	0.2089	0.2000	0.2000
R14	R12	0.203	0.2010	0.2000	0.2000
R21	R12	0.736	0.5575	0.2027	0.2000
R11	R13	0.203	0.2033	0.2002	0.2000
R21	R13	–	0.5749	0.3585	0.3528
R16	R14	0.209	0.2086	0.2000	0.2000
R19	R14	0.586	0.8672	0.2042	0.2075
R13	R15	0.200	0.2055	0.2000	0.2000
R19	R15	0.305	0.6421	0.2665	0.2005
R2	R16	0.202	0.2536	0.2001	0.2000
R17	R16	0.465	1.2469	0.3337	0.3661
R2	R18	0.502	0.9027	0.7907	0.7300
R15	R18	0.780	1.2878	0.9585	0.8557
R13	R20	0.800	1.3466	0.8181	0.7971
R16	R20	0.782	1.1026	0.9161	0.7885
R11	R22	0.800	1.1052	0.8460	0.7392
R14	R22	0.756	1.0015	0.8900	0.8465
R5	R24	0.685	1.1842	0.8491	0.7960
R8	R24	0.800	0.9168	0.9354	0.7896
Sum of CTI (s)		13.108	18.2230	12.4622	11.4736

considered as a meshed subtransmission/distribution system with DG units, as shown in Figure 13.8. A total of 38 DOCRs, with the normally inverse characteristics ($\alpha = 0.14$ and $n = 0.02$), are considered for the system, which are installed on each end of the lines. The short-circuit current values for near-end faults are given in Table 13.10. The range of TDS is from 0.1 to 1.1 and for PCS from 1.5 to 6. The CT ratio for each relay was assumed equal to 1,000/5. The CTI_{min} is set to value of 0.3 s.

The results obtained using proposed hybrid GSA–SQP along with GSA and SQP are presented in Table 13.11. Table 13.12 shows the CTI calculated from the optimized values of TDS and PCS, and there is no miscoordination. It can be seen

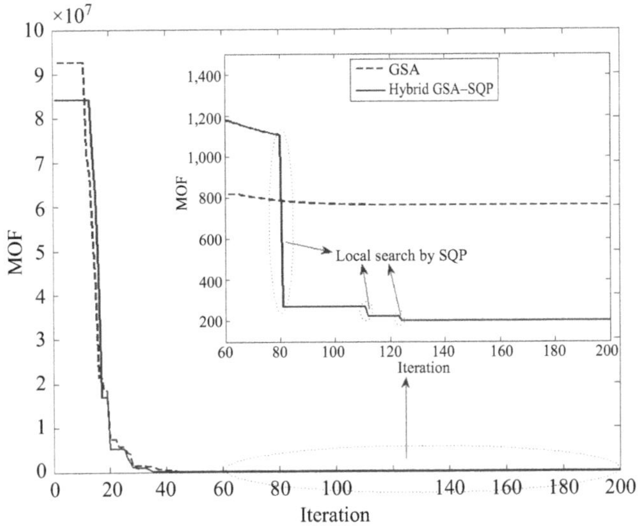

Figure 13.7 Comparative convergence profiles of GSA and hybrid GSA–SQP for test system 2

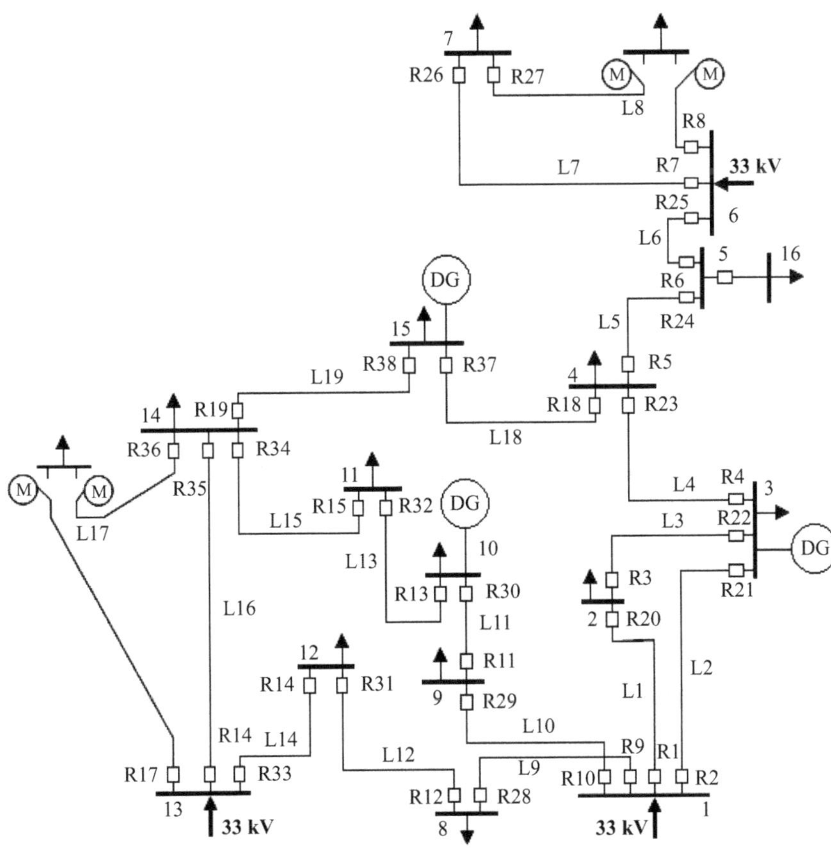

Figure 13.8 Test system 3: IEEE 30-bus system

Table 13.10 Fault currents for test system 3 [32]

Primary relay	Backup relay	$I_{f,primary}$ (A)	$I_{f,backup}$ (A)	Primary relay	Backup relay	$I_{f,primary}$ (A)	$I_{f,backup}$ (A)
R3	R1	4,086.7	4,086.7	R9	R20	7,212.6	1,103.5
R4	R2	5,411.2	2,138.8	R10	R20	7,339.3	1,095.8
R22	R2	4,333.0	2,147.0	R1	R21	7,665.3	698.8
R4	R3	5,411.2	3,272.5	R9	R21	7,212.6	721.2
R21	R3	5,411.8	3,243.6	R10	R21	7,339.3	716.1
R5	R4	4,960.8	3,001.3	R20	R22	3,481.5	3,481.5
R18	R4	4,719.4	3,002.1	R21	R23	5,411.8	2,193.5
R6	R5	2,416.0	2,416.0	R22	R23	4,333.0	2,204.6
R7	R6	5,669.0	1,790.9	R18	R24	4,719.4	1,717.7
R8	R6	5,607.7	1,774.8	R23	R24	3,689.7	1,724.2
R27	R7	1,472.3	1,472.3	R24	R25	2,695.0	2,695.0
R26	R8	1,026.8	1,026.8	R1	R28	7,665.3	1,552.0
R12	R9	5,034.9	5,034.9	R2	R28	7,985.7	1,545.8
R11	R10	3,457.1	3,457.1	R10	R28	7,339.3	1,538.0
R13	R11	3,727.3	2,875.0	R1	R29	7,665.3	1,380.6
R14	R12	2,906.5	2,906.5	R2	R29	7,985.7	1,375.2
R15	R13	2,660.5	2,660.5	R9	R29	7,212.6	1,379.0
R16	R14	6,185.6	1,668.1	R29	R30	2,518.9	2,518.9
R17	R14	7,492.9	1,641.1	R28	R31	2,036.8	2,036.8
R19	R15	5,445.2	1,527.3	R30	R32	2,998.8	2,149.0
R35	R15	4,222.0	1,533.2	R31	R33	3,263.6	3,263.6
R36	R15	6,420.2	1,509.7	R32	R34	2,930.4	2,930.4
R19	R16	5,445.2	3,128.3	R17	R35	7,492.9	1,885.4
R34	R16	5,796.6	3,123.9	R33	R35	6,456.2	1,954.5
R36	R16	6,420.2	3,052.4	R16	R36	6,185.6	490.9
R19	R17	5,445.2	801.3	R33	R36	6,456.2	500.6
R34	R17	5,796.6	800.1	R5	R37	4,960.8	1,961.0
R35	R17	4,222.0	794.0	R23	R37	3,689.7	1,968.5
R38	R18	3,133.2	2,292.2	R34	R38	5,796.6	1,886.8
R37	R19	3,788.9	2,940.9	R35	R38	4,222.0	1,896.7
R2	R20	7,985.7	1,053.9	R36	R38	6,420.2	1,867.7

that the results obtained from the hybrid GSA–SQP are better than those obtained from the original GSA and SQP methods. This clearly indicates that the proposed algorithm can be effectively used to solve the DOCRs coordination problem for large-scale power systems.

13.4.4 Statistical evaluation of the results

A statistical evaluation of the results is presented in this section. Table 13.13 shows the comparison of minimum, maximum, and standard deviation of the results obtained by GSA and hybrid GSA–SQP over 10 runs and the results of SQP under three different initial choices of control variables. In the case of SQP, the problem was solved with different initial choices, such as lower limits of control variables,

Table 13.11 Results for test system 3

Relay	GSA		SQP		Hybrid GSA–SQP		
	TDS	PCS	TDS	PCS	TDS	PCS	$OT_{primary}$
R1	1.0731	1.5393	0.4372	2.6744	0.3038	3.8158	0.9006
R2	0.6752	1.5000	0.5791	1.7575	0.3455	1.5000	0.7131
R3	0.5549	1.5000	0.3472	2.6903	0.2762	2.7562	0.9458
R4	0.2145	6.0000	0.3353	2.8637	0.2584	2.6926	0.7658
R5	0.1670	6.0000	0.3496	2.4616	0.2625	1.6592	0.6611
R6	0.1001	4.5297	0.1769	3.2217	0.1597	1.9809	0.6073
R7	0.1001	4.7929	0.2831	2.1619	0.1732	1.8244	0.4299
R8	0.2648	2.3894	0.2218	1.8887	0.1570	1.5872	0.3718
R9	1.0995	1.5000	0.5090	3.4047	0.3133	3.3528	0.9014
R10	0.4892	5.9977	0.4853	4.1095	0.3123	3.5250	0.9115
R11	0.5000	3.4787	0.6024	2.2426	0.2797	2.7844	1.0530
R12	0.4682	4.2032	0.3627	3.1807	0.2170	3.5937	0.7654
R13	0.6268	1.5000	0.3233	4.4211	0.2494	2.6182	0.8720
R14	0.6755	1.5549	0.2678	2.1714	0.2155	2.1395	0.7723
R15	0.4803	1.5011	0.3311	1.7343	0.2334	1.6058	0.7565
R16	1.0996	1.5000	0.4214	2.1123	0.2678	3.0708	0.7931
R17	0.2929	1.5000	0.1934	1.7176	0.1468	1.6293	0.3176
R18	0.2759	3.8552	0.3870	2.0983	0.2412	2.4930	0.7345
R19	0.6865	1.8317	0.4226	2.1682	0.2405	2.8394	0.7281
R20	0.1069	4.1612	0.2666	2.0309	0.1443	2.3975	0.4993
R21	0.3343	1.5000	0.2160	1.7407	0.1604	1.5000	0.3769
R22	0.4900	1.5000	0.3298	2.4447	0.1702	4.0080	0.6943
R23	0.1558	5.4904	0.2421	3.1497	0.2100	2.5687	0.7311
R24	0.3802	1.5000	0.3247	2.1049	0.2016	2.2364	0.7716
R25	0.4822	1.5000	0.3744	2.4249	0.2523	2.6631	–
R26	0.3695	1.5005	0.1681	1.9096	0.1106	1.5095	0.6248
R27	0.1000	4.1890	0.2080	2.3950	0.1215	1.6362	0.5570
R28	0.1039	5.8654	0.4775	1.7040	0.1932	2.5489	0.9631
R29	0.4439	2.1601	0.3064	2.1555	0.1643	2.6733	0.7305
R30	0.2168	6.0000	0.3448	2.5695	0.1837	3.6716	0.9013
R31	0.6293	1.5000	0.3649	3.1340	0.2191	3.0692	0.9025
R32	0.2744	4.2042	0.2071	4.4850	0.2095	3.2170	0.9527
R33	0.9231	1.5000	0.3544	4.2707	0.2857	3.1774	0.8428
R34	0.4494	3.6290	0.2605	4.5405	0.2516	3.6611	0.8339
R35	0.1905	6.0000	0.3782	1.7569	0.2289	2.4531	0.7285
R36	0.2109	1.5000	0.1240	1.5484	0.1000	1.5000	0.2216
R37	0.5387	1.5736	0.2788	2.3403	0.2138	2.3547	0.7029
R38	0.3608	2.3341	0.3284	1.9759	0.1941	2.8879	0.7901
OF (s)	51.7747		41.0390		26.8258		

upper limits of control variables, and middle values between lower and upper limits. In some cases, it gives the feasible solution, but worse than hybrid GSA–SQP. Moreover, for test system 3 in cases with upper and middle initial choices of control variables, the SQP method terminates without finding even the feasible solution.

Table 13.12 CTI for test system 3

Relays		GSA	SQP	Hybrid GSA–SQP	Relays		GSA	SQP	Hybrid GSA–SQP
Backup	Primary				Backup	Primary			
R1	R3	1.3817	0.3001	0.3003	R20	R9	0.3003	0.3745	0.3001
R2	R4	1.3773	1.1825	0.4415	R20	R10	0.8557	0.3440	0.3003
R2	R22	1.1041	1.1647	0.5106	R21	R1	0.4816	1.0362	0.4157
R3	R4	0.6049	0.3003	0.3003	R21	R9	0.3000	0.5872	0.3672
R3	R21	0.8072	0.7926	0.6946	R21	R10	0.8096	0.5640	0.3675
R4	R5	0.8110	0.3588	0.3736	R22	R20	0.8499	0.3028	0.3001
R4	R18	0.5758	0.3010	0.3000	R23	R21	0.7799	0.8055	0.6213
R5	R6	0.9516	0.5894	0.3000	R23	R22	0.3036	0.3011	0.3005
R6	R7	0.6337	0.4487	0.3001	R24	R18	0.4519	0.5013	0.3001
R6	R8	0.3002	0.6498	0.3627	R24	R23	0.6061	0.6475	0.3005
R7	R27	0.3905	0.3154	0.3000	R25	R24	0.3181	0.3005	0.3000
R8	R26	0.3283	0.3593	0.3004	R28	R1	0.3287	1.0523	0.3009
R9	R12	0.8542	0.5432	0.3005	R28	R2	1.2351	0.9195	0.4928
R10	R11	1.0525	0.3081	0.3002	R28	R10	0.8215	0.6669	0.3000
R11	R13	0.7340	0.6774	0.3015	R29	R1	0.3815	0.7019	0.3001
R12	R14	0.5407	0.6784	0.3000	R29	R2	1.2592	0.5696	0.4927
R13	R15	0.4591	0.9170	0.3000	R29	R9	0.3021	0.3487	0.3009
R14	R16	0.3005	0.3047	0.3005	R30	R29	0.3002	0.3010	0.3002
R14	R17	2.1787	0.9659	0.7894	R31	R28	0.9451	0.3054	0.3002
R15	R19	0.3006	0.4009	0.3034	R32	R30	0.3854	0.3002	0.3001
R15	R35	0.9817	0.4979	0.3005	R33	R31	0.8414	0.3035	0.3000
R15	R36	1.5811	1.2751	0.8178	R34	R32	0.7038	0.3286	0.3000
R16	R19	1.4739	0.3043	0.4048	R35	R17	2.3219	1.1239	0.8565
R16	R34	1.7260	0.4792	0.3000	R35	R33	0.6794	0.3147	0.3004
R16	R36	2.7744	1.1844	0.9290	R36	R16	0.5153	0.8062	0.6214
R17	R19	0.3349	0.4633	0.3857	R36	R33	0.8267	0.5974	0.5173
R17	R34	0.5865	0.6216	0.3000	R37	R5	1.2115	0.3079	0.3731
R17	R35	1.0389	0.5633	0.4153	R37	R23	1.1300	0.3973	0.3003
R18	R38	0.4516	0.4813	0.3000	R38	R34	0.3005	0.4817	0.3005
R19	R37	0.7814	0.6019	0.3042	R38	R35	0.7297	0.4039	0.4009
R20	R2	1.7676	0.6810	0.5591	R38	R36	1.3294	1.1793	0.9228
					Sum of CTI (s)		51.79	35.89	24.55

It is clear that the proposed hybrid GSA–SQP algorithm provides better and more stable solutions than the original GSA and SQP for all three test systems.

13.5 Educational program package *ocdocr*

The *ocdocr* is an educational program package of MATLAB® m-files for solving the problem of optimal coordination of DOCRs. The program package is available at [33]. The simulation results presented in Sections 13.4 are obtained by the *ocdocr* program on a 2.20-GHz PC with 3.0-GB RAM.

Table 13.13 Comparison of the results (OF) obtained with GSA, SQP, and hybrid GSA–SQP

	Statistics for hybrid GSA–SQP			**Statistics for GSA**			**Results for SQP with initial choice of control variables as**		
	Min	**Max**	**Std. dev.**	**Min**	**Max**	**Std. dev.**	**Lower**	**Upper**	**Middle**
Test syst. 1	4.7806	4.7806	0	4.8257	4.9915	0.056	4.7810	4.8383	4.9162
Test syst. 2	8.8923	8.9316	0.0139	14.7384	22.9571	2.509	10.2430	9.8740	9.7940
Test syst. 3	26.8258	29.6896	0.9808	51.7747	66.9562	4.921	41.0390	18.0874*	16.2475†

*Solution with miscoordination between backup/primary relays: R1/R3, R2/R4, R2/R22, R3/R4, R3/R21, R4/R5, R8/R26, R9/R12, R10/R11, R11/R13, R12/R14, R13/R15, R16/R19, etc.
†Solution with miscoordination between backup/primary relays: R1/R3, R3/R4, R3/R21, R8/R26, R9/R12, R10/R11, R11/R13, R12/R14, R13/R15, R14/R16, R16/R19, R16/R34, etc.

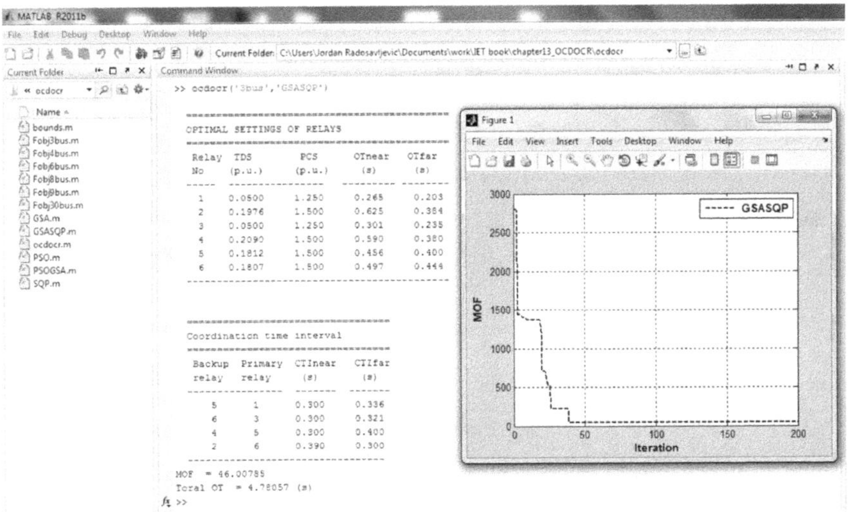

Figure 13.9 Example of the ocdocr program execution

An example of the *ocdocr* program run with the hybrid GSA–SQP solution method is presented in Figure 13.9.

List of the m functions belonging to the *ocdocr* program package is given in Table 13.14.

Table 13.14 Content of the ocdocr program package

ocdocr.m	function []=ocdocr(testsystem,method) %ocdocr – main program %-------------------------------- %Input variables: %testsystem - test system ('3bus', '4bus', '6bus', '8bus', '9bus', '30bus') %method - solution method ('GSA', 'SQP', GSASQP', 'PSO', 'PSOGSA') ...
bounds.m	function [numbervar,lowerbound,upperbound]=bounds(testsystem) % bounds - Define the number of control variables, lower and upper limits %of control variables ...
Fobj3bus.m	function [MOF,dt,top,ti]=Fobj3bus(x) %Fobj3bus - Calculate the objective function for the 3bus test system %Input variable: %x - vector of control variables (TDS and PCS) %Output variables: %MOF - Modified objective function %dt - Coordination time intervals between backup and primary relays %top - Operation time (OT) of primary relays %ti - Total OT of primary relays ...
Fobj4bus.m	function [MOF,dt,top,ti]=Fobj4bus(x) %Fobj4bus - Calculate the objective function for the 4bus test system ...
Fobj6bus.m	function [MOF,dt,top,ti]=Fobj6bus(x) %Fobj6bus - Calculate the objective function for the 6bus test system ...
Fobj8bus.m	function [MOF,dt,top,ti]=Fobj8bus(x) %Fobj8bus - Calculate the objective function for the 8bus test system ...
Fobj9bus.m	function [MOF,dt,top,ti]=Fobj9bus(x) %Fobj9bus - Calculate the objective function for the 9bus test system ...
Fobj30bus.m	function [MOF,dt,top,ti]=Fobj30bus(x) %Fobj30bus - Calculate the objective function for the 30bus test system ...
GSA.m	function [MOF,xbest,dt,top,ti]=GSA(testsystem) %GSA solution method ...
GSASQP.m	function [MOF,xbest,dt,top,ti]=GSASQP(testsystem) %Hybrid GSASQP solution method ...
PSO.m	function [MOF,xbest,dt,top,ti]=PSO(testsystem) %PSO solution method ...
PSOGSA.m	function [MOF,xbest,dt,top,ti]=PSOGSA(testsystem) %PSOGSA solution method ...
SQP.m	function [MOF,xbest,dt,top,ti]=GSASQP(testsystem) %SQP solution method ...

13.6 Conclusion

The coordination between DOCRs is nonlinear and highly constrained optimization problem in which two settings, namely, TDS and PCS of each relay are considered as decision variables. The main objective is to minimize the sum of OTs of all the primary relays, which are expected to operate in order to clear the faults of their corresponding zones. In this chapter, a hybrid GSA–SQP optimization algorithm has been proposed and successfully applied to solve the DOCRs coordination problem. The SQP routine is incorporated in GSA as a local search mechanism to improve the performance of the conventional GSA algorithm. The proposed approach has been tested and investigated on three different test systems. Simulation results show that the hybrid GSA–SQP algorithm provides effective and robust high-quality solution. Moreover, the results obtained using hybrid GSA–SQP are either better or comparable to those obtained using other techniques reported in the literature. The proposed hybrid GSA–SQP algorithm is suitable to find the global optimal solution for the DOCRs coordination problems, because it benefits from the advantages of both GSA and SQP methods.

References

[1] Noghabi A.S., Mashhadi H.R., Sadeh J. 'Optimal coordination of directional overcurrent relays considering different network topologies using interval linear programming'. *IEEE Transactions on Power Delivery*. 2010;**25**(3): 1348–54

[2] Thangaraj R., Pant M., Deep K. 'Optimal coordination of over-current relays using modified differential evolution algorithms'. *Engineering Applications of Artificial Intelligence*. 2010;**23**:820–9

[3] Ezzeddine E., Kaczmarek R., Iftikhar M.U. 'Coordination of directional overcurent relays using a novel method to select their settings'. *IET Generation, Transmission & Distribution*. 2011;**5**(7):743–50

[4] Bedekar P.P., Bhide S.R. 'Optimum coordination of directional overcurrent relays using the hybrid GA-NLP approach'. *IEEE Transactions on Power Delivery*. 2011;**26**(1):109–19

[5] Ezzeddine M., Kaczmarek R. 'A novel method for optimal coordination of directional overcurrent relays considering their available discrete setting and several operation characteristics'. *Electric Power Systems Research*. 2011;**81**:1475–81

[6] Meskin M., Domijan A., Grinberg I. 'Optimal co-ordination of overcurrent relays in the interconnected power systems using break points'. *Electric Power Systems Research*. 2015;**127**:53–63

[7] Moravej Z., Adelnia F., Abbasi F. 'Optimal coordination of directional overcurrent relays using NSGA-II'. *Electric Power Systems Research*. 2015;**119**:228–36

[8] Urdaneta A.J., Nadira R., Jimenez L.G.P. 'Optimal coordination of directional overcurrent relays in interconnected power systems'. *IEEE Transactions on Power Delivery*. 1988;**3**(3):903–11

[9] Zeineldin H.H., El-Saadany E.F., Salama M.M.A. 'Optimal coordination of overcurrent relays using a modified particle swarm optimization'. *Electric Power Systems Research*. 2006;**76**:988–95

[10] Mahari A., Seyedi H. 'An analytic approach for optimal coordination of overcurrent relays'. *IET Generation, Transmission & Distribution*. 2013;**7**(7):674–80

[11] Urdaneta A.J., Ramon N., Jimenez L.G.P. 'Coordination of directional relay timing using linear programming'. *IEEE Transactions on Power Delivery*. 1996;**11**(1):122–9

[12] Birla D., Maheshwari R.P., Gupta H.O. 'A new nonlinear directional overcurrent relay coordination technique, and Banes and Boons of near-end faults based approach'. *IEEE Transactions on Power Delivery*. 2006;**21**(3): 1176–82

[13] Papaspiliotopoulos V.A., George G.N., Maratos N.G. 'A novel quadratically constrained quadratic programming method for optimal coordination of directional overcurrent relays'. *IEEE Transactions on Power Delivery*. 2015;**32**(1):3–10

[14] Sharaf H.M., Zeineldin H.H., Ibrahim D.K., El-Zahab E.E.D.A. 'A proposed coordination strategy for meshed distribution system with DG considering user-defined characteristics of directional inverse time overcurrent relays'. *Electrical Power and Energy Systems*. 2015;**65**:49–58

[15] Correa R., Cardoso G. Jr., De Araujo O.C.B., Mariotto L. 'Online coordination of directional overcurrent relays using binary integer programming', *Electric Power Systems Research*. 2015;**127**:118–25

[16] Singh M., Panigrahi B.K., Abhyankar A.R., Das C. 'Optimal coordination of directional over-current relays using informative differential evolution algorithm'. *Journal of Computational Science*. 2014;**5**:269–76

[17] Yang H., Wen F., Ledwich G. 'Optimal coordination of overcurrent relays in distribution systems with distributed generators based on differential evolution algorithm'. *International Transactions on Electrical Energy Systems*. 2013;**23**:1–12

[18] Moirangthem J., Krishnanand K.R., Dash S.S., Ramaswami R. 'Adaptive differential evolution algorithm for solving non-linear coordination problem of directional overcurrent relays'. *IET Generation, Transmission & Distribution*. 2013;**7**(4):329–36

[19] Chelliah T.R., Thangaraj R., Allamsetty S., Pant M. 'Coordination of directional overcurrent relays using opposition based chaotic differential evolution algorithm'. *Electrical Power and Energy Systems*. 2014;**55**: 341–50

[20] Salazar C.A.C., Enriquez A.C., Schaeffer S.E. 'Directional overcurrent relay coordination considering non-standardized time curves'. *Electric Power Systems Research*. 2015;**122**:42–9

[21] Albasri F.A., Alroomi A.R., Talaq J.H. 'Optimal coordination of directional overcurrent relays using biogeography-based optimization algorithms'. *IEEE Transactions on Power Delivery*. 2015;**30**(4):1810–20

[22] Hussain M.H., Musirin I., Abidin A.F., Rahim S.R.A. 'Solving directional overcurrent relay coordination problem using artificial bees colony'. *International Journal of Electrical, Electronic Science and Engineering*. 2014;**8**(5):705–10

[23] Singh M., Panigrahi B.K., Abhyankar A.R. 'Optimal coordination of directional over-current relays using teaching learning-based optimization (TLBO) algorithm'. *Electrical Power and Energy Systems*. 2013;**50**:33–41

[24] Mansour M.M., Mekhamer S.F., El-Kharbawe N.E.S. 'A modified particle swarm optimizer for the coordination of directional overcurrent relays'. *IEEE Transactions on Power Delivery*. 2007;**22**(4):1400–10

[25] Amraee T. 'Coordination of directional overcurrent relays using seeker algorithm'. *IEEE Transactions on Power Delivery*. 2012;**27**(3):1415–22

[26] Tripathi J.M., Adhishree, Krishan R. 'Optimal coordination of overcurrent relays using gravitational search algorithm with DG penetration'. *Proceedings of 6th IEEE Power India International Conference*; Delhi, India, Dec 2014. IEEE; 2015. pp. 1–6

[27] Alam M.N., Das B., Pant V. 'A comparative study of metaheuristic optimization approaches for directional overcurrent relays coordination'. *Electric Power Systems Research*. 2015;**128**:39–52

[28] Papaspiliotopoulos V.A., Kurashvili T.S., Korres G.N. 'Optimal coordination of directional overcurrent relays for distribution systems with distributed generation based on a hybrid PSO-LP algorithm'. *Proceedings of MedPower 2014*; Athens, Greece, Nov 2014. *IET*; 2014. pp. 1–6

[29] Radosavljević J., Jevtić M. 'Hybrid GSA-SQP algorithm for optimal coordination of directional overcurrent relays'. *IET Generation, Transmission & Distribution*. 2016;**10**(8):1928–37

[30] Rashedi E., Nezamabadi-pour H., Saryazdi S. 'GSA: a gravitational search algorithm'. *Information Sciences*. 2009;**179**:2232–48

[31] Sivasubramani S., Swarup K.S. 'Hybrid DE-SQP algorithm for non-convex short term hydrothermal scheduling problem'. *Energy Conversion and Management*. 2011;**52**:757–61

[32] Mohammadi R., Abyaneh H.A., Rudsari H.M., Fathi S.H., Rastegar H. 'Overcurrent relays coordination considering the priority of constraints'. *Transactions on Power Delivery*. 2011;**26**(3):1927–38

[33] Radosavljević J. *ocdocr—Matlab based program for optimal coordination of DOCRs*. 2017. Available at: https://www.researchgate.net/profile/Jordan_Radosavljevic2/contributions

Chapter 14

Steady-state analysis of self-excited induction generators

14.1 Introduction

The application of induction generators (IGs) follows the development of distributed generation systems. Advantages of IGs in regard to widely applied synchronous generators are a lower unit cost, brushless cage-rotor construction, absence of a separate dc source, good over-speed capability, better transient performances, and inherent overload protection. The inability of a direct regulation of reactive power is a disadvantage of IGs.

An IG can operate in parallel with the distribution network, when its rotor speed is greater than the synchronous speed of the air-gap revolving field. Better utilization of renewable energy may be achieved by developing small-scale autonomous power systems like wind and mini hydropower plants, which use the self-excited IGs (SEIGs) [1]. Those small isolated microgrids are identified as promising concept to energize remote and rural communities [2–4]. However, poor voltage regulation characteristics of IGs require an efficient and reliable control system of voltage and frequency in stand-alone microgrids with SEIGs [5,6]. It is well known that if an appropriate capacitor bank is connected across an externally driven induction machine, an electromotive force (EMF) is induced in the machine windings due to the excitation provided by the capacitor. The induced voltage and current would continue to rise, until the reactive power supplied by the capacitor is balanced by the reactive power demanded by the machine. This results in an equilibrium state being reached, and the machine now operates as an SEIG at a voltage and frequency assigned by the value of the capacitance, speed of the prime mover, and parameters of the machine and the load [7,8].

The key issue that needs to be solved at the SEIG is the voltage and frequency regulation. The voltage and frequency variables are nonlinear dependences of the speed, magnetizing reactance, excitation capacitance, and load. The terminal voltage of SEIG can be controlled by continuously adjusting the reactive power supply to meet the demand of the generator and loads. This is achieved either by simple switched capacitors, switched reactors, or by using power electronics such as static Var compensators (SVC). The generated frequency is highly dependent on the speed of the rotor, so that unless there is a fixed speed prime mover, the microgrid with SEIGs will see a frequency that changes with prime mover and drops down when the load increases. Consequently, generator frequency can be controlled by variation in consumer loads or by using an electronic load controller at the generator terminals in parallel to consumer loads. The operating principle behind the electronic load controller is that they consume the difference between the power output of the SEIG and the power absorbed by the connected consumer loads [2,3].

Any regulation strategy for SEIGs requires an efficient procedure for computing of the frequency and magnetizing reactance, that is, the frequency and EMF for a given rotor speed, loads, and the machines parameters.

A huge number of journal publications have so far been dedicated to the steady-state analysis of an autonomous three-phase SEIG with three-phase balanced static load [8–16]. Analysis of the SEIG feeding power to a dynamic load (induction motor) is presented in [7,17,18]. The performance of a SEIG is usually determined through its equivalent circuits. An overwhelming majority of the aforementioned researchers used either the loop impedance approach [7,8,18] or the nodal admittance approach [9–14] in analyzing the circuits. Irrespective of the representation manner, the analysis of three-phase SEIG requires a computation of the generated frequency (F) and magnetizing reactance (X_m) for the given operating conditions. Knowing the values of X_m and F, the steady-state performances of the SEIG can easily be determined through their equivalent circuits in conjunction with appropriate magnetization characteristic.

Unbalanced operation of a three-phase SEIG may be caused by manufacturing tolerances of excitation capacitances, failure of some excitation capacitance modules, or disconnection of loads by consumers [19]. Moreover, autonomous microgrids with three-phase SEIGs often employ single-phase distribution schemes due to a low cost, easy maintenance, and simplicity in terms of protection [20–22]. The steady-state analysis of three-phase SEIG with unbalanced loads and excitation capacitances is well documented [19–25]. By means of the method of symmetrical components, the complex three-phase generator-load system has been reduced to a simple equivalent passive circuit. In order to determine the excitation frequency and magnetizing reactance, the author used different methods such as pattern search minimization technique of Hooke and Jeeves [19] and genetic algorithm [25]. Valid simplifying assumptions in the sequence equivalent circuit, which are applied in [20], make the final equation convenient in order to obtain the solution for X_m and F.

Usually, the stand-alone renewable energy sources with SEIGS do not go much beyond 15 kW [3]. In order to utilize the full potential of the hydro and wind energy source, a number of SEIGs are required to be operated in parallel. For parallel operation of SEIGs, there was a simplified model described in [26] and steady-state analyzes in [27–31]. These papers are focused on three-phase SEIGs with balanced excitation capacitances and loads. However, in practice, the conditions for symmetrical and balanced operation of three-phase SEIGs are often not achieved. The steady-state analysis and performance characteristics of parallel operated three-phase SEIGs with unbalanced load conditions are considered in [32].

The aim of this chapter is to show how a metaheuristic optimization technique such as GA can be applied in solving a nonoptimization problem, i.e., in determining the steady-state performances of a SEIG for general operating conditions. GA-based optimization procedure is applied to the analysis of SEIG feeding balanced/unbalanced and static/dynamic loads. The GA has been applied to the computation of the unknowns by minimizing the total impedance module of the equivalent passive circuit.

14.2 System configuration

A modified example of SEIG-based small autonomous microgrid system presented in [2] is shown in Figure 14.1. The SEIG is coupled to the small hydro turbine and is feeding both balanced three-phase and unbalanced single-phase consumer loads. As the SEIG is driven by an unregulated small hydro turbine, the mechanical power input of SEIG is constant; therefore, to control system frequency, the balance between mechanical power input and the electrical power output of SEIG has to be maintained by using an electronic load controller. Therefore, all the parameters of the SEIG such as voltages, winding currents, capacitor currents, power output, and frequency remain at same operating point in steady state irrespective of loading condition which is known as "single point operation" [2].

As mentioned before, generated voltage of the SEIG depends on the prime mover speed, connected terminal capacitance, and load. Initially, excitation capacitors are used to excite IG to build up the rated voltage at no-load condition. But as power factor and power of load fluctuate, ideal compensation will require variable reactive compensators such as SVC, static compensator, or dynamic voltage restorer. The controlled reactive power is responsible for keeping the terminal voltage constant with change in load with Var compensator connected in parallel with a fixed capacitor and load. The main function of such controllers would be to provide required variable capacitive reactive power to maintain constant voltage across the load at varying prime mover speed and power of loads. Therefore, an effective capacitive reactive power controller has become central to the success of SEIG system for stand-alone applications [33].

The required reactive power is different for different rating machines for maintaining constant terminal voltage, and it depends on the machine parameters and magnetizing characteristics. The capacitive reactive power required for excitation and voltage control of a SEIG at different load conditions is calculated based on appropriate steady-state equations of the SEIG.

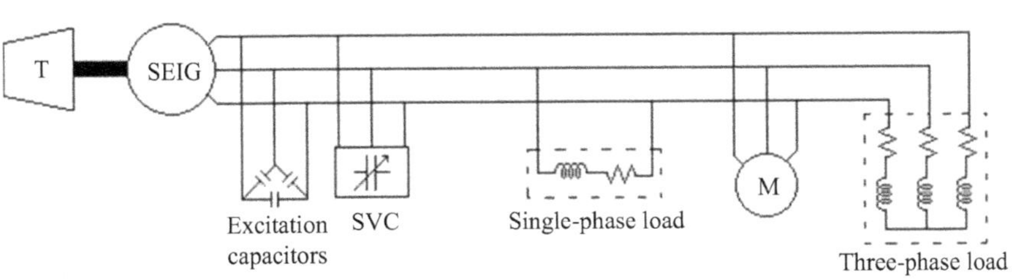

Figure 14.1 SEIG-based microgrid

14.3 Induction generator model

If three-phase induction machines feed an unbalanced/asymmetrical external network, then a natural analysis mode is to resort the system to symmetrical components and rotating field concepts [1,20]. Therefore, the IG is modeled with its positive- and negative-sequence circuits as it is shown in Figure 14.2.

In these circuits, all parameters refer to the rated frequency, and it is assumed that all inductive reactances are proportional to the rated frequency, where F is the per-unit frequency—the ratio of the generated frequency to the rated frequency—and v is the ratio of the actual rotor speed to the synchronous speed corresponding to the rated frequency. In these circuit models, it is also assumed that all parameters, except the magnetizing reactance X_m, are constant and independent of the saturation. The value of X_m depends on the core flux which in turn depends on the ratio of the air-gap voltage to the frequency. The variation of the positive-sequence air-gap voltage with magnetizing reactance X_m can be expressed by an appropriate equation based on the synchronous speed test data fitting.

According to Figure 14.2, the positive-sequence and negative-sequence impedances of the generator, that is, impedances $\underline{Z}_p$ and $\underline{Z}_n$, are as follows:

$$\underline{Z}_p = \frac{R_s}{F} + jX_s + \frac{(jR_c \cdot X_m/(R_c + jFX_m))((R_r/(F - v)) + jX_r)}{jR_c \cdot X_m/(R_c + jFX_m) + ((R_r/(F - v)) + jX_r)} \tag{14.1}$$

$$\underline{Z}_n = \frac{R_s}{F} + jX_s + \frac{(jR_c \cdot X_m/(R_c + jFX_m))((R_r/2F) + jX_r)}{(jR_c \cdot X_m/(R_c + jFX_m)) + ((R_r/2F) + jX_r)} \tag{14.2}$$

where F is the per-unit frequency, v is the per-unit speed, R_r is the stator resistance per phase in p.u., R_r is the rotor resistance per phase referred to the stator in p.u., R_c is the core loss resistance per phase in p.u., X_s is the stator reactance per phase in p.u., X_r is the rotor reactance per phase referred to the stator in p.u., and X_m is the magnetizing reactance per phase at the rated frequency in p.u.

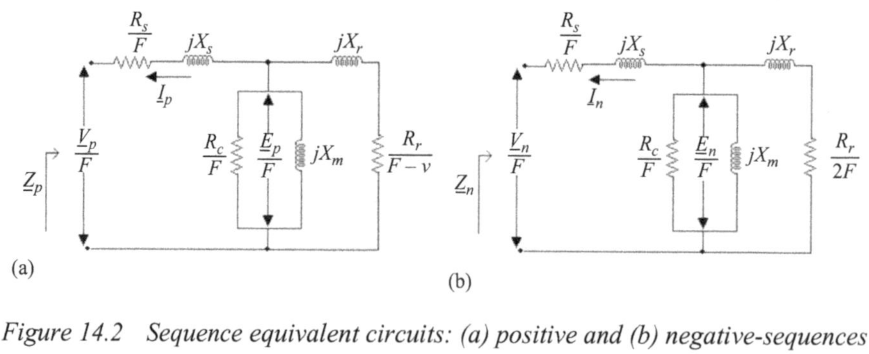

Figure 14.2 Sequence equivalent circuits: (a) positive and (b) negative-sequences

14.4 Steady-state equations of SEIG

The circuit connection of a three-phase delta connected SEIG feeding a three-phase unbalanced load is shown in Figure 14.3. In order to produce the reactive power necessary for initiating self-excitation, at least one of the terminal impedances $\underline{Z}_A$, $\underline{Z}_B$, or $\underline{Z}_C$ must contain a capacitive element. Assigning appropriate values for the terminal impedances, a specific unbalanced operating condition can be simulated.

The delta connection is deliberately selected herein due to the fact that zero-sequence quantities do not exist. However, the presented concept can easily be extended to the star connection with the isolated neutral point applying the star-delta transformation.

The circuit shown in Figure 14.3 has no active voltage sources. Accordingly, the SEIG may be regarded as a passive circuit when viewed across any two stator terminals [19]. For convenience, a phase is chosen as the reference and the input impedance of the SEIG across terminals 1 and 3 in Figure 14.3 will be considered. In this manner, the equivalent scheme shown in Figure 14.3 can be transformed into a simple circuit given in Figure 14.4.

According to the marks given in Figure 14.3, the current and voltage balance equations may be written as follows:

$$\underline{I}_A + \underline{I}_{\ell A} - \underline{I}_B - \underline{I}_{\ell B} = 0 \tag{14.3}$$

$$\underline{I}_A + \underline{I}_{\ell A} - \underline{I}_C - \underline{I}_{\ell C} = 0 \tag{14.4}$$

$$\underline{I}_{\ell A} = \frac{\underline{V}_A}{\underline{Z}_A} = \underline{Y}_A \underline{V}_A \tag{14.5}$$

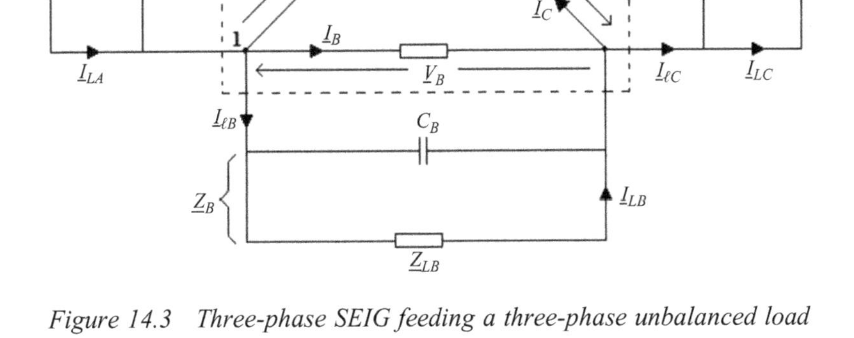

Figure 14.3 Three-phase SEIG feeding a three-phase unbalanced load

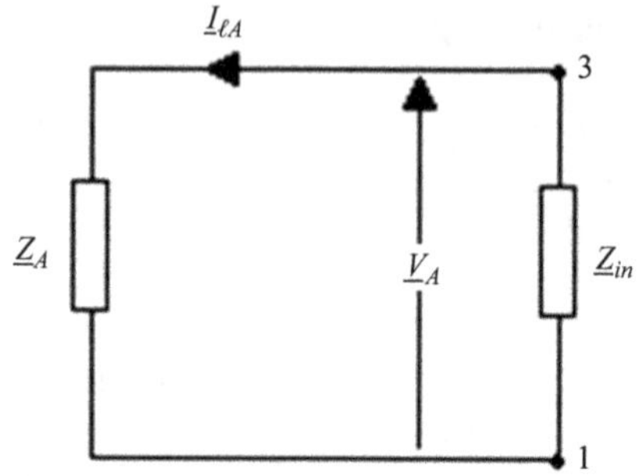

Figure 14.4 Simplified circuit for the three-phase SEIG feeding a three-phase unbalanced load

$$\underline{I}_{\ell B} = \frac{\underline{V}_B}{\underline{Z}_B} = \underline{Y}_B \underline{V}_B \tag{14.6}$$

$$\underline{I}_{\ell C} = \frac{\underline{V}_C}{\underline{Z}_C} = \underline{Y}_C \underline{V}_C \tag{14.7}$$

Based on the symmetrical component theory, the following equations are developed for the delta connected system:

$$\underline{V}_A + \underline{V}_B + \underline{V}_C = 0 \tag{14.8}$$

$$\underline{V}_A = \underline{V}_p + \underline{V}_n \tag{14.9}$$

$$\underline{V}_B = \underline{a}^2 \underline{V}_p + \underline{a}\,\underline{V}_n \tag{14.10}$$

$$\underline{V}_C = \underline{a}\,\underline{V}_p + \underline{a}^2 \underline{V}_n \tag{14.11}$$

$$\underline{I}_A = \underline{V}_p \underline{Y}_p + \underline{V}_n \underline{Y}_n \tag{14.12}$$

$$\underline{I}_B = \underline{a}^2 \underline{V}_p \underline{Y}_p + \underline{a}\,\underline{V}_n \underline{Y}_n \tag{14.13}$$

$$\underline{I}_C = \underline{a}\,\underline{V}_p \underline{Y}_p + \underline{a}^2 \underline{V}_n \underline{Y}_n \tag{14.14}$$

where $\underline{a}$ is the complex operator $e^{(j2\pi/3)}$.

This follows from (14.3) and (14.4)

$$\underline{I}_{\ell A} = -\underline{I}_A + \underline{I}_B + \underline{I}_{\ell B} \tag{14.15}$$

$$\underline{I}_{\ell A} = -\underline{I}_A + \underline{I}_C + \underline{I}_{\ell C} \tag{14.16}$$

Substituting (14.6), (14.7), and (14.12)–(14.14) into (14.15) and (14.16), it is obtained that

$$\underline{I}_{\ell A} = \underline{A} \cdot \underline{V}_p + \underline{B} \cdot \underline{V}_n \tag{14.17}$$

$$\underline{I}_{\ell A} = \underline{C} \cdot \underline{V}_p + \underline{D} \cdot \underline{V}_n \tag{14.18}$$

where

$$\underline{A} = \left(\underline{a}^2 - 1\right)\underline{Y}_p + \underline{a}^2\underline{Y}_B \tag{14.19}$$

$$\underline{B} = (\underline{a} - 1)\underline{Y}_n + \underline{a}\,\underline{Y}_B \tag{14.20}$$

$$\underline{C} = (\underline{a} - 1)\underline{Y}_p + \underline{a}\,\underline{Y}_C \tag{14.21}$$

$$\underline{D} = \left(\underline{a}^2 - 1\right)\underline{Y}_n + \underline{a}^2\underline{Y}_C \tag{14.22}$$

From (14.17) and (14.18), it is obtained that

$$\underline{V}_n = \frac{A - C}{D - B}\underline{V}_p \tag{14.23}$$

According with Figure 14.4, the phase voltage $\underline{V}_A$ is

$$\underline{V}_A = -\underline{Z}_{in} \cdot \underline{I}_{\ell A} \tag{14.24}$$

Finally, by substituting (14.9), (14.17), and (14.23) into (14.24) for the SEIG input impedance, it is obtained that

$$\underline{Z}_{in} = \frac{(A + D) - (B + C)}{BC - AD} \tag{14.25}$$

The positive-sequence and negative-sequence admittances of the SEIG, that is, the admittances $\underline{Y}_p$ and $\underline{Y}_n$, are functions of the F and X_m. Therefore, the SEIG input impedance may be expressed as

$$\underline{Z}_{in} = R_{in}(X_m, F) + jX_{in}(X_m, F) \tag{14.26}$$

In accordance to Figure 14.4, the following voltage balance equation can be written as

$$(\underline{Z}_A + \underline{Z}_{in})\underline{I}_{\ell A} = 0 \tag{14.27}$$

$$\underline{Z}_{loop}\underline{I}_{\ell A} = 0 \tag{14.28}$$

where $\underline{Z}_{loop} = \underline{Z}_A + \underline{Z}_{in}$.

For a successful self-excitation, it is required that $\underline{I}_{\ell A} \neq 0$, hence,

$$\underline{Z}_{loop} = 0 \tag{14.29}$$

14.5 Steady-state equations of parallel operated SEIGs

Microgrid may consist of either wind power plant or hydropower plant or a combination of both constituting the generating unit [33]. Figure 14.5 shows such a microgrid which consists of several parallel operating SEIG with its prime mover and reactive power compensators which are connected together to supply the same load.

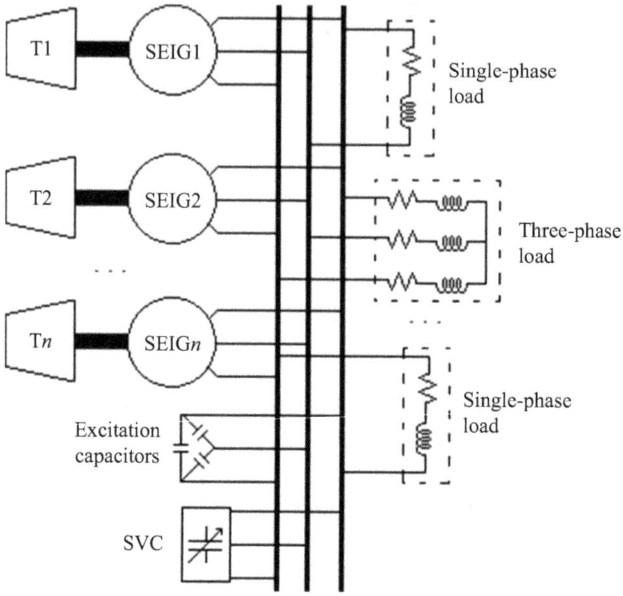

Figure 14.5 Microgrid with SEIGs connected in parallel

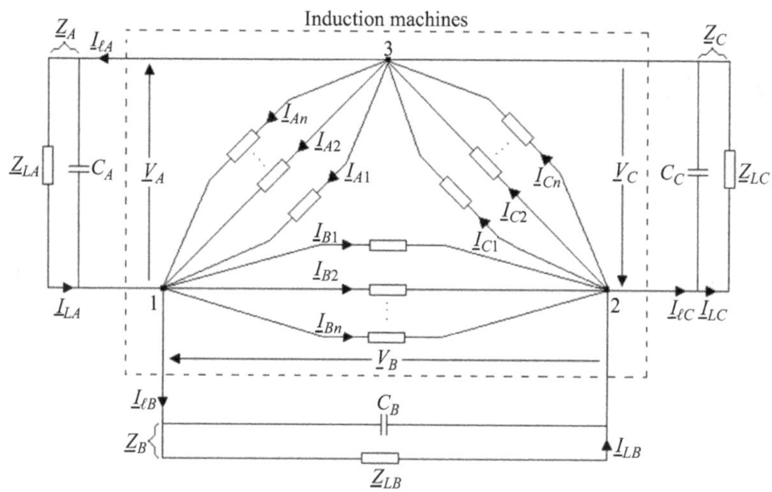

Figure 14.6 Parallel SEIGs feeding unbalanced load

The circuit connection of *n* three-phase delta-connected SEIGs feeding a three-phase unbalanced load is shown in Figure 14.6.

Using analogy with single SEIG considered in Section 14.4, the equivalent scheme shown in Figure 14.6 can be transformed into a simple circuit given in

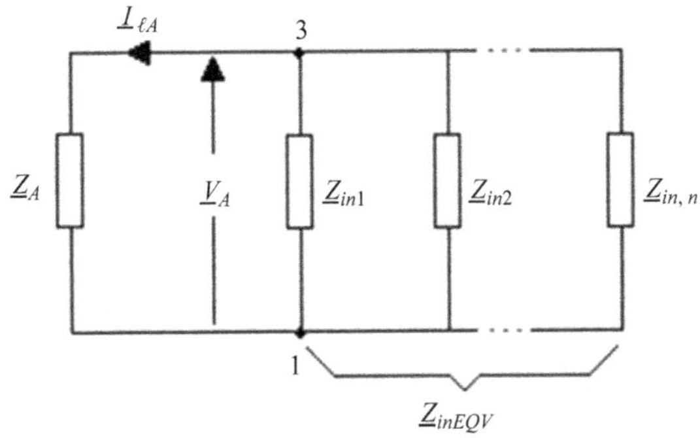

Figure 14.7 Simplified circuit of parallel SEIGs

Figure 14.7. For the parallel SEIGs, input impedance is obtained as follows:

$$\underline{Z}_{inEQV} = \frac{(\underline{A} + \underline{D}) - (\underline{B} + \underline{C})}{\underline{B} \cdot \underline{C} - \underline{A} \cdot \underline{D}} \tag{14.30}$$

where

$$\underline{A} = (\underline{a}^2 - 1) \sum_{i=1}^{N} \underline{Y}_{pi} + \underline{a}^2 \underline{Y}_B \tag{14.31}$$

$$\underline{B} = (\underline{a} - 1) \sum_{i=1}^{N} \underline{Y}_{ni} + \underline{a} \underline{Y}_B \tag{14.32}$$

$$\underline{C} = (\underline{a} - 1) \sum_{i=1}^{N} \underline{Y}_{pi} + \underline{a} \underline{Y}_C \tag{14.33}$$

$$\underline{D} = (\underline{a}^2 - 1) \sum_{i=1}^{N} \underline{Y}_{ni} + \underline{a}^2 \underline{Y}_C \tag{14.34}$$

$\underline{Y}_{pi}$ and $\underline{Y}_{ni}$ are the positive-sequence and negative-sequence admittance of the *i*th IG obtained based on (14.1) and (14.2).

The positive-sequence and negative-sequence admittances, $\underline{Y}_{pi}$ and $\underline{Y}_{ni}$, that is the complex coefficients $\underline{A}$, $\underline{B}$, $\underline{C}$, and $\underline{D}$, are functions of F and X_{mi}. Therefore, the SEIGs input impedance may by expressed as

$$\underline{Z}_{inEQV} = R_{inEQV}(X_{m1}, X_{m2}, \ldots, X_{mn}, F) + jX_{inEQV}(X_{m1}, X_{m2}, \ldots, X_{mn}, F) \tag{14.35}$$

In accordance to Figure 14.7, the following voltage balance equation can be written as

$$\left(\underline{Z}_A + \underline{Z}_{inEQV}\right)\underline{I}_{\ell A} = 0 \tag{14.36}$$

$$\underline{Z}_{loop}\underline{I}_{\ell A} = 0 \tag{14.37}$$

where $\underline{Z}_{loop} = \underline{Z}_A + \underline{Z}_{inEQV}$. For a successful self-excitation, it is required that $\underline{I}_{\ell A} \neq 0$, hence,

$$\underline{Z}_{loop} = 0 \tag{14.38}$$

In order to ensure that all generators have the same terminal voltage (the common bus voltage), the following equation must be satisfied:

$$\left|\underline{Z}_{p1}\right| \cdot \left|\underline{I}_{p1}\right| - \left|\underline{Z}_{pi}\right| \cdot \left|\underline{I}_{pi}\right| = 0 \quad (i = 2, \ldots, n) \tag{14.39}$$

According to Figure 14.2(a), the magnitude of the positive sequence stator current for machine i can be related to the machine magnetizing reactance X_{mi}, provided the relationship between the forward field air-gap voltage $|\underline{E}_{pi}/F|$ and magnetizing reactance X_{mi} is known. The magnetizing characteristic relating the $|\underline{E}_{pi}/F|$ with the X_{mi} can be obtained experimentally by a synchronous speed test. Generally, the variation of $|\underline{E}_{pi}/F|$ with X_{mi} over the practical region of operation can be approximated by linear segments with expressions of the type

$$\left|\frac{\underline{E}_{pi}}{F}\right| = a_i + b_i \cdot X_{mi} \quad (i = 1, \ldots, n) \tag{14.40}$$

Therefore, (14.39) can be expressed as

$$\left|\frac{\underline{Z}_{p1}}{\underline{Z}_{pmr1}}\right|(a_1 + b_1 X_{m1}) - \left|\frac{\underline{Z}_{pi}}{\underline{Z}_{pmri}}\right|(a_i + b_i X_{mi}) = 0 \quad (i = 2, \ldots, n) \tag{14.41}$$

where $\underline{Z}_{pmri}$ is the total positive-sequence impedance of the magnetizing and rotor circuit of the ith machine. In accordance with Figure 14.2(a), it is

$$\underline{Z}_{pmri} = \frac{jX_{mi} \cdot ((R_{ri}/(F - v_i)) + jX_{ri})}{jX_{mi} + ((R_{ri}/(F - v_i)) + jX_{ri})} \tag{14.42}$$

14.6 Solution method

Equation (14.28), i.e., (14.38), should be separated into real and imaginary parts, in order to solve it explicitly. The usual practice is to derive the complex coefficients of the nonlinear equations manually and solve them by using the numerical Newton–Raphson or polynomial method. The necessary mathematical manipulations and the lengthy derivations of nonlinear equations required for the unknown variables in the methods are tedious, time consuming, and liable to human error. They require a tremendous human effort for accurate programing and debugging. The model lacks flexibility as coefficients are valid only for a given circuit configuration. An inclusion of the core loss resistance or load inductance will increase the order of the equations. These disadvantages can be overcome by using the optimization techniques in [12–14,19,25,32]. In this way, the problem of solving a nonlinear complex algebraic equation is reduced to an optimization problem with a predefined minimum. The unknown variables in this equation are regarded as control variables of the optimization problem. To transform the problem of solving nonlinear equation into an optimization problem, the objective function and the control variables should be defined.

14.6.1 Overview of genetic algorithm

The genetic algorithm is one of the techniques, that is, one optimization procedure based on the natural evolution process imitation [34]. They belong to the methods of a directed random search of solution domain with the aim of finding a global optimum. The classical optimization methods start from a single possible initial solution, and they reach the optimum by applying the heuristic rules iteratively. The GA starts from the population, which is a group of individuals. Each individual represents a potential solution of the optimization problem. The individuals are presented in the same way, usually through a column or a string of data. The quality of each solution or individual is determined based on the fitness function values. Through a series of GA operations, a new population is obtained and its individuals are engendered by the individuals from the previous population according to the natural evolution principles: the choice of parents, crossover, and mutation. Basic operations of the GA are:

1. *Representation of individuals*: All data (variables) that make an individual are written in a string. A string is composed of substrings. Each substring represents a binary encoded variable on which the process of optimization is carried out. The number of substrings, therefore the size of a string, depends on the number of variables that are optimized.
2. *Initialization*: Individuals with random strings are generated that set up the initial population.
3. *Fitness function calculation*: It is used to rate the quality of an individual, and it represents an equivalent of the function that should be optimized, that is, objective function.
4. *Selection*: During the selection process, the individuals that will participate in the reproduction (parents) are selected. The point of the selection is to store and transfer good individuals to the next generation.
5. *Crossover*: Crossover is an exchange process of column parts between two individuals, that is, "parents." One or two new individuals are engendered by the crossover, that is, a "child." The possibility of inheriting the first parent's characteristics by a child is introduced during this process.
6. *Mutation*: Mutation is a way to give a new piece of information to an individual. Mutation represents an accidental bit variation of an individual, generally with a constant probability for each bit within a population. The mutation probability can further vary depending on the size of the population, application, and preferences of the explorer. A fixed value which is often kept during the whole genetic algorithm is used for each generation.
7. *Ending conditions*: The process of finding the optimal solution is an iterative process which ends when a maximum number of generations is achieved or

when another criterion is fulfilled, such as a minimum offset from the best fitness value and medium fitness value of all individuals in a current population. If end conditions are fulfilled, the best individual thus obtained is the semioptimal solution in question. Otherwise, return to 3.

14.6.2 Application of GA to SEIG

The self-excitation requirement given by (14.29) can be expressed as follows:

$$\left|\underline{Z}_{loop}\right| = 0 \tag{14.43}$$

This equation can be considered as an objective function of the optimization problem, which control variables and unknown variables of the equation. Therefore, to solve (14.43), the optimization problem is stated as follows:

$$\min\left|\underline{Z}_{loop}(X_m, F)\right| = \min\sqrt{\left(\mathrm{Re}(\underline{Z}_{loop})\right)^2 + \left(\mathrm{Im}(\underline{Z}_{loop})\right)^2} \tag{14.44}$$

where unknowns X_m and F are considered as control variables of the optimization problem.

The feasibility region of the problem control variables is defined as

$$X_m^{\min} \leq X_m \leq X_m^{\max}; \quad F^{\min} \leq F \leq F^{\max} \tag{14.45}$$

The terminal voltage at SEIG is considered as functional operating constraints in this optimization problem.

Depending on the specific problem, in addition to the X_m and F, the unknowns, i.e., control variables, could be the excitation capacitance, rotor speed, and load impedance as well. For example, the voltage control of the SEIG can be achieved by a change in the excitation capacitance for a constant value of the speed and load impedance. In this case, the optimization problem can be expressed as follows:

$$\min\left\{\left|\underline{Z}_{loop}(X_m, F, C)\right| + \left|V - V_{spec}\right|\right\} \tag{14.46}$$

with the constraints

$$X_m^{\min} \leq X_m \leq X_m^{\max}; \quad F^{\min} \leq F \leq F^{\max}; \quad C^{\min} \leq C \leq C^{\max} \tag{14.47}$$

where V is the terminal voltage, and V_{spec} is the specified terminal voltage (usually $V_{spec} = 1$ p.u.).

The flowchart describing the MATLAB® realization of the GA optimization method implemented to solve the above optimization problem is shown in Figure 14.8. The GA enables a simultaneous computation of the X_m and F values for any given circuit configuration and operating conditions such as the speed, excitation capacitances, and load impedances.

The evaluation of the objective (fitness) function is performed by the MATLAB program based on the mathematical model presented in Section 14.4.

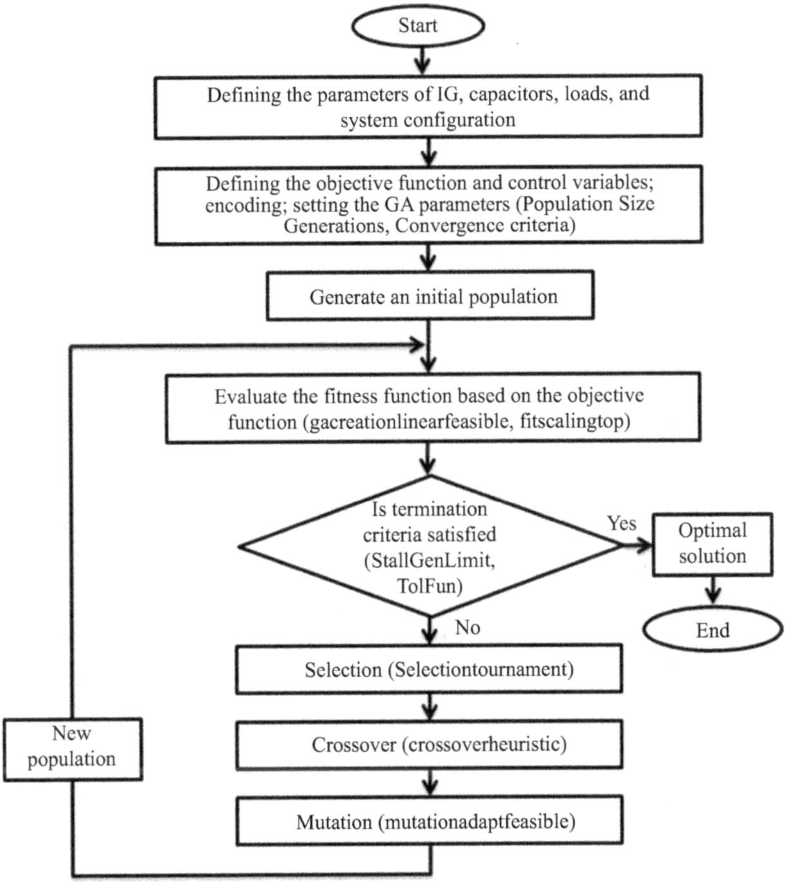

Figure 14.8 Flowchart of GA

Having calculated the X_m and F, the next step is to determine the normalized air-gap voltage E_p/F for the value of X_m based on the relevant magnetizing characteristic of an induction machine. The magnetizing characteristics relating the E_p/F with the X_m can be obtained experimentally by a synchronous speed test. Knowing the air-gap voltage E_p, the positive-sequence current and voltage can be computed in accordance with Figure 14.2(a) as follows:

$$\underline{I}_p = -\frac{E_p}{(FR_r/(F-v)) + jFX_r} - \frac{E_p}{jFR_mX_m/(R_m + jFX_m)} \tag{14.48}$$

$$\underline{V}_p = E_pF - (R_s + jFX_s)\underline{I}_p \tag{14.49}$$

The negative-sequence voltage $\underline{V}_n$ can be determined by (14.23). Finally, the terminal voltages and stator currents of the SEIG can be calculated by (14.9)–(14.14) and the line currents by (14.5)–(14.7).

14.6.3 Application of GA to parallel operated SEIGs

If the self-excitation requirement (14.29) and voltage-balance equation (14.36) are treated as criterion functions and the frequency, magnetizing reactance, excitation capacitance, and speed as control variables within the specified limits, then the optimization procedure can be applied for the determination of the steady-state condition of three-phase SEIGs operating in parallel.

14.6.3.1 Objective function

Equation (14.38) can be expressed as

$$f_1 = \left|\underline{Z}_{loop}\right| = 0 \tag{14.50}$$

The system of (14.41) is equivalent with following expression:

$$f_2 = \sum_{i=2}^{N}\left\{\left|\frac{\underline{Z}_{p1}}{\underline{Z}_{pmr1}}\right|(a_1 + b_1 X_{m1}) - \left|\frac{\underline{Z}_{pi}}{\underline{Z}_{pmri}}\right|(a_i + b_i X_{mi})\right\} = 0 \tag{14.51}$$

The control of the common bus voltage can be achieved by changing in the excitation capacitances as well as changing rotor speeds of SEIGs. By controlling the voltage level, one attains the voltage on the common bus of the SEIGs, that is, on the load which has the value equal to a previously specified one. This condition can be expressed as follows:

$$f_3 = \left|V_{spec} - V_t\right| = 0 \tag{14.52}$$

where V_{spec} is the specified voltage, and V_t is the common bus voltage ($V_t = \left|\underline{V}_A\right|$).

Therefore, the objective function in case of parallel operated SEIGs can be expressed in the following form:

$$\min Fobj = \min\{f_1 + f_2 + f_3\} \tag{14.53}$$

14.6.3.2 Control variables

Depending on the specific problem, the unknown variables, i.e., control variables of the optimization problem, can be magnetizing reactances (X_{mi}), frequency (F), excitation capacitance (C), and rotor speeds (v_i). Accordingly, the inequality constraints which define the solution space of the optimization problem may be expressed as

$$X_{mi}^{\min} \le X_{mi} \le X_{mi}^{\max} \quad i = 1, 2, \ldots, n \tag{14.54}$$

$$F^{\min} \le F \le F^{\max} \tag{14.55}$$

$$C^{\min} \le C \le C^{\max} \tag{14.56}$$

$$v_i^{\min} \le v_i \le v_i^{\max} \quad i = 1, 2, \ldots, n \tag{14.57}$$

General steps for solving the optimization problem using GA are described in Figure 14.8.

14.6.3.3 Performance equations

The GA enables a simultaneous computation of the unknown values of X_{mi} and F. Having thus determined the X_{mi} and F, the next step is to determine the normalized air-gap voltage $|\underline{E}_{pi}/F|$ according to value of X_{mi} based on the relevant magnetizing characteristic for all induction machines. Because the phase angles of the normalized air-gap voltages are not known, it can be supposed that the angle of the SEIG with index 1 is equal to zero. Knowing the air-gap voltage $\underline{E}_{p1} = E_{p1}\angle 0^\circ$, in accordance with Figure 14.2(a), the positive-sequence current of the SEIG can be computed as

$$\underline{I}_{p1} = -\frac{\underline{E}_{p1}}{\underline{Z}_{pmr1}} \tag{14.58}$$

Now, the positive-sequence voltage is

$$\underline{V}_p = \underline{E}_{p1} - (R_{s1} + jFX_{s1})\underline{I}_{p1} \tag{14.59}$$

When $\underline{V}_p$ is calculated, the positive-sequence currents of the other SEIGs can be determined

$$\underline{I}_{pi} = -\frac{\underline{V}_p}{F\underline{Z}_{pi}} \quad (i = 2, \ldots, n) \tag{14.60}$$

The phase angles of the normalized air-gap voltages of the other SEIGs

$$\theta_i = \arctan\frac{\mathrm{Im}\left\{\underline{V}_p/F + ((R_{si}/F) + jX_{si})\underline{I}_{pi}\right\}}{\mathrm{Re}\left\{\left(\underline{V}_p/F\right) + ((R_{si}/F) + jX_{si})\underline{I}_{pi}\right\}} \quad (i = 2, \ldots, n) \tag{14.61}$$

The negative-sequence voltage $\underline{V}_n$ can be determined by (14.23) and the negative-sequence currents in accordance with Figure 14.2(b) by equation

$$\underline{I}_{ni} = -\frac{\underline{V}_n}{F\underline{Z}_{ni}} \quad (i = 1, \ldots, n) \tag{14.62}$$

Accordingly, with Figure 14.2(b)

$$\underline{E}_{ni} = \frac{jX_{mi} \cdot ((R_{ri}/(F + v_i)) + jX_{ri})}{jX_{mi} + ((R_{ri}/(F + v_i)) + jX_{ri})} \cdot \underline{I}_{ni} \quad (i = 1, \ldots, n) \tag{14.63}$$

Finally, the terminal voltages and stator currents of the SEIGs can be calculated by (14.9)–(14.14) and the line currents by (14.5)–(14.7).

The output power (P_{Geni}) and the input power (P_{Inpi}) of the ith SEIG are

$$P_{Geni} = \mathrm{Re}\left\{\underline{V}_p \cdot \underline{I}^*_{pi} + \underline{V}_n \cdot \underline{I}^*_{ni}\right\} \quad (i = 1, \ldots, n) \tag{14.64}$$

$$P_{Inpi} = \mathrm{Re}\left\{\frac{E^2_{pi}}{(R_{ri}/(F - v_i)) - jX_{ri}} + \frac{E^2_{ni}}{(R_{ri}/(F + v_i)) - jX_{ri}}\right\} \quad (i = 1, \ldots, n) \tag{14.65}$$

and reactive power requirement

$$Q_i = \text{Im}\left\{\underline{V}_p \cdot \underline{I}^*_{pi} + \underline{V}_n \cdot \underline{I}^*_{ni}\right\} \quad (i = 1, \ldots, n) \tag{14.66}$$

14.7 Simulation results

14.7.1 Steady-state analysis of SEIG

The proposed GA-based procedure is tested on a three-phase, four pole, 50 Hz, delta-connected, squirrel cage induction machine of 2.2 kW, 230 V, and 8.6 A.

Parameters of the induction machine

$$R_s = 0.07224(\text{p.u.}),\ R_r = 0.03795(\text{p.u.}),\ X_s = X_r = 0.10459(\text{p.u.}) \tag{14.67}$$

Variation of air-gap voltage with magnetizing reactance at rated frequency

$$\begin{aligned}
X_m < 1.77464 &\rightarrow \frac{E_d}{F} = 1.49744 - 0.007X_m \\
2.06096 > X_m \geq 1.77464 &\rightarrow \frac{E_d}{F} = 2.02226 - 0.01338X_m \\
2.32904 > X_m \geq 2.06096 &\rightarrow \frac{E_d}{F} = 2.52129 - 0.0186X_m \\
X_m \geq 2.32904 &\rightarrow \frac{E_d}{F} = 0
\end{aligned} \tag{14.68}$$

In order to verify the mathematical model and methodology, the results obtained herein for the three-phase SEIG with balanced resistive loads and the results of analysis presented in [14] are comparatively given in Table 14.1. Moreover, the proposed approach was implemented to voltage control of the SEIG in case of feeding static and dynamic loads.

14.7.1.1 Balanced/unbalanced operation

The steady-state analysis results of the three-phase SEIG for the different connections of balanced/unbalanced resistive loads and excitation capacitances are

Table 14.1 Verification of the proposed procedure for the balanced resistive loads

Case	C, R (μF, Ω)	Rotor speed ν (p.u.)	Reference [14]				Proposed method results	
			Computed results		Experimental			
			F (p.u.)	V (p.u.)	F (p.u.)	V (p.u.)	F (p.u.)	V (p.u.)
1	36,160	0.95333	0.934899	0.583202	0.943800	0.582608	0.941519	0.573805
2	36,160	1.04666	1.033292	0.942325	1.033800	0.943478	1.033292	0.942324
3	51,220	0.85666	0.848210	0.765084	0.848600	0.756521	0.848209	0.765083

Table 14.2 Case studies of the three-phase SEIG for the different circuit configurations

Case	Note	$\underline{Z}_A = R_{LA}//C_A$		$\underline{Z}_B = R_{LB}//C_B$		$\underline{Z}_C = R_{LC}//C_C$		Results		
		R_{LA} (Ω)	C_A (μF)	R_{LB} (Ω)	C_B (μF)	R_{LC} (Ω)	C_C (μF)	F (p.u.)	X_m (p.u.)	$\lvert\underline{Z}_{loop}\rvert$
1	Bal. load bal. ex. Cap	150	50	150	50	150	50	0.9854	1.4359	$1.29 \cdot 10^{-6}$
2	Bal. load unbal. ex. Cap	150	40	150	50	150	60	0.9854	1.4334	$1.89 \cdot 10^{-6}$
3	Unbal. load bal. ex. Cap	100	50	150	50	200	50	0.9844	1.4558	$1.23 \cdot 10^{-6}$
4	Unbal. load unbal. ex. cap	100	40	150	50	200	60	0.9844	1.4486	$5.41 \cdot 10^{-6}$
5	Single-ph. load bal. ex. cap	100	50	–	50	–	50	0.9919	1.3524	$1.96 \cdot 10^{-6}$
6	Single-ph. load C2C ex. cap.	100	50	–	100	–	–	0.9907	1.2916	$2.90 \cdot 10^{-6}$
7	Steinmetz connection	100	–	–	120	–	–	0.9914	1.5025	$5.74 \cdot 10^{-6}$
8	Plain single-ph. connection	100	120	–	–	–	–	0.9893	1.6226	$2.54 \cdot 10^{-6}$

Table 14.3 Performance of the SEIG for the circuit configurations listed in Table 14.2

Case	Phase voltages			Phase currents			Voltage unbalanced factor	Electric power output	Efficiency η (%)
	V_A (p.u.)	V_B (p.u.)	V_C (p.u.)	I_A (p.u.)	I_B (p.u.)	I_C (p.u.)	VUF (V_n/V_p) (p.u.)	$P_{out.}$ (p.u.)	
1	1.0675	1.0675	1.0675	0.8342	0.8342	0.8342	0	1.0414	84.9735
2	1.0467	1.0800	1.0786	0.7710	0.8164	0.9314	0.0202	1.0434	84.7754
3	1.0533	1.0445	1.0740	0.8841	0.7568	0.8709	0.0166	1.1017	85.2691
4	1.0348	1.0591	1.0869	0.8307	0.7384	0.9677	0.0284	1.0999	84.9691
5	1.0970	1.0971	1.1614	0.8615	0.6461	0.9856	0.0387	0.5535	77.0888
6	1.1790	1.1787	1.0652	0.9067	1.2000	0.6433	0.0654	0.6386	75.4552
7	1.0418	1.1674	0.9723	0.4659	1.2217	0.8049	0.1097	0.4989	73.2919
8	1.1154	0.8660	1.0512	1.3242	0.6621	0.6621	0.1454	0.5707	73.3721

presented in Tables 14.2 and 14.3. It is considered that the generator is driven at a constant speed of 1.0 p.u. (i.e., 1,500 rpm) by a regulated prime mover.

14.7.1.2 Voltage control of the SEIG feeding a static load

Figure 14.9 shows the variation of the excitation capacitance required to maintain the constant terminal voltage of $V_t = 1$ p.u. with the change in the resistive load current at the constant generator speed of $v = 1.0$ p.u. The excitation capacitance has to vary continuously in order to keep the machine terminal voltage at a constant

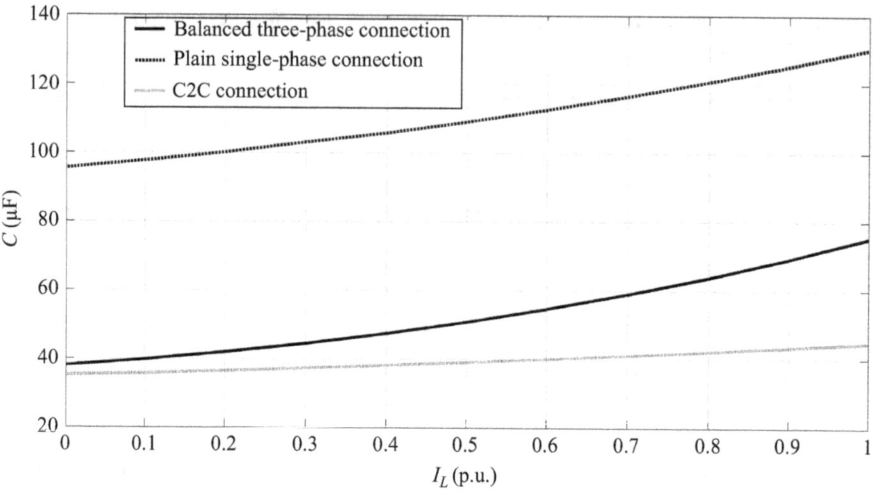

Figure 14.9 Excitation capacitance versus load current (p.f. = 1) for a constant terminal voltage of $V_t = 1$ (p.u.)

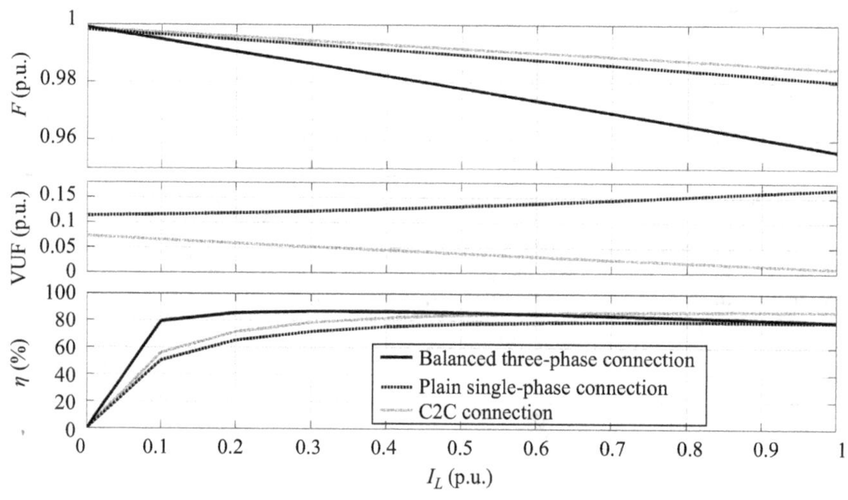

Figure 14.10 Frequency, voltage unbalance factor, and efficiency versus load current (p.f. = 1) under constant terminal voltage of $V_t = 1$ (p.u.)

value. The excitation capacitance requirement increases (i) from 38 μF at no load to 75 μF at 1 p.u. load current for the balanced three-phase connection, (ii) from 96 to 130 μF for the plain single-phase connection, and (iii) from 35.3 to 45 μF for the C2C connection.

The variations of frequency, voltage unbalance factor, and efficiency with load current at the corresponding values of the excitation capacitance for the constant terminal voltage of $V_t = 1$ p.u. are shown in Figure 14.10. With an increase in load, the frequency decreases from 0.9991 p.u. at no load to 0.9561 p.u. at the nominal load for the balanced three-phase connection, from 0.9982 to 0.9805 p.u. and from 0.9987 to 0.9848 p.u. for the plain single-phase, and C2C connection, respectively. The change in frequency is the highest in the case of the balanced three-phase connection due to the highest value of the load. The voltage unbalance factor increases from 0.1135 p.u. at no load to 0.1672 p.u. at the nominal load for the plain single-phase connection and decreases from 0.0732 p.u. at no load to 0.0092 p.u. at the nominal load for the C2C connection. It is obvious that the C2C connection has better performances than the plain single-phase connection.

14.7.1.3 Voltage control of the SEIG feeding a dynamic load

It is considered that the SEIG is loaded by a three-phase, 1.0-kW, four pole, delta-connected, squirrel cage induction motor, which is modeled through equivalent circuit with following parameters: $R_s = 0.0918$ (p.u.), $R_r = 0.0635$ (p.u.), $X_s = X_r = 0.2112$ (p.u.), $X_m = 1.95$ (p.u.), and

$$\underline{Z}_{im} = \frac{R_s}{F} + jX_s + \frac{(jR_m \cdot X_m/(R_m + jFX_m))((R_r/sF) + jX_r)}{(jR_m \cdot X_m/(R_m + jFX_m)) + ((R_r/sF) + jX_r)} \tag{14.69}$$

where $\underline{Z}_{im}$ is the per phase equivalent impedance of the induction motor (positive sequence), and s is the induction motor slip.

Figure 14.11 shows the variations of the capacitance requirement, frequency, and magnetizing reactance of the SEIG versus the induction motor slip at a constant generator speed for the constant terminal voltage of $V_t = 1$ p.u. The excitation

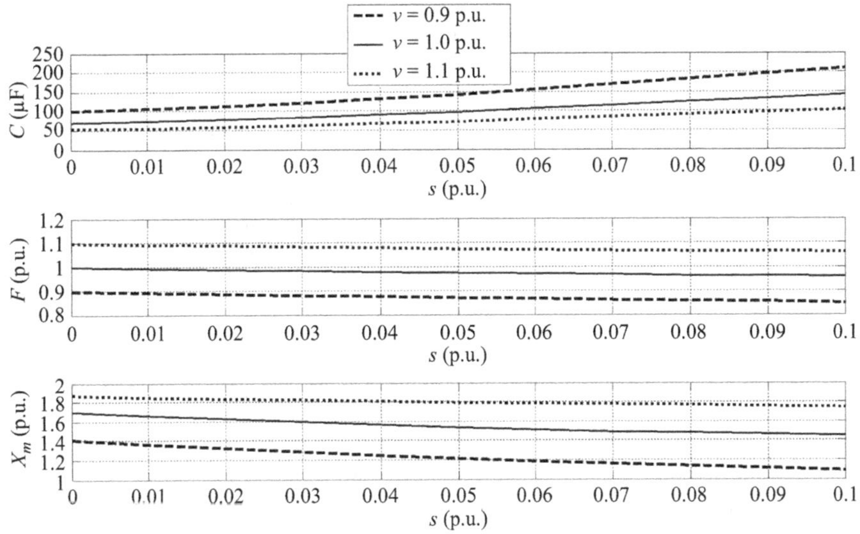

Figure 14.11 Excitation capacitance, frequency, and magnetization reactance versus induction motor slip at the constant SEIG speed for a constant terminal voltage of $V_t = 1$ p.u.

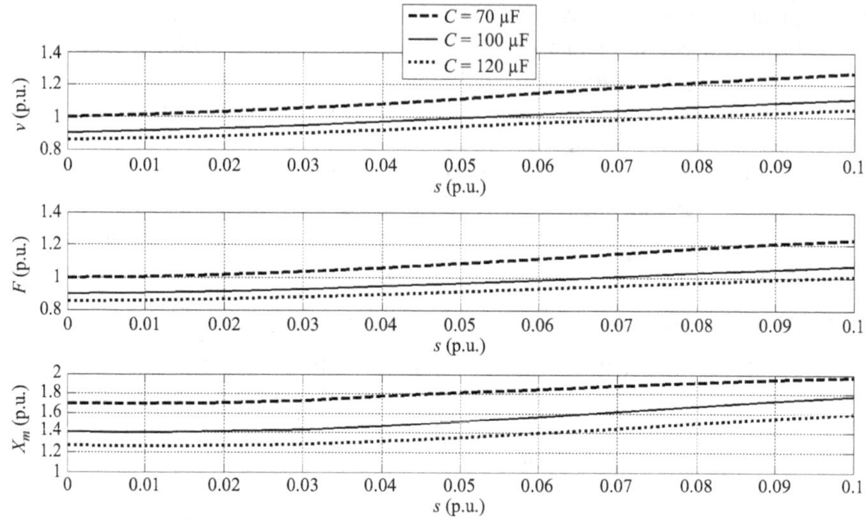

Figure 14.12 Generator speed, frequency, and magnetization reactance versus induction motor slip at the constant excitation capacitance for a constant terminal voltage of $V_t = 1$ p.u.

capacitance has to vary continuously in order to keep the machine terminal voltage at a constant value. The C value increases with the increase in induction motor slip. The higher speed of the SEIG corresponds to the lower value of the excitation capacitance for the same value of the slip. The values of F and X_m decrease with the increase in slip.

The variations of the speed, frequency, and magnetizing reactance of the SEIG versus the induction motor slip at the constant excitation capacitance for the constant terminal voltage of $V_t = 1$ p.u. are shown in Figure 14.12. The generator speed has to vary continuously to keep the machine terminal voltage at a constant value. The value of the speed v increases with the increase in the slip. The higher value of the excitation capacitance corresponds to the lower value of the generator speed for the same value of the slip. The values of F and X_m increase with the increase in the slip.

14.7.2 Steady-state analysis of parallel operated SEIGs

The proposed GA-based procedure for parallel operated SEIGs is tested on two three-phase, 380-V (line), four pole, delta connected, 1-kW induction machines, namely, a wound rotor machine (WRM) and a squirrel cage machine (SCM). The per-unit measured parameters of the two machines are given in Table 14.4.

The measured per-unit variations of the air-gap voltages $|\underline{E}_p|/F$ based on a 220 V with X_m for the two machines are approximated as follows over the practical range of interest.

Table 14.4 Machines parameters [28]

	R_s	R_r	X_s	X_r	X_{max}
WRM	0.09175	0.06354	0.2112	0.2112	2.00
SCM	0.16543	0.09324	0.1060	0.1060	2.84

For the WRM

$\left|\underline{E}_p/F\right| = 1.0007 - 0.1741X_m$ for $X_m \leq 1.3666$

$\left|\underline{E}_p/F\right| = 1.4298 - 0.4881X_m$ for $1.3666 < X_m \leq 1.7707$

$\left|\underline{E}_p/F\right| = 3.0192 - 1.3857X_m$ for $X_m > 1.7707$

For the SCM

$\left|\underline{E}_p/F\right| = 2.4408 - 0.6080X_m$ for $X_m \leq 2.3839$

$\left|\underline{E}_p/F\right| = 3.4592 - 1.0352X_m$ for $2.3839 < X_m \leq 2.8116$

$\left|\underline{E}_p/F\right| = 9.7841 - 3.2848X_m$ for $X_m > 2.8116$

14.7.2.1 Basic approach

For known values of the excitation capacitances (C_A, C_B, C_C), the load impedances ($\underline{Z}_{LA}$, $\underline{Z}_{LB}$, $\underline{Z}_{LC}$), the SEIGs speeds (v_1, v_2, v_3) and machines parameters, unknown values of the magnetizing reactance's ($X_{m1}, X_{m2}, \ldots, X_{mn}$), and frequency ($F$) can be determined by minimizing the following objective function:

$$\min Fobj = \min\{f_1 + f_2\} \tag{14.70}$$

$$\text{subject to: } \begin{cases} X_{mi}^{\min} \leq X_{mi} \leq X_{mi}^{\max} & i = 1, 2, \ldots, n \\ F^{\min} \leq F \leq F^{\max} \end{cases} \tag{14.71}$$

where f_1 and f_2 are defined by (14.50) and (14.51), respectively.

The steady-state analysis results of parallel operated SEIGs for the different connections of balanced/unbalanced resistive loads and excitation capacitances are presented in Tables 14.5–14.7. It is considered that the SEIGs have constant speeds of 1.0 p.u.

As shown in Table 14.7, WRM output power is less than the output of SCM. This is due to the rotor impedance ratio of the SEIGs. On the other hand, the reactive power of the WRM is significantly higher than the reactive power of the SCM due to smaller value of the magnetizing reactance. The result of it is that the phase current of the WRM is significantly higher than the phase current of the SCM, as can be seen in Table 14.6. This clearly shows the importance of the parameters of the individual SEIGs in parallel operation.

The poor generator performance, such as phase imbalance and poor efficiency, will result when single-phase loads are supplied by three-phase SEIGs (cases 6 and 7). This disadvantage is overcome to a large extent by the use of the C2C connection (case 8).

Table 14.5 Case studies for the different circuit configurations of parallel SEIGs

Case	Note	$\underline{Z}_A = R_{LA}// C_A$		$\underline{Z}_B = R_{LB}// C_B$		$\underline{Z}_C = R_{LC}// C_C$		Results			
		R_{LA} (Ω)	C_A (μF)	R_{LB} (Ω)	C_B (μF)	R_{LC} (Ω)	C_C (μF)	F (p.u.)	X_m (p.u.) *WRM*	X_m (p.u.) *SCM*	*Fobj*
1	Bal. load; bal. cap.	100	60	100	60	100	60	0.9578	0.9314	2.3093	$1.32 \cdot 10^{-7}$
2	No load; bal. cap.	–	40	–	40	–	40	0.9964	1.1409	2.4590	$3.82 \cdot 10^{-7}$
3	Bal. load; unbal. cap.	100	50	100	60	100	70	0.9576	0.9313	2.3087	$3.48 \cdot 10^{-7}$
4	Unbal. load; bal. cap.	50	60	100	60	150	60	0.9522	1.0554	2.3729	$4.43 \cdot 10^{-7}$
5	Unbal. load; unbal. cap.	50	50	100	60	150	70	0.9522	1.0334	2.3593	$7.49 \cdot 10^{-7}$
6	Single-ph. load; bal. cap.	33	60	–	60	–	60	0.9620	0.9417	2.3283	$2.96 \cdot 10^{-7}$
7	Plain single-ph. Connection	33	180	–	–	–	–	0.9386	1.8268	2.7899	$7.07 \cdot 10^{-7}$
8	C2C connection	33	60	–	120	–	–	0.9551	0.9356	2.3043	$8.99 \cdot 10^{-7}$

14.7.2.2 Controlling terminal voltage by excitation capacitance

It is well known that the voltage control of the SEIG can be achieved by a change in the excitation capacitance for constant values of the speeds and load impedances. In this case, the multiobjective function can be expressed as

$$\min Fobj = \min\{f_1 + f_2 + f_3\} \tag{14.72}$$

$$\text{subject to:} \begin{cases} X_{mi}^{\min} \le X_{mi} \le X_{mi}^{\max} \; i = 1, 2, \ldots, n \\ F^{\min} \le F \le F^{\max} \\ C^{\min} \le C \le C^{\max} \end{cases} \tag{14.73}$$

where f_1, f_2, and f_3 are defined by (14.50)–(14.52), respectively.

This approach has been applied for C2C connection. Figure 14.13 shows the variation of excitation capacitance C required to maintain the constant terminal voltage of $V_t = 1$ p.u. with the change in the resistive load P_L at the constant generators speeds v for the C2C connection. The excitation capacitance has to vary continuously in order to keep the terminal voltage at a constant value. From this figure, it is clear that C increases with P_L for fixed values of V_t and P_L, C increases as v is decreased. The variations of frequency F and output powers of generators P_{Gen} with load P_L at the corresponding values of the excitation capacitance are shown in Figure 14.14. With an increase in load, the frequency decreases. The range of change in frequency is approximately the same for all three speeds. The power load must be equal to the sum of output powers of parallel operated SEIGs. Figure 14.14 clearly shows that the power output of SCM generator is greater than the output power of WRM generator, although they have the same rated power. For fixed values of V_t and P_L, the ratio of the output powers increases as v decreases.

Table 14.6 *Phase voltage, voltage unbalance factor, and currents for the circuit configurations are listed in Table 14.5*

Cs.	V_A (p.u.)	V_B (p.u.)	V_C (p.u.)	VUF (V_n/V_p)	I_A (p.u.)		I_B (p.u.)		I_C (p.u.)		$I_{\ell A}$ (p.u.)	$I_{\ell B}$ (p.u.)	$I_{\ell C}$ (p.u.)
					WRM	*SCM*	*WRM*	*SCM*	*WRM*	*SCM*			
1	0.9715	0.9715	0.9715	0	1.1194	0.6645	1.1194	0.6645	1.1194	0.6645	1.7584	1.7584	1.7584
2	0.9460	0.9460	0.9460	0	0.7050	0.3734	0.7050	0.3734	0.7050	0.3734	1.0782	1.0782	1.0782
3	0.9437	0.9946	0.9764	0.0305	1.0464	0.6257	1.1458	0.6172	1.1756	0.7696	1.4773	1.8000	2.0130
4	0.8881	0.9017	0.9718	0.0572	1.0665	0.8260	0.9398	0.5091	1.1753	0.7375	1.9802	1.6244	1.6622
5	0.8700	0.9321	0.9844	0.0712	1.0193	0.8116	0.9822	0.4576	1.2452	0.8401	1.7735	1.6791	1.9414
6	0.8636	0.9596	1.1070	0.1486	1.0198	0.9729	0.8662	0.1190	1.4382	0.9714	2.4440	1.5840	1.8274
7	0.5672	0.3412	0.6403	0.3439	0.9977	1.0313	0.3530	0.6793	0.6779	0.3959	3.0350	0	0
8	0.9939	0.9733	0.9364	0.0346	1.1665	0.6327	1.2020	0.8065	1.0551	0.6468	2.8059	3.1905	0

Table 14.7 Output powers and efficiency for the circuit configurations are listed in Table 14.5

Case	Power load P_L (p.u.)	P_{Gen} (p.u.)		Efficiency		Q_i (p.u.)	
		WRM	*SCM*	*WRM*	*SCM*	*WRM*	*SCM*
1	0.7159	0.3239	0.3920	70.6876	80.7330	−1.0381	−0.5129
2	0	−0.0092	0.0092	25.0589	28.3383	−0.6669	−0.3531
3	0.7164	0.3247	0.3917	70.5678	80.3282	−1.0416	−0.5159
4	0.7637	0.3620	0.4017	73.9621	79.0921	−0.9075	−0.4793
5	0.7659	0.3617	0.4042	73.2110	78.3864	−0.9388	−0.4937
6	0.5715	0.2756	0.2958	67.3593	70.9987	−1.0499	−0.5414
7	0.2465	0.1541	0.0924	71.4587	46.8577	−0.3032	−0.2150
8	0.7569	0.3430	0.4139	70.7812	79.8785	−1.0504	−0.5243

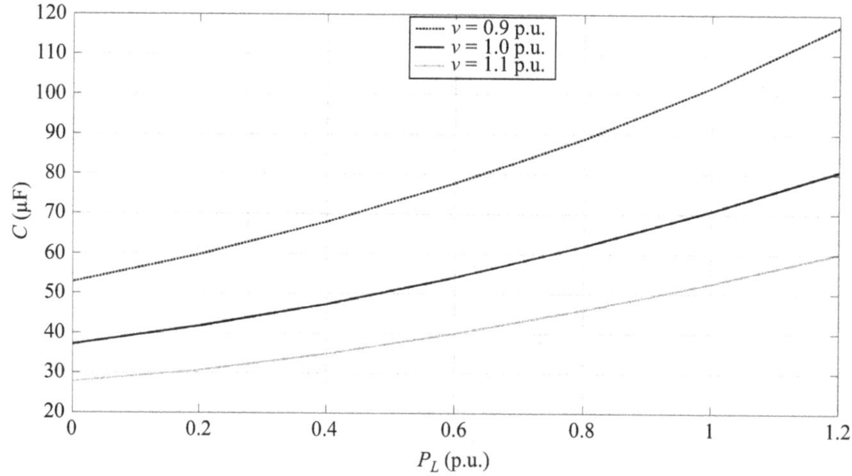

Figure 14.13 Variation of C with P_L and v for a constant terminal voltage of $V_t = 1$ p.u.

14.7.2.3 Controlling terminal voltage by SEIGs speeds

For a fixed value of *C*, it may be possible to achieve a certain degree of voltage regulation by varying the speed of one or more machines [25,26]. In this situation, two cases can be analyzed: (i) all SEIGs have common but variable speeds and (ii) the speed of only one SEIG is varied while keeping the speeds of the remaining $(n-1)$ SEIGs fixed. There can also be used multiobjective function.

$$\min Fobj = \min\{f_1 + f_2 + f_3\} \tag{14.74}$$

$$\text{subject to:} \begin{cases} X_{mi}^{\min} \le X_{mi} \le X_{mi}^{\max} & i = 1, 2, \ldots, n \\ F^{\min} \le F \le F^{\max} & \\ v_i^{\min} \le v_i \le v_i^{\max} & i = 1, 2, \ldots, n \end{cases} \tag{14.75}$$

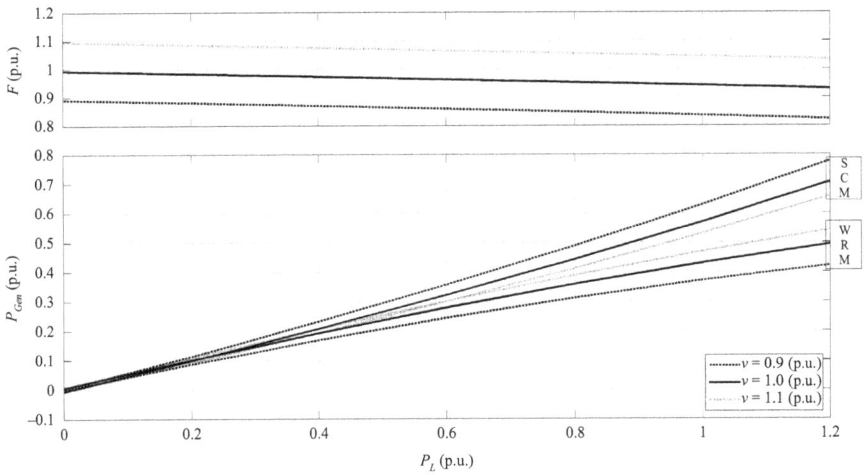

Figure 14.14 Corresponding variations of F and P_{Gen} with P_L and v in accordance with Figure 14.13

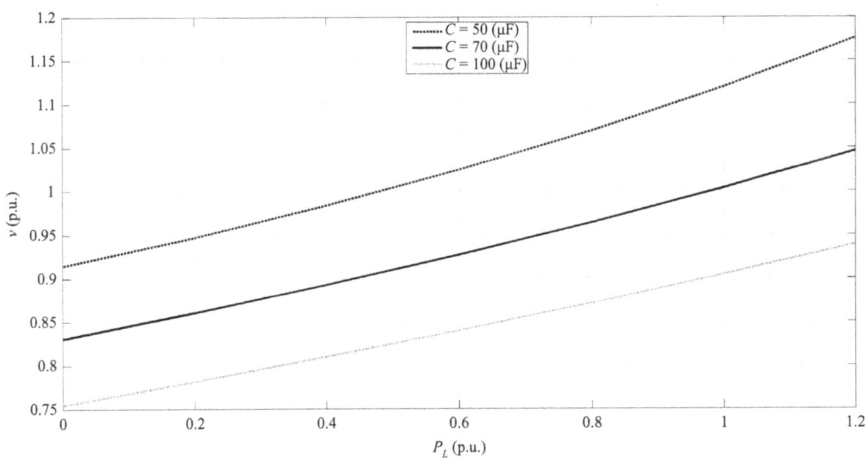

Figure 14.15 Variation of v with P_L and C for a constant terminal voltage of $V_t = 1$ p.u.

1. ***C2C connection, both SEIGs have common but variable speeds***
Figure 14.15 shows the variation of SEIGs speeds v required to maintain the constant terminal voltage of $V_t = 1$ p.u. with the change in the resistive load P_L at the constant excitation capacitance C. The SEIGs speeds have to vary continuously in order to keep the terminal voltage at a constant value. From this figure, it is clear that v increases as P_L increases.. For fixed values of V_t and P_L, v decreases as C is increased. The corresponding variations of frequency F and

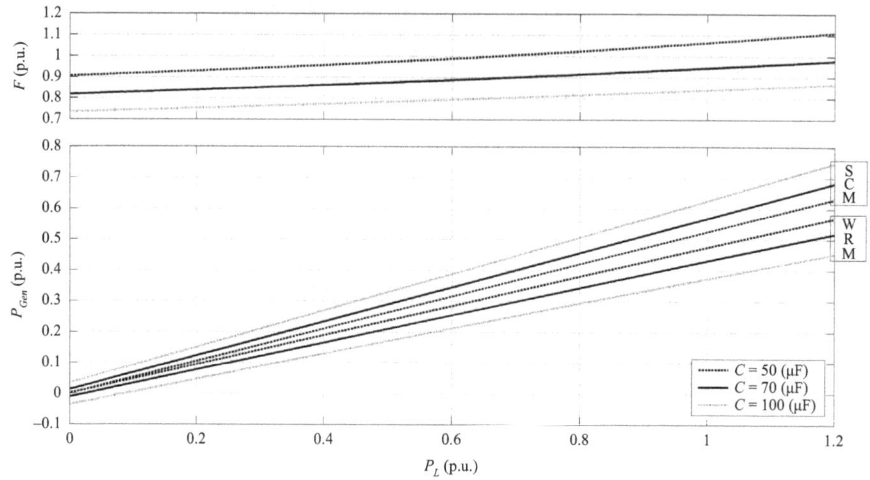

Figure 14.16 Corresponding variations of F and P_{Gen} with P_L and C in accordance with Figure 14.15

output powers of generators P_{Gen}, with load P_L are shown in Figure 14.16. With an increase in load, the frequency increases. The range of change in frequency is approximately the same for all three excitation capacitances. As shown in Figure 14.16, the power output of SCM generator is higher than the output power of WRM generator. The ratio of output powers SCM/WRM decreases as P_L increases. For fixed values of V_t and P_L, the ratio of the output powers increases as C is increased.

2. ***C2C connection, v_{SCM} is variable and v_{WRM} is kept constant***
   For the C2C connection SEIGs parallel system of an SCM and a WRM, voltage control by the speed of SCM is illustrated in Figures 14.17 and 14.18. Based on these results, it is clear that voltage control by the speed of only one machine is generally possible, but very limited.

   Figure 14.17 shows the variation of the speed of the SCM (v_{SCM}) required to maintain the constant terminal voltage of $V_t = 1$ p.u. with the change in the resistive load P_L at the constant excitation capacitance C and the constant speed of the WRM (v_{WRM}). The v_{SCM} has to vary continuously in order to keep the V_t at a constant value. From this figure, it is clear that the v_{SCM} control characteristics largely depend on values of P_L, C, and v_{WRM}. Generally, the control characteristics improve as v_{WRM} and C is increased. However, as can be seen from Figure 14.18, in cases $v_{WRM} = 0.9$ p.u. and $v_{WRM} = 1$ p.u., the WRM output power is negative in over range of P_L, which means that the WRM operate as motor. For $v_{WRM} = 1.1$ p.u., the SCM operate as motor or generator depending on P_L and C. For fixed value of P_L, the operating mode of SCM greatly depends upon the C. In addition to the values of P_L, C, and v_{WRM}, the control characteristics is highly dependent on the SEIGs parameters.

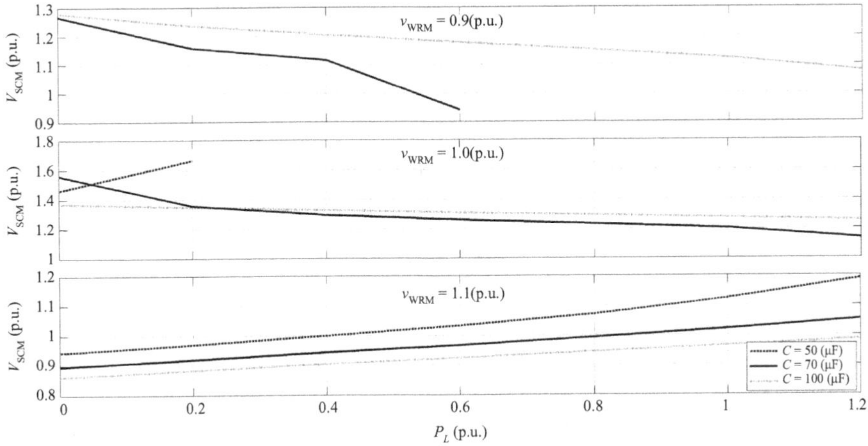

Figure 14.17 Variation of v_{SCM} with P_L, C, and v_{WRM} for a constant terminal voltage of $V_t = 1$ p.u.

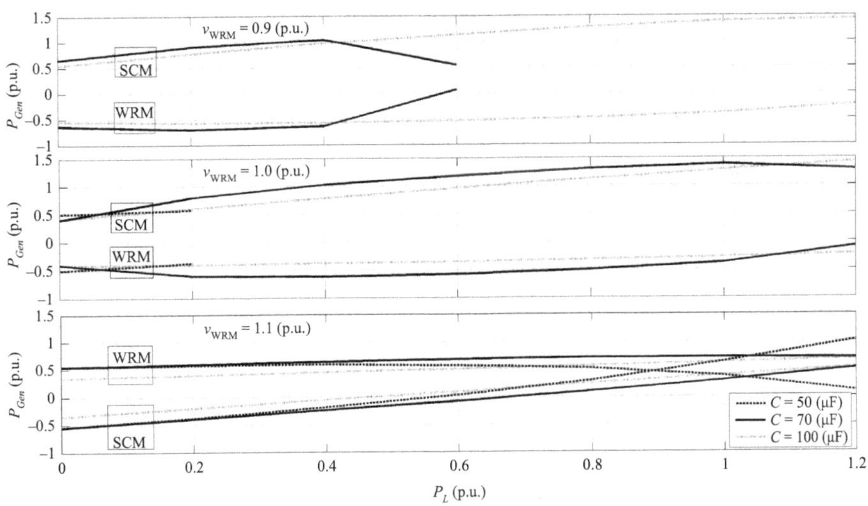

Figure 14.18 Corresponding variations of SEIGs power outputs with P_L, C, and v_{WRM} in accordance with Figure 14.17

14.8 Conclusion

In this chapter, a GA-based procedure for general steady-state analysis of three-phase SEIGs feeding balanced/unbalanced and static/dynamic loads has been presented. The complex three-phase generator-excitation capacitance-load system is transformed to a simple equivalent passive circuit by using the symmetrical component theory. The MATLAB realization of the GA has been applied to the computation of the unknowns by minimizing the total impedance module of the equivalent passive circuit. The proposed methodology is general since it can be applied to different configurations of excitation capacitances and consumers. Application of GA gives great freedom in the choice of objective function and control variables in the analysis of different modes of AG, which is the main advantage of the proposed procedure over other methodologies to deal with this problem. The voltage control of the SEIG feeding balanced/unbalanced and static/dynamic loads has been investigated. It is demonstrated that the voltage control of the SEIG can be achieved by changing the excitation capacitance and generator speed.

The voltage control of the parallel operated SEIGs feeding an unbalanced load has been investigated as well. The methodology for voltage control has been implemented for C2C connection under single-phase load. From these results, it is clear that the voltage control can be achieved by varying the excitation capacitance or by varying the common speeds of the SEIGs. The voltage control by varying the speed of only one SEIG is generally possible, but very limited. In addition to the excitation capacitance and speeds of the SEIGs, the voltage control characteristic is highly dependent on the induction machine parameters.

The methodology proposed in this study can be exploited in planning, development, and operation of autonomous small-scale SEIGs-based microgrids.

References

[1] Simoes M.G., Farret F.A. *Renewable energy systems – design and analysis with induction generators*. New York: CRC Press; 2004

[2] Singh B., Murthy S.S, Chilipi R.R., Madishetti S., Bhuvaneswari G. 'Static synchronous compensator-variable frequency drive for voltage and frequency control of small-hydro driven self-excited induction generators system'. *IET Generation, Transmission & Distribution*. 2014;**8**(9):1528–38

[3] Thomsen B., Guerrero J.M., Thogersen P. 'Faroe islands wind-powered space heating microgrid using self-excited 220 kW induction generator'. *IEEE Transactions on Sustainable Energy*. 2014;**5**(4):1361–6

[4] Ambia M.N., Al-Durra A., Caruana C., Muyeen S.M. 'Islanding operation of hybrid microgrids with integration of wind driven cage induction generators'. *Sustainable Energy Technologies and Assessments*. 2016;**13**:68–75

[5] Chae W.K., Lee H.J., Hwang S.W., Song I.K., Kim J.E. 'Isolated microgrid's voltage and frequency characteristic with induction generator based wind turbine'. *Smart Grid and Renewable Energy*. 2014;**5**:180–92

[6] Jain M., Gupta S., Masand D., Agnihotri G. 'Soft computing technique-based voltage/frequency controller for self-excited induction generator-based microgrid'. *Journal of Circuit, Systems, and Computers*. 2016;**25**(2):1–29

[7] Shridhar L., Singh B., Jha C.S., Singh B.P. 'Analysis of self excited induction generator feeding induction motor'. *IEEE Transactions on Energy Conversion*. 1994;**9**(2):390–6

[8] Chan T.F. 'Steady-state analysis of self-excited induction generators'. *IEEE Transactions on Energy Conversion*. 1994;**9**(2):288–96

[9] Chan T.F. 'Self-excited induction generators driven by regulated and unregulated turbines'. *IEEE Transactions on Energy Conversion*. 1996;**11**(2):338–43

[10] Chakraborty C., Bhadra S.N., Chattopadhyay A.K. 'Excitation requirements for stand-alone three-phase induction generator'. *IEEE Transactions on Energy Conversion*. 1998;**13**(4):358–65

[11] Alghuwainem S.M. 'Steady-state analysis of a self-excited induction generator including transformer saturation'. *IEEE Transactions on Energy Conversion*. 1999;**14**(3):667–72

[12] Alolah A.L., Alkanhal M.A. 'Optimization-based steady state analysis of three phase self-excited induction generator'. *IEEE Transactions on Energy Conversion*. 2000;**15**(1):61–5

[13] Nigim K.A., Salama M.M.A., Kazerani K. 'Identifying machine parameters influencing the operation of the self-excited induction generator'. *Electric Power Systems Research*. 2004;**69**:123–8

[14] Joshi D., Sandhu K.S. 'Performance analysis of three-phase self-excited induction generator using genetic algorithm'. *Electric Power Components and Systems*. 2006;**34**:461–70

[15] Haque M.H. 'Comparison of steady state characteristic of shunt, short-shunt and long-shunt induction generators'. *Electric Power System Research*. 2009;**79**:1446–53

[16] Haque M.H. 'A Novel method of evaluating performance characteristics of a self-excited induction generator'. *IEEE Transactions on Energy Conversion*. 2009;**24**(2):358–65

[17] Kuo S.C., Wang L. 'Steady-state performance of a self-excited induction generator feeding an induction motor'. *Electric Power Components and Systems*. 2002;**30**:581–93

[18] Alolah A.I., Alkanhal M.A. 'Analysis of three phase self-excited induction generator under static and dynamic loads'. *Proceedings of the Electric Machines & Drives Conference*; Antalya, Turkey, May 2007. IEEE; 2007. pp. 1783–6

[19] Chan T.F., Lai L.L. 'Steady-state analysis and performance of a stand-alone three phase induction generator with asymmetrically connected load impedances and excitation capacitances'. *IEEE Transaction on Energy Conversion*. 2001;**16**(4):327–33

[20] Murthy S.S., Singh B., Gupta S., Gulati B.M. 'General steady-state analysis of three-phase self-excited induction generator feeding three-phase unbalanced load/single-phase load for stand-alone applications'. *IEE Proceedings—Generation, Transmission and Distribution*. 2003;**150**(1):49–55

[21] Alolah A.I., Alkanhal M.A. 'Excitation requirements of three phase self-excited induction generator under single phase loading with minimum

unbalance'. *Power Engineering Society Winter Meeting*; Singapore, Singapore, Jan 2000. IEEE; 2002. pp. 257–9

[22] Kumaresan N. 'Analysis and control of three-phase self-excited induction generators supplying single-phase AC and DC loads'. *IEE Proceedings Electric Power Applications*. 2005;**152**(3):739–47

[23] Wang Y.J. 'Analysis of a self-excited induction generator supplying unbalanced loads'. *2004 International Conference on Power System Technology—POWERCON 2004*; Singapore, Singapore, Nov 2004. IEEE; 2005. pp. 1457–62

[24] Idjdarene K., Rekioua D., Rekioua T., Tounzi A. 'Performance of an isolated induction generator under unbalanced loads'. *IEEE Transactions on Energy Conversion*. 2010;**25**(2):303–11

[25] Radosavljević J., Klimenta D., Jevtić M. 'A genetic algorithm-based approach for a general steady-state analysis of three-phase self-excited induction generator'. *Revue Roumaine des Sciences Techniques, Série Électrotechnique et Énergétique*. 2012;**57**(1):10–9

[26] Watson D.B., Milner I.P. 'Autonomous and parallel operation of self-excited induction generators'. *International Journal of Electrical Engineering Education*. 1985;**22**:365–74

[27] Al-Bahrani A.H., Malik N.H. 'Voltage control of parallel operated self-excited induction generators'. *IEEE Transactions on Energy Conversion*. 1993;**8**:236–42

[28] Al-Bahrani A.H., Malik N.H. 'Steady state analysis of parallel operated self-excited induction generators'. *IEE Proceedings C—Generation, Transmission and Distribution*. 1993;**140**(1):49–55

[29] Lee C.-H., Wang L. 'A novel analysis of parallel operated self-excited induction generators'. *IEEE Transactions on Energy Conversion*. 1998; **13**(2):117–23

[30] Chakraborty C., Bhadra S.N., Chattopadhyay A.K. 'Analysis of parallel-operated self-excited induction generators'. *IEEE Transactions on Energy Conversion*. 1999;**14**(2):209–16

[31] Sandhu K.S. 'Steady-state analysis of multimachine system with self-excited induction generators'. *Electric Power Components and Systems*. 2006;**34**:483–96

[32] Radosavljević J., Klimenta D., Jevtić M. 'Steady-state analysis of parallel-operated self-excited induction generators supplying an unbalanced load'. *Journal of Electrical Engineering*. 2012;**63**(4):213–23

[33] Jain M., Gupta S., Angihotri G. 'Design of voltage controller for parallel operated self excited induction generator-microgrid'. *Proceedings of International Conference on Energy, Automation, and Signal (ICEAS)*; Bhubaneswar, Odisha, India, Dec 2011. IEEE; 2012. pp. 1–6

[34] Lee K.Y., El-Sharkawi M.A. *Tutorial on modern heuristic optimization techniques with applications to power systems. IEEE Power Engineering Society 2002*; IEEE Catalog Number 02TP160; 2002

Index

acceleration coefficients 81, 90, 92–4
active power loss, minimization of 189, 191
adaptive acceleration coefficients
  particle swarm optimization (PSO) with 92–3
adaptive differential evolution (ADE) 450
adaptive gbest-guided GSA 131
  improving the exploitation of 131–3
  slow exploitation of 131
adaptive inertia weight
  particle swarm optimization (PSO) with 91–4
adaptive modified firefly optimization algorithm (AMFA) 410
adaptive particle swarm optimization (APSO) 93, 95, 100, 366
agent 114
ant colony optimization (ACO) algorithm 9, 182, 235, 268, 367
ant colony optimization (ACO)–artificial bee colony (ABC) algorithm 367
arithmetic crossover 52–3
arithmetic crossover operation (ACHS) 268
artificial bee colony (ABC) algorithm 9–10, 182, 235, 268, 367, 450

backtracking search optimization algorithm (BSA) 8, 182, 367
bacterial colony foraging optimization (BCFO) system 17–18, 367
bacterial foraging algorithm (BFA) 184
BAT search algorithm 185
binary encoding 42
binary gravitational search algorithm (BGSA) 125–6, 139, 149
binary integer programing 450
biogeography-based optimization (BBO) 18–19, 184, 235, 246, 268, 450
bird flocking 9, 76, 194, 417
black hole (BH) algorithm 13–14, 147
black-hole optimization (BH) 183
B-loss matrices 271
Brownian-like random-walk movement 24
Broyden–Fletcher–Goldfarb–Shanno (BFGS) quasi-Newton updating method 457
bus location coefficients (*BLCs*) 368–9, 371–2, 381–4, 391–2

C2C connection 492–3, 499–500
ceedgui (educational software) 277, 286–92
center of inertia (COI) 193
chaotic krill herd algorithm (CKHA) 246, 248
chemotaxis 17–18
classical heuristic methods 2
classical optimization methods 1, 98, 149, 378, 485
cloud theory 64, 366
clustered-gravitational search algorithm 139–40
colliding bodies optimization (CBO) algorithm 12–13, 183

combined economic and emission dispatch (CEED) 60–1, 98, 146–7, 267
  ceedgui (educational software) 286–92
  gravitational search algorithm (GSA) 274–5
  hybridization of PSO and GSA 275
  particle swarm optimization (PSO) 273–4
  problem formulation 269
    emission function 270
    fuel cost function 270
    generation capacity constraint 271–2
    power balance constraint 271
    slack generator calculation 272–3
  PSOGSA implementation to the CEED problem 275–6
  simulation results 277–85
completely connected topology 88
comprehensive learning PSO (CLPSO) 97
constriction coefficients 81–2
control variables, defined 177
coordination time interval (CTI), minimization of 454–5
Coriolis force 11
criticism of metaheuristics 31–3
crossover 39, 50
  arithmetic 52–3
  heuristic 53
  multipoint 51
  one-point 51
  shuffle 52
  uniform 51–2
Cuckoo search (CS) algorithm 15, 183, 268, 367
cumulative density function (CDF) 311–13, 420–1
current optimum opposition-based gravitational search algorithm (COOGSA) 130
current transformer (CT) 452
decay curve 118
decoding 42, 44
description of metaheuristics 2–4
deterministic optimal power flow 299–300
  constraints 301–2
  objective function 300–1
diesel electric generators (DEGs) 297
diesel generator 303, 415
differential evolution (DE) 7, 181, 246, 268, 450
differential search algorithm (DSA) 24, 184–5, 235
directional overcurrent relays (DOCRs), optimal coordination of 449
  ocdocr (educational program package) 468–70
  optimal coordination of 65
  problem formulation 450
    coordination criteria 454–5
    limits of relay operation time 453
    limits of the settings 452–3
    modification of objective function for minimization of CTI 454–5
    objective function 451–2
  simulation results 459
    statistical evaluation of the results 466–7
    test systems 459–66
  solution method 455
    gravitational search algorithm (GSA) 455–7
    hybrid GSA–SQP algorithm 457–9
    sequential quadratic programing (SQP) 457
distributed generation (DG) 338, 363, 407, 451
  bids calculation 414
    diesel generator 415–16
    microturbine and fuel cell 414–15
  genetic algorithm (GA) 377
    optimal DG placement and sizing by using GA 378–81

opsdg (educational program package) 399–400
partial search of variants 372
optimal DG placement by using partial search of variants 373–5
optimal DG sizing by using partial search of variants 375–7
penetration 297
preliminary locations of 369–72
simulation results 381
distribution network Zajecar 391–9
IEEE 31-bus system 381–91
units modeling, for OPF 302
diesel generator 303
electric grid 306
fuel cell 304
microturbine 304
mini hydropower plants 306
photovoltaic 305–6
wind turbine 304–5
distribution management system (DMS) 407–8
distribution network operators (DNO) 331, 368
distribution networks
optimal placement and sizing of distributed generation in 63–4
optimal power flow in 61–3
power-flow calculation in 165, 170
backward/forward sweep power-flow algorithm 170–4
pfdngui (power-flow software) 174–5
Stara Planina 346–7
Zajecar 391–9
distribution system operator (DSO) 297

economic dispatch (ED) problem 145, 267
solution 98–9
educational program package
ocdocr program 468–70
opsdg program 399–400
ORPD program 257–63
efficient evolutionary algorithm (EEA)-based OPF 59, 179
electric grid 306
electromotive force (EMF) 475–6
elitist selection 49
emigration process 18
emission function 270
energy and operation management (EOM) 148
of microgrids: *see* microgrids (MGs), energy and operation management (EOM) of
energy distribution company (EDC) 381
energy-storage devices 412
evolution 39
evolutionary optimization techniques 377
evolutionary programing (EP) 7–8, 181–2, 268, 339
exact optimization methods 1–2, 5
excitation capacitance 476, 486, 488, 494–6
exploitation 3
exploration 3

fast decoupled power-flow method 163–4
firefly algorithm (FA) 14–15, 183
fitness 5, 44, 77, 114, 194
linear scaling 46
relative fitness 45–6
flexible AC transmission systems (FACTS) devices 59, 145, 179, 183
fuel cell (FC) 297, 304, 407, 414–15
fuel cell power plants (FCPPs) 148, 367
fuel cost 181
minimization of 188
fuel cost and active power loss, minimization of 191, 207
fuel cost and voltage deviation, minimization of 191
fuel-cost and voltage-stability index, minimization of 191
fuel cost function 270

fuzzy adaptive PSO (FPSO) 91–2, 99
fuzzy logic 180
fuzzy logic-controlled genetic algorithm 268

Gaussian distribution 15, 92
Gauss–Seidel method 159–60
gbest 9, 76–82, 88–93, 131–5, 141, 182, 194–6, 199–200, 238, 240–1, 273–6, 417–19
generation capacity constraint 271–2
genetic algorithm (GA) 6, 39, 90, 178, 181, 307–8, 339, 365–6, 377–81, 409, 450, 476–7, 485–6
  application to parallel operated SEIGs 488
    control variables 488
    objective function 488
    performance equations 489–90
  application to power system problems 57
    combined economic and emission dispatch (CEED) 60–1
    optimal coordination of directional overcurrent relays 65
    optimal energy and operation management of microgrids 64–5
    optimal placement and sizing of distributed generation in distribution networks 63–4
    optimal power flow 57–60
    optimal power flow in distribution networks 61–3
    optimal reactive power dispatch (ORPD) 60
    self-excited induction generator (SEIGs), steady-state analysis of 66
  application to SEIG 486–7
  basic structure of 40–1
  control parameters 54–5
  crossover 50
    arithmetic crossover 52–3
    heuristic crossover 53
    multipoint crossover 51
    one-point crossover 51
    shuffle crossover 52
    uniform crossover 51–2
  fitness function 44
    linear scaling 46
    relative fitness 45–6
  GA-based tabu search (TS) method 366
  implementation, to the deterministic OPF problem 308
  MATLAB realization of 502
  multiobjective optimization using 55–7
  mutation 53–4
  population size and initial population 44
  representation of individuals (encoding) 41
    binary encoding 42
    Gray coding 43
    real-value encoding 43
  selection 46
    elitist selection 49
    k-tournament selection schemes 50
    linear ranking selection 49
    simple selection 48
    simple tournament selection 50
    stochastic universal sampling 48
global best: *see* gbest
glowworm swarm optimization (GSO) algorithm 25–6
gradient-based method 450
Gram–Charlier expansion series 299, 315, 423, 430, 444
gravitational constant 11, 117–18
gravitational search algorithm (GSA) 10–11, 113, 181, 196–8, 235, 238–9, 255, 268, 274–5, 298, 309–10, 339, 367, 410, 450, 455–7, 459, 471
  adaptive gbest-guided GSA 131
    improving the exploitation of 131–3
    slow exploitation of 131

applications, to power system problems 145
economic dispatch using GSA 146–7
optimal coordination of overcurrent relays 149
optimal DG placement and sizing in distribution networks 147–8
optimal energy and operation management of microgrids 148–9
optimal power flow 145–6
optimal power flow in distribution networks 147
optimal reactive power dispatch 146
binary version of 125–6
Clustered-gravitational search algorithm 139–40
convergence of 124
example usage of 123–4
hybrid PSO and GSA algorithm 140–4
implementation, to the deterministic OPF problem 310–11
interaction of agents in 116
MATLAB code of 120–2
modified GSA 126–8
nondominated sorting GSA 135
acceleration of agents, updating 137–8
external archive, updating 136–7
list of moving agents, updating 137
mass of moving agents, updating 137
mutation operator use 138
update and mutate position of agents 138–9
opposition-based GSA 128–9
current optimum opposition-based GSA 129–31
original GSA, description of 115, 117
parameters of 117–19
self-adaptive GSA 133–5
Gray coding 43
gray wolf optimizer (GWO) algorithm 14, 183

habitat suitability index (HSI) 18–19
Hamming distance 42–3, 126
harmony memory accepting rate 22
harmony search (HS) 22–3, 185, 268
harmony search algorithm (HSA) 235
heuristic crossover 53
heuristic optimization techniques 1, 26, 298, 476
heuristics, defined 2
high-voltage direct current (HVDC) links 179
Holland's algorithm 39
hybrid algorithms 185, 268, 450
hybrid GSA–SQP algorithm 255–6, 457
application to ORPD problem 256–7
implementation of 458–9
simulation results of 257
hybrid PSO (HPSO) method 90–1, 96
hybrid PSO and GSA (hybrid PSOGSA) algorithm 140–4, 146, 199, 273
application of, to OPF problem 199
gravitational search algorithm 274–5
hybridization of PSO and GSA 275
implementation, to the CEED problem 275–6
optimal reactive power dispatch (ORPD) using 238
gravitational search algorithm 238–9
hybridization of PSO and GSA 240
particle swarm optimization (PSO) 238
particle swarm optimization 273–4

IEEE 30-bus test system 200, 242–6, 465
comparative study 208–11

critical analysis of 246–9
minimization of fuel cost (case) 201–5
and active power loss (case) 207
with considering the valve-point effect (case) 205–6
and voltage deviation (case) 207
and voltage-stability index (case) 207–8
IEEE 31-bus system 355, 381–91
IEEE 118-bus test system 211–13, 249–53
critical analysis of 254–5
immigration process 18
imperialist competitive algorithm (ICA) 23–4, 184–5, 368
improved particle swarm optimization (IPSO) 94, 96–8, 101, 180
individual best: *see* pbest
induction generators (IGs) 475
inertia weight 81, 90
informative differential evolution (IDE) 450

Jacobian matrix 369–70
JANA algorithm 28–9
Jaya algorithm 27–8, 185

Kbest 11, 117–18, 126, 139, 197, 239, 274, 309, 456
krill herd algorithm (KHA) 16–17, 183, 235
k-tournament selection schemes 47, 50

lbest 88–9
league championship algorithm (LCA) 19–20, 184
Lévy flight behavior 15
L-index 190, 207
linear programing (LP) 365, 450
linear ranking selection 49
linear scaling 46
local controllers (LCs) 408–9, 411
low-voltage (LV) distribution network 337
low-voltage (LV) grid-connected MG 411

MATLAB code 380
of gravitational search algorithm 120–2
of particle swarm optimization 83–5
medium-voltage (MV) distribution network 337
metaheuristic optimization methods 2–3, 353
metahopt 33–4
microgrids (MGs), energy and operation management (EOM) of 64–5, 101, 148, 407
probabilistic EOM of 419
procedure for solving 422–3
statistical characterization of the input random variables 420–1
statistical evaluation of the output variables 421–2
problem formulation of EOM 411
constraints 412–14
distributed generation bids calculation 414–16
objective function 412
simulation results 424
MATLAB program eom used for deterministic EOM 437–43
microgrid MG1 424–30
microgrid MG2 430–6
solution method 416
application of PSO to EOM 418–19
particle swarm optimization (PSO) 417–18
microgrid system central controller (MGCC) 408–9, 411
microhydro power plants (MHPs) 346
microturbine (MT) 297, 304, 414–15
mine blast algorithm 20–1
mini hydropower plants (MHP) 297, 306, 391

modified adaptive θ-particle swarm optimization (MA θ-PSO) 268
modified bacterial foraging algorithm (MBFA) 268
modified differential evolution (MDE) 450
modified objective function (MOF) 455
Monte Carlo simulation 101, 313, 410, 421
moth swarm algorithm (MSA) 15–16, 183
multiobjective differential evolution (MODE) 268
multiobjective evolutionary programing (MOEP) 268
multiobjective optimization problem (MOP) 55, 63
multiobjective optimization using GA 55–7
multiobjective particle swarm optimization (MOPSO) 268
multiobjective performance index 63, 365
multipoint crossover 51
mutation 8, 39, 53–4, 485

nature-inspired metaheuristics 3, 113
negative loads 340
network power losses 366
Newton gravitation theory 197, 456
Newtonian gravity 10, 115, 274, 455–6
Newton–Raphson method 161–3, 246
Newton's second law of motion 11–12
'new' algorithm, creating 28–31
niched Pareto genetic algorithm (NPGA) 268
nominal operating cell temperature (NOCT) 306, 416
nondominated sorting genetic algorithm (NSGA) 55–7, 135–9
  flowchart of 56
  NSGA-II approach 57, 59, 64–5, 179, 268, 366, 450
nondominated sorting gravitational search algorithm 135
  acceleration of agents, updating 137–8
  external archive, updating 136–7
  list of moving agents, updating 137
  mass of moving agents, updating 137
  mutation operator use 138
  update and mutate position of agents 138–9
nonlinear programing (NLP) problem 255, 450, 457
numerical Newton–Raphson/ polynomial method 484

ocdocr program 468–70
off-voltage tap-changing transformers (OVTCTs) 337
one-point crossover 51
onlookers 16
operating times (OTs) 452
opfdngui (solution software) 324–7
opposition-based chaotic differential evolution (OCDE) 450
opposition-based GSA (OGSA) 128–31, 246, 248
opsdg (educational program package) 399–400
optimal coordination of directional overcurrent relays 65, 101–2, 449–70
optimal DG placement by using partial search of variants 373–5
optimal DG sizing by using partial search of variants 375–7
optimal distributed generation 145
optimal energy and operation management 64–5, 101, 148–9
optimal placement and sizing of distributed generation in distribution networks63–4
optimal power flow (OPF) 57–60, 94–7, 145, 177

optimal power flow (OPF), in distribution networks 61–3, 99, 147, 297
  deterministic OPF analysis 318–21
  deterministic optimal power flow 299–300
    constraints 301–2
    objective function 300–1
  distributed generation (DG) units modeling for OPF 302
    diesel generator 303
    electric grid 306
    fuel cell 304
    microturbine 304
    mini hydropower plants 306
    photovoltaic 305–6
    wind turbine 304–5
  opfdngui (solution software) 324–7
  probabilistic OPF analysis 321–3
  probabilistic optimal power flow 311
    procedure for solving probabilistic OPF 314–16
    statistical characterization of the input random variables 311–13
    statistical evaluation of output variables 313–14
  solution methods 307
    genetic algorithm (GA) 307–8
    gravitational search algorithm (GSA) 309–11
optimal power flow (OPF), in transmission networks 177
  equality constraints 186
  gravitational search algorithm (GSA) 196–8
    application of, to OPF problem 198
  hybrid PSOGSA 199
    application of, to OPF problem 199
  IEEE 30-bus test system 200–11
  IEEE 118-bus test system 211–13
  inequality constraints 186–8
  literature overview 178–85
  multiobjective function 191–2
    fuel cost and active power loss, minimization of 191
    fuel cost and voltage deviation, minimization of 191
    fuel-cost and voltage-stability index, minimization of 191
  objective function 188
    active power loss, minimization of 189
    fuel cost, minimization of 188
    voltage-profile improvement 189
    voltage-stability enhancement 189–90
  OPF graphical user interface (*opfgui*) 213–23
  particle swarm optimization (PSO) 194–5
    application, to OPF problem 195–6
  transient-stability-constrained OPF 192–3
optimal reactive power dispatch (ORPD) 60, 97–8, 145, 235
  application of PSOGSA to 240–1
  ORPD program 257–63
  problem formulation 236–7
  program execution 259
  simulation results of PSOGSA 241
    IEEE 30-bus test system 242–9
    IEEE 118-bus test system 249–55
  using hybrid GSA–SQP algorithm 255–6
    application of hybrid GSA–SQP to ORPD problem 256–7
    simulation results of hybrid GSA–SQP 257
  using hybrid PSOGSA 238
    gravitational search algorithm 238–9
    hybridization of PSO and GSA 240
    particle swarm optimization (PSO) 238

optimal Volt/Var control in distribution networks 337
decomposition of voltage-control problem 339–49
using metaheuristic optimization 350
problem formulation 350–2
simulation results 355–8
solution method 352–5
optimization, defined 1
optimization algorithms 141, 246

parallel operated SEIGs
application of genetic algorithm to 488–90
steady-state analysis of 494–501
steady-state equations of 481–4
parcel's velocity 12
Pareto-optimal solutions 55, 136, 145–6
particle 76, 194, 417
particle swarm optimization (PSO) 9, 75, 180, 194–5, 235, 238, 268, 273–5, 298, 366–7, 410, 417–18, 430, 450
with adaptive acceleration coefficients 92–3
with adaptive inertia weight 91–2
and acceleration coefficients 93–4
application 195–6
to EOM 418–19
to OPF problem 195–6
applications, to power system problems 94
economic dispatch (ED) problem solution 98–9
optimal coordination of directional overcurrent relays 101–2
optimal energy and operation management of MGs 101
optimal placement and sizing of distributed generation 100–1
optimal power flow 94–7
optimal power flow in distribution networks 99
optimal reactive power dispatch 97–8
discrete binary PSO 89–90
elements of 417
example usage of 85–7
general flowchart of 79
general remarks about 82
hybridization of PSO and GSA: *see* PSOGSA (hybrid PSO and GSA algorithm)
hybrid PSO (HPSO) method 90
MATLAB code of 83–5
parameters of 79
acceleration coefficients 81
constriction coefficients 81–2
inertia weight 81
maximum velocity 80
population size and initial population 80
population topology 88–9
PSO with an aging leader and challengers (ALC-PSO) 246, 248
particle velocity 76, 194, 417
passive congregation 94, 180
pathfinders 16
pattern search minimization technique 476
pbest, 9, 76–8, 81–2, 88–92, 194–6, 238, 273–4, 417–19
pfdngui (power-flow software) 174–5
pfgui (power-flow software) 164–9
phenotype 44
photovoltaic (PV) systems 297, 305–6, 407, 415–16
pickup current setting (PCS) 449
population 5, 76, 113–14, 194, 417
population-based metaheuristics 3–4, 352–3, 450
ant colony optimization (ACO) algorithm 9
artificial bee colony (ABC) algorithm 9–10

backtracking search optimization algorithm (BSA) 8
bacterial colony foraging optimization 17–18
biogeography-based optimization (BBO) algorithm 18–19
black hole (BH) algorithm 13–14
classification of 4
colliding bodies optimization (CBO) algorithm 12–13
criticism of metaheuristics 31–3
cuckoo search algorithm 15
differential evolution (DE) algorithm 7
differential search algorithm (DSA) 24
evolutionary programing (EP) 7–8
firefly algorithm 14–15
genetic algorithm 6
glowworm swarm optimization (GSO) algorithm 25–6
gravitational search algorithm (GSA) 10–11
gray wolf optimizer (GWO) algorithm 14
harmony search (HS) 22–3
imperialist competitive algorithm 23–4
Jaya algorithm 27–8
krill herd (KH) algorithm 16–17
league championship algorithm 19–20
metahopt 33–4
mine blast algorithm 20–1
moth swarm algorithm (MSA) 15–16
'new' algorithm, creating 28–31
particle swarm optimization (PSO) 9
shuffled frog-leaping algorithm (SFLA) 17
sine cosine algorithm (SCA) 21–2
spiral optimization algorithm (SOA) 26–7
teaching–learning-based optimization (TLBO) 19
wind-driven optimization (WDO) 11–12

population size 44, 80
population topology 88–9
power balance constraint 271–2, 279, 412–13
power flow analysis 57, 155, 157
power-flow calculation 155
  in distribution networks 165, 170
    backward/forward sweep power-flow algorithm 170–4
    pfdngui (power-flow software) 174–5
  in transmission networks 156
    bus classification 158
    pfgui (power-flow software) 164–9
    power-flow equations 157–8
    solution methods 158–64
power-flow software
  pfdngui 174–5
  pfgui 164–9
power system problems
  genetic algorithm applications to 57
    combined economic and emission dispatch (CEED) 60–1
    optimal coordination of directional overcurrent relays 65
    optimal energy and operation management of microgrids 64–5
    optimal placement and sizing of distributed generation in distribution networks 63–4
    optimal power flow 57–60
    optimal power flow in distribution networks 61–3
    optimal reactive power dispatch (ORPD) 60
    self-excited induction generator (SEIGs), steady-state analysis of 66
  gravitational search algorithm (GSA) applications to
    economic dispatch using GSA 146–7
    optimal coordination of overcurrent relays 149

optimal DG placement and sizing in distribution networks 147–8
optimal energy and operation management of microgrids 148–9
optimal power flow 145–6
optimal power flow in distribution networks 147
optimal reactive power dispatch 146
particle swarm optimization (PSO)
applications to 94
economic dispatch (ED) problem solution 98–9
optimal coordination of directional overcurrent relays 101–2
optimal energy and operation management of MGs 101
optimal placement and sizing of distributed generation in distribution networks 100–1
optimal power flow 94–7
optimal power flow in distribution networks 99
optimal reactive power dispatch 97–8
primary relay 461
probabilistic energy and operation management of microgrid 419
procedure for solving 422–3
statistical characterization of the input random variables 420–1
statistical evaluation of the output variables 421–2
probabilistic load model 313
probabilistic optimal power flow 311
procedure for solving 314–16
statistical characterization of input random variables 311
probabilistic load model 313
solar irradiance modeling 312–13
wind speed modeling 311–12
statistical evaluation of output variables 313–14
probability density function (PDF) 299, 420
prohibited operating zones (POZs) 272
proportional selection 47–8, 56
prospectors 16
proton-exchange membrane fuel cell (PEM-FC) 415
PSOGSA (hybrid PSO and GSA algorithm) 141–4, 147, 199, 211, 235–6, 238, 241, 254, 263, 269, 273, 283, 286, 292
application
to OPF problem 199–200
to ORPD problem 240–1
different steps of 141
IEEE 30-bus test system 242–9
IEEE 118-bus test system 249–55
MATLAB code of 142–4

quadratically constrained quadratic programing 450
quadratic programing (QP) 298
quasioppositional DE (QODE) 246, 248

ranking selection 47, 49
reactive power compensation devices 337
real power generation capacity 412–13
real-value encoding 43
relative fitness 45–6
reliability index 366
ring lattice topology 88
roulette-wheel selection 48

seasonal control of voltage 340
first stage 340–4
second stage 344
simulation results 346–9
with ULTCTs 344–6

security constraints 180, 187, 352
seeker optimization algorithm (SOA) 450
selection 39–40, 46
- elitist selection 49
- k-tournament selection schemes 50
- linear ranking selection 49
- simple selection 48
- simple tournament selection 50
- stochastic universal sampling 48

self-adaptive charged system 410
self-adaptive GSA 133–5
self-excited induction generators (SEIGs) 475
- induction generator model 478
- sequence equivalent circuits 478
- simulation results 490
  - steady-state analysis of parallel operated SEIGs 494–501
  - steady-state analysis of SEIG 490–4
- single point operation 477
- solution method 484
  - application of GA to parallel operated SEIGs 488–90
  - application of GA to SEIG 486–7
  - genetic algorithm 485–6
- steady-state analysis of 66
- steady-state equations of 479–81
- steady-state equations of parallel operated SEIGs 481–4
- system configuration 477

sequential quadratic programing (SQP) 255, 450, 457, 471
shuffle crossover 52
shuffled frog-leaping algorithm (SFLA) 17, 183–4
sigmoid function 89
sign and reordering mutations 138
simple selection 48
simple tournament selection 50
sine cosine algorithm (SCA) 21–2, 184
single-solution-based metaheuristics 3
slack generator calculation 272–3
solar irradiance modeling 312–13
space of possible solutions 5, 114
spinning reserve, in microgrids 413
spiral optimization algorithm (SOA) 26–7, 268
squirrel cage machine (SCM) 494–5
static Var compensators (SVCs) 62, 182, 338–9, 475
steady-state analysis of SEIG 66, 490–4
steady-state equations
- of parallel operated SEIGs 481–4
- of self-excited induction generators (SEIGs) 479–81

stochastic universal sampling 48
stopping criteria 8, 77, 194, 417
Stott method 163–4
suitability index variables (SIVs) 18–19
Swarm 76, 194
Swarm intelligence 75

teaching–learning-based optimization (TLBO) 19, 184, 235, 367, 450
thyristor-controlled series capacitors (TCSCs) 59
time dial setting (TDS) 65, 449
total harmonic distortion (THD) control 62
tournament selection procedure 50
trajectory methods 3
transient-stability-constrained OPF 180–3, 192–3
transient-stability index (TSI) 193
transmission networks, power-flow calculation in 156
- bus classification 158
- *pfgui* (power-flow software) 164–9
- power-flow equations 157–8
- solution methods 158–64
  - Gauss–Seidel method 159–60
  - Newton–Raphson method 160–3
  - Stott method 163–4

under-load tap changer (ULTC) 61–2
underload tap-changing transformers (ULTCTs) 337–8
uniform crossover 51–2

valve-point effect 188, 270
variant search 372–7
  optimal DG placement by using 373–5
  optimal DG sizing by using 375–7
velocity of an agent 117, 310, 456
voltage-control problem, decomposition of 339–49
voltage deviation (VD) 298, 337
voltage deviation index 366
voltage-profile improvement 189
voltage regulators (VR) 337
voltage-stability enhancement 181–2, 189–90
voltage stability index 191–2, 366
von Neumann topology 88–9

Weibull distribution 311–12, 420, 430
  cumulative density function (CDF) for 420
wind-driven optimization (WDO) algorithm 11–12, 286
wind-speed modeling 311–12, 420
wind turbines (WTs) 297, 304–5, 407, 415–16
wound rotor machine (WRM) 494–5

Zajecar 391–9

www.ingramcontent.com/pod-product-compliance
Lightning Source LLC
LaVergne TN
LVHW011857200926
848484LV00002B/473
9781785615467